W9-BKW-175

THE SAS URBAN SURVIVAL HANDBOOK

How to protect yourself from domestic accidents, muggings, burglary, and attack

JOHN "LOFTY" WISEMAN

Skyhorse Publishing

Skyhorse Publishing books may be purchased in bulk at special discounts for sales promotion, corporate gifts, fund raising, or educational purposes. Special editions can also be created to specifications. For details, contact Special Sales Department, Skyhorse Publishing, 555 Eighth Avenue, Suite 903, New York, NY 10018 or info@skyhorsepublishing.com.

www.skyhorsepublishing.com

Library of Congress Cataloging-in-Publication Data

Wiseman, John, 1940–
    The SAS urban survival handbook : how to protect yourself from domestic accidents, muggings, burglary, and attack / John "Lofty" Wiseman.

    p. cm.
  ISBN-13: 978-1-60239-216-8 (pbk. : alk. paper)
  ISBN-10: 1-60239-216-1 (pbk. : alk. paper)
  1. Survival skills. I. Title.

GF86.W57 2008
613.6—dc22

                    2007046887

10 9 8 7 6 5 4 3 2 1

Printed in the United States of America

WARNING

The publishers cannot accept any responsibility for any prosecutions or proceedings brought or instituted against any person or body as a result of the use or misuse of any techniques described or any loss, injury or damage caused thereby. All relevant local laws protecting certain species of animals and plants and controlling the use of firearms and other weapons must be regarded as paramount.

For my mother
EDIE
who brought me up through
difficult times with love and
affection, and gave me a
sense of responsibility to
my fellow man

CONTENTS

FOREWORD

My first book, The SAS Survival Handbook, deals with survival in the wild away from all resources, so you may be surprised to see me turning my attention to the problems and dangers of the urban environment.

Unemployment and the high cost of accommodation contribute to serious social problems—the number of homeless sleeping rough in London, New York and other major cities is powerful evidence of the stresses that exist. For the newcomer, in particular, the going can be tough and frightening. It really is a jungle out there.

Everyone is exposed to an enormous number of risks and dangers, which they have to try to cope with on top of all their own personal, health and financial stresses. These range from ordinary domestic accidents to burglary, muggings and even the terrorist threat.

In 27 years as a soldier I was always glad to return home, where I could relax in the 'safe' domestic environment. How wrong I was! I once returned after four months in a fierce war without a scratch. I'd only been in the house for two minutes when I stood up in the kitchen and cut my head badly on a cupboard door. That started me thinking. Every year there are thousands of fatal accidents in the home—more than on the roads, and many more than in the great outdoors.

The home has all the ingredients for disaster under one roof—fire, electricity, water, gas, sharp knives, power tools, heights, poisons and chemicals. In the wild you rarely have two dangerous things to contend with at the same time. At home you also have people from every age group. I have watched my seven children growing up and have seen the dangers they were exposed to—not just in and around the house, but at school and on the streets.

Safety is largely a matter of common sense and following basic rules. This is easiest when you're not under pressure. If you're in unfamiliar surroundings and under threat, your judgement is likely to be clouded. If you're late for work or school, you may forget to lock the house and take a short cut through a dangerous area. You must recognize when you're taking chances and learn to make decisions quickly.

The pace of urban life is sometimes too fast to keep up with. The world is what we have made it—if we want to enjoy all of its pleasures, we must be aware of the dangers. I didn't write this book to scare anyone. I want to show you how to avoid these dangers. The same awareness and readiness that are required when surviving in the wild really are necessary in the city, too. That's why this book is called The Urban Survival Handbook.

I would like to thank Christopher MacLehose of Harvill for first encouraging me to write this book, and Tony Spalding for helping me to shape it. Also thanks to all the people who helped me gather information and my mates with whom I learned my own basic skills. Most of all, thanks to my family—concern for whom first awakened me to the dangers of the urban jungle.

John "Lofty" Wiseman

We are constantly reminded that the world is a dangerous place—pollution, food processing, violence and natural disasters are all beyond our direct control. City life adds further stresses, which may become too much to bear.

Essentials

BE A SURVIVOR Natural enemies • Pollution • Radiation • Processed Foods

BODY MATTERS Vitamin overdose • Exercise • Chemical risks • Substance abuse

STRESS Dealing with stress • Working out your problems • Relaxation • Counselling • Fear • Violence in the city • Catastrophes

CITY SURVIVAL KIT

BE A SURVIVOR!

Every man, woman and child has to face a multitude of dangers in their lives. The added pressures of urban life can often make those problems seem impossible to deal with. It's VITAL to understand the dangers at home, at work, on the streets (at least) and to get to grips with as many as possible. Take control! The safer you make your day-to-day existence, the better able you will be (and the more time and energy you will have) to deal with major crises. How can you expect to deal with a serious health, emotional or financial set-back, if you're totally preoccupied with surviving the dangers of your own home?

Your strategy must be to avoid unnecessary risks in the urban environment and minimize the damage caused by those you can't avoid. You must develop an attitude to urban life that will help you cope with all kinds of situations.

Safety and security

The first place to make safe and secure MUST be your home. Make sure that the structure of the building is safe and keep it that way. Prepare it to withstand both the pressures of natural forces and the added hazards of modern cities. Older houses, in particular, may be affected by damp, insect infestation, rot, settlement and subsidence. All fixtures and fittings should be carefully checked to ensure that they do not invite accident— and immediate action taken if they do.

Everything must be made secure against intruders— primarily because YOU might be at risk. People have been attacked and even held hostage in their own homes. If you decide to put things right yourself, make sure you really CAN do the job. In the end, it may not be worth the aggravation. Unskilled DIY is the cause of an enormous number of injuries—bodged jobs can lead to more stress and further accidents.

The workplace needs to be as safe and secure as the home. It may include machinery, chemical hazards or processes that are highly dangerous. Many jobs are made difficult enough by heavy competition and stress. At home, you may be able to sort things out yourself, but it's more complicated to deal with dangers at work and in public places.

Leisure activities, too, bring their own risks. The element of danger may give a 'thrill' which is an essential part of some hobbies and sports—but a severe injury could be disastrous.

Travel, for business or for pleasure, brings new risks to be assessed and prepared for. There have always been the problems of adapting to a change of diet and climate, and having to cope with a society whose language and social rules you do not understand. These are made worse by jet lag, the strain of airport high-security procedures and the fear of terrorism. Wherever you are, there is a possibility of danger from natural phenomena and from human violence—you could get caught up in an earthquake or be mugged.

AGAINST ALL ODDS

Take control of yourself, your health and your immediate surroundings. On a simple practical level, there's a lot you can do to make life safer and less complicated—but you must do so against a backdrop of constant natural and man-made dangers. You can't change the world, but you can HELP! Minimize the risks to yourself by

understanding some of the natural dangers. DON'T contribute to the world-pollution crisis.

Natural enemies

Some hazards exist as part of nature. There is a huge number of poisonous plants commonly found in gardens or indoors as pot plants. Some are deadly if eaten, others cause irritation when they come into contact with the skin. DO NOT eat any plant, unless it is already a recognized 'food'. Even some everyday food plants are poisonous if eaten in their raw state, including the familiar potato. Its flowers, leaves and stems are poisonous, as are the potatoes themselves if eaten raw—especially if they are green!

Cockroaches, fleas, rats and other common vermin help to spread disease. Rats alone have been responsible for the spread of plague (transmitted by rat fleas) and a serious illness called Weil's disease is carried in their urine. And, of course, there is a multitude of poisonous and dangerous animals, reptiles, fish and insects the world over—from Australia to India and America to the Arctic Circle.

There is even a risk from the life-giving sun itself. Over-exposure, especially of skin that hasn't seen the sun for months, can lead to skin cancers if no protective barrier is worn.

Radon gas is a major hazard, although not specifically a city problem. It occurs mainly in areas built on granite, where the rock contains uranium 238 and thorium. Official figures indicate that radon gas is responsible for 20,000 deaths in the USA alone, every year. The carcinogenic (cancer-producing) gas rises through the ground and up into buildings. Sealing floors and improving ventilation is one of the recommendations for reducing exposure.

Pollution

Many of the environmental problems of our cities can be traced directly to ourselves. In most densely-populated parts of the world, industrial, domestic and traffic emissions so pollute the environment that large numbers of the population—especially children—suffer from respiratory problems and various forms of chemical poisoning.

It is only really in this century that people began to realize the link between industrial and domestic smoke and air pollution. Avoiding the filthy air was one reason why cities spread away from their industrial hearts.

Clean-air legislation has made some progress, but traffic emissions and the burning of fossil fuels still pollute the air of most of the world's cities. Often the atmosphere is so contaminated that even the buildings are being slowly eaten away by the chemical-laden air and rain.

Of course, it is not just our cities that are affected by air pollution. When major atmospheric pollution combines with moisture, sulphuric acid, sulphites and sulphur dioxide are produced. These spread around the globe, falling as acid rain. This affects people, trees, plants and animals thousands of miles away. It also leaches mercury, iron, cadmium and magnesium out of soil to pollute streams and rivers, further poisoning fish and other wildlife.

The acid content of the water increases the risks of poisoning the drinking supply. Many harmful chemicals also escape the filters of the water-treatment systems.

Gases such as carbon dioxide have built up in the atmosphere, contributing to a so-called global warming—the 'greenhouse effect'—which may dramatically change the world's climate and sea levels. Inadequate controls and accidents have led to horrific spillages of radiation, poisonous gases and chemicals, often in densely populated areas.

CHEMICAL SPRAYS

If you must use pesticides, fungicides or weedkillers in your own garden, be careful to avoid skin contact or inhalation of droplets or vapours. Even those of low toxicity can do you harm. Toxic effects vary considerably, but most of the chemicals you will use will persist in the soil. DDT is now banned, in most countries, but even a newborn baby has traces in its body tissues. 2,4,5-T was found to cause horrific disorders, several years ago, but is still available.

Radiation

There has always been a background level of radioactivity from natural sources, but the development of nuclear power—whether for peaceful or military purposes—has greatly increased the risks. Incidents such as those at Chernobyl and Three Mile Island have made our worst fears REAL. Fallout from Chernobyl has disastrously affected an enormous number of lives in many countries.

We are what we eat

Much of the food we eat now often carries a heavy dosage of chemicals absorbed from pesticides and artificial fertilizers (which also leach into waterways and pollute them). Many foods are also over-processed or in some way 'interfered with' before they reach us.

For the city dweller, growing your own food—however organically—is no longer safe. Research has suggested that home-grown vegetables in many city gardens absorb even more pollutants from the air than are present in most non-organically grown farm produce. Lettuce is the worst—it absorbs more nitrates and lead than any other vegetable.

It may not be safe to eat food grown within a seven-mile radius of a major city. Large airports also produce toxins, causing (for instance) very high levels of aluminium in the soil.

Processed foods

Food marketing today seems to try to make all food look appealing. Food may have been treated with bleach or colourings. Various additives may be used to provide or increase flavour or aroma, preservatives to make the food keep longer. Fruit may be sprayed or painted with wax. Irradiation is also being used in some countries to kill organisms and increase shelf life of 'fresh' products. It is claimed this process has no ill effects. Experts differ on this, but there is serious concern—not least because it is not detectable in most foods. All these additives and processes are IN ADDITION TO the chemicals used in growing produce.

ALL fruit and vegetables must be THOROUGHLY washed. It is advisable that they should also be peeled, unless you are quite sure that they have been organically grown without the use of pesticides. If you like cooking with the zest of citrus fruit or baking potatoes in their jackets, you should always be very careful.

It's up to us

We have got to live with these problems and do everything we can to improve things. It will mean a dramatic change in attitudes—especially in our cities—to conserve energy and produce it in less harmful ways. We are all aware of most of the problems, but action on both the personal and governmental level still lags far behind what's needed. In many cases we have learned too late—the damage has already been done. In some places the countryside, its wildlife and even its people, are sick and dying. We are all at risk. This is the background against which you have to solve all your other problems.

REMEMBER

The growth of knowledge and understanding, followed by positive action, which has occurred over environmental issues is an example of the pattern which YOU should apply to every aspect of your life. Ignorance is definitely NOT bliss! It is a killer. The more you know and understand, the better you can cope with any situation. Knowledge will give you confidence and experience will increase your abilities.

BODY MATTERS

It's up to YOU to make sure that you have a balanced diet and proper exercise. Monitor your own health and have regular medical check-ups—particularly if you smoke cigarettes, drink a lot of alcohol or have a hazardous/high-stress job. How much you need to eat depends on your physique and the kind of life you lead. Exercise will be more necessary if you have a sedentary job than if your work involves physical exertion.

Most people have a pretty good idea of when they are over or under weight. If you're worried about being fat or skinny you should check with your doctor, before adopting a particular diet. There could be non-dietary reasons for your condition or you may be worrying needlessly. There are a great many cases of severe dietary disorders each year, due to people having distorted ideas of what is a 'perfect' physique.

In the 1980s, many people—worried by all the media coverage on diet, sport, fitness and longevity—took to eating pasta and salads. Both are good for you, but don't constitute an adequate diet on their own. Many people became 'hooked' on exercise as well—doctors found that some of their patients were literally exhausted!

Vitamin and other dietary supplements should not be necessary, unless you feel particularly 'run down' or you've been ill. A mixture of vegetables, fruit, meat and fish and dairy products is generally believed to provide all the proteins, carbohydrates, vitamins and minerals which you need. In recent years, many people have cut down on the amount of meat they eat—many have stopped eating meat altogether. A vegetarian or vegan diet needs more careful planning to ensure the body gets all the nutrients it needs when animal fats and proteins are excluded.

VITAMIN OVERDOSE

Excessive vitamin intake can do as much harm as having too little—especially for the very young, the elderly and pregnant women. It's quite common to reach toxic levels of some vitamins, especially with today's super-high-dose supplements. A boost of vitamin C, for example, has long been thought to help fight common colds (although there is little scientific evidence for it). An excess increases the risk of forming kidney stones. Excessive doses during pregnancy can lead to withdrawal symptoms in newborn infants, producing symptoms very like scurvy. Effervescent tablets often consist largely of sodium bicarbonate and this would be a risk to patients on a low sodium diet for heart or blood pressure problems. Natural vitamin C from citrus fruits, for example, should be adequate.

Exercise

Don't suddenly take up strenuous exercise if you have been inactive for a long time. You should have a medical check-up first. You have to build up your stamina gradually and match the kind of exercise you do to your age and condition. Most injuries in amateurs occur in the first few weeks of a season when people are out of practice—evidence enough that sudden exercise can be damaging.

In fact, every sport and exercise session should start with a warm-up (see WORK & PLAY: **Sport**). Walking and swimming provide excellent exercise for all ages and are safe for most people.

! WARNING

If you find that ANY exercise you undertake puts you under uncomfortable strain, causes chest pain or shooting pains down the arms, STOP AT ONCE. You may be overdoing it. Have a medical check-up and then opt for a gentler form of exercise. You can increase the work your body has to do, as it develops greater strength and stamina. NEVER take strenuous exercise classes in conditions which are very hot or very humid compared with those in which you normally live.

Chemical risks

In addition to the toxic substances in the environment and in our food, there are a huge amount of poisons lying around in every home or in daily use at work.

There are great dangers in presuming that all over-the-counter medicinal preparations are totally safe—and appropriate for the condition YOU think you have. Many people dose themselves up to toxic levels.

Children are probably at greatest risk—young developing bodies won't process chemicals efficiently. In some countries (including Britain) it has been decided that aspirin and products containing aspirin are NOT suitable for children under twelve years of age. As we get old, our bodies become more delicate again—it's wise to take a smaller amount and reduce the maximum dosage period of drugs like paracetamol.

The threat of poisoning is ever-present, though—with the bewildering range of toxic and potentially-toxic substances in the house, garden and garage/shed (see POISONS).

Substance abuse

It should not be necessary to warn about hard drugs—in most countries they are illegal—but drug-taking has long been common among young people and those under stress. Quite apart from the physical and mental damage which misuse of drugs can do, obtaining drugs requires 'stepping outside the law', in most countries. Funding a serious habit sometimes leads to further involvement in crime.

Drug tolerance (larger doses are gradually needed for the same effect) and dependency can develop very rapidly—DON'T RISK GETTING HOOKED!

It has often been claimed that the occasional social use of cannabis does much less harm than smoking cigarettes. Prolonged use of cannabis may carry a higher risk of lung cancer when mixed with tobacco. Animal tests have indicated that the immune system may be weakened and that the sperm count in males may be lowered. Cannabis does not create physical dependence—there are no definite withdrawal effects.

Greater danger may be in contaminants which are likely in cannabis—powerful weedkillers have been found in some samples. In a purified form, cannabis is being considered for various medicinal purposes.

The effects of amphetamines, barbiturates, cocaine, heroin and heroin derivatives, solvents, glues and other 'abused' substances are much more quantifiable—and much more harmful. All have been associated with countless deaths. But while you sit shaking your head with dismay that so many people seem to abuse their bodies with so many chemicals, how much alcohol do you drink? Although legal and socially acceptable in most countries, alcohol IS poisonous. It causes millions of deaths directly by poisoning and indirectly through accidents, violence and loss of judgement.

Tranquillizers are taken by thousands of people every day and can lead to the same 'addictive' pattern of tolerance and dependence as 'illegal' drugs. They should, if absolutely necessary, only be taken for short periods. As soon as you find yourself desperately wanting or needing a repeat prescription to be able to cope with life, you are on a dangerous path. You MUST try other methods of relieving stress. You need to talk about your problems, solve some of them, find a counsellor or help-group. These will give you long-term help—tranquillizers WON'T. Investigate relaxation techniques.

WARNING

If you use any relaxant, hallucinogenic or mood-changing drugs—tranquillizers, alcohol, cannabis, LSD, cocaine, heroin, for example—do NOT drive a vehicle or operate machinery while under the influence. You may feel perfectly capable, but your reactions and decision-making abilities are affected. If ALCOHOL is combined with other drugs it can cause serious problems, even with prescribed drugs. For pregnant women these are even more serious since the foetus can be affected.

ARE YOU AN ADDICT?

There are other 'social addictions' that can produce adverse effects. How many cups of tea or coffee do you drink in a day? Tea contains tannin, both contain caffeine—as do soft drinks such as 'cola'. Caffeine acts first as a stimulant and then a depressant. It can cause insomnia, nausea, gastric disorders, palpitations and nervous anxiety. Links have been suggested with cancer and benign lumps in the breast. There may be risks to foetuses during pregnancy and to breast-fed babies, if large quantities are consumed.

STRESS

Stress is an unavoidable part of living and it's VITAL that you learn to minimize it—and to cope with it or reduce its harmful effects. Every life has its problems, but city life tends to magnify stress. Travelling to and from work in crowded public transport or driving in rush-hour conditions can be a considerable twice-daily strain. Fear of attack in the streets, dodging traffic, being buffeted by crowds—it all adds up.

At work—and work problems rate very highly amongst causes of stress—pressures may be enormous. Some people live in constant fear of being sacked or made redundant. There may be delivery dates or deadlines to meet and aggressively competitive colleagues. Management may be difficult or unsympathetic. There may be sexual or racial harassment. You may have to deal with ill-tempered customers or clients. Strangely enough some people seem to 'thrive' on such pressures, but for most of us it's quite impossible. Sometimes, if anything else happens to us in other areas of our lives, the pressure just gets too great to bear.

Being unemployed can be just as stressful—it affects both public- and self-image and deprives you of MONEY—which is unfortunately vital for urban survival.

Stress is really a blanket term for excessive emotional pressure. It can be provided by (or focused on) something quite minor—a dripping tap, a persistent smell, or constant background noise, such as that produced by air-conditioning.

Poor living conditions are not only physically unhealthy, but depressing and stress-inducing. Fear of illness, attack or robbery can all prey on the mind. Research increasingly supports the contention that people are more susceptible to depression and stress at certain times of the year. When days are shorter and light levels lower, more people succumb to stress.

Worries about sexual matters and relationships can be very destructive, but the most stressful events in most people's

RESTORE THE BALANCE

Stress puts extra demands on the body, burning up nutrients to fuel your nervous energy. That is why people under stress tend to lose weight and anxiety produces a drawn and gaunt appearance. To help the body cope with the physical effects of stress, make sure it has the particular nutrients that are important for the nervous system and body functions.

Calcium and zinc are burned up under stress and need replacement. Vitamin B1 is needed to control nerve conduction and reaction. Vitamin B2 is needed to control the breakdown of fats and carbohydrates in the liver to ensure wastes and toxins do not accumulate, making you more tired and sluggish. Vitamin B3 is needed to help control the amount of oxygen-carrying red cells in the blood. Since B vitamins are NOT stored in the body, because they are water-soluble, you need a constant supply.

lives are those which create sudden changes. Death of a family member or a friend, divorce, redundancy, unemployment, moving house, starting a new job are all at the top of the list for causing stress. Even 'happy' events such as a marriage or the birth of a child can cause great anxiety.

Parents may worry—sometimes excessively—about their children. Often these worries don't subside when the children are grown into adults themselves!

Dealing with stress

Recognizing the causes of stress in YOUR life is the first step. Make it a challenge to identify and sort out problems—don't let things pile on top of you. Stress really drags you down mentally and puts you at great risk physically.

You need to find something better to do than worry. If you can sort out your day-to-day life to reduce stress, you can start to look ahead to possible problems and try to avert disasters before they happen. If you're overcome by your present problems, how will you cope if anything major goes wrong? You are starting from the wrong place!

Overcoming stress requires great self-control. You need to be able to clear your thoughts and make both body and mind relax. This also helps create a calm and balanced attitude which will greatly reduce the impact of stress situations.

CAN YOU REDUCE STRESS?

Are you ALWAYS rushing about? Do you NEED to place yourself under so much stress? Can you have the occasional quiet day or evening? Can you travel at a different time or use a different route or another form of transport? Is it really necessary to be so competitive? Is that difficult relationship worth the damage it causes? Take a look at your lifestyle, review it and assess what you want. Think about changing your job for something less demanding. Let life centre around something else!

Different people have different ways of relieving tension. Tranquillizers, alcohol, smoking, drugs, eating and violence are NOT solutions. They are the traps people fall into. If you're lucky and have a creative hobby, the concentration involved can be very therapeutic. When a project is completed, you may feel genuinely proud of yourself. Anything which can build confidence and self-esteem is worthwhile.

Work out your problems

Finding solutions to your problems is obviously the best way of removing stress. When that seems impossible you can free yourself for a while by concentrating on physical exertion to help the mind and body relax. Exercise is a great way of overcoming stress. After a good workout your problems don't seem so bad. The tension and anxiety go from your body and you see things in a different light.

Repetitive exercises can be calming and the burning up of energy is like releasing a pressure valve to rid you of built-up tension. Don't get hooked on exercise, though, and make sure your diet compensates for the energy expended. Try to 'train' with a friend—you can pace each other and it's a lot less boring than doing it on your own.

Relaxation

Proper relaxation is not just putting your feet up and watching television—though that can have its role in taking your mind off worries. Learn to relax properly. There are numerous relaxation 'techniques'—each person needs to find one that works for them. Try this:

1 Put on some loose clothing and lie on your back on a bed or on the floor. Reach out with your toes and the top of your head, making yourself as 'tall' as possible. Release and relax.

2 Try to empty your mind. If you can't, focus on something peaceful and restful—the sea, the sky, a tree. If you can't keep your problems from popping into your head, mentally grab hold of something simple and familiar—an egg, a cup, a button. Don't let anything else into your mind. Imagine you are floating on air.

3 When your mind is clear, begin to tense and release all the body's muscle groups in turn. Start with the toes, one foot at a time. Follow with the legs, buttocks, abdomen, back, chest, fingers, hands, arms and neck. Do one group at a time, relax that group and move on to the next. Finally tense the whole body then relax.

Try a variation on this relaxation exercise. Imagine each part of your body is growing very heavy and sinking through the floor. You should soon feel the tension drain away. Investigate local relaxation classes or try yoga or meditation.

Learn to breathe!

Controlled breathing is another way of combatting stress. When you are tense your breathing tends to be short and shallow. Slow it down and make it deeper. Pay attention to how you exhale. As you do, let your whole body relax.

Stand with your hands on your hips, elbows slightly forward, shoulders slightly rounded. Now breathe in for a count of three, hold for three and exhale for three. Keep this going until you have full control and the rhythm becomes natural. Breathe in through the nose and out through the mouth.

Use your lungs completely. Let your abdomen rise and fall—don't let your chest heave up and down. Try it lying down in a darkened room. Imagine the room is filled with pure clean white light—your body is filled with dark smoke. Each time you breathe in you take in the white light. Each time you exhale you release the dark smoke.

REMEMBER

Relaxation and meditation techniques can be learnt, but just being alone and calm from time to time may give your mind a chance to relax. At the very least, this sort of 'private' time enables you to devote your thoughts to sorting out your problems.

MASSAGE

Relieving physical tension is the first step to relieving mental stress. If you have a partner get them to massage your neck muscles, your shoulders and the top of your spine. Even better, go to a masseur and enjoy the luxury of a full massage. To release neck tension on your own move the head rapidly backwards a few times in a very gentle motion. Push it slightly further than it wants to go with a bouncing action. Now bounce it forwards and to each side—again, very gently. Let it loll relaxed to one side and roll it through a full 360 degrees, keeping it relaxed. There, did that help?

Counselling

A problem shared is a problem halved. DON'T let problems build up inside you. DON'T hold grudges and brood about things. Hard though it may seem, it's better to 'bite the bullet' and get the problem out into the open with the people concerned rather than bottle everything up to explosion point. Discuss worries with a close friend. If you don't feel you can share your problems with anyone you know, seek professional counselling and help-groups. It is often easier to get a new perspective on problems with people who don't know you.

Fear

Fear is an element in most forms of mental stress—fear of the unknown and the known, fear of not being able to cope, fear of violence. It is not something to be ashamed of, but you must NOT let it build up out of proportion. Fear is a natural emotion and not necessarily harmful. When people are suddenly very afraid, they can run faster, jump higher and do things they never thought possible. This is due to the adrenalin the body produces in a stressful situation. It can make feats of superhuman strength achievable.

Fear can even be an asset. It makes you wary and more alert on the streets, for instance. Don't try to eliminate it with false courage or deny its existence.

Panic is fear that gets out of control. You MUST control panic—it can be disastrous. You must learn to use fear and anxiety—not let them use you. Don't let your imagination run riot and conjure things to be scared of. Assess the real situation and confront it.

Learn to think and plan when you are afraid. One way is to enrol on a survival course and be put under pressure while being carefully monitored. You must get to know yourself.

The violent city

For many people the greatest threat in city life is that of violence. Media reports suggest that muggings and other violent assaults have increased considerably in recent years. At the same time, people seem less and less prepared to go to the help of others. There is not much point in exposing yourself to serious physical risks if you can't really help. You can, however, at least call the police or an ambulance. Too many people don't even do that—they don't want to be 'involved'.

Avoid violence. Don't make yourself vulnerable—use your common sense—but don't shut yourself indoors. What is the point of being in a city if you cannot enjoy city life? Be sensible about where you go alone, especially at night. Know your city, avoid any areas known to be dangerous. Be aware of local problems and don't provoke violence by your behaviour. Wearing a football team's colours and shouting abuse about the other side's supporters is a sure way to get your nose flattened. Don't take short cuts through dark alleys.

Sometimes attack comes out of the blue. That's when it is useful to have some knowledge of self-defence. Follow the basic advice given later (see SELF-DEFENCE), but think about taking a proper course. You'll learn much better from an instructor than from any book—but don't expect an attacker to keep to the rules. You may need to let fly with everything you've got. Learn how to use your natural weapons.

Terrorism and armed violence

The use of terrorism for political ends has been a growing problem in recent years. We have to learn to live with it and to be vigilant. Whilst you cannot deal with bombs and suspect packages and vehicles personally, you can report any suspicions and let the experts sort them out. Be alert. Learn what the procedures are for hostile action against your workplace, whether by terrorists or armed robbers, so that you will be able to respond rapidly and calmly.

Catastrophes

There are some situations that really are beyond our control, from train crashes to natural disasters. That does not mean that you cannot be prepared for them. Developing your own skills and reactions will help you to avoid accidents and enable you to assist others who are involved in them.

If you live in places where earthquakes, floods or hurricanes could strike, you must learn to prepare for them, evacuate as necessary or deal with the aftermath. As with everything else in life, awareness, knowledge, confidence, training and experience will set you up to be a good survivor.

Being prepared

You are what you think you are. You will achieve what you think you will—at least, you have little chance of achieving ANYTHING if you don't believe you can! Other people will have widely different opinions about you, but it is what YOU really think of yourself that matters. Confidence in yourself and self-knowledge must be backed up by what you have learned about your city and how to deal with situations.

Of course, no two problems are identical. There is no advice that can cover EVERY situation. Develop the ability to take in and assess a situation quickly, then apply a calm and logical solution based on what you know and what you have learned. No one can hope to do better than that.

Equip yourself to deal with small day-to-day problems. If you know what your programme is for the day, you will carry what you need—but you should also carry an emergency kit to cope with some of those unforeseen occurrences. It need not take up much space—a small tin like a tobacco tin is ideal, but you shouldn't smoke the tobacco! Always carry your kit with you and it will see you through many a crisis. With it you can (at least):

- Repair a tear
- Sew on a button
- Pin a broken zip
- Remove a splinter from your finger
- Read a telephone book/street map in poor light
- Attract attention
- Make notes
- Leave a message
- Get rid of a headache
- Pay for a taxi-ride
- Make an urgent phone call
- Cover a cut or graze

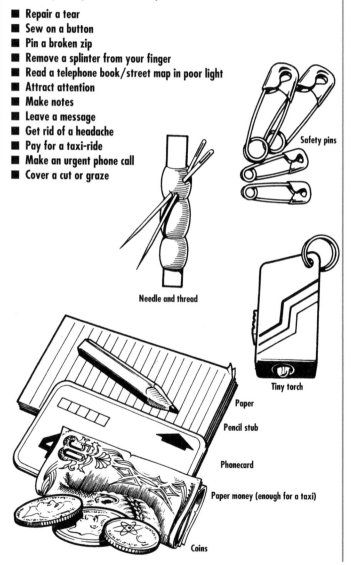

Safety pins

Needle and thread

Tiny torch

Paper

Pencil stub

Phonecard

Paper money (enough for a taxi)

Coins

ESSENTIALS BE A SURVIVOR ■ STRESS

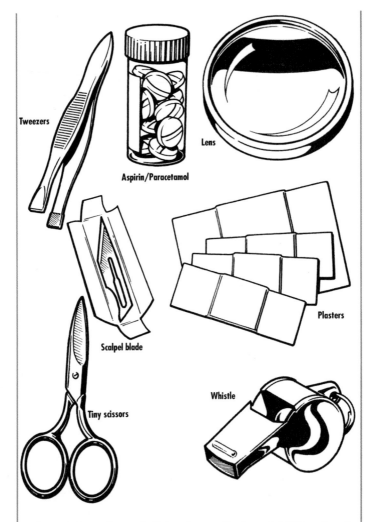

Tweezers

Aspirin/Paracetamol

Lens

Scalpel blade

Plasters

Tiny scissors

Whistle

Start with this as the standard kit and personalize it to suit YOUR needs and YOUR lifestyle. Cyclists and motorists could have their own versions. You may need a tiny screwdriver (especially if you wear glasses), antihistamine tablets (if you get hay fever), other medicines, tampons, condoms, a spare pair of contact lenses or eye drops, matches, breath fresheners, something to tie long hair back—you name it!

Keep the kit you carry small and easy to slip into a pocket or your handbag. If it becomes too bulky or inconvenient to carry, you will leave it at home on just the day you need it! It is like wearing a seat belt in a car. You wear it every time you drive—but the day you forget it is the day you have an accident! It is the same with your emergency kit. Carry it ALL THE TIME.

2

Y ou may think that you're safe at home, but statistics indicate otherwise. The number of serious injuries and fatalities is staggering. Take steps to prevent yourself, your family and visitors from coming to harm.

Safety first

SAFE AS HOUSES? Structure • Subsidence & settlement • Outside checks • Inside checks • Damp/condensation • Rot/infestation • Asbestos risks • Repair priorities

ELECTRICITY Know the enemy! • Electrocution • Earthing • Domestic supplies • Voltage • Power consumption • Replacing a fuse • RCDs/RCBs/ELCBs • Outdoor circuits • Low-voltage installations

GAS Carbon monoxide • Gas leaks • Bottled/cylinder gas

LIQUID FUEL Paraffin • Portable heaters

SOLID FUEL

WATER Drinking water • Lead • Purification • Filtering water • Hard & soft water • Domestic water supply • Waste systems • Burst pipes • Tank floods

COMMON ACCIDENTS Falls • Suffocation • Choking • Burns/scalds • Bleeding

ROOM CHECK Kitchen—gas/electric/microwave ovens, fridges/freezers, food mixers, washing machines • Living room—fires, televisions/VDUs, children's toys • Bedroom—electric blankets, nursery • Hallway/stairs/landing • Loft—access, insulation • Garden—pesticides, bonfires, electricity • Garage/shed—storage, tools, car repairs

LEAVING THE HOUSE Out for the day • Going away

PETS Safety & hygiene • Unusual pets

SAFE AS HOUSES?

 Most people think of the home as the one place where they can feel safe and secure—but statistics show that in industrialized societies there are as many accidents in the home as on the roads. In Britain alone, one person in 30 will have a serious accident in the home. Hospitals treat at least three million domestic injuries—and, since some result in visits to the doctor or self-help, the actual number of accidents may be much higher. Even in the garden, over 90,000 accidents are reported, ranging from slight to fatal.

Types of accident

Everyone is at risk, but children and old people particularly so. The young are inquisitive and full of energy, while older people can quite often be handicapped by poor eyesight, bad hearing and slower reactions.

A quarter of a million British children require hospital treatment for accidents which result from the structure of the house itself—falling and head injuries from hard edges and corners of furniture.

Children at play around the home will have quite surprising accidents—most of which cannot be foreseen. Most hurt themselves by falling, tripping or slipping on a flat surface. Statistics are lower for falls from one level to another, such as out of a window. Next come injuries against objects such as the corners of tables, followed by cuts and deeper wounds, suffocation, poisoning and burns.

It's up to you

People spend an average of 16 hours a day at home, so make sure yours is safe. Keep it properly maintained—put right any problems that increase risk. Make sure that you do not create new dangers as you make changes or improvements.

Most occupations and areas of work are covered by stringent safety laws. These should always be followed and it is not totally up to you to ensure that you are protected from injury. But in the home, safety is largely a matter of common sense and personal discretion.

There are rules and regulations regarding the structure itself, but these have changed over the years and older dwellings may (quite literally) be deathtraps. Correctly installed and maintained gas, electrical and other services should be safe—but wear and tear on these, added to the possibility of human error, could be lethal.

Surveyors will tell you if a structure is safe or not, but no one will come to your house to check for worn electrical flexes, loose stair carpets and slippery floors. You are unlikely to be arrested for not paying due attention to an old and weak ceiling which could collapse. It is up to YOU to judge these things for yourself or call in an expert when in doubt. Don't wait for something to happen!

Many household accidents could be avoided with a better understanding of the risks involved—an increased awareness of dangers which could seriously affect the lives of the occupants of your home.

PLEASE NOTE

All **EMERGENCY!** and **SAVE A LIFE!** panels are intended only to guide you through emergency procedures, and to avoid the need to search through the pages of this book in a crisis. More-detailed instructions appear in the HEALTH chapter. Learn the procedures BEFORE you need to use them.

THE STRUCTURE

Keeping your house in good repair is not just a matter of maintaining market value. Damage, defects and decay could create a serious risk to life and limb. When you buy a house it is usually thoroughly inspected by a qualified surveyor, but no one can spot every potential problem, nor predict exactly how a building will behave as time passes and external conditions change.

Every structure 'settles' throughout its lifetime. This settlement is caused by the sheer weight of materials used in the construction, but is also affected by changes in the use of the building—floors and walls are affected by heavy traffic and the positioning of unusually heavy furniture. This process can be accelerated by environmental and local conditions. If you are worried about the safety of your home, it would be worth seeking professional advice.

There are several ways in which the very location of your home may be a severe threat:

Vibration

Trains, heavy traffic, roadworks and even low-flying aircraft can progressively weaken the fabric of a building. If you can feel vibration—and it's not uncommon to 'hear things rattling' as a train passes—then your home is under threat.

Numerous factors determine the seriousness of such vibration. If a building has been next to a railway for 75 years and shows no serious damage, it may never do so. However, the type or intensity of rail traffic may change over the years and new problems may develop. Expect problems in new buildings when local conditions change for the worse.

Bombs, other explosions and earthquakes may do great damage, although the real extent of it may not be visible.

Clay soils

The main problem with clay subsoils is that they expand when wet and contract when subjected to drought. Shrinkage can be substantial in dry seasons, causing foundations to sink. When a period of wet weather follows, the clay expands, eventually lifting the foundations back up again. This may not be too serious if the shrinkage and heave occur evenly under the house, but uneven movement can be disastrous. A leaking water pipe on one side of a structure would cause local swelling of the clay. More surprisingly, a large tree near a house will 'drink' up vast quantities of water, causing local shrinkage.

Obviously a leak in a pipe must be located and fixed, but dealing with trees is more difficult. Remedies range from cutting down the tree, to severe root pruning. Sometimes the remedy only makes matters worse. With the cause of the shrinkage removed, the clay takes on water and begins to heave. This process could take several years, but will cause damage. Always consult an expert.

REMEMBER

If you are planting a tree near a house, and you expect the tree to achieve any appreciable size, it should not be planted within 10 m (over 30 ft) of the foundations. Even when not on clay soils, the roots of a tree can cause havoc. Willow will target the slightest leak in a buried water or drainage pipe and could cause damage.

Chalky/sandy soils

In severe winters, chalky and sandy soils may be subject to 'frost heave'. As they take on water and freeze, they expand—with enough power to lift foundations. This problem is fairly rare and mainly affects unheated buildings.

Marshy ground

Whether the building is built on known marshy ground, possibly on a valley floor or foot of a hillside, or in close proximity to a river or lake, it must be expected that problems will occur. New buildings should be built with the possibility of gradual sinking taken into account. Underground rivers or bodies of water may be harder to detect.

Mining

Mining or tunnelling. Dramatic settlement may be expected when mining has been close to the surface. This may not be a gradual process—numerous instances of sudden subsidence have been reported. You should know if your area was once or is still being mined.

Sloping ground

Building on hill and mountain sides may be no problem at all. All over the world there are buildings which have stood in precarious positions for centuries. However, there are also numerous instances of sudden collapse, usually associated with mudslides after heavy rainfall.

Landfill

Much depends on the quality of the landfill. All sorts of problems have been reported, from sudden subsidence to underground explosions of decaying matter. These are impossible to predict. There is also the possibility of toxic substances percolating up through the soil.

SUBSIDENCE AND SETTLEMENT

 All buildings settle in response to movement in the earth and expansion and contraction of their own structure. This can produce sloping floors with doors, windows—even walls—out of alignment. Many old buildings survive like this for centuries—think of the Leaning Tower of Pisa.

Any new or sudden movements, generally indicated by a crack of growing length or width, or creases or splits appearing in wallpaper at corners, should be investigated. Recent very dry, hot summers in temperate countries have dried out soils and greatly increased the number of properties affected by subsidence damage.

Any crack, especially a new one, should be investigated—or even monitored over a period of weeks or months to ascertain the cause. Don't assume that the building is falling down—but don't ignore the cracks either. Monitoring may involve marking the crack in some way to work out the direction of movement. A crack in an external wall will let in damp and this could lead to further problems.

Consult an expert such as a surveyor or a professional builder. Your insurance company may be able to advise, but many policies do not allow for the effects of subsidence and settlement.

> **! WARNING**
>
> Climbing plants, particularly ivy, should not be allowed to scramble unchecked over the outside of a building. If the walls are in good condition or the bricks and mortar are new, there should not be much problem. The aerial roots will invade bad joints or cracked rendering, expand and severely weaken a wall. Eventually, the weight of old plants (stems can grow as thick as your neck!) could pull a wall down.
>
> Care should be taken to avoid ivy invading the spaces around the eaves of the house, spaces between slates, tiles, flashing and weatherboarding. As the roots and stems expand they will open up any crack they can find. Remedial treatment may be difficult if the situation has got out of hand.

OUTSIDE CHECKS

 Take a critical look at your home from outside and in. Make use of a neighbour's upper windows to help you see the roof. Use binoculars if you can. The most serious damage could lead to collapse of all or part of the building, but anything which lets in rain or damp can start a chain of problems. Paint and other decorative treatments have a vital function in protecting wood and other materials, and require regular upkeep.

Roof

No roof covering is expected to last forever. The effects of the weather and conditions inside the roof space will take their toll. Look for sagging (drooping) roof lines. This may only indicate that, over a long period of settlement, the external walls have stood firm while the internal structure has 'sunk' slightly. This may not be serious. However, timbers may be decaying because of inadequate weatherproofing, condensation in the roof space or even infestation of wood-boring insects. Sometimes a new roof covering may be too heavy for the old timbers, or new work—such as a loft window—may be inadequately braced.

From the outside, try to find a good vantage point such as a neighbour's upper window—use binoculars from the ground. Look for deformation of the shape and slipped tiles or slates. The latter may be due to failure of the nails holding them in place and could be progressive.

From the inside, working with a good light (and strong boards across the joists when there is no attic flooring) look for damp and rot. Its location may help pinpoint external problems. It may also be due to condensation in the roof space. All roofs are designed with some ventilation, usually at the eaves, to allow the space to 'breathe'.

Wood-boring insects can do very considerable damage if allowed to colonize the roof space. Badly-infected timbers may have to be cut out and replaced. However serious the infestation, all internal timbers should be treated.

Lead roofs are vulnerable to the extremes of climate. Repeated expansion and contraction will crack the lead, resulting in further problems. Avoid long runs of lead which accentuate the problem.

All flat roofs must have good drainage. Most felt roofs will blister and trap water, but more damage is done by walking on them, or resting ladders on them, than is done by the weather. Cracked or leaky felt may only need to be painted with a new protective coating. Consult a local supplier.

WARNING

When insulating the loft space, do not carry the insulating material too far into the eaves. The loft needs to 'breathe' and you must not cut off or restrict its air supply. To do so could result in a build-up of condensation, which could lead to rot. This problem is particularly aggravated by 'magic' roof paints, used to seal and secure old slate roofs. This thick coating seals in the slates (trapping moisture which was present at the time of application) and could lead to far more serious problems than those you were trying to solve.

Chimneys and flashings

Flashings round roof edges and the base of chimneys always cause problems, but chimneys are at risk from inside as well. The soot contains acids, which eat away at the fabric of the stack. Pots on the top are set in mortar (known as flaunching). This may break up over the years, allowing pots to fall. Chimney stacks have been known to fall, sometimes onto the roof and thereby down into the house. Check your chimneys!

Gutters and downpipes

Leaks are easiest to see when it's raining—you may be able to use a hose to produce the same effect. Look for stains on the wall, algae growing round brackets and joins. Plastic guttering is notorious for drooping or failing if the runs between retaining

Pay particular attention to the sides of your house which face the prevailing wind or receive the severest 'baking' from the sun. Heavy rain, driven by strong winds, will find its way into all sorts of cracks and crevices. Flashings may be lifted by particularly strong gusts. Rain may be driven underneath roof coverings.

Strong sunlight gradually cooks paint and breaks it down. You will always have the worst problems with woodwork on the 'sunny' side of the house. Black and dark paints suffer most, because they absorb the heat.

The weight of a heavy snow fall may be too great for weak roofs and gutters to bear. Freezing temperatures will freeze water which has penetrated cracks and crevices, causing them to expand and let in more water. Remember, water bursts a frozen pipe indoors and can do the same to blocked downpipes outside.

clips are too great. Drooping can cause otherwise secure seals to fail. Cast iron and aluminium guttering are subject to rust and corrosion (respectively).

Blockages, such as leaves, twigs and birds' nests can cause water to sit in the gutters for long periods of time instead of racing away as is intended. Settlement of the house may mean that gutters, originally sloping to allow water to escape, no longer do so.

Leaky overflows, if they have persisted for a long time, will usually make a stain on the wall or the ground below. In winter a leaky overflow may produce a deadly patch of ice. Check the tank!

Walls

Look for horizontal or vertical distorting, eroded mortar between brickwork, cracked rendering, missing tiles or other cladding, soil or other obstructions covering the damp-proof course, blocked airbricks. Sulphates in some kinds of brick (and in chimney soot) can cause mortar, rendering and concrete to expand and crack. Areas looking damp when the rest of the wall has dried may indicate where old bricks, which have become porous, are located.

Doors and windows

Check for severe distortion and paint or varnish in bad condition (plastic frames and some metal frames may not need painting). Ferrous metals and most woods must be protected from the weather. Look for gaps between frames and walls and cracked putty round windowpanes. Clear drip grooves under sills, which are designed to allow water to drip harmlessly off the edge. When blocked or bridged, water will penetrate where the sill joins the wall.

Woodwork

All external woodwork should be weatherproofed, usually with paint, varnish or preservative. Most problems are easy to spot. Bodged repairs never last. It's far better to get the job done properly. Check for wet rot—usually apparent from the 'sponginess' of the wood, which may also break away very easily. This rot must be treated because it will spread or lead to dry rot—which is more serious (see **Rot**).

INSIDE CHECKS

Structural distortion

Movement of the building may produce interior cracks in the plasterwork, distortion of door and window frames, slipped cracks in the corners of rooms (most noticeable when wallpaper looks stretched or torn). If the structure is actually sound, remedial work is mostly cosmetic. In severe cases, doors and windows will stick, glass in windows will break, and floors will slope. Changes in use of a building, heavy furniture or a whole library of books in an upstairs room will take their toll. So will changes in climate, which cause the internal timbers to expand and contract.

Support walls

Home 'improvers' sometimes knock out a wall between two small rooms, to create a larger one. Careful checks must be made to see if you are removing a support wall.

If the joists supporting the floor above (or in the loft space) run parallel to the wall, you are probably all right. However, if they run at right angles to the wall—or travel at right angles over the wall—then you are going to have to insert a special reinforced steel joist. In some cases a support pillar may be required to give extra strength.

When you consider that the load on this wall may include the support of a roof brace, it is easy to imagine what would happen if the wall was removed!

Check thoroughly. A wall which is not load-bearing on the ground floor might be positioned to help carry one of the upper floors of the building.

DAMP

Damp conditions can be disastrous to your health, and distressing to live with. There are three main problems:

Rising damp

 Rising damp is caused by water soaking up from the ground into the walls. Most houses have a dampproof course to prevent this, but this barrier may be bridged on the outside (perhaps by a pile of earth), or on the inside (the plaster on the wall may have been taken down too far). In houses with cavity walls, debris may have accumulated at the bottom of the cavity.

This problem must be checked very early on. Look for damp plaster and peeling wallpaper at or near floor level—although in severe cases there is no limit to how high the damp will rise. Skirtings, floor joists and floorboards may be damp or even rotten.

Penetrating damp

Penetrating damp is more likely in older homes—or at least, buildings with solid walls. Cavity walls don't tend to favour damp penetration, unless the cavity has been bridged. A full check must be made outside to try to find the point of entry.

Condensation

Condensation—evident as water droplets on walls and ceilings, windows heavily laden with moisture, damp furnishings—is caused by warm, moisture-laden air coming into contact with cold, uninsulated surfaces. It is aggravated by bad ventilation. In severe cases mould may form, plaster walls may disintegrate, wood and furnishings may rot.

TESTS FOR DAMPNESS

A moisture meter, which you may be able to hire, will not only tell you if a surface is damp—it will also tell you where the greatest concentration of moisture is located. This could be invaluable if you are having trouble locating the source of the damp. The meter has two pins which penetrate the damp surface and produce a reading.

If you can't decide whether a surface is damp or not—perhaps it is a cold external wall or a concrete floor—tape a sheet of polythene on a selected area. If there is a lot of moisture present in that area, droplets will form on the underside of the polythene sheet.

Raising the temperature in a room increases the capacity of the air to hold moisture without condensing. But it's a major battle against cold walls and windows which need insulating. Cold pipes should be generously lagged. In the desire to cut off draughts, there is the danger of sealing all ventilation from a room (see **Gas**).

Condensation also occurs when a chimney has been closed off or removed without the insertion of an airbrick, or in the roof when loft insulation is carried too far into the breathing spaces round the eaves.

REMEMBER

Just as we need fresh air to survive, good ventilation is vital to keep your home healthy. Airbricks—basically bricks with holes in—are used to vent the spaces under floors or in sealed chimney stacks. These prevent the build-up of moisture-laden air. Don't block them!

In rooms with high levels of condensation, such as kitchens and bathrooms, extractor fans should be employed—either on an outside wall or through a window. If neither is available it may be necessary to run ducting to a suitable escape point.

WARNING

Don't seal off every little draught from a room, particularly if it's regularly occupied or contains a fuel-burning appliance. Moisture build-up from human breath and burning fuel will produce condensation. Most fuel-burning appliances also produce moisture, along with carbon monoxide and carbon dioxide—which are highly poisonous, especially in concentration.

ROT/INFESTATION

 **All woodwork, especially softwoods (which comprise a large amount of the inner framework of most houses), should not be exposed to persistent wet conditions, whatever the cause. The occasional spillage of water is harmless, but constant wetting will lay the way open for wet rot or dry rot.**

Rot

Areas most likely to develop rot are roof timbers (either from damp penetration or condensation), bathrooms (round the bath, shower, toilet, bidet), kitchens (round the sink), window frames, door frames, floors and skirtings (particularly in the presence of rising damp).

Dry rot is the more serious. It lives on timber, favouring damp unventilated spaces. Usually it goes undetected until it is well under way, the first sign often being a floorboard beginning to weaken. It gives off a distinctive mushroomy smell which should be noticeable in severe cases.

The wood darkens in colour and cracks across and along the grain, producing an uneven network of squares, often with grey furry patches of mould. The wood becomes very light and crumbly. If the air is very moist round the fungus, it may produce larger fluffy whitish growths. After a year or so it produces thick bracket-like fruiting bodies with rust-red spores clearly visible.

Dry rot is fast and deadly. It will travel across brickwork to reach unaffected timber. The spores travel well, and if they land on moist timber the whole process is accelerated.

Wet rot needs damp wood to survive. The fungus usually only penetrates damp wood. It doesn't spread in the same way—it's unlikely to produce fruiting bodies indoors.

It grows into the wet wood, softening it, and producing cracks along the grain. More than one fungus can produce wet rot, but the wood may appear lighter in colour and very fibrous, sometimes with brown strands on the surface.

Wet rot tends to die if the wood dries out, but the path may be open for dry rot to develop.

Any case of rot must be treated as a top priority. Affected timbers may need to be cut out and replaced. At any rate, whatever the scale of the emergency, expert help should be sought. All wood should be treated with chemicals to inhibit further mould growth, after all possible sources of damp have been eliminated. Any wood being cut out and removed should be in sealed polythene bags and disposed of finally by burning.

! WARNING

New timber used for skirting boards, window frames and door frames may be 'green wood'. Traditionally timber was seasoned before use—left for a couple of years to expand, contract or warp and adjust to atmospheric moisture. With increased demand, wood is sometimes kiln-dried to speed up the process.

Much new wood may still be 'green', containing a substantial amount of water. It is quite common for such wood to rot, once installed and painted.

Wood-borers

Groups of small holes in timber, or small clouds of powdered wood falling from them, indicate the presence of wood-boring insects (see colour pages). In most cases, it is the larvae of various beetles which do the damage, tunnelling through the wood until they emerge as adult beetles.

The main culprits in temperate climates are **woodworm** (common furniture beetles), which favour the sapwood of softwoods—which comprise most of the wooden structure of most houses—plus the sapwood of oak. They also have a taste for beech, birch, elm, mahogany and older-style plywood.

Small piles of wood dust round clean 1.5–2 mm (1/16 in) holes indicate recent activity. These are the flight holes from which the adult beetles emerge after the larvae have eaten their way around the inside of the timber for about three years. The best time to treat woodworm is in the spring, before the warm weather arrives and the beetles emerge. The pupae are usually just below the surface.

Deathwatch beetles prefer partly decayed oak and are usually found in older buildings, producing more damage and 3 mm (1/8 in) flight holes. They are quite rare.

House longhorn beetles are also quite rare, but represent a very large family of beetles around the world—many of which can do considerable damage. The flight holes are oval, usually about 10 × 6 mm (3/8 × 1/4 in), filled with bore dust. House longhorns tend to attack the sapwood of seasoned softwoods. The larval stage can last eleven years, so by the time the flight holes appear, timbers may be 'eaten away' inside.

Powder post beetles tend to inhabit timber yards and sawmills, favouring the sapwood of new hardwoods. They don't like old wood. Their flight holes are about 1.5 mm (1/16 in) across.

Wood-boring weevils tend to prefer decaying hard or soft wood—they are most commonly found in damp cellars and buildings close to water. Once the rot is treated the weevils tend to die out.

In sub-tropical and tropical climates, **termites** are a major problem. They look like large white ants—and are often so called. They attack almost anything that contains cellulose—wood, paper, cardboard, chipboard. They don't live in wood—they eat it or take it back to their nest, which may be underground. Attack by termites should be treated as an emergency. Chemical and biological controls are used, but termites are very determined.

It's common to try sheet metal at ground level to protect posts and structural timbers—but even sheet metal dampcourses don't stop them. They can create a route through most thin substances, including metal, glass and concrete. They may even attack the sheathing of cables, causing short circuits and power failures, and make holes in piping.

The **lead cable borer** in the US favours hardwoods like oak and laurel—also called 'short circuit', it bores into lead-lined cables. The **bamboo borer**, although it only breeds in tropical climates, may be exported to any location in bamboo furniture. It may attack wood in the home—not just bamboo—if conditions are favourable.

Carpenter ants (mainly in the US) create galleries inside timbers, depositing excavated wood outside their nests. **Carpenter bees** bore 20 mm (3/4 in) diameter holes, which lead to large galleries which run with the grain of the wood.

Avoiding infestation

Woodworm, in particular, like old furniture—the thin plywood backs of wardrobes, all-plywood 'utility' furniture and basket work. Treat any secondhand or imported furniture before installing it in your home as a matter of course. Treat all timbers if there has been any evidence of attack. Treat all new wood that is brought into the house. Don't store lots of old furniture or wood in the loft.

SAFETY FIRST   SAFE AS HOUSES?   ROT/INFESTATION

WOOD PRESERVATIVES/INSECTICIDES

 Protective clothing (including boots, gloves, filter mask and goggles) MUST be worn when applying fungicides, insecticides and preservatives. They—and the fumes given off by them—are toxic for up to a week after application. During this period:

- Do NOT smoke or use naked flames near chemicals or treated areas—keep a fire extinguisher nearby
- Turn off gas pilot lights in treated areas
- Turn off electricity or isolate circuits in the area by removing fuses for at least 48 hours
- Temporarily remove loft insulation and securely cover water tanks for at least two weeks, if treatment is in loft
- Protect any rubber cables close to timber to be treated with sealer (shellac)—or cover with polythene sheet
- If spraying, use a coarse spray, rather than a fine mist
- Keep area well ventilated
- Do NOT sleep in a treated area and try to avoid any use for at least a week after application—longer if possible
- Do not eat, drink or smoke during treatment
- Take a shower or bath immediately after finishing the work

See **DIY/CRAFT HAZARDS: Protective clothing**

The main object of treatment is to safeguard the main structural timbers of the house—the roof timbers and floor joists. Treatment is a hazardous process. Don't hesitate to contact experts to deal with large infestations.

Rotting wood—fences, outbuildings and piles of timber— may harbour any number of wood pests. Some pests don't travel very far—others will seek out certain timbers. DON'T allow a large amount of wood to rot near your house—even fungal spores may spread.

Shipping furniture around the world seems a very likely way of introducing new pests to a country. This has already happened with several of the species listed.

Termites (in sub-tropical and tropical climates) may initially make their presence known in fences and outbuildings. DON'T wait for them to discover your house. If you see evidence of attack, or large 'white ants', the time has come to act.

ASBESTOS

Asbestos is now considered a dangerous mineral. Inhalation of the tiny fibres (thousands would fit into a millimetre cube) may eventually lead to:

Asbestosis: a scarring of the lung tissue, leading to restricted breathing and sometimes death.

Mesothelioma: a form of cancer affecting the lining of the abdomen or chest. It is usually fatal and has been linked to exposure to asbestos dust.

Lung cancer: although smoking is considered the primary cause of lung cancer, reports suggest that exposure to asbestos increases the risk considerably.

> # REMEMBER
> Not even a small piece of asbestos, such as a simmering mat for a stove, should be thrown into the wastebin. Waste disposal carts often crunch up rubbish and this would release fibres into the air. Dampen asbestos and double-wrap in polythene bags. Consult local government health departments for instructions on safe disposal locally.

Types of asbestos

Most old asbestos looks greyish in colour, but in fact there are three main types: brown, blue and white—white being the most common. Blue and brown have most often been linked with illness and are seldom used now. To be on the safe side, treat ALL forms as extremely dangerous, with NO 'safe' level for exposure to dust and fibres.

Common uses for asbestos
- Oven gloves
- Old fire blankets, fire gloves
- Toasters, hairdryers, irons
- Tumbledryers
- Insulation round hot water pipes
- Ironing board iron rests
- Wall cladding
- Roofing
- Guttering, downpipes, flues
- Simmering pads
- Car brake linings and underseals
- Fire doors
- Old electrical storage heaters
- Old textured wall and ceiling coatings
- Ceiling tiles
- Flooring

What to do
Sheet asbestos in good condition under layers of paint is best left where it is. Asbestos is at its most dangerous when friable (flaky and crumbly)—as is often the case with oven door seals and pipe insulation. NEVER drill, saw, sand or scrub asbestos. ALWAYS keep an eye open for signs of wear and tear. You should NEVER attempt large-scale removal yourself. Call an expert for advice.

IF YOU MUST TACKLE SMALL-SCALE REMOVAL
- Keep other people away
- Wear disposable mask, overalls (preferably disposable), rubber or washable boots and disposable gloves
- Wet asbestos
- Do NOT use power tools—cut slowly by hand
- Remove as large pieces as possible—the less it breaks up the better
- Put all pieces in at least two heavy-duty plastic bags and mark clearly: DANGER! ASBESTOS DUST! DO NOT INHALE!
- Clean up with a damp cloth and put cloth and disposable clothing in same bag before closing
- Wash non-disposable clothing
- Shower, washing all over including your hair

PUTTING THINGS RIGHT

House repairs take time and money, quite apart from the upheaval and stress that may be caused. Try to think, during every day of major work, that—if done properly—many jobs may not need doing again in your lifetime. Bodged repairs will only lead to more work, more stress and more expenditure.

Outdoor maintenance may not seem as rewarding as tackling interior decoration—but it is much more important. You may find yourself redecorating very soon if you haven't had a proper job done on a leaky roof, or had rotten wood removed and made good.

Priorities

Give priority to making the house watertight—bad weather may force this decision upon you. Whatever the weather, treat the following as urgent:

- Serious cracks in walls, rendering
- Defects in the roof
- Rising damp, wet rot, dry rot
- Problems with gutters, downpipes or external drainage
- Unsafe chimney stacks

Inside: Do something about:

- Faulty or inadequate electrical systems
- Leaky, bodged plumbing and heating systems
- Inadequate lighting on stairs and changes of floor level
- Ventilation
- Blocked or damaged chimneys and flues

Annual inspections

Once a year, or immediately after a bad storm or very high winds, check the condition of the roof. Look for loose tiles and slates, displaced flashings where roofs abut a wall or round the base of the chimney, blocked gutters and downpipes, signs of water penetration through walls and ceilings, deteriorating external woodwork, blocked airbricks.

REMEMBER

When large-scale remedial work is undertaken it may be your chance to make major improvements to your home. Bear in mind that you are likely to need planning permission for certain types of building work.

BASIC SERVICES

Every house is different, as is every lifestyle, but most homes rely on water supplies, electricity, gas, oil, or solid fuels. Remember it is your home and you are in control. All the basic services, while making life easier or more comfortable, can also be sources of great danger.

You SHOULD know where to go in your home to turn off any or all of the supplies—both for routine maintenance and in the case of emergency. DON'T wait until the storage tank in the loft overflows and brings down a ceiling. DON'T allow someone to be asphyxiated by gas before checking that a gas appliance is correctly installed.

It is NOT enough to know where the levers or stopcocks are. Check that you can operate them and that they do the job they are supposed to do. It is not impossible that someone has, at some time, bypassed a meter or rerouted a supply for convenience. The stopcock on the water inlet may NOT completely turn off all the water in the house. CHECK! When you switch off the electricity, is ALL the power off?

If you are in any doubt about the efficiency of the controls, or if they are very hard to get at, get them sorted out.

In a multi-occupied dwelling, some of the shared services may have only one turn-off point. Is it accessible to you at any time—day or night? If not, consider having your own controls installed. This is obviously vital in the case of gas or electricity.

ELECTRICITY

Most of us take electricity completely for granted. We flip a switch and a light comes on. We want toast, and pop the bread in the toaster. We want food quickly and microwave ovens offer just that. There is an electrical gadget for every job, and the demand on domestic supplies is enormous.

In most houses, particularly older ones, the supply is barely up to the demand. More and more sockets are added—or worse, people start overloading sockets and adding make-do extension leads.

There are two hazards which must always be remembered when dealing with electricity—FIRE and ELECTROCUTION. Almost half of all domestic fires, some causing loss of life, are caused by electrical faults. Electrocution is less common, but still claims hundreds of lives every year.

Most common-sense precautions will protect you from both hazards. Because electricity is so dangerous, it is well worth understanding as much about it as possible.

INTRODUCTION

Electricity existed in nature before man ever thought of generating it himself and using it to his advantage. Lightning is probably the most violent and dangerous manifestation—although simple 'static' produced when combing your hair or the charge built up in garments during tumbledrying is also electricity. Most of us have walked across a man-made-fibre carpet and then touched something, only to receive a jolt of pain in the fingertips. This is a minor electric shock.

Even the human body is operated by tiny electrical impulses sent through the brain, nerves and muscles. This is partly why electrocution is so serious.

Man-made electricity is only useful to us when it is channelled through an appliance. When current flows through a lightbulb, the filament glows. When connected to an electric motor, the motor turns. In all appliances heat is generated—and a magnetic field. The heat may be a problem, although adequate ventilation and careful use of appliances should take care of this. Sometimes the heat generated is the whole point—as in an electric fire.

The magnetic field is usually too small to matter, unless you live in close proximity to a group of pylons or a generating station, when the powerful magnetic field may have a damaging effect on your health.

The circuit

Think of electricity as energy being pushed under pressure along a wire. The electric current flows because the pressure is greater on the side it enters an appliance—the live wire—than on the side where it leaves the appliance—the neutral wire.

The electric current moves in a circuit. If that circuit is broken, it cannot flow. By switching off a switch or unplugging an appliance, you are breaking the circuit.

The supply of electricity you receive flows from the incoming mains, through your meter and fuse box/consumer unit, along the fixed cables which are (usually) hidden in the walls, ceiling and floors. Via socket and plug or lampholder, the current flows through the live wire to the appliance, causes the appliance to do whatever it does, and then returns to the mains supply again down the neutral wire.

> ▶ The live wire—in some countries—is known as the active, positive or hot wire. The earth—especially in the US—is called the ground.

Conductors/insulators

The current flows through some substances more easily than through others. Substances which allow electricity to flow are called conductors. Most metals are conductors—copper, brass and aluminium are commonly used for wiring.

Substances which resist or prohibit the flow of electricity are called insulators. Most non-metals fall into this category—but in practice the most common insulators are plastics, rubber and ceramic.

For these reasons most domestic wiring is copper, which is a good conductor and flexible enough to work with. It is sheathed in waterproof plastic to restrict the flow of electricity to the copper.

Test of strength

Conductors have to be thick enough to take the amount of current passing through them. Generally, the stronger the current, the thicker the conductor has to be. The longer the run of cable, the thicker too. DON'T expect a thin cable used for wiring a battery-operated doorbell to carry a full-strength mains current. The wire will heat up, melting the insulating plastic coating and then melting itself at the weakest point—possibly starting a fire in the process!

Trying to pass a heavy current in this way is causing the wire to act like a fuse (see **Fuses**). It is also the principle of the light bulb—but the filament does not burn out because air, which is needed for combustion, is excluded from the bulb.

High voltages

Very high voltages carried in overhead cables are insulated by the air. Ceramic insulators are used to protect the pylons and the points at which the current comes down to earth. The currents carried may be as high as 400,000 volts, and are reduced at substations to be used domestically.

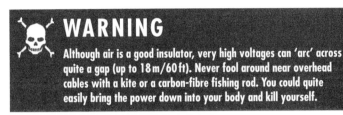

WARNING

Although air is a good insulator, very high voltages can 'arc' across quite a gap (up to 18 m/60 ft). Never fool around near overhead cables with a kite or a carbon-fibre fishing rod. You could quite easily bring the power down into your body and kill yourself.

Earthing/grounding

Electricity always wants to return to earth. This process of 'earthing' is what happens when you receive an electric shock from an appliance, or from a live wire. The electricity is trying to return to earth through you. The human body is not a very good conductor, but so strong is the tendency for electricity to earth itself, it will take any route offered to it. Effectively, you form a circuit.

Wearing rubber boots, which cut off the route to earth and therefore break the circuit, should protect you. Standing on wet ground makes you a more efficient circuit and is therefore deadly.

For this reason, mains circuits and most appliances carry a third wire (apart from the live and the neutral)—the earth wire. In appliances, particularly those with metal casings, the earth wire will be fixed to the inside of the casing. It is intended to offer a swifter route to earth for the electricity than you do—thereby protecting you. In reality, it does more to protect the appliance—the fuse does more to protect you (see **Fuses**).

REMEMBER

The earthing of your domestic wiring is designed to save your life and prevent fires. Do NOT tamper with it or disconnect it. It is advisable to have your earth checked every few years or after work has been done on your wiring.

The electricity should pass down the earth wire to the socket. The socket is connected to the mains supply, which is also earthed. All the earth wires on the domestic circuit should run to one heavy earth cable at the meter. It used to be quite common for the mains supply's earth to be connected to the water pipes to guarantee a quick and easy journey for the electricity, but the introduction of plastic plumbing has rendered this method unreliable. You will still find that water and gas pipes are earthed to protect anyone touching part of these systems should any part of the electrical wiring come into contact with either of them.

> ## ☠ WARNING
>
> Plastic plumbing is becoming more common. This is not effective as a means of earthing your domestic electricity. If all or even part of your plumbing has been replaced with plastic, your mains supply may no longer be earthed. CHECK!

ELECTROCUTION

 It is VITAL to understand electrocution, and what to do if it occurs. Through touching an appliance which has become 'live' or by touching a live wire, the current is carried through the body. Since many materials, including water and even urine, are effective conductors, it is possible to receive a shock without realizing that you are in any immediate danger.

It is quite safe to touch properly-insulated wiring in dry conditions, but use electrical screwdrivers with properly-insulated handles and stand on non-conductive surfaces—a rubber mat will do, or thick rubber boots. Do NOT touch or lean on pipes or scaffolding as there is a possibility that these could give the current an earth again.

A shock is very painful and frightening. The effects and seriousness depend on the circumstances and the individual. If the body is wet, or in wet conditions, a shock is more likely to be fatal. Most fatal shocks at domestic

> ## ☠ WARNING
>
> If someone is electrocuted by extremely high voltages from overhead cables, do not even get close to them. If you attempt to sever the contact with a 'non-conductor' such as a stick or a piece of clothing, the current will 'jump' the gap. Call for help.

SAVE A LIFE!

ELECTROCUTION

You must act quickly. If someone else is with you, they should call an ambulance while you apply first aid

- ► **CHECK! Is current still flowing?**
  DON'T touch victim or source of current
- ► **CHECK! Is there water about? You may be at risk**
- ► **SWITCH OFF at consumer unit or socket OR**
- ► **Pull victim free with leather belt, dry towel, stick**
  Knock victim's hand free with a stick IF victim is dry, grasp clothing WITHOUT touching body

If victim has fallen there may be other injuries—don't move the person unless you have to

IS VICTIM BREATHING? IS HEART BEATING?
If not breathing: Give artificial respiration—either mouth-to-mouth or mouth-to-nose. DON'T GIVE UP!

If heart not beating: Check for pulse. Victim may look 'blue', especially round lips. Give cardiac compression.

If not breathing AND heart not beating: Give six compressions alternated with two lung inflations.

If you have help: One person concentrates on breathing, the other on the heart.

WHEN BREATHING AND HEARTBEAT ARE OK
Place victim in recovery position.

BURNS
As soon as possible—at once, if there is more than one first-aider, reduce the temperature of the burns with cold water/ice. If working alone, attend to the heart and breathing first.

voltage happen when the current passes from one hand to the other—and therefore through the heart. Expect any of the following:

❍ Muscle spasms which cause the victim to grip the source of current. You must NOT touch the victim or the source of the current—you will also receive a shock. You MUST sever the contact with a non-conductor—a leather belt, a stick, a piece of furniture, a piece of clothing (avoiding body contact)

❍ The person may have fallen and hurt themselves, or may fall when the current is broken

❍ The heart may have stopped beating

❍ Breathing may have stopped

❍ Severe muscle spasms have been known to break bones

❍ There may be severe burns

❍ The victim may be unconscious or even 'dead' (but revivable with urgent medical attention)

Don't take chances

If someone has received a shock, especially if they have lost consciousness as a result, they should have an immediate medical check-up. Even if the victim appears well, they should be kept warm and quiet until the effects of the shock have definitely passed.

Burns, even when the shock does not appear to have other adverse effects, may be quite deep. Reduce the temperature as quickly as possible. Seek medical attention.

> ☠ **WARNING**
>
> If you attempt to work on any part of a domestic circuit, then the whole circuit must be rendered harmless at the consumer unit. If you have a fuse box, take the fuse out and keep it in your pocket. That way you definitely KNOW no one will switch the current back on while you are working. DON'T play games with electricity. DON'T take chances with electricity. DON'T GUESS. If a problem develops, go to the meter/consumer unit/fuse box and turn off the supply.

No earth wire?

Not every appliance has an earth wire. You can see from the flex (especially when connecting a plug) if there is only a live and a neutral wire.

Older lighting circuits were rarely earthed. This was because it was considered unlikely that people would regularly touch ceiling fittings.

Some appliances have plastic casings which cannot become live, or may claim to be 'double insulated'. Such appliances rarely have an earth wire.

With appliances which have no earth wire, or which claim to be 'double insulated', make very sure you use the recommended fuse, both in the plug for the appliance and at the fuse box/consumer unit by the meter.

☠ WARNING

On domestic lighting circuits which do not include on earth wire, be very careful when using metal light fittings. If a fault develops, you are NOT protected. Always turn off the switch when changing the bulb. Put a piece of tape over the switch to avoid someone else turning on the light from habit when entering the room. To be completely safe, turn off the power at the fuse box/consumer unit (or remove the fuse until you have changed the bulb).

NON-EARTHED SYSTEMS

Remote areas may receive mains power on a non-earthing system, whereby current trying to run to earth is fed back into the mains along the neutral wire. These systems are rare and are governed by strict regulations. Do NOT attempt to make changes to the system yourself, unless you are a qualified electrician. This system is sometimes known as a protective multiple earth.

WHAT THE TERMS MEAN

Current is the rate of flow of electricity, and is measured in amperes, known as amps (usually seen as **A**).

Voltage is the pressure difference between the live and neutral wires (usually expressed as volts or **V**).

The **resistance** of an appliance to the flow of electricity, or the amount it impedes the flow (impedance), is measured in ohms (usually Ω).

Watts (**W**) are used to measure the power consumption of an appliance. Actually the number of watts indicates the amount of electrical energy the appliance consumes in a set period of time. This sounds complicated, until you remember that a 15 W lamp is pretty much at the lower end of the scale. It doesn't get very hot or glow very brightly. A 100 W lamp gives out a lot of light and heat, using more electrical energy. An electric fire with one straight radiant element is likely to be 1000 W (1 kW). Most of the electrical energy is turned into considerable heat to make the fire effective.

Simple maths

You may be unsure about the power rating of an appliance, but in practice most electrical equipment is well-labelled to tell you the voltage, impedance or power consumption. Often the instruction booklet gives you more information.

If you know two of the values, it is possible to calculate the third:

W = V × A indicates the consumption of power
A = W ÷ V indicates a safe fuse or flex rating

AC/DC

When voltage is expressed on an appliance as, say, 240 V, it's quite usual to see the letters AC immediately following. This refers to the type of supply.

Electricity is produced by a conductor (usually a coil of copper) rotating in a magnetic field. This process produces what appears to be a steady current, but one half of each revolution produces a positive charge, the other results in a negative charge. The current alternates between positive/negative at approximately 50 times a second, in Britain. This is called alternating current—AC.

Direct current (DC) does not alternate positive/negative—and is usually achieved by electronically rectifying the alternating current. In some situations a DC current is necessary—usually in industry. DC current is often used for controlling motors, lifts, trains and underground trains. Many motorized household appliances convert AC to DC—including turntables, some food mixers and some fans.

DOMESTIC SUPPLIES

Electricity for domestic use is not supplied at the same voltage throughout the world. This doesn't really matter unless you intend to travel with electrical equipment. There may even be regional variations within a country. There are adaptors and transformers available to enable electrical equipment to be used safely. These step up or decrease the voltage accordingly. You should find out in advance if this is likely to be a problem—KNOW YOUR DESTINATION!

Major differences

Power supplies, electrical sockets and plugs vary the world over. In the UK and Australia, for example, voltages are very similar and both use three-pin plugs. The important difference is that plugs are not fused in Australia—whereas they always are in the UK. In the US, not only is the voltage lower, but plugs are two-pin and unfused. Switches on lamps may be turn buttons, not rocker switches or press buttons. Room switches may operate in the opposite direction to the UK. 'Up' may be ON, while 'down' is OFF. All could lead to confusion and mistakes could be dangerous.

If travelling or emigrating, be aware of national variations. KNOW YOUR DESTINATION!

Types of domestic circuit

The electricity supply enters a premises through a sealed service fuse box (to which only the electricity company has access), passes through the meter (which records the current used) and travels through a set of fuses, before being distributed about the building.

Very old systems had separate fuses for each outlet. Power (socket) and lighting circuits, even quite recently, had separate fuse boxes. A messy fuse board may indicate old, messy or dangerous wiring.

A modern system usually has a single fuse box, or consumer unit, linked to a smaller number of circuits, with each circuit protected by a fuse or circuit-breaker. Each appliance is protected by a fuse in its plug. Alternatively the actual socket (or special switch) may also be fused.

Heavy cable is used to serve a number of outlets, which are usually arranged as a ring—starting and ending at the consumer unit. This is known as a ring main.

A less-common domestic arrangement—a radial circuit—passes the current from one outlet to another, but ends at the last outlet.

Spurs or extensions may have been taken off either of these types of circuit at any point, to serve another outlet.

SAFETY FIRST ■ ELECTRICITY ■ DOMESTIC SUPPLIES

45

Special circuits may be run off to serve rooms such as kitchens, where demand for power is high, or for individual appliances such as cookers. It is unlikely that you will overload a circuit, unless you use a lot of demanding equipment such as electrical heaters.

Power limits

A 13-amp socket will be able to supply up to, but not more than, 3000 watts power if safely and correctly installed. Most ring mains carry a fuse of 30 amps. This should tolerate a total load of 7200 watts—on the whole circuit.

The following box should help you work out if you are within safe limits:

With apparatus, or even an entire installation, that creates a very heavy demand for electricity, three-phase power is commonly used. Generally speaking, this makes available three times the 240 volt (AC) supply. The three phases are balanced so that they are used equally, and in most cases separately fused and arranged so that the alternations of each supply are 'out of sync' with one another.

The commonest uses of three-phase power are kilns, hoists (reversible motors), factories, photographic/film studios, workshops, schools, hospitals and other large buildings with a high number of rooms, sockets and lights.

ADDED DANGER

Three-phase equipment should be regularly checked for safety. In large buildings it should be arranged so that you can never come into contact with more than one phase at a time. To do so could result in a shock of 415 volts. Usually three phases are each assigned different tasks—NOT one phase to sockets, one to lights etc—so the risk of making contact with more than one phase is eliminated. In most cases the three phases each feed an equal number of rooms.

NO NEUTRAL WIRE?

Three-phase motors usually have no neutral wire. The motor is fed by all three phases, and the consumption is balanced so that there is no power to be returned via a neutral wire.

WARNING

Never attempt any work on a three-phase circuit, unless qualified to do so. You would be risking a shock of 415 volts.

FUSES

 When a wire carries a current, heat is generated. If the current is high and the wire is too thin to cope with it, the wire will get hot—and probably 'burn out' or melt. A fuse is a deliberate 'weak link'. If the current rises (usually because a fault has developed) above the intended limit for a fuse, the fuse wire gets very hot, melts and breaks the circuit. This may be caused by:

◐ Electricity leaking to earth, either directly, or through the earth wire. The earthed metal casing of an appliance may have become live and the current is returning to earth. This causes an increase in the current because there's no resistance.

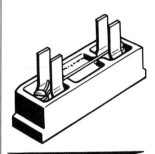

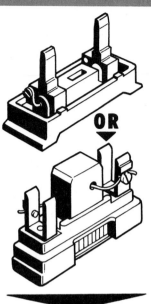

OR

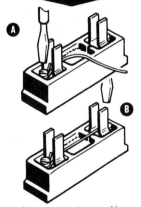

▲ Loosen screws. Remove old wire. Fix new wire at one end (A) with screw. Take wire across, tighten second screw (B). SNIP OFF SPARE WIRE!

▲ With some types of fuse, it is necessary to thread the new piece of fuse wire through (C). Follow path of old wire or compare to another complete fuse.

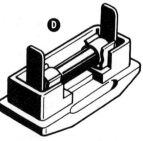

► Cartridge fuses (D) need to be removed and replaced. With mini-circuit breakers (E), flip switch or press the reset button.

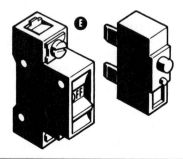

SAFETY FIRST ELECTRICITY ■ FUSES

The live and earth wires may have touched in a faulty plug or socket. Spilled water or another conductor may have formed a new connection.

◗ Similarly, live and neutral wires may have become connected either directly or through a fault in an appliance.

◗ Too heavy a load may have been placed on a circuit—perhaps too many high-powered appliances have been connected to it. The fuse won't allow the circuit to be endangered by too much current passing through it.

WARNING

When a fault develops and a fuse 'blows', a dangerous amount of heat may have been generated in an appliance, a plug or a socket. Check the circuit for signs of overheating.

When a fuse 'blows'

First check the fuse in the plug or socket. The easiest way to check them is to replace them, since a 'blown' fuse may not look any different from a good one.

If these are working, go to the fuse box/consumer unit and check the fuse there or press the reset button (see **Replacing a fuse**).

It's usually more obvious when a fuse has blown at the fuse box (if you know which outlets are on a circuit) because a whole circuit will be rendered inoperative. If a light on one floor's lighting circuit blows a fuse, then all the other lights on that circuit will go out.

Replacing a fuse

If you know that switching on a light or a particular piece of equipment has blown a fuse, switch it off before attempting to replace the fuse. If a circuit fuse has blown, test something else in the socket to make sure that the fault is not in the circuit. Check for possible obvious causes before retrying the equipment or you will just keep blowing fuses.

An expansion fuse or 'circuit breaker' is enclosed and usually all you have to do to re-establish the circuit is to press the reset control button or flip a switch (E). You will often find circuit breakers on sink disposal units and other equipment which is designed for short spells of usage and can get overheated if run for too long (or are prone to jamming).

Cartridge fuses (D), which you will find inside plugs and special sockets as well as in fuse boxes, are sealed capsules containing a fuse wire. A few are transparent and enable you to see the fuse, but most give no visible sign that they have blown.

You must simply replace the fuse and if the appliance works you will know that the fuse was the problem. You can always test an appliance fuse by fitting it into the plug of equipment which you know is still operational.

Old style fuses in a fuse box each consist of a short piece of wire between two screws or terminals, passing over or through an insulated pad. You will sometimes see that the middle of the wire has melted. If it looks complete, try lifting it with your screwdriver. Replace by fitting another piece of fuse wire of correct value between screws or terminals (A, B and C).

❗ WARNING

Always use the correct value of fuse in a plug for an appliance, or at the main fuse box. Don't simply copy the fuse you found there. It may not have been the correct fuse.

Fuse ratings for appliances

Using the correct fuse is VITAL. The lower-rated fuses, such as 3-amp, are used in the plugs of appliances with low power consumption. They are intended to 'blow' more easily, protecting delicate equipment. Using a fuse of a higher rating may allow a fault to cause overheating, possibly starting a fire.

On a 240 volt supply, 3-amp fuses are used for appliances up to 750 watts—this includes computers, televisions, clock radios and fridges.

A 5-amp fuse is suitable for appliances which consume 760–1000 watts. A 10-amp fuse suits appliances 1000–2000 watts.

RCDs/RCBs/ELCBs

Residual current devices (also called residual current breakers and earth leakage circuit breakers) come in various forms. Some plug into a socket, and the appliance is then plugged into them; some may be fitted into an extension cable; some replace sockets altogether; or they may be part of the consumer unit. If anything goes wrong with the current, such as a sudden leak to earth (which might be you getting a severe shock!), they instantly cut the power and need to be reset. They are very responsive and are designed to give you protection, in addition to that given by fuses and earthing. ELCBs were so called because they respond to even small leakage of current to earth—a leakage which might not blow a fuse.

RCDs are VITAL life insurance. Users of power tools, electrical gardening equipment or kitchen gadgets where liquid is involved should ALWAYS use RCDs. If upgrading your domestic wiring, consider installing these safety devices on the whole system.

Above this and up to 3000 watts maximum, a 13-amp fuse is safe.

Any electrical dealer or electrician will be able to advise you on the correct fuse ratings. Often the appliance is labelled, or the instruction manual will indicate the amount of power consumed by the appliance. (For a general guide only, see **Power consumption.**)

Main circuit fuses

On a 240-volt supply, and especially on special circuits for cookers, showers and storage heaters, the correct fuses MUST be used. In general, these are:

Immersion water heater: 15–20 amp
Lighting circuits: 5 amp
Ring power (socket) circuits: 30 amp
Radial power (socket) circuits: 20–30 amp
Showers: 30 amp
Oven (up to 12 kW): 30 amp
Oven (over 12 kW): 45 amp
Storage heaters: 20 amp

REMEMBER

Always keep a flashlight near your electricity meter. If it is in a dark cupboard, this will help you see what you're doing. If you need to switch off the power at night, you will need to avoid fumbling around. DON'T use a candle if the fuse box/consumer unit is near your gas meter!

CABLES/FLEXES

 Live, neutral and earth wires are colour-coded for identification to make sure that the correct electrical connections are made. Unfortunately, colour coding is not standard around the world, but it is beginning to be. This can never take account of strange 'bodged' wiring jobs people may have created.

Main circuit wiring (cable)

Old rubber sheathing is still common and should be treated with suspicion. Rubber degrades with age and crumbles. But even modern wiring (on supply cables) will have a red live, a black neutral and a green-and-yellow earth. In practice, the earth wire is often unsheathed in the centre of the cable. Special green-and-yellow sheathing is sold for you to cut to length, to sleeve sections of the earth which are exposed when making connections. DON'T use spare pieces of black or red sheathing. DON'T use electrical insulating tape.

Appliance wiring (flex)

The live is always brown, the neutral is always blue, the earth is always green-and-yellow (when there is an earth wire).

Old appliances—which should really be rewired and updated by an expert (or discarded)—will probably have a red live, a black neutral and a green earth. If the wires are old, the colours of the sheathing may have faded or discoloured. DO NOT GUESS! Old rubber insulation is, by now, not safe.

> **! WARNING**
>
> When connecting wiring with screw fittings, DON'T over tighten the screws. To do so may sever individual strands of multi-stranded flex. These may become dislodged and form an unwanted and dangerous connection. With cable or single-stranded flex, over tightening the screw may flatten (or even sever) the wire. Heavily-flattened wire may act like a fuse, with a reduced capacity for carrying current. It can overheat, may burn out and may start a fire.

Types of domestic cable

On a 240-volt supply, the lighting circuits are generally formed by 1.0 mm² twin-and-earth (a plastic casing housing three single-strand wires—one sheathed red, one black and one unsheathed). Each wire is about 1 mm in diameter.

The power cable for the main power (socket) circuits is very similar, but of a heavier weight.

On special circuits for ovens, electrical heating installations or shower heaters, even thicker cables may be used.

> **! WARNING**
>
> No mains wiring alterations or installations should be attempted by anyone who is not qualified to do so. Because of the dangers, in some countries such as Australia, severe restrictions are imposed on the type of electrical work that you can do yourself.

Types of flex

Flexes may contain two sheathed wires—brown (live) and blue (neutral)—in an outer casing **OR** three wires, if a yellow-and-green earth wire is included. Sometimes a thin two-core flex may have no outer casing—but this is only really suitable for wiring battery-operated doorbells.

Most flexes, inside the internal coloured sheathing, are made up of several thin strands of wire, which gives the flex FLEXibility. When stripping the sheathing from the ends, for wiring a plug, take care not to sever individual strands. Losing one or two may not matter, but if you lose too many the current-bearing capacity of the wire is reduced at that point. This could cause overheating—or even fire.

When you buy flex, ask the dealer which is the correct one for a particular appliance. There is a rising scale of thickness to cope with different power consumptions. The flex is usually marked with its recommended maximum load.

When using longer flexes—perhaps an extension lead to a garden tool—you will need a thicker cable to cope with the voltage drop over the distance.

If the flex needs to withstand a lot of movement and handling (an iron, a power tool, a hairdryer), a heavier flex should be used than is suggested by the power consumption.

For pendant light fittings, the flex needs to be heavy enough to support the lampshade, even if only a 25-watt lamp is used. NEVER expect a heavy pendant fitting to be carried by the flex alone. A chain, or similar, must be anchored into a joist above the ceiling. The flex is for carrying power—not weight.

REMEMBER

Always use the flex grips on plugs to hold the flex. Make sure ALL the flex, including the outer sheath, is held. NEVER pull plugs out of their sockets by the flex.

MISWIRING A PLUG

DANGER! Miswiring a plug, by swapping the live and neutral wires, is extremely dangerous. The fuse in the plug is totally ineffective, because the live wire is not connected to it.

Normally, what happens when a table lamp (for instance) is switched off, is that the switch breaks the circuit—stopping the live current at the switch. When the live and neutral wires are swapped, the switch effectively breaks the neutral wire and the lamp appears to behave correctly—by going off. However, all you have done is to cut off the return path for the electricity, so that the current cannot flow.

THE LAMPHOLDER IS STILL LIVE! A slip when changing the bulb could lead to ELECTROCUTION.

Old wiring

Wiring circuits should be checked at least every 20 years. Old wiring can be recognized by primitive bakelite plugs and fittings, fabric-covered flex hanging from ceiling fittings and more than one fuse box.

In Britain two-pin sockets and plugs should be replaced with modern three-pin 13-amp fittings, when rewiring is done. Replacing the sockets alone is not good enough.

> ## ⚠ WARNING
>
> Keep all flexes and leads as short as possible. Many falls and accidents are caused by tripping over trailing flexes! Taping cable to the floor is a temporary safety measure, but you should consider installing new sockets where you actually want to use the equipment.

Adaptors

Overloading a socket with an adaptor which can take several other plugs is NOT safe. Worse still, other adaptors may be joined on, allowing still more plugs to be fitted! Not only will you overload the socket electrically, but literally as well. The adaptor will work loose in the socket from the sheer weight of plugs and flexes. This can weaken electrical connections. Exposed pins may be touched, risking electrocution. Have you got enough sockets to avoid this?

Extension leads

If you have to use extension leads inside your house, either you have not got enough sockets or the arrangement of furniture and appliances means that your sockets are in the wrong positions. Multi-socket extension leads are undoubtedly safer than adaptors, but:

◑ **DON'T** leave the lead coiled when in use. It will generate heat and possibly start a fire.

◑ **DON'T** use the sockets if there is any sign of charring anywhere, even on the flex.

◑ **DON'T** get portable sockets wet.

◑ **DON'T** exceed the 3000-watt maximum that the socket on the wall could cope with.

Extension leads that bring power from sockets to hard-to-reach places like lofts or gardens, temporarily, should be made of a flex suitable for the power consumption and the length of run involved. **Some cheap multi-socket extension** leads have been found to be extremely dangerous. If any problem occurs, discontinue use.

REMEMBER

If you must extend the wiring on an appliance, NEVER join two wires by simply twisting the ends together—even if you bind the join with electrical insulating tape. THIS IS UNSAFE. Use proper shielded protectors or junction boxes. Even an in-line switch means that a new piece of flex could be joined on. It is preferable to fit a new longer flex. Since this means opening the appliance and tampering with it, consult an expert.

WARNING

If a two-part connector or a plug/socket extension is used to extend a flex, NEVER NEVER **NEVER** arrange this so that the live flex ends with the exposed pins. It must end with the socket. If outdoors temporarily, wrap the connection in polythene for protection against moisture. NEVER pull on an extended flex whilst in use. If you do, go back and check the connection.

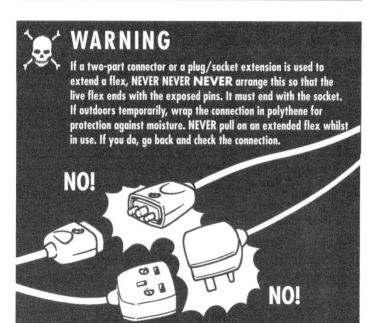

KITCHENS AND BATHROOMS

The kitchen with its numerous gadgets and the power-hungry oven creates a greater demand for electricity. Consider installing a separate circuit. There is also the presence of water to consider—electricity and water are a very dangerous combination. Some gadgets, such as irons, are handled a lot and flexes and connections can wear out.

In bathrooms, the main danger is the presence of water and the metal sanitary fittings, even if you're only temporarily using power tools. Protect yourself with an RCD/RCB/ELCB. Both rooms deserve special consideration, electrically speaking (see ROOM CHECK: **Kitchens** and **Bathrooms**).

OUTDOOR CIRCUITS

If you require electricity outside the house for a garage, greenhouse, lighting or a water pump, proper weatherproof exterior fittings and cables must be used.

All outdoor wiring should be fitted with RCDs/RCBs/ELCBs, which should protect you immediately if a fault develops.

External wall lights

Porch lights can be run from the household supply, but the exposed wiring must be protected by a conduit. All connections MUST be weatherproof. The lamp inside the fitment must be protected by a rubber skirt or cup which prevents contact with the connections.

Electricity for outbuildings

Supplies to outbuildings may be run off the domestic supply.

They can be carried overhead (up to 5 m/15 ft): Provide posts and a support wire (galvanized catenary wire). The wire should slope slightly to encourage water to run off it. At the lower end the cable should have some slack arranged as a drip-loop to prevent water following the cable all the way to the end connections. Keep overhead cables above head height and be aware of the possibility of wind damage.

They can also be carried underground (over 5 m/15 ft): Cables must be armoured (such as MICS—mineral insulated copper sheathed) and run in a strong conduit. They should be run at the base of walls and fences if possible—or buried low enough to avoid damage from digging.

The outdoor supply should be earthed, both from the main earth connection at the meter/consumer unit and also in the outbuilding. The circuit should be fitted with an RCD/RCB/ELCB. If the use of the outbuilding involves the need for more than one light and one socket, it should be fitted with its own fuse box/consumer unit.

Garden sockets

It is possible to fit sockets in the garden, preferably in a sheltered location. Two main kinds of socket are available (and which claim to be weatherproof). One has a hinged plastic cover which protects the socket when not in use. The other has a screw-on cover—each plug that fits into it also has a cover fitted round the flex. WHEN the plug is inserted, the cover is screwed onto the socket, covering the plug.

Gas is a relatively cheap, clean fuel for cooking and domestic heating. Whether produced as a product of the coal industry (coal gas), 'natural gas', or in cylinders (propane and butane), it is extremely DANGEROUS—and should never be taken for granted. Coal gas is highly poisonous—fatal cases of asphyxiation are very common. Some 'natural' gas, such as North Sea gas, is a lot less poisonous, but will still make you very ill in sufficient concentration. Both are EXPLOSIVE—the tiniest spark could trigger a devastating explosion.

The supply

The gas enters your premises via the service pipe, which ends with an isolating (on/off) valve, just before the meter. This valve is operated by turning the spanner-like handle. It should move without too much effort—but it shouldn't be loose either. If you have any concerns about it, if you doubt you have the strength to turn it on and off, call the gas company and ask them to send someone to sort it out.

Pipes from the meter carry the gas around your home to each appliance. Each appliance will probably be connected by a fixed coupling, with or without an on/off key—in some cases to an outlet point, with its own on/off key. It is usual for a flexible hose to be joined onto one of these—the commonest being a gas cooker. The flexible hose allows the cooker to be eased forward for cleaning without disrupting the supply.

All internal gas plumbing is made of copper, these days. If you find any pipes which appear to be lead composition or iron, they are likely to be rather old. Get them checked!

! WARNING

Do **NOT** tackle even the simplest of gas installations or maintenance jobs yourself! Gas plumbing is a job for experts only. Any problem with gas supplies or appliances should be reported immediately.

VENTILATION

To burn safely, all gas appliances MUST have a supply of fresh air. This may be supplied by normal room ventilation or through a balanced flue, a system which

OTHER OPTIONS

Low-voltage installations

Consider low-voltage lighting—even in the garden. Not only does it use a lot less power, it also works on such low voltages that there is no danger of electrocution from the fittings. It has become common to use low-voltage lighting in shops, in gardens and (more recently) in homes. Low-voltage pumps are also available for garden ponds and fountains.

Generally speaking, the mains supply is reduced by a transformer. If used for outdoor lighting, the transformer must be kept in the house, garage or outbuilding.

Lighter cables may be used in conduits on walls, or buried—sleeved in a length of ordinary garden hosepipe. It is still important to check that all connections are weatherproof.

> **! WARNING**
>
> A transformer should be properly housed and subjected only to the loads specified by the manufacturer. Transformers generate heat while operating—even those which plug into a socket to operate electronic equipment. Ensure adequate ventilation and switch them off when not in use.

Cordless devices

More and more cordless devices are available. These are mainly irons and kettles. There is a base unit which is plugged into a socket and carries full mains power. The 'cordless' appliance only receives power when it's sitting in the base unit. The very nature of these types of equipment means that a careful eye should be kept on them for signs of wear and tear, especially i water is involved. Follow safety instructions.

Rechargeable devices such as power tools, food mixe electric screwdrivers—even electric toothbrushes—are cord' when in use, but must be returned to their recharging u Check carefully for wear and tear. DON'T try to use the rec ing unit for anything except the tool it was intended for.

Rechargeable batteries

There are so many portable, expensive-to-run, operated devices on the market—especially persona' that rechargeable 'NiCad' (nickel cadmium) ba becoming very common. These batteries do NOT and overcharging can be dangerous. Always follow draws air directly into the apparatus from outside the house and discharges fumes directly out again. This is also called a 'room-sealed' system.

Never obstruct airbricks or attempt to seal off all draughts in a room where gas equipment is used. Beware double glazing!

Failure to ventilate could lead to the appliance using up all the available oxygen, possibly causing unconsciousness and subsequent suffocation. It also increases greatly the risk of a build-up of carbon monoxide—a very poisonous gas which has no smell. You are only likely to be aware of it when you begin to be subjected to its poisoning effects.

NEVER cover or lean anything against the external vent of a balanced flue. Obstructing the air flow could cause inefficient burning and the production of poisonous fumes.

Gas appliances MUST be checked and serviced annually. Blocked jets and the accumulation of deposits may make the apparatus dangerous as well as inefficient.

All gas burners give off carbon dioxide and water vapour. With permanent appliances these are taken away by the flue or chimney, but a fault may sometimes discharge them into the room. Unvented appliances may cause severe condensation problems if room ventilation is poor.

Carbon monoxide

Explosions caused by leaking gas are frighteningly common, but four times as many people die from carbon monoxide poisoning. Unburnt gas does not contain carbon monoxide—but it is produced if a gas appliance is not burning efficiently.

You cannot smell carbon monoxide, but breathing it for even fairly brief periods of time can be FATAL. Coal gas produces higher concentrations of carbon monoxide than 'natural' gas. If gas appliances are installed by a professional and regularly serviced (once a year)—and ventilation is adequate—the risk is minimized. Most cases of carbon monoxide poisoning involve gas appliances with blocked flues or chimneys, or dangerous old gas water heaters in bathrooms.

Avoid cheap gas equipment from non-specialist dealers and only buy secondhand equipment if you are willing to pay someone to completely overhaul it to make sure it is safe.

Carbon monoxide is also produced by exhaust fumes from vehicles, oil burners, wood/coal/charcoal fires and bonfires. Always ensure adequate ventilation. NEVER run the engine of a vehicle in a sealed garage!

Look for the danger signs

A yellow/orange flame indicates inefficient burning of the gas. **Stains/soot/discolouration** around the appliance may indicate that exhaust gases are leaking into the room. **A slightly gassy smell** of burning may also indicate the presence of exhaust gases.

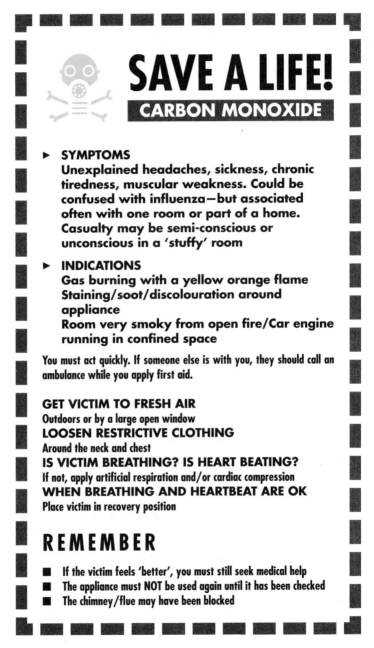

SAVE A LIFE!
CARBON MONOXIDE

► **SYMPTOMS**
Unexplained headaches, sickness, chronic tiredness, muscular weakness. Could be confused with influenza—but associated often with one room or part of a home. Casualty may be semi-conscious or unconscious in a 'stuffy' room

► **INDICATIONS**
Gas burning with a yellow orange flame
Staining/soot/discolouration around appliance
Room very smoky from open fire/Car engine running in confined space

You must act quickly. If someone else is with you, they should call an ambulance while you apply first aid.

GET VICTIM TO FRESH AIR
Outdoors or by a large open window
LOOSEN RESTRICTIVE CLOTHING
Around the neck and chest
IS VICTIM BREATHING? IS HEART BEATING?
If not, apply artificial respiration and/or cardiac compression
WHEN BREATHING AND HEARTBEAT ARE OK
Place victim in recovery position

REMEMBER

■ If the victim feels 'better', you must still seek medical help
■ The appliance must NOT be used again until it has been checked
■ The chimney/flue may have been blocked

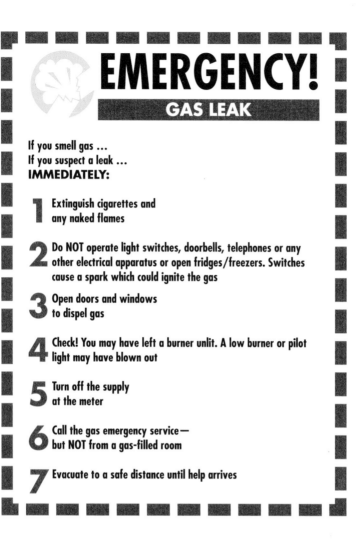

EMERGENCY!
GAS LEAK

If you smell gas ...
If you suspect a leak ...
IMMEDIATELY:

1 Extinguish cigarettes and any naked flames

2 Do NOT operate light switches, doorbells, telephones or any other electrical apparatus or open fridges/freezers. Switches cause a spark which could ignite the gas

3 Open doors and windows to dispel gas

4 Check! You may have left a burner unlit. A low burner or pilot light may have blown out

5 Turn off the supply at the meter

6 Call the gas emergency service— but NOT from a gas-filled room

7 Evacuate to a safe distance until help arrives

Gas safety rules

- All gas installations and maintenance MUST be done by experts—NEVER attempt to do any of the work yourself.
- NEVER use, or allow to be used, an appliance which is known to be or suspected to be faulty in any way.
- If you suspect a leak, turn off the supply at the meter and inform the gas company.
- Gas must have a ready supply of air to burn efficiently. NEVER seal off all ventilation to a room.
- NEVER fool around with gas. NEVER!

> # REMEMBER
>
> Water is likely to damage your property when pipes leak or installation work is shoddy—but leaks are visible and easily sorted out. Gas leaks can kill you, even before you have detected any problem.

Old gas water heaters

Old gas water heaters which do not have a 'balanced' flue or 'room-sealed' flue (which takes its air supply and discharges its exhaust through the wall) are EXTREMELY DANGEROUS. In some countries, such as Britain, it is now illegal to instal them. It may seem to work well but—**IF** there is inadequate ventilation, **IF** the appliance is not serviced at least every six months, **IF** the flue is leaky, damaged or blocked—**THIS TYPE OF APPLIANCE CAN KILL.**

- **NEVER** seal off ventilation or airbricks.
- **NEVER** use the heater while you are in the bath.
- **ALWAYS** open the door and windows when you are using the heater.

Gas alarms

Unfortunately, there are no domestic gas alarms or detectors—although there are quite simple devices in use in industry. Some can even detect the presence of carbon monoxide. Most people already have smoke alarms and, as soon as gas detectors become available, 'every home should have one'!

BOTTLED/CYLINDER GAS

 For those without a mains supply, two kinds of liquefied petroleum gas (LPG) are widely available: **propane** and **butane.** Neither has a noticeable smell so a stenching agent is usually added. The smell is similar to that of mains gas.

Propane is not widely used indoors, because it is stored under great pressure, and this increases risk. When it is used to fire domestic central heating, it is stored outside either in a bulk tank (which is replenished by a tanker) or in large cylinders beside the house. In Britain, a maximum of four cylinders of 47 kg capacity is permitted.

Butane cylinders are mainly used to fuel portable heaters. The cylinders of both gases should be treated with care. DON'T knock, handle them roughly or bang them to see if they are empty. DON'T subject them to heat or leave them exposed to sunlight or rain.

Both propane and butane have other uses. Both may be used for blow-torches to burn off paint or braze metal. Butane is common as lighter fuel, for camping gas and even cordless hair curlers (see POISONS). REMEMBER that these gases are **POISONOUS.**

Changing cylinders

Follow manufacturer's and supplier's instructions to the letter. If these are incomplete or confusing, DON'T guess! Call in an expert to show you how to change cylinders safely and easily.

> ## ❗ WARNING
>
> Good ventilation is required to allow exhaust gases to escape—and to avoid the build-up of condensation. NEVER block established vents in walls and windows or attempt to seal every draught from a room. Beware double glazing! You may need to provide extra ventilation.

Portable heater dangers

FIRE! There are thousands of domestic fires every year caused by heaters which have been placed too close to furnishings, or knocked over.

Some portable heaters are fitted with a special safety device which shuts off the gas if there is insufficient air for combustion. SOME DO NOT! Most shut off the supply if the pilot light goes out. SOME MAY NOT—or may not be working properly. Have the appliance serviced regularly (once a year).

❍ **NEVER** place anything over a heater—especially clothes for drying—many fires start like this!

❍ **NEVER** position a portable heater in a corridor. It may be knocked over or block an escape route.

❍ **NEVER** store gas cylinders below ground in a cellar or basement. Gas is heavier than air and could be undetected if a leak occurred, creating a severe fire risk.

❍ **AVOID** inhalation of gases. Breathing low concentrations will cause dizziness and nausea. Higher concentrations may asphyxiate or lead to death.

❍ **AVOID** contact with liquid gases. They can cause 'burns' similar to frostbite. Wash areas affected with warm water for at least ten minutes. Flush out eyes with warm running water for at least ten minutes. If in doubt—always seek medical attention.

❍ **KEEP CYLINDERS UPRIGHT**—in case the valve leaks.

SAFETY FIRST GAS ■ BOTTLED/CYLINDER GAS

63

- **DON'T** knock or kick containers—there is enormous pressure inside. Faults in the cylinder could lead to explosion.
- **NEVER** use portable gas heaters in bathrooms or other small rooms where there is too little ventilation.
- **LEAKS:** Turn off the cylinder immediately if you suspect a leak. Open doors and windows to ventilate. Take heater outdoors. If leak not immediately traceable, brush liquid detergent over connectors: it will bubble at the escape point.
- **NEVER** leave a heater on when you intend to go to bed or go to sleep.
- **NEVER** knock a heater over, especially when it is on. Liquid gas may escape causing a fire/explosion risk.

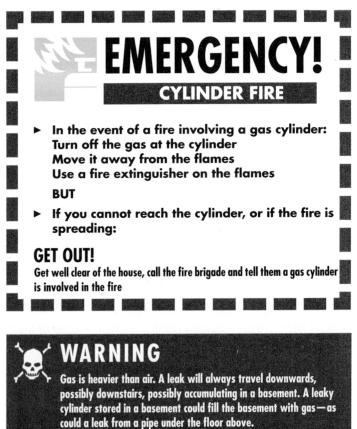

EMERGENCY!
CYLINDER FIRE

▸ **In the event of a fire involving a gas cylinder:**
  Turn off the gas at the cylinder
  Move it away from the flames
  Use a fire extinguisher on the flames
  BUT

▸ **If you cannot reach the cylinder, or if the fire is spreading:**

GET OUT!
Get well clear of the house, call the fire brigade and tell them a gas cylinder is involved in the fire

WARNING
Gas is heavier than air. A leak will always travel downwards, possibly downstairs, possibly accumulating in a basement. A leaky cylinder stored in a basement could fill the basement with gas—as could a leak from a pipe under the floor above.

If you smell gas, do not enter the basement or switch lights off or on. Extinguish all naked flames/cigarettes. If you have a mains supply, turn it off at once—and report the problem (see EMERGENCY!: GAS LEAK panel).

LIQUID FUEL

Two grades of fuel oil—paraffin and gas oil—are common in domestic use. Both are a fire risk and should be treated with respect. Both produce carbon monoxide when burning (see Gas). If a fire starts, the result may be the spillage of burning liquid—which is extremely dangerous. Use a fire extinguisher—a dry extinguisher is suitable—aimed at the base of the fire and into the appliance.

PARAFFIN

Paraffin is also known as kerosene. It is used for portable heaters and blow lamps, although many people still use old paraffin lamps (oil lamps). In some appliances—heaters and oil lamps—the fuel soaks up a wick. Blow lamps and storm lanterns require the paraffin to be pressurized by pumping air into the sealed paraffin reservoir.

Paraffin heaters

Paraffin heaters are safe when used properly, but are the cause of many fires. The heater must be placed on a level surface—away from draughts. If you set it in a fireplace, partially block the chimney to reduce draughts. Check the flame is burning blue (not yellow or orange) and not flickering in a draught.

It's not advisable to buy secondhand heaters. They may be dangerous. New ones supply a special trimmer for the wick. Use it every time you fill the reservoir.

Make sure there is plenty of air coming into the room. Paraffin heaters will produce deadly carbon monoxide as well as water vapour, which can lead to condensation problems. Don't seal off every draught. Don't block airbricks or fixed vents.

WARNING

When going to bed, you MUST extinguish a paraffin heater. Do so at least 20 minutes before retiring, and give the heater a final check before you leave it for the night. Make sure the wick is not still smouldering.

Refilling tank

Always use paraffin—NEVER petrol. If reservoir is removable take it outdoors to refill. Use a funnel and wipe any spillages from reservoir before replacing. It is easier to refill a detachable reservoir from a small storage tank with a draw-off tap than to pour paraffin from a can—but keep the tank/can in the garden shed, if you have one, not the house.

EMERGENCY!
HEATER FIRE

▶ **Do NOT use water or a water extinguisher (red) to put out a fire, even if there is a spillage of burning fuel**

DO NOT MOVE HEATER
On no account try to carry a burning heater

SMOTHER THE FIRE
Use a fire blanket or a dry powder (blue) extinguisher. Aim it right into the appliance or at base of flames
EVACUATE and call the fire brigade

GAS OIL

Gas oil is usually stored outside in large tanks and is piped to oil-fired boilers. All the safety and maintenance advice for gas and propane boilers applies here (see **Gas**).

Oil-fired boilers and heaters need efficient chimneys or flues. When a boiler is serviced (annually), make sure that the chimney/flue is cleaned out too. If the boiler has a balanced/room-sealed flue (which takes in air for combustion and expels exhaust gases through an outside wall), make sure nothing obstructs the flow of air around the vent outside.

Oil tanks

The tank for the oil must be well maintained. The main problem with these tanks is that any water that enters, sinks to the bottom and can cause rusting. In Britain, the tanks may be used to hold a maximum of 3500 litres (770 gallons)—which is a lot of oil. Imagine the problems you will incur if the tank is not well looked-after and the base of the tank suddenly gives way! There is a drain-off cock on some tanks, at the bottom, to let any accumulated water out.

Once upon a time open fires were the only source of heating. They are inefficient and wasteful. A lot of the heat goes up the chimney—polluting the atmosphere with several harmful by-products. In some areas only special smokeless fuels may be used to reduce atmospheric pollution.

Closed stoves and cooking ranges are more efficient, radiating more of the heat into the room and making it possible to burn fuel more slowly.

All solid fuels produce ash, dust and soot, acids and tars when they burn. Some of this material collects in the chimney, reducing its efficiency—which could be a very serious fire hazard. An inefficient chimney will also allow smoke and gases to fill the room where the fire is burning. This is very dangerous. One of these gases will be carbon monoxide—inhalation of which, even for a short amount of time, can be FATAL (see **Gas**).

The chimney should be swept annually, and once during the burning season. Apart from the risk of blockage and fire, acids in the soot and tar will attack the lining of the chimney during the summer months.

> **REMEMBER**
>
> For a fire to burn well and safely, the chimney must be kept clear of obstructions and the room must have a good source of ventilation. As warm air rises up the chimney, it takes with it potentially dangerous gases and the fire is able to draw on new supplies of air.

Burning wood

Wet or unseasoned wood is no good for burning—if you can get it alight it will be smoky and possibly cause dangerous sparks. Seasoning takes about two years—at the end of which time the wood looks drier and will have drying splits in the ends.

Provided they are well guarded and no flammable materials are left close by, stoves and heaters can be left burning slowly overnight, but get a good blaze going in the morning to warm up the chimney.

Servicing

Apart from making sure the chimney is clean and efficient, check all flues and stoves for signs of smoke leakage. Firebacks in open fires may need attention. They protect the chimney from the intense heat of the fire, so don't allow large cracks to go unattended. Renew any fire cement as it deteriorates.

WATER

Water is one of our basic human needs, but attitudes to safe drinking water and levels of sanitation vary remarkably around the world. The water system in most homes is prone to all sorts of problems, which may not be immediately life-threatening, but which can do a lot of damage to your home and your peace of mind.

A leaky pipe may go ignored for months or years, especially if the outward signs are not very dramatic. But this leak may lead to structural problems and rot. When the damage begins to show, the remedial work may be expensive or very difficult. Your home could become damp and smell damp. Your health could suffer.

PLUMBING DECISIONS

Simple plumbing jobs, such as changing the washer in a leaky tap or in the valve in a ball-cock assembly, are not very difficult and could be attempted by the beginner. Larger jobs, which might result in having to turn the water off for days (while you feel your way through the process) are not such a good idea. The work can also be quite strenuous.

Because plumbing for water is fairly simple in principle, with fairly low risk factors involved, many people 'have a go'. The plumbing in older homes and multi-occupied homes may be very difficult to follow as a result. There are numerous books on the market which describe simple procedures, and the plumbers' supply shop may be able to advise you. What neither of these can take into account are the idiosyncrasies of your existing plumbing.

If money and the cost of plumbing work are your main worry, do-it-yourself plumbing might be the answer. You may save money simply by understanding what you are asking a plumber to do—how much work it is and the sort of material costs involved. You may make his job quicker if you have already identified a problem before he arrives. If you attempt a difficult job and get in a mess, it may cost you more to have a plumber sort it out.

Remember that several special tools are required. Some homes have virtually no tools at all. Do you want to buy an adjustable spanner or a pipe bender for one job? Borrowing or hiring tools is a possibility.

WATER SUPPLY

Urban water supplies come from many sources—rivers, wells, reservoirs and underground springs. All of it is processed to offer safe drinking water to the public, despite the fact that much of it is used for washing, flushing toilets and other things for which pure water is not necessary.

Modern demands for water are putting increasingly heavy demands on supplies, especially in times of drought. Processing is costly and uses up valuable energy resources. Economy in use is therefore a first priority in reducing pressure on the environment and on the pocket.

Pure water?

Water treatment is a very complicated business, involving settlement tanks, filtration beds and the addition of chemicals. Most water authorities prefer the water going into the mains to be slightly alkaline—in order not to corrode the system. To ensure this, lime is added, but only at about 20 parts per million. Other chemicals are used too, like ferrous sulphate to make impurities cling together so that they can be filtered out more easily (about 18 parts per million).

A small amount of chlorine (about 5.5 parts per million) is used to kill off bacteria which may be present. By the time the water leaves the treatment plant, the level of chlorine will have been reduced to less than one part per million.

Most chemicals added to water, such as the chlorine or copper from your own pipework, may affect the taste, but are harmless at the concentrations at which they occur.

Other chemicals, which are known to be present in some supplies, are far from harmless. These include:

Aluminium: Sometimes added during water treatment. May be carried into reservoirs from the soil. May be naturally present. Links are being considered with Alzheimer's disease (premature senility). Water is unlikely to be a major source of aluminium—but it has been suggested that water-borne aluminium is in a readily absorbed form.

Nitrates: Increase the risk of a rare blood disease which affects bottle-fed babies (whose feed is made up with tap water). Suspected (not yet proven) link with cancer. Leached into supplies from fertilizers and manure in intensive farming. Very difficult to remove.

Lead: Most likely source is your own plumbing. Old pipework was made of lead (see panel). Very serious. Can damage the brain and nervous system, cause anaemia and affect the muscles, and stunt mental development in children.

Polycyclic aromatic hydrocarbons: From coal tar pitch, used to coat the inside of mains pipes (taken out of use in the UK in the mid-1970s). Suspected link with cancer.

Industrial chemicals: Such as industrial degreasing agents and dry cleaning fluid. Suspected link with cancer. These are

Plumbing for water used to be made entirely of lead. In old houses this may still be the case. In updated systems, some lead piping may still be found as the connection to the water mains. Lead poisoning is extremely serious and in some cases debilitating or fatal. It attacks the brain and the nervous system. It affects the muscles. It causes anaemia. The effects can be devastating in young children.

Some of the lead we absorb is from atmospheric pollution, but levels found in water samples indicate that water-borne lead is a major problem.

The problem is worse in areas with soft water, because the pipes do not develop a protective coating of limescale (see **Hard and soft water**). Soft water may even dissolve small quantities of lead as it travels through the pipes. No lead plumbing should be considered safe, however.

The long-term solution is to have ALL lead plumbing removed and replaced—but be prepared for a major upheaval.

The short-term solution is to filter all drinking water (see **Filtering water**). Never drink water that has been standing in the pipes. Run the tap for a few minutes before drinking.

among the numerous chemicals pumped into streams and rivers when pollution controls are flouted.

Bacteria: Those of diseases spread by contact with human excreta are the most serious risk, though contamination is comparatively rare. In areas where bacterial diseases such as cholera are endemic, local water should be avoided in favour of a safer alternative even though you may be likely to be more at risk when you go swimming.

Fluoride: Most supplies contain minute amounts. There is disagreement about the safety of adding sodium fluoride to water supplies in some countries/areas. This is done at a level of about one part per million, primarily to increase children's resistance to dental decay—and is effective at doing so.

Higher doses of fluoride during early tooth formation (possibly also because of fluoride dental treatments and fluoride supplements given to children) have been shown to produce a mottling of the teeth. This is common and only a cosmetic problem.

Opinion as to whether there is any long-term health risk is divided—even scientific assessments have come up with conflicting results. In Britain, very few water authorities add fluoride to their supplies. There may be a health risk if you ingest quantities of fluoride from other sources as well (toothpaste, for instance). In higher doses (upwards of ten parts per million), fluoride is more widely believed to be a health hazard.

Filtering water

Various filters are available. Some plumb directly into the drinking water pipe, others fit onto a tap. There are also two-part jug systems—tap water runs from a top section, through a filter, into a jug below. Most contaminants, including lead (but excluding nitrates), can be reduced or removed—but the replaceable filters work best when new. Decide for yourself whether the benefits outweigh the cost.

Hard and soft water

Hard water basically contains more calcium and magnesium than soft water. The water filters described are good at removing 'hardness' when the filters are fairly new, yet there is evidence to suggest that hard water is better for you in some ways. Hard water dissolves less lead from pipes—it coats them with limescale. Evidence suggests that there are more cases of heart disease in areas with soft water. If this is true, the reasons have yet to be made clear.

Hard water can damage plumbing systems by 'furring' up the inside of pipes and appliances. There are softeners which can be fitted to your system. The main one uses common salt to remove the 'hardness', another uses a magnetic field to prevent limescale forming. Again, these are not cheap options. If you only want a 'softer' bath, soda or bath salts will reduce scum and the amount of soap needed.

There are various commercial treatments to de-scale systems and appliances. Consult a dealer or plumber who will tell you if your system needs such attention.

> **! WARNING**
>
> If you instal a sodium-based water softening device, it is advisable NOT to include the main drinking water tap. Drinking sodium-softened water all the time could add marginally to your body's sodium levels. Also, removing all calcium from the water you drink may not be a good idea.

Boiling water/purifying tablets

Boiling water for several minutes should kill off most bacteria, which are not normally present in correctly-processed drinking water. This should only be necessary if there is a declared emergency, or if supplies are from a primitive source.

Water-purifying tablets are available, but these are only really intended for short-term use—while on holiday for example. Follow manufacturer's instructions.

DOMESTIC SYSTEMS

Before the water enters your premises, there is usually a heavy-duty stopcock (on/off valve) which is intended for use by the water company. Inside your premises, there should be another stopcock/gate valve so that the entire supply to your home can be isolated. Make sure you can operate this valve. In some areas there is also a meter, which registers the amount of water you have used.

Indirect systems

The supply will feed at least one tap directly—this is your drinking water. The rest of the taps or appliances in your home are usually fed by a large tank containing 189 litres (50 gallons) or more. This tank is actually a large heavy cistern and is often situated in the loft space.

TANK/CISTERN

The height of the cistern above your taps and other outlets determines your internal water pressure. This is known as the 'head of water'. The cistern is so heavy that it is important to check that the supporting timbers are strong enough and not deteriorating from rot or insect infestation.

The cistern should be protected from freezing by lagging. Loft insulation is not carried under the tank. Any warmth filtering up from the rooms below is useful to help prevent freezing.

Like most cisterns, this tank contains a floating ballcock which cuts off the intake when the desired level of water is reached. When you use water, the level falls allowing the ballcock to drop—opening the inlet valve—so more water enters the tank to refill it. Keep the tank covered and check that it's clean and in good repair. Everything from rotting wood to dead birds and rats have been found in loft cisterns.

In indirect systems there is usually a sealed **hot-water tank** (keep well lagged)—usually in a cupboard, bathroom or kitchen. This heats water, either by an electric immersion heater (like an electric kettle), or by heat exchange from a coiled pipe inside the tank. This coiled pipe is fed with hot water by the central heating boiler.

Often, also in the loft, will be a **smaller cistern** which feeds water to water-filled central heating systems. Both hot water tanks and water-filled central heating systems require '**expansion pipes**', which are open-ended to release bubbles and steam. These are usually allowed to drip back into the respective feed cistern. DON'T block these pipes. If a fault develops, water may boil in the pipes. Pressure MUST be allowed to escape.

All cisterns are fitted with an **overflow** or **warning pipe** (which drains through an outside wall) in case they overfill.

Direct systems

As water enters the premises, it feeds all the outlets in turn. Hot water is achieved by taking the water through a boiler—or, in some cases, there may still be a hot-water tank with an overhead cistern.

Scaling of pipes and appliances is more of a problem with direct systems in hard water areas. Water-filled heating systems are usually still fed by a cistern, and protected by an expansion pipe.

Which is best?

In indirect systems, most of the water is isolated from the mains supply—so if the water in the house becomes contaminated, it cannot syphon back and affect the mains supply. The pressure from the tank is constant—which is ideal for showers and other appliances which need to receive hot and cold water at roughly the same pressure.

Direct systems are easier to understand and may even be cheaper, but—apart from the possibility of limescale damage—special appliances and controls may be needed to avoid contamination of the mains supply.

You're in control

You MUST know where every stopcock/gate valve is, to be able to isolate all or part of your system. If you trace your pipes, you will find valves to isolate feeds to and (in most cases) from loft cisterns. At the lowest point on most systems—not only central heating systems—there is usually a special drain-off tap, to which a hose may be fitted for speedy draining.

REMEMBER

It's not enough to know where these stopcocks/gate valves are. You must check regularly that they are working efficiently BEFORE an emergency arises—which is exactly the wrong time to start learning about plumbing!

It's a good idea to fit an isolating valve under EVERY tap or before every appliance to make it possible to perform routine maintenance—such as curing leaks and changing tap washers—without having to drain down the relevant parts of the system. Pumps on showers and central heating systems should always have isolating valves—or even proper gate valves—to allow regular servicing.

WASTE SYSTEMS

Waste pipes are now usually made of plastic, although older houses may still have lead or galvanized piping. All pipes carrying water away incorporate a trap—the U-bend, S-bend, P-trap or bottle-trap—which partially fills with water to prevent smells from the drains coming back into the house. In the toilet this is built into the fitment, but on other outlets it forms the first section of the piping.

In most houses all the wastepipes are connected to a single vertical soil pipe, which carries the water down to the drains and into sewers. This pipe is extended upwards to roof level where its top is left open to act as a vent for smells. It also prevents suction (produced by the pull of water running through the pipes) from drawing water out of any of the individual traps.

Sometimes WCs are connected to a soil pipe, while baths, basins and other wastes feed into a separate waste pipe. These pipes go to the sewers via a gully trap, into which external drains and rainwater run-off pipes also discharge.

WARNING

Cracked or broken drain inspection covers should be replaced. There is a risk of disease-carrying bacteria being released or carried away by flies. Debris may fall in and block the drain.

Cesspools and septic tanks

Even in cities there are sometimes premises not connected to a main sewer, either because they are very old properties predating the sewage system or because topography makes mains sewerage impossible.

A cesspool is a lined hole in the ground where sewage collects. When full it has to be emptied. It must be sealed so that sewage cannot escape and water cannot get in from surrounding ground. It should be childproof!

A septic tank is a small-scale sewage treatment system. Two chambers are usual. In the first tank, bacteria break down waste into a harmless liquid. The second is a filter bed. Sludge from a septic tank requires emptying in a city environment, but not as frequently as a cesspool (perhaps once a year instead of once a week or once a month).

DON'T use excessive quantities of disinfectants, bleaches or household detergents. If the bacteria in the chambers are killed you might slow down the reaction—clogging the tank.

COMMON PROBLEMS

Blocked wastepipes: First attempt to remove the blockage with a suction-cup plunger, if there is water which won't run away. **If the water has run away**, but very slowly, and there is a trap in the first section of pipe, open the trap. Hair, grease, food and other debris should be cleaned out and the trap reconnected. **If there is no trap**, just a loop in the pipe, and the water has gone, try a chemical drain clearer—following the manufacturer's advice (see POISONS).

Sometimes it is necessary to disconnect the waste pipe from the appliance and push something down the pipe to clear the blockage. A length of hosepipe may work, particularly if connected to a running tap—but BEWARE of the backwash, particularly if you have tried using a chemical drain clearer. If nothing works, call in an expert.

Tank/cisterns overflowing: A stream of water is issuing from an external overflow/warning pipe. The ballcock is allowing the tank to overfill. It may have jammed. It may be adjusted too high. It may have filled with water so that it is not floating. Most likely the washer or diaphragm on the inlet valve needs replacing—to make it cut off the water supply when the water level is correct.

> **! WARNING**
>
> Overflowing water in winter may create a slippery ice patch on the ground below. Accidents do happen! The water may also freeze as it leaves the pipe. This freezing may travel back to the cistern. Fix the simple problem—and save having to deal with even worse ones.

Freezing pipes/tanks/cisterns: The whole water system should be lagged to prevent freezing. Don't carry loft insulation under cisterns—they need all the warmth they can get. If a pipe freezes, the expansion of the ice may split the pipe or force a joint apart. When it thaws you will get very wet! If freezing occurs, turn off hot water and central heating systems—the water is not flowing and this equipment may overheat.

Thawing should be done with great care, and only on identified areas of freezing. If the whole system is frozen you may need to call in a specialist. Don't use blow torches or hot-air strippers. You may melt plastic pipes—or the solder at (capillary) joints. A hairdryer is fine, used with care—but be careful of spurting water if dealing with a split pipe (turn off the stopcock!). An electric fire, used with great care, may be all right to gently warm a cistern—but supervise constantly.

Leaky joints: There are two main types of plumbing joint—soldered (capillary) joints and larger compression fittings which have to be tightened with spanners or wrenches. Problems at soldered joints mean that the whole system has to be drained of water. Compression fittings may be gently tightened (if they are recent fittings)—but beware of over tightening. Consult a specialist.

Leaky taps: When taps leak from the spout, the washer needs replacing. If they leak from the top, the packing or O-ring seal is faulty. Don't keep tightening a leaky tap, you may wear down the seating for the washer and then the whole tap may need to be replaced.

Central heating overheating: Can be very serious, may be for any of a number of reasons:

❍ The chimney or balanced flue (room-sealed flue) may be blocked. Clear all obstructions.

❍ The system could have become loaded with limescale. It will need draining down and treating by a specialist.

❍ The thermostat on the boiler may be faulty. Shut the system down and call a specialist.

❍ The main thermostat or programmer may be faulty. Shut the system down and call a specialist.

❍ The pump may not be working (you can hear it running, or feel a slight vibration if you touch it). Shut the system down. The pump must be isolated and serviced.

❍ A motorized valve may be stuck. Most systems include at least one electrically-operated valve which controls water flow. This is quite a common problem. Shut the system down and have the valve replaced. In most cases, only the tiny motor needs replacing—which means you won't have to drain the system.

Radiator problems: If a radiator stays cool at the top, there is air in the system. You could be damaging the system. Buy a key and gently 'bleed' the radiators. Slowly and carefully undo the valve, holding a cloth ready to stem the flow of water. Air hisses as it escapes, bubbles follow—when water flows, quickly close the valve. Bleed all radiators at the beginning, middle and end of the heating season as part of simple routine maintenance.

If a radiator is cool in the middle and cold at the bottom, you have a build up of sludge, corrosion particles and debris which need flushing from the system. Consult a specialist, before the system is damaged.

If either of the above is allowed to continue for too long, radiators may be damaged beyond repair. In the worst cases, a radiator may pinhole—spurting water under pressure!

EMERGENCY!

BURST PIPE/TANK FLOOD

- ► **Locate main stopcock and turn it off**
- ► **If water has reached an electrical appliance or light fitting:**
  DON'T touch the appliance
  Go to the fuse box/consumer unit and cut the power
- ► **Turn on all the taps, to drain the pipes and tank**
- ► **Turn off your central heating/immersion hot water heater**

BURST PIPE
If you can locate the actual source of the leak, a do-it-yourself repair may be possible. Repair 'kits' include hose clips (A) or epoxy putty (B). Otherwise call a specialist.

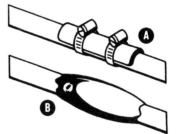

BURST TANK
If the tank is leaking, turn off the supply to the tank—or tie up the ballcock to prevent the tank refilling. You may now be able to turn the main stopcock back on again to get drinking water.

YOU COULD TRY
If the patch of damp is on a ceiling and is dripping, it could be worth making a small hole in the centre with a small screwdriver. Stand on a stepladder, NOT a chair. Don't forget to have a couple of buckets ready to catch the water.

If the leak is very bad and the ceiling is drooping—and dripping in the centre—be VERY careful. Piercing the bulge may be possible. It would be safer to evacuate the room if the ceiling looks likely to fall.

COMMON ACCIDENTS

Casualties from accidents in the home are more numerous than those on the roads. Everyone is at risk, particularly children under four—with peak risk at two—and the elderly. The kitchen is the most dangerous room, it seems, with up to 25 per cent of accidents related to it. First-aid advice given here is abbreviated as a reminder for emergencies only.

PLEASE NOTE

All **EMERGENCY!** and **SAVE A LIFE!** panels are intended only to guide you through emergency procedures, and to avoid the need to search through the pages of this book in a crisis. More detailed instructions appear in the HEALTH chapter. Learn the procedures BEFORE you need to use them.

FALLS

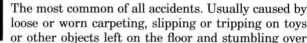

The most common of all accidents. Usually caused by loose or worn carpeting, slipping or tripping on toys or other objects left on the floor and stumbling over appliance flexes. The majority of accidents involve vacuum cleaners—caused by tripping over or stepping backwards on to them! Bad lighting is often a contributory factor.

Children under two try to walk or crawl up- or downstairs before they are physically ready to do so. Even as they get older, children are easily thrown off balance if they catch a toe on a piece of carpet or slip on a shiny floor. Rushing round a corner, they may fail to judge their speed sufficiently well to stop.

Falls from windows account for many deaths and injuries, especially for children. Low windows, especially those with

SAVE A LIFE!
FALLS

There may be no cuts and bruises, but internal injuries are possible. Concussion is likely after a blow to the head. Even worse, compression may develop—which is life-threatening. ALWAYS seek medical attention.

climbable objects in front of them, are the most dangerous. Balcony rails which can be squeezed through or climbed are another serious risk.

Worn and loose floor coverings can be a real hazard. Non-slip mesh is available in two versions to stabilize rugs on polished floors or to prevent them creeping on carpets.

◑ Repair any rents, which could catch either toe or heel, or replace.

◑ Ensure rugs don't curl up at the edges.

◑ Don't polish the floor beneath rugs.

◑ Keep stair coverings close fitting.

◑ Avoid loose mats at top or bottom of stairs.

Steps up and down, especially shallow ones, are a frequent site for falls, as are changes in width of the treads of stairs (when the stairs turn a corner). Though interior decoration usually seeks to unify a space, this is a case for different treatments of the surface to make the change more noticeable. Good lighting is essential in all such situations.

SUFFOCATION

Plastic bags, wrappings, clear kitchen film and even uninflated balloons commonly cause suffocation. Plastic bags are convenient and reusable—so less wasteful of energy than paper if you get a lot of use out of them—but punch a hole or two in them as air vents, just in case a child sticks its head inside. Most products today are overwrapped—thin plastic films can easily suffocate if placed over the face.

SAVE A LIFE!

SUFFOCATION

Remove cause—if it is a plastic bag, tear it away from the mouth and nose. If breathing has stopped or is very weak, apply artificial respiration. If heartbeat has stopped, apply cardiac compression. DON'T give up if suffocation has just occurred. Seek urgent medical attention.
For choking see SAVE A LIFE!: Choking (next page).

Dispose of them as soon as possible. Tell suppliers you don't need their wrappings! Make sure a child doesn't raid the waste bins!

It's not necessary to cover the whole head to cause suffocation. Plastic films may be sucked into the mouth, small objects can easily block a windpipe. Never let young children play with small or breakable toys—and choose sweets carefully. Even peanuts are dangerous to young children.

SAVE A LIFE!
CHOKING

If you believe that there is a blockage in the windpipe, you must act quickly. Remove any matter, including loose dentures (leave fixed ones), from the mouth or throat.

CHILD OR BABY
Lay the child or baby over your lap, with the head hanging downwards. Slap sharply — the younger the casualty, the gentler you should be — between the shoulder blades three to four times to dislodge the obstruction.

ADULT
Try to get casualty to sit, bending over so that head is lower than lungs. Strike sharply between the shoulder blades three to four times to dislodge the obstruction.

IF THIS DOES NOT WORK
Apply Heimlich manoeuvre/abdominal thrust.

IF BREATHING STARTS
Get casualty comfortable and reassure. Sips of water may be given.

IF UNCONSCIOUS
Place in recovery position. Seek URGENT medical assistance.

IF BREATHING DOES NOT START
Apply artificial respiration. An unconscious casualty's throat may have relaxed sufficiently for you to get air past the obstruction into the lungs. **CALL AN AMBULANCE.**

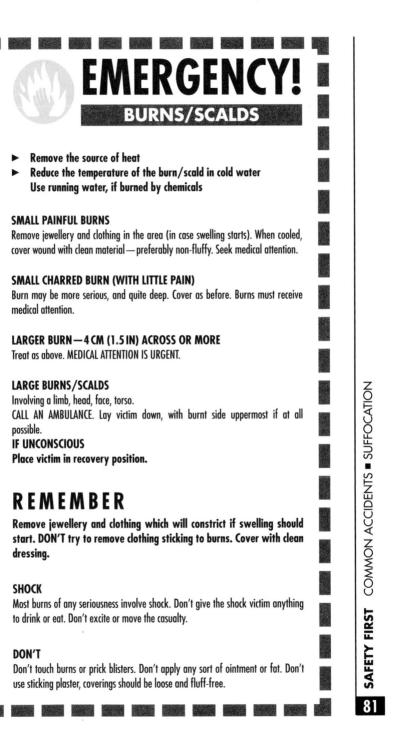

EMERGENCY!

BURNS/SCALDS

► Remove the source of heat
► Reduce the temperature of the burn/scald in cold water
  Use running water, if burned by chemicals

SMALL PAINFUL BURNS
Remove jewellery and clothing in the area (in case swelling starts). When cooled, cover wound with clean material—preferably non-fluffy. Seek medical attention.

SMALL CHARRED BURN (WITH LITTLE PAIN)
Burn may be more serious, and quite deep. Cover as before. Burns must receive medical attention.

LARGER BURN—4 CM (1.5 IN) ACROSS OR MORE
Treat as above. MEDICAL ATTENTION IS URGENT.

LARGE BURNS/SCALDS
Involving a limb, head, face, torso.
CALL AN AMBULANCE. Lay victim down, with burnt side uppermost if at all possible.
IF UNCONSCIOUS
Place victim in recovery position.

REMEMBER
Remove jewellery and clothing which will constrict if swelling should start. DON'T try to remove clothing sticking to burns. Cover with clean dressing.

SHOCK
Most burns of any seriousness involve shock. Don't give the shock victim anything to drink or eat. Don't excite or move the casualty.

DON'T
Don't touch burns or prick blisters. Don't apply any sort of ointment or fat. Don't use sticking plaster, coverings should be loose and fluff-free.

BURNS/SCALDS

 The main causes of injury are when children, especially toddlers, tip over saucepans/frying pans on cooker hobs, or when children pull kettles off work surfaces. Some sort of safety fence should be fitted round the hob. Handles should NEVER be left sticking out from the hob area, as these are a risk for adults, too. Jugs, teapots and cups of hot drink cause many injuries. NEVER drink hot drinks with a child sitting on your lap.

Everyone, especially children, should note how hot the external casing of ovens and cookers become. Irons and kettles, when switched off, do not cool down for some time—a child might think it's safe to touch these. Steam from a boiling kettle can scald.

Young children should not supervise the running of their own baths. Nasty scalds could happen, quite apart from any subsequent accidents which might occur in the panic when the child realizes its mistake.

Switch off irons and move them to a safe place to cool down when you have finished using them or if the doorbell/phone rings when there are children about—or take the child with you to answer the door/phone.

FIRE

 The most common cause of accidental death in the home. Nearly half of all fires are known to be started by electrical faults (see FIRE!).

POISONING

 Information about chemical hazards and poisoning—their effects and how to treat them—can be found in POISONS.

ELECTRICAL DANGERS

 Children are inquisitive. They will stick things into empty sockets—use childproof covers. You MUST understand the risks of electricity (see **Electricity**).

GLASS

Ordinary window glass shatters easily, breaking into sharp-edged, often dagger-like pieces. Areas which are more vulnerable to impact should be fitted with safety glass instead. There are two main kinds.

Toughened glass is processed to be five times stronger than ordinary glass, withstanding both rapid temperature change and impact. Instead of breaking into large pieces, it shatters into small granules. It must be ordered to size.

Laminated glass consists of layers of glass bonded together with a strong clear plastic film between them. It absorbs some of the energy of impact, and reduces the danger of flying fragments of glass. However, this sort of glass should never be used in the home for a window which might need to be used as an escape route—in a fire, for example. It can be VERY difficult to break through.

Ordinary glass can be made safer by applying a special adhesive film to the surface. This stays clear once fixed and holds the glass together if it is broken.

Many accidents involve people walking into doors or even windows which they have not seen. Glass partitions, including windows onto patios and balconies, should have barriers which prevent people walking right up to them or decoration on the panel which will draw attention to its surface. Glass doors should always carry some form of decoration. Though the glass may not shatter, the impact can do considerable injury.

EMERGENCY!
BLEEDING

Small cuts may be cleaned and dressed. If bleeding does not stop, cover with a dressing and apply pressure to control it.

LARGE WOUNDS/SERIOUS BLEEDING
Call for an ambulance. Cover with a dressing and apply pressure, elevating the bleeding limb. If necessary, hold the wound closed. If there is a foreign body in the wound, do not attempt to remove it—it may stem some of the bleeding.

ROOM CHECK

Make a close inspection of your home and consider the risks which each room presents to you and your family. Do not forget the hazards posed for the very young and the elderly. Even though they may not be part of your live-in family, your home should also be safe for visitors.

KITCHEN

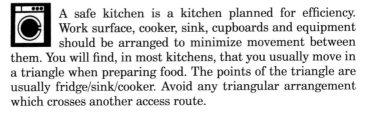

A safe kitchen is a kitchen planned for efficiency. Work surface, cooker, sink, cupboards and equipment should be arranged to minimize movement between them. You will find, in most kitchens, that you usually move in a triangle when preparing food. The points of the triangle are usually fridge/sink/cooker. Avoid any triangular arrangement which crosses another access route.

Watchpoints

COOKERS Avoid placing the cooker under a window or next to a door. It would be easy to burn yourself reaching across to open the window. Pots could be knocked when coming through the door. In both positions draughts could blow out low flames on a gas cooker. With built-in ovens check the position of air-vents which may blow hot air out at you. Beware of external cooker surfaces, they can get hot enough to give you a bad burn. Don't stand over the oven door when opening it—the blast of hot air and steam could be nasty.

Have heat-resistant working surfaces next to the hob and oven to take hot pans and dishes. Have the sink within easy reach for draining/straining and for transferring heavy pans for washing. Clean the cooker regularly—the build-up of grease and dirt around an oven can become a fire hazard. Cookers on wheels (with a flexible attachment to the mains in the case of gas supplies) enable a thorough cleaning job to be done behind as well as all over the oven itself.

STORAGE Store all food and equipment close to the place where it will be used: for instance, place vegetable storage near the sink. Avoid placing cupboards or shelving over the hob, or hanging anything there. Reaching over it could easily set clothes alight. Do not warm or store dishes above the hob for the same reason. If you have to use high cupboards, do not keep frequently-used objects in them. Get a set of kitchen steps rather than standing on chairs or stools.

DANGER Do not have plugs for kettles or appliances near the hob—flex trailing across the burners or hotplates

SAFETY FIRST ■ ROOM CHECK ■ KITCHEN

could easily be charred or burned. Do not have television or mains radios near the sink or where leads have to pass near the cooker. Keep electric flexes well away from water and places where they could constitute an ELECTROCUTION hazard.

FLOORS Keep floors clean and mop up spillages immediately—spills can make them dangerously slippery—but avoid unnecessary polishing. Carpets are not usually practicable in a kitchen—though types are now available which are easily cleaned—NEVER use loose rugs in a kitchen.

PLUS Many people feed pets in the kitchen but cats or dogs under the feet, especially if pestering for food, can be a hazard. Do not allow cats on work surfaces. Can you find a more appropriate pets' corner? Keep the kitchen well-ventilated. Cooking produces a lot of steam and a washing machine in the kitchen will add to that. Condensation not only steams up windows but can make floors slippery too.

CHILDPROOFING

- If you need to have small children in the kitchen, keep them out of harm's way by constructing a partition so that they are out of the work area.
- Fit child-resistant catches on low-level cupboards—especially if you keep bleaches and cleaners underneath the sink. Ideally, store potential poisons out of a child's reach.
- Keep knives and other dangerous equipment out of sight of children.
- Fit a guard round the hob so that pans cannot be pulled down from it. If possible, cook on back rings, where handles are out of reach.
- Split-level, high-set ovens will keep one danger area out of reach of very small children.
- Avoid flexes hanging from kettles, irons or other electrical equipment, which children could pull. Coiled flexes are safer than trailing ones. Keep kettles at the back of work surfaces—out of reach and sight of toddlers.
- Do not use long tablecloths which children could tug, pulling hot food and sharp utensils down on top of them.
- Teach children that cookers—not just hobs but oven doors and sides—are hot and can burn.
- Children use drawer units as stepladders. Fit latches to lock drawers.
- Don't store children's drinks where you keep alcohol.

Electric cookers

Some small cookers, including microwaves, are below 3000W rating, and can be connected to a 30-amp ring main circuit. Most ordinary electric cookers are over 3000W and require a separate circuit, run directly from the consumer unit, via a cooker control unit which must be within 2m (6ft 6in) of the cooker. This unit has an isolating switch, but is sometimes combined with a socket for use with an electric kettle or other appliances.

The gauge of cable used and the fuses required will depend upon the power consumption of the cooker, which should be indicated on the appliance (see **Electricity**).

Some ovens use a cleaning system in which the surfaces are specially coated and the oven is run at a high temperature to burn dirt off. Follow the instructions carefully. In all other cases ensure that the oven and/or rings are switched off before starting to clean.

Microwave ovens

Microwave ovens cook food through the action of high-frequency radio waves which quickly penetrate the food, vibrating the water molecules within it to produce heat and rapid cooking. Leakage of microwaves can have harmful effects, but ovens have to meet official safety requirements. If the door or seals of an oven appear damaged, or its performance changes, get it checked—do NOT take risks.

DON'T allow grease or dust to accumulate around the seal or door frame. The door must close properly. DON'T peer closely through the door at regular intervals. Leakages are more dangerous inches from the oven.

 WARNING

Microwave radiation is very dangerous. In perfect condition, ovens are designed so that the radiation is entirely contained and is switched off when the door is opened. Nevertheless any tampering with the system or inefficient cleaning could result in radiation being released. NEVER allow an unqualified person to attempt to repair a microwave. If you suspect any malfunction take it to the suppliers or their service centre.

Ovens MUST cut out when the door is opened and some are now equipped with cut-out mechanisms if they become overheated—but there is a considerable risk of fire if food is allowed to cook for too long or instructions are not followed.

NEVER put oven foil, metal dishes or dishes decorated with metal in a microwave. Since many ceramic colour glazes and paints contain metallic elements, it is safest to use only dishes which are designed for microwave ovens.

Always check that the time you have set reads accurately on the timer digital display. It is particularly easy to mis-key touchpad controls. Adding an extra digit could have serious repercussions.

Set to the minimum cooking time—you can always reset for a little longer but you can't uncook a burned-up plateful.

Clean the oven carefully after use, following the maker's instructions. Spillages which are recooked can easily overheat and burst into flame

Gas cookers

Older-style, self-igniting gas rings using a pilot light make siting of a gas cooker particularly important. If the pilot jet blows out, the kitchen could fill with gas. ALWAYS check to see that a ring has lit—and stayed alight.

If matches are used to light the cooker, do NOT leave them on the top of the stove. They could all catch fire—or fall down beside the cooker and ignite there later.

Match the pot to the size of the ring and do not turn flames high enough to reach up the sides of pans.

Always close oven doors gently. A slam could create a draught of air which blows out the gas jets if they are turned low by the thermostat.

Fat fires

A commonly-occurring cooking hazard—a pan of hot fat catches alight. This can happen with both shallow and deep frying. Use a pan with a lid—and keep the lid close by so that you can replace it, cutting off the oxygen supply to the flames and extinguishing the fire. If you have a fire-blanket—a must for every kitchen—approach with the blanket raised to protect your face, and smother the flames. If you have no lid or blanket use a chopping board or damp towel. Don't forget to turn off the heat source. Don't move the pan until the fat cools.

Keep careful control of fat temperature and level and do not place wet food into the fat. Dry it first on kitchen paper. Wet food may make fat spit and increase the risk of igniting. Watch out how much fat you use in a pan—as soon as food is added, the level will rise. This is because of the volume of the food itself and the bubbling action of the fat as it is added. Most fires start at this point during cooking.

Automatic deep fryers with a thermostat to prevent overheating avoid the need to monitor the frying so closely—but do not stand the fryer immediately in front of the socket into which it is plugged. A fire is unlikely unless the control breaks down, but you need to be able to switch it off easily. When you open the lid, watch out for a cloud of steam!

Automatic fryer or not, a large volume of hot oil can take a couple of hours to cool down!

▶ **TURN OFF THE HEAT**

IF YOU HAVE A LID FOR THE PAN
Replace it immediately to smother the flames

IF YOU HAVE A FIRE BLANKET
Approach with the blanket held up to protect your face and smother the flames

IF YOU HAVE NO LID OR BLANKET
Cover the pan with a damp towel or chopping board

DO NOT
Move the pan until the fat has cooled

REMEMBER

Low cooker extractor hoods may draw flames up into them. Smother the flames around the pan first. Then switch off the extractor and aim a dry extinguisher into it.

Refrigerators/freezers

Refrigerators have made life much safer by enabling us to keep perishable foods in good condition, at least for short periods. Freezers give the convenience of being able to prepare meals far in advance and to take advantage of seasonal price differences. However, they must be kept clean, free of ice build-up and at the correct temperatures if they are to do their job efficiently and safely.

When defrosting fridges, put the contents into a coolbox or an insulated pack—the kind used for bringing frozen food back from the frozen-food store. When defrosting the freezer, you may be able to transfer some of the contents to the frozen food compartment of a fridge—otherwise use the insulated bags again. If they are not available, wrap frozen items in layers of newspaper. **Do not allow frozen food to thaw (or even partially thaw) and then refreeze. Depending on the type of food, this can be dangerous.**

Fridges and freezers usually have a thermostat with a numbered scale, but it cannot give you an accurate idea of their temperature. A fridge should be kept at 2–5°C (35–40°F), while a freezer should be –18°C (–0.4°F) or colder. Make regular checks by leaving a thermometer in the 'warmest' part for at least a couple of hours, then taking a reading. In a fridge put the thermometer in the middle of the top shelf. In an upright freezer place at the top front edge on the side the door opens, or in the top basket of a chest freezer. It is a good idea to leave a thermometer permanently attached in these positions.

REMEMBER

Fridges should be kept at a temperature below 5°C (40°F) if they are to be cold enough to inhibit the growth of bacteria — which will still develop slowly as long as the temperature is above freezing point.

WARNING

Fridges and freezers can be coffins, especially chest freezers with latches which hold the lids tightly closed. It can be impossible to open them from the inside. When defrosting lock the room or garage they are in — a child might not freeze to death but could suffocate, since the air supply would soon be used.

Food processors/mixers

Modern electric food processors, with their rapidly-turning blades, look much more dangerous than the old-fashioned kinds. However, most of them will not work until a safety cover is securely in place and offer no danger in use—provided that you follow all the operating instructions and don't overload the motor. Old-fashioned mincing machines were more likely to trap the fingers and graters can more easily take a chunk out of them. However, modern blades should be handled with caution, especially when washing them, and still stored out of the reach of children.

Always unplug electrical appliances before wiping or removing parts for washing. NEVER put the motor unit in water. Wipe it with a slightly damp cloth.

Follow the makers' instructions regarding running time and the type and quantity of food it can process. If the motor-casing becomes hot, switch off and allow it to cool down. If it seems to be getting hot regularly, or smells hot, or if the motor sounds under strain, you may be misusing it in some way, or it may need servicing.

Electric kettles and jugs

ALWAYS unplug before filling. Keep and use the kettle on a firm surface, well away from the edge. Do not leave the flex dangling—or plugged in and switched on, while not connected to the kettle. Live flex hanging over a sink, or even on a surface where liquids could be spilled, could easily make contact with water and short-circuit—at worst cause ELECTROCUTION.

It's best to shorten the flex, so that there is just enough to comfortably reach the kettle from the socket. Or fit a curly cable instead, to avoid trailing flexes which a child could grab.

Descale the kettle regularly. Scale build-up could affect the automatic cut-out device—or cause overheating.

> ## ! WARNING
>
> All electrical appliances generate heat in use. Follow manufacturers' instructions carefully concerning running times and safe ventilation. Fitted-in appliances must have adequate ventilation and insulation to avoid a fire risk.

Toasters

Few people switch off toasters when not in use. You should, to avoid children 'toasting' things—like paperback books! Make sure you regularly empty out all crumbs/debris from the toaster. If this accumulates, overheating may occur. NEVER place the toaster so that anything can interfere with the push-down operating lever.

Tests have shown that some toasters can melt dramatically and burst into flame in these circumstances. Some toasters only stop heating the elements when the lever returns to the 'up' position. NEVER fiddle about inside a toaster with a metal utensil. If your toast gets stuck, first switch off and unplug. Then free the toast with a plastic or wooden spatula—but avoid touching the elements.

Washing machines

NEVER leave switched on, especially if the machine is 'plumbed in'. If it is, ALWAYS turn off the water when the machine is not in use. The water-intake valves could fail under constant pressure of mains water. If the waste pipe is hooked over a sink, ensure water in the sink never covers the end of the pipe. Water could be syphoned back into the machine.

ALWAYS follow manufacturers' instructions about safe loading and operation. If a violent wobbling takes place during spinning, switch off the machine and try the last part of the programme (final rinse to spin) again to try to redistribute the

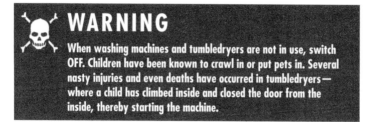

load. If this doesn't work, you may have to let the machine 'empty' the water (on front loaders) and find a part of the cycle that will allow you to open the door.

If you use a lot of bleach in your machine regularly, the rubber seals may begin to degrade a lot sooner.

Using the wrong washing powder (using manual in automatics) can produce violent results, with more foam being produced than you or the machine can handle. The safe operating of the machine may also be affected.

Cupboards and drawers

Some of the simplest mishaps can lead to worse accidents. As you are distracted by a minor misadventure, you are less able to react efficiently to something else. A slip in the kitchen can bring you into contact with flame, hot liquids and sharp implements as well as sharp corners and hard surfaces.

Fix stops on the inside of drawers to prevent them being pulled out too far and spilling their contents.

Get into the habit of ALWAYS shutting cupboard doors. Standing up from stooping or turning round, it is easy to catch against the door or knock your head against the bottom edge— a nasty injury if it is metal!

Knives

Never put sharp knives into washing-up water with other items. It is easy to touch or grasp a blade unintentionally. Handle each knife individually. Wash, wipe and put away, in a childproof place—not 'on display' where children (or even intruders!) may misuse them.

Pass knives with care! Hold the blade with the cutting edge AWAY from the palm or fingers and offer the handle to the other person. DON'T walk about holding a knife.

ALWAYS cut away from yourself. A sharp knife cuts smoothly and is therefore less likely to inflict injury than a blunt one (which has to be used with more pressure so cannot be halted as quickly).

NEVER use knives, or other cutlery, for tightening screws, removing bicycle tyres or other jobs. It will not only damage blades, but produce burrs which could cause cuts if they catch against the hand or lips in normal use.

Sharpening a knife consists of wearing away some of the metal to produce a thin cutting edge. The traditional method is to hone it on a whetstone and then finish the edge on a sharpening steel—a grooved metal rod. These techniques require practice to do them safely and efficiently.

There are several other kinds of sharpener, some electrically-powered, which offer metal discs, small steels, tungsten cutters or ceramic grinders to remove metal from both sides of the blade at once. In all cases, they operate by drawing the knife between the cutters. If they are easy to keep stable, they are probably safer than a steel.

Accidents in sharpening knives usually occur when the sharpening device slips—not the knife, so choose the one you feel is solid and stable.

Some individual knives also come with their own holster, which not only, protects the blade when not in use but sharpens it each time it is drawn in and out. They only sharpen that knife and cannot be used on others unless they exactly fit the holster.

Wash knives before sharpening—and afterwards clean both knife and sharpener to remove the small particles of metal which will still be attached to them.

REMEMBER

It is well worth investing in a few good basic knives, instead of lots of gadgety ones. Good knives are usually of better steel, and honed to a longer-lasting edge. Sometimes, good kitchen equipment shops may be able to recommend specialist companies who re-sharpen good knives when the edge needs major work. Treat these good knives with care, and they could last for years.

LIVING ROOM

The majority of living room accidents involve tripping over something—so ensure there are no trailing cables from lamps and electrical appliances, or small obstructions. Never leave things lying around on the floor. Teach children not to leave toys scattered about where they or others could trip over them.

The living room is the area where there is most likely to be an open fire, a gas fire or an electric heater—whether or not the house is centrally heated. Do not leave newspapers or fuel close to open fires.

Fit a fireguard

A sparkguard, consisting of a screen of fine wire mesh, should be placed round an open fire, whenever it is left unattended and especially when you go out of the house or up to bed. An even better practice would be to rake the embers so that the fire goes out completely. The guard should be anchored to the wall, and enclosed at the top, sides and front—if all sparks are to be safely contained.

CHILDPROOFING

For a home with young children, a secure fireguard fixed to the wall is needed with any sort of fire—not just an open one. Children must not only be kept away, but also prevented from poking things through into the flames or onto the electric element. An open mesh will allow you to enjoy the full benefit of the fire, but closer mesh is needed for an efficient sparkguard.

Keep alcohol away from youngsters—do not leave drinks out on a table and always return the bottle to a secure cupboard or cabinet, which should be locked before you leave the room. NEVER keep children's drinks in the same place as alcohol—don't tempt them to help themselves.

WARNING

NEVER use fireguards for airing or drying clothes, nor hang things to dry above the grate. ENSURE that portable oil and gas heaters are on a stable base and out of draughts. Follow the instructions already given. Never cover the ventilation grills of these or of any electric fires. DON'T use an overmantel mirror for brushing hair, shaving or applying make-up. It is easy to get too close and, if wearing loose clothes, to set yourself alight.

Other fire hazards

Upholstered furniture often burns easily and foam, plastic and synthetic fabrics produce thick smoke and toxic gases. Modern furniture should be fire resistant and tested to withstand a smouldering cigarette—and labelled so. Look for items that have also been tested as resistant to a lighted match.

Candles can create a very cosy and romantic atmosphere, but use them with caution. People who are used to lighting their homes with oil lamps and candles know how much heat is generated above them. Pass your hand above a naked candle flame and you will realize how carefully it must be placed so that it cannot ignite anything overhead or nearby. Always set candles securely in holders with a firm base. **NEVER put lighted candles on top of a television set.**

Cigarettes not only damage your health, they are a serious fire risk. Make sure all ashtrays are stable—the lidded kind that extinguish cigarettes are best. Do not put sweet papers in them—they could ignite. Never balance ashtrays on the arms of chairs or place them in the middle of a sofa seat. A fallen cigarette end could start a fire.

NEVER smoke in bed! However many times this advice is repeated, people still do it and start serious fires.

FIRE FACTS

Some furnishings burn at a temperature of over 1000°C (1832°F), and produce deadly choking fumes and smoke. In some cases a whole house has been engulfed in under TWO MINUTES (see FIRE!).

Over twice as many people are asphyxiated by fumes and smoke, than are killed by the flames. Keeping low to the floor as you escape, covering your mouth and nose, will help protect you from the worst effects. Apart from the smoke, heat and carbon monoxide dangers—polyurethane foam gives off hydrogen cyanide when burnt. PVC gives off hydrogen chloride. Both are highly toxic. Note that some foam furnishings may still be made of polyurethane foam. PVC is very common—even electrical cables are insulated with it—and any appliance, especially a television, contains enough PVC to give off a roomful of fumes.

If a fire gets out of control—EVACUATE. Call the fire brigade.

Television sets/computer screens

ALL electrical equipment should be given plenty of ventilation. Don't block the air vents provided in the appliance.

Televisions involve many hazards. They operate using very high voltages—at least 20,000 volts for colour, 10,000 volts for black-and-white. Their circuits take the main voltage and amplify it. These high voltages are still present for a while after the television has been switched off. NEVER take the back off the set, or poke anything through the holes. NEVER place fish tanks, flower vases, potted plants, drinks or any other liquids on top of or close to televisions. The danger of electrocution is very real and very serious.

TV/computer fires

Because of fire risks, televisions should be switched off at the mains and unplugged when not in use—not left on 'standby', which is possible with some sets.

If you smell burning, or if you see smoke or flames, switch off at the mains immediately. If the TV is blocking your route to the socket, switch off at the fuse box/consumer unit.

Use only a dry extinguisher or fire blanket—NEVER liquid extinguishers. A damp blanket (when the power is off) covering the entire set should contain flames and some fumes—it may also help contain glass fragments in case the tube 'explodes' (actually implodes). Do NOT approach the set from the front, because of the danger of flying glass fragments.

Evacuate because of the risk of highly toxic fumes, and ventilate the room as soon as the fire is fully extinguished. If there is any sign of the fire getting out of control, evacuate and call the fire brigade.

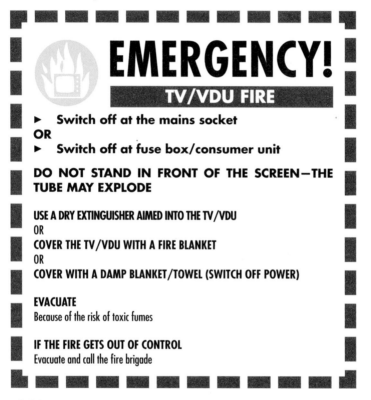

EMERGENCY!

TV/VDU FIRE

▶ **Switch off at the mains socket**
OR
▶ **Switch off at fuse box/consumer unit**

DO NOT STAND IN FRONT OF THE SCREEN—THE TUBE MAY EXPLODE

USE A DRY EXTINGUISHER AIMED INTO THE TV/VDU
OR
COVER THE TV/VDU WITH A FIRE BLANKET
OR
COVER WITH A DAMP BLANKET/TOWEL (SWITCH OFF POWER)

EVACUATE
Because of the risk of toxic fumes

IF THE FIRE GETS OUT OF CONTROL
Evacuate and call the fire brigade

Children's toys

Check all toys for sharp edges, finger traps or pieces which could be swallowed. Look especially for eyes on spikes in old soft toys. Do not let small children play with anything small enough to become lodged in nose, ear or throat.

Lead paint was regularly used on old toys and young children should not be allowed to play with them, since they might put them in their mouths or suck fingers which have picked up flakes of paint.

All toys for sale in shops have to pass stringent safety tests. This is particularly important with toys for young children. Beware cheap imported toys (often seen in markets) which may evade the testing procedures.

Allow children to use only non-toxic paints and crayons—make sure they are so labelled. Many artists' colours contain poisonous chemicals which could be sucked from brush or hands (see DIY/CRAFT HAZARDS: Artists' paints).

KEEP TOYS TIDY! Make sure children have a cupboard or box in which to keep their toys and a shelf for their own books. Toys left lying around are a common cause of falls—especially on stairs. Small round objects and wheeled toys are most dangerous, but anything lying on the stairs could send you headlong down them.

BEDROOM

 The main fire risk in bedrooms is caused by people smoking in bed. Another common cause of fire is the electric blanket. Some mattresses are now made of foam and risks are as serious as with foam-filled furniture in the living room. Toxic fumes given off during burning are a major cause of loss of life (see FIRE!). Nightclothes should be made of flame-retardant fabric.

The bedroom seems to be a room where people stumble around a lot in the dark—usually getting up in the night to go to the toilet. DON'T leave shoes or clothing where they can be tripped over. DON'T leave drawers and doors open to be walked into. DON'T leave trailing flexes around.

WARNING

Don't be tempted to drape a cloth or place a paper bag over a light to dim it. Use a low-wattage bulb, a dimmer switch or a safety-approved nightlight.

Electric blankets

Some new blankets have good thermostatic controls, or devices to prevent overheating. The more safety features, the better. Older electric blankets may represent a serious hazard.

❍ ALWAYS switch off before going to bed—unless a special type made to operate on extra-low voltage (ELV).

- Have your electric blanket checked by the manufacturer if there is any sign of fraying, scorching, worn flex, loose connections or tie tapes damaged or missing—every year, even if nothing appears to be wrong. Do so in the summer when you don't need to use the blanket.
- NEVER use an electric overblanket as an underblanket or vice-versa.
- ENSURE underblankets are flat and securely tied to mattress. Check regularly. NEVER secure with safety pins.
- Do NOT use a hot-water bottle with an electric blanket.
- NEVER use an electric blanket that is wet for any reason. Even if an electric blanket is only slightly wet, do not switch it on in an attempt to dry it out.
- Roll rather than fold an electric blanket when in storage. Keep it on top of a pile so that it's not squashed by other bedding or store as an extra blanket on a spare bed.

THE NURSERY

- Ensure cot mattresses fit well and leave no room at head or foot.
- Cot bars should be not more than 6cm (2.5in) apart so that a child cannot trap its head.
- Baby bumpers (to keep child from banging against rails) are not really necessary. If you use them, ensure ties are as short as possible—use a double knot and trim off the ends.
- Babies under 12 months old should NOT have pillows.
- Lie babies on their side or front. On their backs, if they are sick, it is possible that they may choke.
- Do not leave babies to sleep in quilted bag 'baby nests' all night.
- Use anti-burglar window catches to prevent children opening windows and falling out. Alternatively fit bars or railings, at least until the child is old enough to understand the dangers.
- Audio baby alarms are essential to detect any problems that may occur—and to avoid you constantly looking in to check the infant is all right. Video (CCTV) versions are also available. It's worth using these until the child is at least four, especially during periods of illness.

! WARNING

Some pillows and duvets/eiderdowns are washable or may be dry cleaned. If dry cleaning is preferred, take the item to a specialist—not a do-it-yourself dry cleaners. After dry cleaning, the bedding MUST be properly aired to prevent the residual poisonous fumes from the cleaning fluid making you very ill—possibly causing lung and eye irritation, possibly death in infants.

BATHROOM

Electricity and water are an extremely dangerous combination. In the bathroom there is a lot of well-earthed metal. When you become wet, especially in the bath or shower, you become an excellent conductor. There should be:

◑ NO ordinary sockets in the bathroom.
◑ NO ordinary switches—unless the pull-cord type.
◑ NO way to touch any electrical appliances from the bath.
◑ NO portable electrical equipment in bathrooms.

The light should be operated, if not by a pull-cord switch, then by a switch OUTSIDE the bathroom altogether. NEVER switch it on or off with wet hands. The same applies to over-mirror lights and heaters. Light fittings should be fully enclosed to avoid the risk of dense steam or water splashes reaching them.

Electric showers should only be fitted by an expert, and protected by an RCD/RCB/ELCB (see **Electricity**).

NEVER take any mains-operated appliance, not even a radio or hairdryer, into the bathroom on an extension lead. NEVER NEVER NEVER allow anything like this near the bath when you are in it!

If you have to use power tools in the bathroom, make sure you use an RCD/RCB/ELCB.

The bath

Falls in the bath account for one third of bathroom accidents. DON'T stand in the bath, especially not when drying yourself. People tend to stand on one foot, resting the other on the side of the bath. One slip is all it takes!

If you must stand in the bath, because the shower is there perhaps, use a non-slip rubber bath mat. Fit grab handles using solid fixings—a grab handle attached to a hollow plasterboard wall may come away in your hand. The same goes for anything else you may clutch at when falling, although slipping in the bath is like slipping on ice. You don't realize what is happening until you land! If you need to fix to a hollow wall, screw into the wooden batons behind the plasterboard.

Test water temperature before getting into the bath. It shouldn't emerge from the taps at nearly boiling point—although many people seem to adjust hot-water cylinder thermostats until this happens. It should not be necessary to set the thermostat much above 64°C (147°F).

Avoid highly polished floor surfaces, and mop up any spills. Don't splash water around. Check that the shower and splashed bathwater don't regularly run down walls or trickle down onto the floor. Small leaks can do more damage than big ones. If you have let the bath overflow, it could be serious enough to bring down a ceiling in the room below. Small leaks on a regular basis could be just as serious.

BATHROOM ROT

The conditions around the shower, bath and toilet, bidet and washbasin are ideal for producing rot in timbers. This room could be the site for the beginning of a devastating problem. Avoid constant leaks, puddles and splashes — avoid rot (see **Damp**).

If condensation is a problem, fit a good extractor fan. If no outside wall is available, it may be necessary to fit ducting to expel the unwanted moisture. The trouble involved in fitting an extractor could be preferable to having to deal with rot and mould later.

REMEMBER

If running a bath steams up your bathroom excessively, the problem could be partly solved by warming the air in the bathroom. Warmer air can hold more moisture without producing condensation. Running some cold water into the bath first, followed by hot — piped UNDER this cold water with a shower attachment — will also help cut down on steam.

Sealants

Check round the wall edges of your bath, shower cubicle and washbasin. There should be a good seal, to avoid water constantly trickling down. It may seem unimportant, but a regular trickle could do great damage to ceilings and electrical fittings below— quite apart from the possibility of rot (see BATHROOM ROT). A long-lasting seal is possible with correctly applied silicon sealant (available in a range of colours). If the gap is wide, it may be necessary to span it with a plastic strip fixed with the same sealant.

Bathroom glass

If a bath is beside a window, ensure that the window is fitted with safety glass to prevent a serious accident. Use safety glass for shower screens and mirrors (see **Common accidents**).

Do not leave parabolic shaving or make-up mirrors on windowledges, especially if there are net curtains at the window. Sunlight focused by the mirror could set fire to the net or any other inflammables on which the beam is concentrated.

Showers

If other taps are turned on in the house you may find shower water temperature suddenly changing. Cold water may be a shock, but hot could scald. Installing thermostatic flow controls should reduce the risk.

Instant showers, in which the water is heated as it is drawn (rather than coming from a hot-water storage cylinder) are of high wattage and require a separate circuit protected by an RCD/RCB/ELCB. They activate when the water is turned on, but should also have a separate on-off switch which should be cord-operated or outside the bathroom. Don't place the shower head where it will spray the control unit but put the unit to the side away from the main jet.

WARNING

DON'T mix toilet cleaners. Various combinations of toilet cleaner can react violently with one another. In some cases this results in the production of highly toxic fumes which—if they don't knock you out altogether—may produce uncontrollable choking and lung irritation.

CHILDPROOFING

- The bathroom is usually the one room that can be locked. Small children can often close a sliding bolt, but have some difficulty unlocking it. Set it out of reach of tiny hands.
- If the door has an ordinary lock, remove the key and fit a bolt higher up. Don't use too strong a bolt. If an accident happens, you may need to break the door down.
- Make the medicine cabinet lockable and keep it locked. Hide the key. Children can otherwise investigate it at their leisure.
- Similarly, lock up toilet cleaners and other chemicals or keep them out of children's reach.
- Some small children like to investigate toilets. Not very hygienic, but not likely to do serious harm—unless a toilet cleaner has been used. Use cleaners only after children have gone to bed.
- Children are particularly likely to get into a bath without realizing that the water is too hot.
- Non-slip mats and baby supports can make bathing easier and safer.
- Never ever go away to answer the door or the phone leaving young children in the bath. Either ignore the caller or wrap the child in a towel and carry them with you.
- Childproof medicine containers may NOT be childproof to a really determined child. Treading on plastic containers may release the contents.

HALLWAY/STAIRS/LANDING

A huge number of reported home accidents involve injury to the feet. You should have a pair of slippers—not wander round in bare feet or only wearing socks or stockings. Also, although you may become 'attached' to your favourite old slippers, you must replace them as soon as they become worn or slippery.

! WARNING

When dealing with polished floors, particularly polished stairs, try to keep the use of waxes to a minimum. There are 'non-slip' waxes available which are not the total answer. The biggest danger comes from over-use of 'silicone wax' furniture polishes. The silicone reduces friction, making furniture feel 'polished'. Even the spray-drift off the edge of a table could cause a dangerous slippery patch on the floor. Don't polish handrails, where a good grip may be needed.

◑ Stairs are a particularly risky place for falls, each step can deliver an injury as you come down. Check the stairs themselves for structural soundness, rot and woodworm.

◑ Check that lighting is effective and never put off changing light bulbs. Two-way switches or timed switches will help ensure that stairs do not have to be negotiated in the dark.

◑ Keep carpets well fitted and in good repair.

◑ Avoid loose rugs in halls.

◑ Avoid placing furniture, plants, bicycles or any obstruction at the foot of stairs and on landings.

◑ Fit a secure handrail the full length of the stairs and on each flight. If there are elderly members of the household, fit a rail up BOTH sides. Continue them onto the landings.

◑ NEVER put things down on the stairs—dustpan and brushes, books, trays, clothes to be put in drawers, towels for the bathrooms, the vacuum cleaner with its flex. It seems the obvious place to put something to remind you to carry it up when next you go upstairs, but by then someone may have tripped on it—yourself, if you forget to pick it up!

◑ Open risers can look attractive, but can be a hazard for the elderly and children. Consider boarding up the gaps until children are old enough not to stick their heads through.

◑ Windows or glass doors opening from stairs should be glazed with toughened glass. Avoid having windows/glass doors at the foot of the stairs.

Fire risks

Do not store flammable materials such as old newspapers, oil-based paint, paraffin, or gas cylinders in a cupboard under the stairs. A fire starting here will spread rapidly, the stairs acting as a chimney, taking the fire to the rest of the house.

Keep doors leading off halls and landings closed. This will slow down the spread of fire.

CHILDPROOFING

Fit gates at top and bottom of the stairs, and at the bottom of the upward flight if there is another floor above the children's rooms. Gates should be easily removable (by adults) in case of fire or another emergency.

If balusters (vertical handrail supports) are more than 6 cm (2.5 ins) apart or there are climbable horizontal rails, board them over until small children have grown old enough to make the risks negligible.

Discourage children from sliding on banisters or playing on stairs and in hallways. Teach them never to leave toys, books, schoolbags or any other objects on stairs or in hallways.

LOFT

Use a proper ladder or stepladder to get into your loft, and make sure that it is firm and stable before climbing it. For a long-term solution, fit a proper loft ladder that can be folded up and pulled down as required.

Take up a torch with you, or an inspection light in a wire cage—never a naked flame. In the long term extend the lighting circuit to the loft, it will make inspections very much easier and help prevent freeze-ups in cold climates. Fit the switch on the landing below so that you do not have to climb into the dark.

Loft entry covers are often unattached boards. Move them gently so that they do not crash back. Fitting hinges, plus a block to support them when open, will avoid this problem—but ensure that they open in a direction which will not obstruct your access. Concertina, folding and sliding ladders are all available ready made, sometimes complete with a built-in hatch cover. These covers often open downwards.

Access and storage

Lofts are not built as rooms, but as air spaces beneath the roof and to house water tanks. They may have no floorboards and if you step off the rafters you will probably go straight through the ceiling into the room below. Fix strong boards across the rafters to give you easy access in places you may need to reach. Do NOT use loose boarding.

Consider putting a proper floor down. In most cases the joists, where you would fix floorboards, are not load-bearing. The floors below have stronger joists to take the loads necessary. In the loft, the joists were only intended to support the ceiling immediately below. That means that if you're considering conversion of the loft-space into a room, this is one of the first problems that needs to be solved.

Although not intended as storage areas, that is how many lofts get used. Lay boards across the joists, to spread the load, wherever you want to place objects. DON'T balance boxes on joists or rest anything directly on the 'ceiling' between the joists—they could end up in the room below!

Do not store heavy objects over the centre of large rooms—it could cause ceilings to sag. Place heavy furniture against the outside walls and the party walls of terrace and semidetached houses where the load will be more readily transferred to the walls—but take care not to add too much weight!

Do not fill the loft with inflammable material. A fire could become established there without you being aware of it below.

Insulation

Check that water tanks and pipes are properly lagged. If the loft is insulated with mineral-fibre blanket, wear gloves and a dust mask if you are likely to disturb it to prevent irritation to skin, lungs and mucous membranes. Insulation is economically and ecologically desirable, but the roof must be ventilated to avoid the risk of condensation in an unheated roof space.

Modern houses often have ventilation holes at the eaves which may become blocked with insulation. You may need to insert airbricks in the gables or drill holes in the soffit boards. Do not insulate over electric cables since they could overheat. Do not insulate under water tanks—warmth from the lower floors helps to prevent freezing.

MIND YOUR HEAD

Are there low doorways, beams, shelves and cupboards which could deliver a nasty bump to the head? Don't put shelves and cupboards where this can happen, especially if the stairs are involved.

If there is an overhead obstruction as you come down the stairs (more likely in an older property), you may have to pad it to prevent people from having a nasty accident. Otherwise you could paint it to highlight the danger, or pin a strip of fabric across so that people duck slightly lower than they would have to normally.

GARDEN

The garden can be a real danger zone—especially for children, who should be able to play there unsupervised. Make sure paths and steps are safe and even. If they become slippery with moss or choked with weeds, they could become dangerous. Ice and snow should be cleared and a thin layer of sharp sand/grit should be laid in frosty weather. A sand-and-salt mixture will clear thin ice and snow, if you don't mind the salt running off into flowerbeds.

The gate (if there is one) should be strong enough to deter a young child from running out into the road. Keep the catch high enough or difficult enough to discourage tiny fingers.

Garden chemicals

Chemical pesticide sprays, slug pellets and similar products are developed to kill. Few of them are selective—they will also do harm to your pets, your children and yourself if ingested, inhaled or absorbed (see POISONS).

Long before pesticides were developed by the petrochemicals industry, there were effective organic ways of dealing with most of our garden pests. Plants that were not dosed with artificial fertilizers and sprays were stronger and more resistant to pests and diseases. Investigate organic gardening and you remove much of the danger from your own garden.

If you MUST use pesticides, take care not to breathe in the spray/mist yourself. You should wear a face mask to help avoid inhalation, goggles for your eyes and gloves to prevent skin contact—at the very least.

Keep the poison well away from garden ponds with fish, frogs or other useful animals in them and from any other animals in the garden—whether pet tortoises or friendly visitors. Unfortunately there is little you can do to protect birds. If they eat the slugs that eat your pellets or the flies you spray with insecticide, the poisons will be passed on through the food chain even if not by direct contact. Keep children, adults and animals out of the garden after spraying.

Do not overuse poisons. Most of the weedkillers and insecticides in common use today are extremely potent, designed for impatient gardeners.

Bonfires

Bonfires are frequently forbidden in urban areas because of air pollution and fire hazards. Composting is the easiest way of disposing of garden waste—providing a useful extra for the flower beds. If fires are permitted keep them well away from trees and fences. Avoid any possibility that the house, garage or shed could be endangered if the fire should get out of control. NEVER throw fuel on if the fire isn't burning as fast as you would like. Do not light them with petrol or paraffin. Build them properly and they will burn with ease (see FIRE!).

Avoid large unstable bonfires. If you have a lot to burn, don't make one massive pile and then set light to it! Have a medium-sized controllable fire, which you can add to gradually as it burns down. Start with newspaper, kindling and fire lighters to get the fire going.

Incinerators will contain a small fire. If you buy one, read the manufacturer's instructions carefully. ALWAYS keep a bucket of water, a fire extinguisher or fire-beater handy when having a bonfire OR a barbecue.

WARNING

Do not burn rubber or plastics. The fumes may be deadly. DON'T inhale too much smoke, even without fumes it contains CARBON MONOXIDE. NEVER put aerosols on the fire — they could blow up. Only build a fire on earth — never tarmac, and never concrete or stone which might explode.

CHILDPROOFING

- Make sure fences are secure and gates lockable, so that children cannot go out or squeeze through by themselves.
- Garden ponds and swimming pools should be fenced off if young children are allowed in a garden unsupervised. In some countries they must be fenced by law. Consider filling in ponds with earth until children are older.
- Water butts should be kept covered.
- Empty paddling pools after use and do not let children play in them unsupervised. A plastic pool left out in the rain can rapidly fill to a couple of inches — quite enough for a small child to drown in, if they fall face down.
- Teach children not to eat any part of plants or to suck their fingers after touching them. Bought seeds may have been coated with toxic chemicals to make them keep.
- Keep sand pits covered when not in use — this will keep out rainwater and mess from any visiting cats or dogs.
- Site swings and play equipment on or over grass, not on hard surfaces.
- Do not let children play near greenhouses, garden frames or any structures which have glass panes.
- Don't let children use garden tools. Always keep children away when you are mowing the lawn.

In the garden:

❍ Make sure your gardening footwear is sturdy enough to withstand crushing—or piercing from above or below.

❍ Use each tool for the job it was intended to do.

❍ Clean and oil tools after use.

❍ Don't use tools with loose or 'bodged' handles.

❍ Put away all tools after use, especially if there are children or pets about.

❍ Wear gloves—don't handle soil excessively.

❍ Always wash your hands, when you finish gardening.

❍ Avoid hand-to-mouth contact—which includes smoking, eating or drinking.

❍ Check when hedge-trimming that there are no hidden posts, wires—or birds' nests.

❍ Avoid lifting heavy weights which are a strain.

> ## ❗ WARNING
>
> DON'T attempt major work on trees, especially if you have to climb to do so. Get professional advice—to avoid injury and making a tree unsafe.

Electricity in the garden

Most accidents happen (and most fatalities are caused) when using electric lawnmowers and hedgetrimmers. You MUST protect yourself with an RCD/RCB/ELCB (see Electricity). In most cases it's a good idea to drape the flex from an appliance over your shoulder to keep it out of harm's way.

DON'T wander about with electrical mowers. Try to mow in more-or-less straight lines or with some sort of method—not up and down slopes, but along them. That way you stand less chance of 'mowing' the cable or your feet, than you do using the mower like a vacuum cleaner!

Extension cables are only temporary solutions. You should choose ones designed for outdoor use—with rubber fittings for extra protection against breakage and moisture. Coiled extension cable MUST be fully unwound before use, to prevent overheating and the possibility of fire.

ALWAYS connect the extension to the appliance FIRST, then—making sure the appliance is switched off—make the connection at the mains socket and switch on. Don't switch on the appliance until you have reached the place where you intend to start working. When you have finished, ALWAYS switch off the appliance, then go to the main socket and switch it off and pull out the plug. Then you can remove the extension cable from the appliance.

NEVER 'tinker' with electrical appliances when they are plugged in. NEVER use electrical tools if it's raining or the ground is wet/damp. NEVER clean or adjust blades of mowers or hedgetrimmers while the power is switched on—it's a risk that's just not worth taking.

REMEMBER

It is possible to have special outdoor sockets fitted, to avoid the need for long extension cables trailed from inside the house (see **Electricity**). You may not see a cable in long grass!

WARNING

ALWAYS ALWAYS ALWAYS wear thick rubber boots when using electrical equipment in the garden. NEVER use electrical tools when it's raining or the ground is wet/damp.

GARAGE/SHED

 In a perfect world, garages and sheds would be tidy and safe with a place for everything. Some garages have to be a toolshed, a gardening shed and a storage area for household overspill all rolled into one—sometimes there is no room for the car!

DON'T clutter up the garage or shed. You'll trip over things or hurt yourself sifting through everything.

Chemicals

All chemicals should either be stored high up, or locked up. Most garages/sheds are not insulated—check that the chemicals can tolerate heat or freezing conditions. Don't store chemicals on the hottest side of the shed. NEVER store them in direct sunlight. Check packets, bottles and cans for signs of leakage, damp and rust from time to time.

NEVER store chemicals where a child could reach them. Either ban children from the garage/shed altogether, or be prepared to lock up the chemicals. Some of the more dangerous household chemicals could be stored in the garage/shed—oven cleaners, caustic soda, acids, solvents etc. But remember the possibility of frost or heat damage.

REMEMBER

Don't store large quantities of petrol, paraffin, turpentine (or substitute) or any other flammable substances. In some countries, including Britain, there are legal limits to the amount of petrol you are allowed to store.

Lighting

If you intend to work in the garage/shed, good lighting is essential. Either have the bench under the window, or consider running power to the outbuilding (see **Electricity**). If you use gas or oil lamps, take extra care. Keep them well away from wood shavings, sawdust, rags and all chemicals/fuel, on a stable surface. NEVER work in a bad light with power tools.

Storage

If you've got rafters or enough headroom, you could build a safe storage area to keep things out of the way. DON'T balance objects on overhead beams, though—unless you want boxes to fall on your head! People often store lengths of timber, pipe and conduit by laying them across the open rafters—usually because they're too long to stack against a wall. Sooner or later, the ends of these droop down—and cause very nasty accidents.

ANYTHING GOES?

Try to think of your garage/shed as another room in your house—especially if you intend to spend any length of time in there. So often 'dodgy' household appliances such as outdated lightfittings and portable heaters are relegated to the garage/shed—and used as if safety were no longer important.

People who would never tackle household electrical work often try to instal sockets and extension cables, breaking all the 'rules' in the process. ELECTROCUTION can be fatal (see **Electricity**).

Tools/garden tools

Most garden tools could inflict quite an injury if stepped on or tripped over. It's best to have stout reliable hooks screwed into the wall (or into a baton on the wall). Have the business ends where you won't brush against them—or, worse still, get them in the face. Above head height would be a good idea—but not so high that they could fall on you.

If you care about your woodworking tools, you probably store them safely. A toolbox is a good investment and saves time looking for 'lost' equipment. Don't search through piles of tools and risk cutting your fingers on knives or chisels.

Pick up nails, screws and small fittings WHEN YOU DROP THEM—they could easily go through a shoe. Don't store reusable wood with nails sticking out of it. Pull them out. If you really haven't time, it's better to drive them in.

Clear up shavings soon after work has finished—or during working, if there's a lot. One spark could start a fire.

! WARNING

Don't let children play with bench-mounted vices. With some kinds, if they are unscrewed too far, the winding gear and half the vice could fall off—possibly onto someone's toes.

REMEMBER

A first-aid kit and a fire extinguisher are a good idea if you intend to work in the garage/shed.

IF YOU DO CAR REPAIRS

- NEVER run a car engine in a closed garage. The exhaust fumes may kill you (see **Carbon monoxide**).
- ALWAYS use axle stands while working on a jacked-up car.
- If there is an inspection pit, take care! Pits often don't have drainage and could easily become waterlogged.
- NEVER use electrical equipment while standing on wet ground!
- Drain the pit regularly and don't let debris, spilt fuel and oil accumulate in the bottom.
- Put duckboarding or a wooden pallet in the bottom to stand on.
- Keep the pit covered when the car is not over it, to avoid anyone falling in.
- Oil on the floor could cause someone to slip over. Shake sand or sawdust on the spillage, wait a while and then sweep it up.
- Watch out for fuel leaks from cars in the garage. A very slight leak might get worse. If you smell strong petrol fumes—remember, petrol vapour is heavy and will be denser nearer the floor (especially in a pit)—DON'T SMOKE. DON'T turn light switches on or off—it could cause a spark. Ventilate the area immediately.

☠ WARNING

If you keep your freezer in the garage, it may produce a tiny spark when the cooling mechanism switches on and off. There is evidence that this is enough to ignite a spillage of petrol or dense vapour—causing a very serious fire or explosion.

BEDTIME CHECK LIST

You should make a nightly, room-by-room check, before retiring, to make sure that all is safe and secure:

■ Turn off all gas and oil appliances, except for pilot-lights
■ Extinguish all fires, naked flames and cigarettes—DOUBLE CHECK!
■ Unplug electrical appliances—especially televisions
■ Make sure there are no obstructions on stairs or in passageways
■ Check everyone is home before securing entrance door bolts or locks which cannot be opened from outside
■ Lock all external doors and secure all windows in unoccupied rooms
■ Set any alarms

OUT FOR THE DAY

If leaving the home for the day, there are several safety and security considerations, which should become automatic. If necessary, make your own check list.

◑ Gas, oil and electrical appliances, such as central heating systems, may be left so long as they are working efficiently on a self-timer programme.

◑ If you have automatic air-conditioning, follow the instructions regarding leaving the system for any length of time.

◑ Extinguish portable gas and oil fires.

◑ Either extinguish open fires, or allow to burn very low if protected by an efficient fireguard.

◑ If anyone in the house smokes, check there are no cigarettes left burning in ashtrays or waste bins.

◑ Unplug all electrical appliances—especially televisions.

◑ Close all internal doors.

◑ Lock all windows.

◑ Check you have switched off any appliances you have been using—the oven or iron for instance.

◑ Set lights on timers to 'come on' as if the house is occupied.

◑ Lock garage and make sure contents are secure.

◑ Check that no ladders are usable and tools are locked away.

◑ Set alarms.

◑ Set telephone answering machine.

◑ Make provision for pets.

◑ If car is in driveway, ensure all doors are locked.

◑ Do not leave a key in ANY hiding place—however 'clever' or 'safe' you think it is.

◑ Lock doors as you leave.

GOING AWAY

If leaving the house for longer than a day—say, a week or two
for a holiday—there are other points to consider.

In warmer weather, central heating/hot water systems
should be completely shut down. In colder weather, especially
when there is a possibility of water pipes freezing, set the thermostat to a low position. In this way, the central heating will
only be activated when room temperatures become very low.
Hot water in a lagged cylinder will stay fairly warm, so the
boiler will only need to fire up briefly to keep it hot. Turn down
the thermostat on the hot water cylinder to be safe. Electric
immersion heaters MUST be switched off.

All electrical appliances should be switched off and
unplugged—except for those lights activated by timer switches.
Try to arrange lights to operate in a realistic pattern at likely
times of the day.

In the morning and afternoon the kitchen would normally
be occupied. In the evening the living and sleeping areas are
normally in use. DON'T just have a hallway light showing.
Some timers allow you to programme several on/off periods so
that you can vary the sequence. An observant burglar keeping
watch over several nights might notice repeating patterns.

If you are in doubt about the safety of your water system,
turn off stopcocks and partially drain down large cisterns (indirect systems). Only do so AFTER you have switched off water-
filled central heating systems and immersion heaters.

It's up to you to make adequate provision for pets and
plants. A trusted neighbour may be able to help you out
and—although you should stop any regular deliveries to your
home—they can make sure no unexpected parcels are left on
your doorstep.

Tell at least one neighbour you have gone and leave a key
with them. It could be an idea to tell the police in case the worst
happens. Let them know who has the key. A neighbour should
know how to switch your alarm on and off. Lock garages/sheds
to prevent access to tools which might be used to gain entry.

PETS

 Although much good can come from owning pets, there are serious diseases/disorders which can be transmitted from pets to humans (see HEALTH). There are also major safety considerations, which are outlined below. Remember that you can be legally as well as morally responsible for your pet's actions. If your dog bites a neighbour's child or runs in front of a car and causes an accident, you could be sued for compensation.

General hygiene

- ❍ A pet should have its **own** bowls, utensils, can-opener, washing and mopping-up equipment.
- ❍ Keep the area where the animal is fed, or where it regularly sleeps, **clean**. Vacuum and scrub surfaces as necessary.
- ❍ **Regularly** wash or discard bedding.
- ❍ Cat litter should be emptied regularly. There are **health risks** to humans, with severe health risks to a pregnant woman and her foetus.
- ❍ Never allow an animal to lick your face, or the faces of children. **NEVER!**
- ❍ If your pet has diarrhoea, be **scrupulous** about cleaning up—but wear gloves and wash afterwards. There could be a danger of cross-infection.
- ❍ **DON'T** allow children—especially toddlers—to play on the ground where dogs regularly defecate. There are severe health risks.
- ❍ Don't garden or work with soil without gloves. If you do, wash **thoroughly** afterwards.
- ❍ Bird keepers should be scrupulous about cleaning out cages. Do it outside or keep people away. Wear breathing protection—to **avoid** inhaling dust from feathers or debris. There is a serious health risk.

General safety

- ❍ Keep pets **away** while you are cooking—both for hygiene and the possibility of nasty accidents.
- ❍ **Remember:** Animals under your feet can cause a serious fall, especially on the stairs. Pets cause a lot of accidents to children and the elderly.
- ❍ **Don't** allow a child to boisterously handle a pet—especially if the animal belongs to someone else. Even animals have tempers—and may retaliate.

- Make sure your fence/gate is **secure** to prevent your dog running free.
- Legally **you** may be responsible if your dog causes distress, damage to property, or an accident.
- You may have to consider a muzzle on large dogs, especially guard dogs which are also pets. This may be a legal requirement in some countries.
- If you have an exotic or dangerous pet, especially a snake, **you** can be responsible for injuries it may cause.
- **Don't** allow your dog to chase pedestrians, cyclists and cars.
- **Take care** when a bitch has a litter to protect. She may be overly-defensive and snap at strangers and children.

AQUARIA

What could be more harmless than a colourful aquarium full of fish? But BEWARE—many of them are carnivores and some have vicious teeth! A moray eel, for instance, may rise out of the water and literally bite the hand that feeds it. Other species may be highly venomous—make sure you KNOW the potential hazards—don't just guess!

Fish aquaria pose serious electrical risks. If you can use appliances which operate on reduced voltages, do so. Since most tanks have to have heaters and pumps, keep fuses to 3amps. Switch off everything before touching the water, and DON'T touch electrical fittings with wet hands. DON'T bodge wiring jobs—you AND your fish may be fried!

UNUSUAL PETS

It is illegal to keep some animals—often a country's own indigenous species. Trade in many endangered species is prohibited by international agreement, so they cannot be exported/imported as this often leads to extreme cruelty or infringement of quarantine laws. It is only thanks to such legislation that Britain and other countries have managed to more-or-less eradicate rabies and other serious diseases (see SELF-DEFENCE: **Animal attack** and HEALTH: **Zoonoses**). Ownership of some animals is prohibited because they are poisonous or otherwise dangerous. To obtain a licence to keep them, it is necessary to show that they will be kept securely with no risk to the public.

In Britain a licence is needed to keep a poisonous snake, but a tarantula or a boa constrictor can be kept without infringing the law. However, it is obvious that they must be housed so that they cannot roam free or pose any risk to others. Owners must know how to handle them and the animals may come to recognize such handling. DO NOT encourage visitors to handle them without supervision and instruction—you could be held responsible if any mishap occurred.

3

More and more DIY tools and products are appearing on the market—statistics indicate that do–it–yourself is increasing likely to become a case of do–it–TO–yourself! Knowing the risks should help you to avoid serious injury.

DIY/craft hazards

USING TOOLS

BASIC TOOLS Hammers/mallets • Screwdrivers • Chisels • Awls • Saws • Files • Knives

POWER TOOLS Drills • Sanders • Saws • Power planers • Electric screwdrivers • Angle grinders • Wood dust

HIRE TOOLS General safety • Chainsaws • Soldering/brazing • Welding

PAINT Emulsion • Gloss • Primers • Stripping

CRAFTS Photography • Ceramics • Enamelling • Stained glass • Artists' paints

LADDERS Platform towers

PROTECTIVE CLOTHING

DOING IT YOURSELF

 There is a growing interest and participation in 'having a go' at home improvements of one sort or another. The number of possible dangers is very high—depending on each individual's level of skill, the complexity of tasks undertaken and the levels of risk involved for the amateur.

Some people will never lift a screwdriver in their lives, others will attempt to build extensions to their homes—or complete new homes.

Most common injuries

In Britain alone this 'have a go' enthusiasm results in an estimated 120,000 casualties a year. Cuts to the hands and fingers from knives, saws, screwdrivers and chisels are the most common injuries—accounting for almost a quarter of the total. An enormous number of injuries are caused by nails, screws and tacks!

Falls from tripping over cables and dust sheets or falling downstairs, falling off stools/chairs/trestles/boxes account for almost as many casualties. Careless or casual use of ladders and stepladders results in some very serious falls.

Less common—but still far TOO common—(in decreasing order of frequency) is the number of people who hit themselves with hammers, injure themselves with power tools and get things in their eyes, sometimes dust and sometimes major debris.

Why do it yourself?

Most people who have been injured while indulging in do-it-yourself—even only at the hobby level—give the same answers. The factors that seem to drive the enthusiasts on are quite understandable:

❍ It is difficult to find reliable professional workers.
❍ The cost of using professionals may be prohibitive.
❍ Thousands of DIY products are available.
❍ Hundreds of handy 'gadgets' are available.
❍ There's a thrill to learning a new process.
❍ There is great satisfaction in a job well done.

It's very true that today's paints and wallpapers are better made and easier to apply. Kit-form furniture is common, but even without it there are lots of fittings and gadgets around which have made traditional joinery unnecessary for the amateur. But easier doesn't mean SAFER.

Danger factors

Many home improvers learn the hard way, practising new skills with each job they undertake. Inadequate safety precautions, not reading safety instructions with tools, the lack of safety instructions with traditional tools and an incomplete grasp of the dangers involved are a frightening combination.

Add to these the fact that new tools, equipment and substances are introduced into the market on a regular basis—and there has to be a first time for every individual with each tool or process, however generally experienced the individual.

ALWAYS ask suppliers or even manufacturers for advice. ALWAYS read the safety instructions. Approach each new job with CAUTION, considering possible hazards.

SAFETY CHECKLIST

Many injuries are caused by:

- Incorrect use of tools and power tools
- Not using/not buying protective guards for power tools
- Not using/not buying dust and debris collectors, which are attached directly to power tools or enable a vacuum cleaner to be connected
- Lack of maintenance of power tools
- Incorrect use of hire tools or inadequate safety briefings
- Underestimating the strength required to operate some hire tools
- Lack of precautions against/understanding of electrocution
- Lack of protection: eyes, ears, nose, lungs, hands, feet, limbs, head
- Inadequate lighting while working
- Slippery or uneven floors, or carelessness with dust sheets
- Lack of fire precautions
- Inadequate ventilation
- Incorrect use of ladders/scaffolding/improvised ladders
- Misuse or misunderstanding of chemicals
- Incorrect storage of chemicals and equipment
- Lifting/carrying too heavy a load or lifting incorrectly
- Lack of care with glass
- Rushing to complete laborious tasks
- Not supporting or anchoring workpiece

REMEMBER

Always work on a firm surface. A collapsible/portable workbench is a good idea—especially with adjustable grips to hold your work steady.

☠ WARNING

Walls, floors and ceilings conceal electric cables and pipes carrying water and gas. Take great care when nailing, screwing or drilling. Use a battery-operated detector to avoid nasty accidents. Piercing a live cable could be FATAL (see SAFETY FIRST: **Electrocution**).

DIY/CRAFT HAZARDS   DOING IT YOURSELF

117

USING TOOLS

 It is not possible to give safety guidelines to cover every do-it-yourself process, nor is this intended to be a do-it-yourself manual. The information about basic tools is intended to assist the inexperienced worker, to help prevent accidents. The prime concern is not to focus on correct DIY techniques, nor to avoid damage to a piece of work (perhaps from a slip with a screwdriver), but to try to ensure the safety of the user of the tool.

Experienced handypersons may find some of the information a little obvious, but should still benefit from a few reminders on safety!

The work area

DON'T leave tools lying around. Try to be methodical about putting them away. A toolbox or special cupboard will do, if you don't fancy the idea of special racks. Put protective covers on (if you have them), especially on piercing tools like bradawls—a cork would do, jammed on the end—and knives and chisels. If knife blades are removable, discard them at the end of working. You'll probably want a new blade next time you use the knife.

Try to be tidy. You'll find things more easily, lower the injury risk of searching through piles of tools or tripping over them and make your tools last longer.

The floor

Keep floors as clear as you can, especially the immediate area where you expect to be working. Keep dust sheets flat and make sure they don't slip on the floor below. Don't leave flexes and extension cables trailing about. If you spill any liquids, mop them up at once.

Lighting

Keep the area well lit, with natural light when possible. Supplement this with inspection lights, spotlights or domestic lighting, bearing in mind the possibilities of electrocution and fire. If fixing permanent work-area lighting, consider low-voltage options. Fixed appliances could have their own separate illumination nearby. Don't work in your shadow!

Ventilation

Ventilation is **vitally** important in several ways. You must NOT expose yourself or others to clouds of dust or fumes while working. Dust is a long-term health hazard and breathing protection should be worn. Damp the floor before sweeping up, or use a mop. DON'T sweep the dust when dry, you will only recirculate it—and

DON'T use a domestic vacuum cleaner. Vacuum cleaners trap larger particles of dust, but smaller ones can pass straight through the machine and become airborne again. If you have a permanent work area, investing in an extractor fan would be a good idea.

Fumes also pose severe health risks. The fumes of some substances may be carcinogenic (cancer-causing) in the long term—in the short term they may make you drowsy and incapable of using power tools safely.

Open windows and doors and try to get air moving through the work area. An accumulation of certain dusts and fumes may also present a risk of explosion.

Noise

Avoid excessive levels of noise, and wear hearing protectors whenever necessary. Don't operate noisy machinery for long periods of time without a break. Remember: if the noise is loud enough, you won't be able to hear whatever else is going on around you. The doorbell or phone may ring. Someone may call for help. The vibration could cause an object to fall from a shelf, out of your line of vision, and onto a heater. Accident records show that almost anything is possible.

> **! WARNING**
> Children and pets should be excluded from the work area. All tools and chemicals should be locked out of the reach of children, when not in use.

Clearing up

At the end of the work period, try to clear the floor and tidy up tools and materials. Fix lids on paint and chemicals. Switch off and unplug tools, appliances and heaters. Look round to see that everything is safe to be left.

Depending on how dirty the job has been, have a thorough wash or shower—not forgetting your hair and fingernails. Even the smallest traces of some toxic substances could be injurious over a period of time.

> **REMEMBER**
> If disposing of any DIY rubbish in household rubbish collections, make sure it is all safe. Young children and pets seem to like rubbish bags. You should NOT use the domestic waste system for toxic or dangerous substances. Wrap up broken glass or knife blades. Sharp objects may pierce the refuse sack and badly cut hands or legs—maybe yours!

BASIC TOOLS

 The use of even traditional workshop tools is risky. As new tools are added to the toolbox, the only way to familiarize yourself with them is to use them. Suppliers, manufacturers and regular or professional tool-users may be able to advise you.

Hammers/mallets/stone hammers

Hammers are for driving in nails and tacks. Depending on the style of hammer, one side of the head performs the basic nailing task. The other may be a narrow rectangular head (A) for coping with small nails, too delicate for normal hammer blows, or for nailing into mouldings (cross-pein hammer).

A rounded head (B) is really for use with sheet metal (hemispherical-pein). The more common claw head (C) is for removing protruding or badly-driven nails (D).

DON'T use hammers to bash wooden- or plastic-handled tools such as chisels or screwdrivers. Cheap or thin-handled claw hammers should be avoided—they may snap when you are trying to pull out deep or stubborn nails.

WARNING

If you miss badly with a hammer, and the handle takes the blow behind the head, the handle may snap—and the head may fly off. Don't raise the hammer excessively. Several medium blows are better than a couple of very hefty ones. Always check that the handle is sound and the head well-fixed.

Mallets (E) are for tapping wood- or plastic-handled tools such as chisels OR for tapping joints together when assembling joinery. Rubber- or leather-faced hammers (F) are available for more delicate work.

Heavy **club hammers** are used with special (cold) chisels to chip into concrete, bricks or paving slabs.

WARNING

If using a club hammer and cold chisel on stone or very hard surfaces, ALWAYS wear eye protection and strong gloves. Debris has an uncanny knack of finding your eyes. A misdirected blow could damage your hand. Watch out for the 'burrs' which occur on the chisel top after a while. These could cut a hand or break and fly off. File them down.

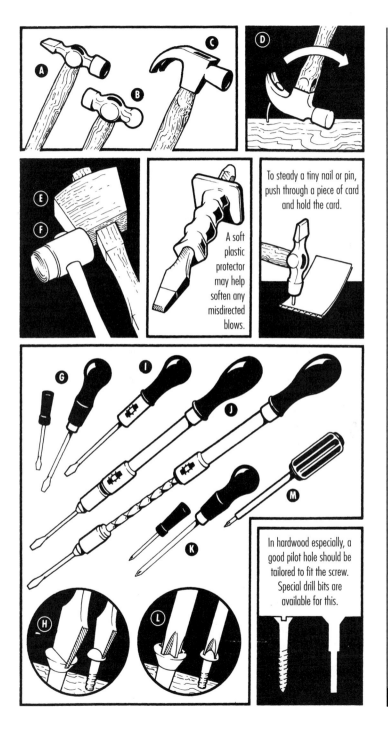

A soft plastic protector may help soften any misdirected blows.

To steady a tiny nail or pin, push through a piece of card and hold the card.

In hardwood especially, a good pilot hole should be tailored to fit the screw. Special drill bits are available for this.

Screwdrivers

Square-ended screwdrivers (G) should fit the screws you are using them for (H). It is worth having more than one size—or a GOOD QUALITY set with interchangeable bits. Ordinary screwdrivers can be hard work—blisters are common.

Rachet screwdrivers (I) are handy—you don't have to release your grip. Most are adjustable to provide the rachet action for screwing or unscrewing, plus a fixed position to allow for use as a conventional screwdriver.

Spiral-rachet screwdrivers (J) are a must for the serious DIY enthusiast or professional, but USE WITH CARE. When released, these tools shoot out to full length and could easily take out an eye or pierce the skin. NEVER point towards your face or at anyone. Slips with these tools can often be very nasty. Keep an eye on where the bit will go if you DO slip. With spiral-rachet screwdrivers, close the tool for final tightening of screws. Most have interchangeable bits.

Cross-headed screwdrivers (K) should also fit the screws that they are used for (L)—in practice, only one small 'electrical' one, and one larger type are needed.

Electrical screwdrivers (M) should offer genuine protection from accidents with electricity. Handles should be shielded with rubber. Some double as 'testers'—a neon, or similar, in the handle lights up if the screwdriver touches a live terminal or wire. **NEVER** work on appliances which are connected to the main supply. **NEVER** work on a live circuit.

Using screws

Avoid undue pressure, which might cause an accident, when driving in screws by:
❍ Drilling correct pilot holes.
❍ Lubricating the threads only of screws with oil or grease—NOT the head—this increases the risk of slipping.
❍ Not using too long a screw.
❍ Using a screwdriver that fits the screw.

> # ❗ WARNING
>
> If a square-ended screwdriver shows wear and tear—reshape the tip, or ask a dealer to have it done for you. A worn tip increases the chance of slipping or fouling the head of the screw, producing sharp burrs.

Chisels

If you intend to use chisels, perhaps to cut a slot for a hinge or to fit a lock, one is not enough. You will need several. Store them safely—either with the ends covered, or in a rack—and keep them very sharp. A blunt chisel requires force and accidents are common. Sharpening is a critical process involving an oilstone—and can take a lot of practice.

Always remove wood PATIENTLY in small amounts, NOT by digging and gouging. If chisel is sharp it should do the work for you.

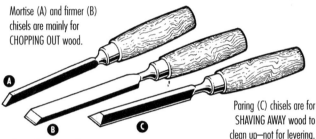

Mortise (A) and firmer (B) chisels are mainly for CHOPPING OUT wood.

Paring (C) chisels are for SHAVING AWAY wood to clean up—not for levering.

Bradawls/awls

Whether square-ended (bradawl) or pointed (awl), they may be good enough to make pilot holes for screws in softwood. Keep them sharp. DON'T use a lot of pressure. DON'T get in the 'line of fire' if you slip—NEVER work into your hand or on your lap. Protect tips and your fingers by jamming a cork on the points when not in use.

Bradawl

Awl

Saws

There is a wide variety of saws. They must be kept sharp (a dealer can arrange this for you) and rust-free. Saws with big teeth are intended to rip through softwood quickly—cutting on the downward stroke. Smaller teeth imply finer work—possibly for plywood or chipboard. The shape of the teeth may mean that more wood is removed on the downward stroke. A dealer will advise you for your specific needs.

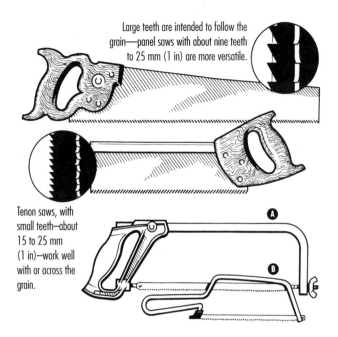

Large teeth are intended to follow the grain—panel saws with about nine teeth to 25 mm (1 in) are more versatile.

Tenon saws, with small teeth—about 15 to 25 mm (1 in)—work well with or across the grain.

Ⓐ

Ⓑ

Tenon saws are for cutting joints in joinery (mortise-and-tenon joints!) and fine work on smaller jobs requiring more care. Most have a strengthening bar along the top edge of the blade, which restricts the depth of cut, so you won't be able to saw sheet materials.

REMEMBER

- KEEP saws sharp, clean and oiled
- Avoid using saws which have been bent or buckled
- Hang up saws when not in use
- The saw is supposed to do most of the work—if you're using excessive pressure, either the saw is blunt or it's the wrong saw for the job
- ALWAYS saw on a steady surface and anchor the workpiece
- ALWAYS support as much of the wood as possible. When the waste wood falls (which you should avoid), the rest may topple off supports
- DON'T saw between two supports, allowing the wood to droop down at the cutting point. This will close the cutting line, gripping the saw
- Keep your fingers well away from the cutting line
- When starting a cut, you may need a thumb to stop the blade 'jumping off' the cut line. Start very slowly with light controlled strokes, until the blade is established in the wood

Hacksaws

A hacksaw (A) is a basic tool for cutting metal bars or strips, angle-iron, metal and plastic pipes. The frame holds the blade taut—don't overtighten, you may cause it to snap. Some types allow for the blade to be set at different angles. The frame limits the depth of cut possible—when it hits the workpiece, you have gone as far as you can. It's not suitable for sheet metal, unless you are making small straight cuts at the edges.

The blade should point downwards—away from the top frame—not upwards. Cutting pressure should be on the forward stroke (away from you). Keep your free hand on top of the frame, at the front (away from you) to steady and guide the saw. Try to use long controlled strokes. Try to anchor the workpiece in a vice.

Small hacksaws (B) are called 'junior' hacksaws. If you are cutting hard steel, they aren't really strong enough.

Change blades regularly and ask a dealer for advice. There are different types and qualities of blade—usually with different size teeth—for different jobs or degrees of finish. NEVER use if the blade is twisted or cracked.

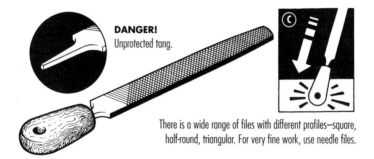

DANGER!
Unprotected tang.

There is a wide range of files with different profiles—square, half-round, triangular. For very fine work, use needle files.

Rasps/files

Most are designed to remove material on the forward stroke. Some have removable handles—**NEVER** use without the handle! It's quite possible to drive the unprotected point (tang) right into your hand. Tap the rasp/file handle downwards on the bench (C) to make sure the handle is secure. Hold the handle firmly, steadying the other end with the tips of your fingers. Use a steady even pressure. Don't work fast—be patient and let the tool do the work. If the face of the tool becomes clogged, give it a good stiff brushing.

Craft knives

Basically a sturdy handle, which can be opened to allow a new blade to be fitted. Choose one that suits your grip. Retractable blade options are a good idea, if the knife is good quality. Blade covers that come with non-retractable types get lost, or don't fit the other blade shapes that are available. The blade cover could also come off—in a pocket, perhaps!

ALWAYS change the blade as soon as it becomes inefficient. You shouldn't have to use excessive pressure. The blade

Some retractable knives accept ordinary reversible blades. Others allow you to snap off the blunt end as you go. For very fine work, scalpels offer precision and control.

is supposed to do the work. ALWAYS buy the correct blades for the make of handle—a sloppy fit is no good. Some of the fancy/ specialist blade shapes are useful—but also quite deadly.

NEVER twist or gouge with the blade. The tip might break and fly off. Remove material slowly—not in large chunks. ALWAYS remove the blade from non-retractable types when putting them away.

Smaller craft knives are advisable for smaller, or more delicate jobs. There's a wide range available. You hold the handle more like a pencil—less like a weapon—and use a lot less pressure. Use the right tool for the right job!

REMEMBER

If throwing away knife blades, make sure you wrap them well to save nasty accidents. Picking up a rubbish bag, or brushing a leg against it, could cause a severe injury.

TOOL RULES

- Always use a tool for the job it was designed to do
- Always use tools correctly—if it feels very awkward, STOP
- Keep hands and other parts of the body out of direct 'line of fire'
- DON'T use excessive force
- Keep 'sharp' tools SHARP
- Check all tools for strength and SAFETY
- Store all tools safely—out of the reach of children

POWER TOOLS

 There's a wide choice of makes and types of tool. The more expensive, top-of-the-range ones are usually intended for heavy-duty or professional use. The lower-priced should be fine for occasional home use—as long as they are of good quality. Often the first thing to fail, especially on drills, is the on/off switch from constant stopping and starting!

Power tools MUST be cleaned and serviced—how often depends on how often you use (or abuse) them. There are service departments of all the major manufacturers practically everywhere.

PLEASE NOTE

Read the following safety notes in conjunction with later information on specific power tools. These notes apply to almost all power tools.

Before you begin

○ DON'T even start to use the tool if you are tired, drunk or feeling ill.

○ If you're working outside, check it is not raining or likely to rain while you are working. DON'T work standing on wet or damp ground. Wear rubber-soled boots. The risk of ELECTROCUTION is always present.

○ Make sure you are protected by an RCD/RCB/ELCB (see SAFETY FIRST: **Electrocution**).

○ Check that the drill bit/saw blade/attachment is firmly fixed.

○ Consider whether you are using the correct tool for the job.

○ If the appliance has a chuck key, remove it before switching on—likewise, spanners and Allen keys!

○ Clear the area where you intend to work. Prepare a solid surface to work on. Clamp the work if possible.

○ Don't wear jewellery, loose clothing—gloves should fit and be kept away from anything which might entangle them. Tie long hair back out of the way.

○ Get ready the correct forms of protection: safety glasses, a face mask or respirator, ear protectors (at least).

○ Clear children and pets out of the work area.

REMEMBER

Always choose tools which are 'double insulated'—with plastic casings. The double-insulation symbol in most countries is two small squares, one inside the other.

- Check that the flex, plug and extension lead are in good condition. Fully unwind extension leads. Don't carry the tool by the flex.
- Make sure (if drilling into a wall, for example) that there are no concealed electrical cables or water or gas pipes.
- If a tool has a fitment for dust/debris extraction, fit it—or use the connector to attach a vacuum cleaner. Use only a cylinder or drum vacuum cleaner—pointing the 'exhaust' out of the work area, preferably outside. Small dangerous particles travel straight through domestic vacuum cleaners (see **Wood dust**).
- If the tool has a guard, make sure it is attached.
- Ensure the area is WELL VENTILATED.
- Make sure the power tool is switched off (or locked off—if that option is available) BEFORE plugging in and switching on at the socket.

WARNING

Lock-on buttons on power tools are good for allowing you to ease your grip slightly—BUT ... if you do have an accident, the tool will carry on working and could do a lot of damage. If you're too tired to hold the on/off trigger down, perhaps you're too tired to be using the tool!

While you work
- DON'T block the ventilation holes on the tool.
- DON'T force the tool, let it dictate the pace of working.
- DON'T use the tool for jobs it was not intended to do.
- Keep an eye on where the flex is at all times. It should NOT be anywhere near the cutting/drilling/sanding site.
- Keep fingers away from the site—even gloves may become entangled, causing you to lose control of the tool.
- If you need to replace the bit/blade/attachment—SWITCH OFF and unplug the tool first.
- Make sure nothing around you can fall—vibrations can cause such an accident. Noise may prevent you from hearing what is happening.
- Never work over your head.
- Never work with the flex stretched tight. Use an extension cable (fully unwound) or move nearer the socket.
- Don't do anything which feels very awkward.
- Don't operate leaning at an angle on a ladder—or balancing on chairs, stools or trestles.
- Keep the area well ventilated to let dust out.

- Don't work in your shadow!
- Keep your fingers away from the 'lock-on' button. On most tools these buttons are a very BAD idea.
- DON'T move about holding the tool while it is switched on.
- **CONCENTRATE!**

When you've finished
- Remove dusty goggles so that you can see what you're doing—but not your mask.
- Switch off the tool and lay it down carefully when all moving parts come to rest.
- Go to the socket, switch off and unplug flex.
- Coil extension cables safely.
- Remove the bit/blade.
- Brush the tool over to remove dust and debris (keeping your mask on).
- Damp down any dust/shavings before sweeping up. A vacuum cleaner will recirculate dangerous particles—even wood dust should not be breathed in (see **Wood dust**).
- When the air is clear remove your mask.
- Store tool in a safe place away from children.

POWER DRILLS

Drills are the most popular power tools. Most accidents occur when a drill bit slips or snaps, or when a bit is too blunt and requires heavier pressure to make it do the job. Accidents also occur when the bit 'breaks through' the material being drilled. ALWAYS switch off the drill when changing bits.

There are many attachments for drills, but these are not recommended—especially circular saw and lathe attachments (these may cause overheating, or damage the chuck bearings). It is advisable to buy specific power tools for specific jobs. Drills work best as DRILLS.

When drilling wood, avoid using too much pressure and ALWAYS prepare for the moment the drill breaks through the other side of the workpiece. A cleaner exit hole—and a way of avoiding the 'surprise'—is to have a piece of waste wood under the drilling site. You should still feel when you have broken through to the other side.

Always anchor the workpiece—use clamps if possible—or use a drill stand, which gives you a lot more control.

When drilling metal, use toughened steel (high-speed steel/tungsten) bits. Work at slower speeds, despite the 'high-speed' label on the bits. Always drill a pilot hole. Keep the bit oiled. Wear eye protection and avoid the sharp hot swarf.

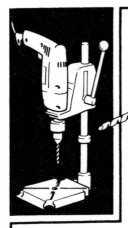

▲ Choose a drill with a sturdy steadying handle and variable speeds. For finer control of vertical drilling, and safety, always use a drill stand.

► The only exception to the 'no-attachments-on-drills' rule—rotary and drum sanders (safer still with a variable speed drill). For fine work, there's no substitute for hand finishing (unfortunately).

C

A

B

D

Orbital (A) and belt (B) sanders generate a LOT of dust—ALWAYS wear a mask and goggles. Circular saws (C) and jigsaws (D) demand concentration, especially at high speeds. ALWAYS switch off when changing blades.

Breaking through may be sudden and tear the metal. Always clamp a block of wood underneath the drill site to absorb the impact, or make a 'sandwich' of wood either side of the metal you wish to drill.

If drilling small pieces of metal, they MUST be anchored firmly—clamped or in a vice. Follow advice above to avoid the metal seizing the bit and spinning round, possibly causing a severe injury to your hands.

You will need a full range of drill bits. There are types for—at least—wood, metal, glass and tiles. DON'T believe that one set will do all jobs—it won't.

Choosing a drill

◗ Look for a drill with more than one speed. At slower speeds, the drill has more turning power (torque)—at higher speeds it produces cleaner holes in wood.

◗ 'Hammer action' is a useful option. You need it for drilling masonry, but read the instructions carefully and be sure to wear eye protection.

◗ 'Reverse action' can be a useful feature, especially at slow speeds for 'unscrewing'.

◗ Hold the drill, feel its weight and balance. If other or additional handle options are available, try those too. At least one extra steadying handle is a good idea.

◗ Lock-on buttons may NOT be a good idea, especially if placed where they could be activated by accident. Lock-OFF buttons are a better idea.

SANDERS

The ONLY attachments recommended for power drills would be the simplest sanders. These are very useful when conventional sanders are too bulky to do a job safely. But they are only as good as the drill they are attached to. A variable-speed drill would make them safer.

Rotary sanders consist of a circular rubber plate, to which you attach sandpaper discs, with a central shaft which is gripped by the drill chuck. Avoid options with a metal backing plate which is almost as big as the rubber disc. These could be more dangerous. Rotary sanding discs should not be used flat—they would be impossible to control—nor at too steep an angle, when they might 'run away with you'.

Drum sanders are similar and should be used with equal care. In this case, a small plastic 'drum' has a strip of sandpaper wrapped round it.

Sanding tools generate a LOT of dust and debris, including grit particles from the sandpaper. You MUST wear eye and breathing protection (at least)—wood dust is a long-term health hazard (see Wood dust). There is also a fire/explosion risk from dust in the air. Ventilate the work area—preferably have dust/debris collectors fitted directly to the tool—and an extractor fan.

DON'T use sanders with wet waterproof sandpapers unless specifically recommended by the manufacturer.

Belt sanders

There are numerous sizes available. Try the weight and see if you think you could manage to work with it. The belt will move at very high speed, possibly taking you with it—or swerving off into a cable or YOU. You will need both hands to operate it. Don't exert a lot of downward pressure. The weight of the tool and the grade of sandpaper should do the work. Only buy a large belt sander if you intend to use it professionally—these are better hired (large ones are called 'floor sanders').

Don't use a belt sander if you want a 'perfect finish'—they are specifically for removing large amounts of material. The belt may wander on the rollers. If worn too thin, or snagged on a nail, the belt may shred and foul the machine. A fitted dust-collection apparatus is essential—as are safety glasses and good breathing protection.

> **! WARNING**
> If you plug in a belt sander and it is switched on, it will 'run away'. Never put it down or abandon it until the belt has totally stopped moving.

Orbital sanders

Also called finishing sanders, the flat sanding pad moves in small circles—it does not rotate. The work is done by using various grades of sandpaper—coarse for heavy smoothing, fine for light finishing—NOT by exerting downward pressure on the tool. This will not make the sander more efficient, it will tear the paper, cause a lot of vibration and possibly overheat the motor. Orbital sanders can produce a better finish than belt sanders.

> **! WARNING**
> Fitting new sanding papers can be very fiddly—NEVER do so when the machine is plugged in or switched on.

A fitted dust/debris collector is ESSENTIAL—as are safety glasses and adequate breathing protection.

SANDER PROBLEMS

Mechanical sanders never quite get 'into' corners—and don't produce perfect results without a lot of time and patience. Flexes tend to be short. Assess whether this will be a problem before you start and use an extension lead (fully unwound) if necessary. If you work at the top of a ladder, the extension socket may swing about in mid-air. Avoid this!

If you never intend to tackle large sanding jobs, consider using ordinary sandpaper (or, in some cases, steel wool) instead. If you only want to sand paintwork for repainting, sugar soap, strong detergent or 'liquid sander' (all used with care) could be a lot easier. Most sanders will not cope with curved surfaces and wooden mouldings.

SAWS

Jigsaws

Jigsaws cause a lot of vibration. The blade, which is exposed, moves up and down rapidly and—because most cut on the upstroke—throws up a lot of debris. Some blow the dust away in front of themselves, so that you can see the cut line. You will find yourself leaning over and peering closely, especially on finer work—and getting a faceful. You MUST wear eye and breathing protection.

NEVER put your hand or any other part of your body underneath the workpiece. The blade protrudes and could easily slice a finger at high speed.

There should be no need to exert much pressure—let the blade do the work. Gently guiding the saw should be enough, unless the blade is blunt or unsuited to the type/thickness of material you are trying to cut. When you push, the blade twists or moves out of alignment, so the cut line wanders and you risk breaking the blade. Jigsaws require a lot of concentration to follow a line accurately—especially at speed. Buy one with

REMEMBER

If the jigsaw has an adjustable sole plate for angled cutting—take care. Always make sure the sole plate is well tightened in position. Don't work too quickly—it's harder to control an angled cut. Let the blade stop moving before you put the jigsaw down.

variable speeds. You should work more slowly with plastics. Cutting metal requires the use of a lubricant.

Only use the 'lock-on' button if you really must—otherwise, if you have an accident, the saw will carry on working by itself.

Some jigsaws allow guides to be fitted, which enable you to follow the edge of a workpiece. These can help you control the tool. Never overreach, but travel with the saw if you can.

Blades

Blades must be chosen with care. You will find many kinds available. They vary greatly for hard and softwoods, plywoods, coarse or fine work, metal, plastic, straight or curved cutting. Replace blades when they become worn—that means as soon as they demand more pressure to do the job. A lot of heat is generated by friction and sawdust may smoulder or ignite. Unplug the tool BEFORE attempting to replace blades. A screwdriver/ Allen key is usually required to do the job. Make sure the new blade is well fixed in.

Circular saw/power saw

Circular saws (also called power saws) are a lot more dangerous than jigsaws. Cutting is done, as the name suggests, by a circular blade. Much care and concentration are required to maintain and use a circular saw.

Circular saws cut on the up stroke and guards should ALWAYS be fitted. **NEVER** operate a circular saw without a bladeguard AND another guard which swings back as the cut is made (and moves back at the end of the cut). **NEVER** fix any guard in any position to 'get it out of the way'. **NEVER** put your hand or any other part of your body underneath the workpiece. The blade penetrates and could easily cut off your hand. Always wear eye and breathing protection.

There is a variety of different blades available, for different jobs. Don't push the saw through the workpiece. If you feel you have to, the blade may be blunt or incorrect for the material you are trying to cut. The blades must be sharpened by a professional.

Make sure the blade is well fixed. Let it reach full speed before starting to cut—and let it stop before lifting the saw away from the workpiece.

NEVER NEVER NEVER touch the blade, or attempt to replace it, while the tool is plugged in. There should not be a lock-on button—if there is, do NOT use it. NEVER. If an accident occurs, a runaway circular saw could do a lot of damage. If faced with a choice when buying, a lock-OFF button is by far the safer option.

If you are left-handed, you may find using a circular saw difficult. With most models, it is hard to see the cutting line.

REMEMBER

With jigsaws and circular saws, it is easier to cut large sheet materials if you drape the flex over your shoulder.

OTHER OPTIONS

Power planers

Use with great care. These machines eat wood! They also spew out a great quantity of debris. Always wear eye and breathing protection. Always try to arrange debris collection at source with a collection bag, or vacuum cleaner attachment.

❍ Make sure **NOT** to connect the power while the tool is switched on.

❍ NEVER leave the tool resting in an upright position, resting on the blade. If it starts up, it may run away, possibly causing a serious accident.

❍ NEVER switch on until in position to commence planing.

❍ NEVER put the tool down until the blades have finished rotating completely.

❍ NEVER rest the tool on the palm of your hand, or on your lap—this is an invitation to disaster.

❍ NEVER plane a piece of hand-held wood!

❍ ALWAYS clamp the workpiece, or use a vice.

❍ ALWAYS work on a clear firm surface.

The cylinder of blades can usually be sharpened, but may need to be replaced in some models. Don't expect to be able to plane nail and screw heads.

NEVER NEVER NEVER touch or attempt to replace a blade while the tool is plugged in. The blade (in domestic models) revolves at up to 20,000 revolutions per minute—more than enough to inflict serious damage.

Electric screwdrivers

Rechargeable, battery-operated and conventional versions are available. They are NOT magic screwdrivers, and all the safety rules for screwdrivers apply—as do power-tool safety rules, in mains-operated versions. You are unlikely to have all the tip styles to suit all screw sizes and types.

The turning force (torque) varies from model to model—but, unless you observe all the normal rules (drilling pilot holes, for example), you might find that the electric screwdriver wants to turn in your hands while the screw stands still!

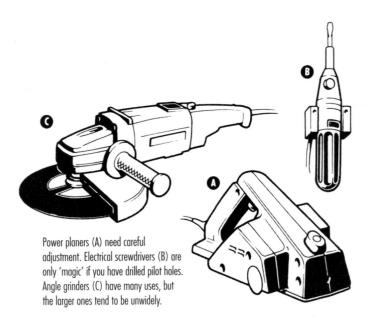

Power planers (A) need careful
adjustment. Electrical screwdrivers (B) are
only 'magic' if you have drilled pilot holes.
Angle grinders (C) have many uses, but
the larger ones tend to be unwidely.

NEVER work 'into' your hand, but always on a firm surface. Don't work on your lap. Keep all parts of your body out of
the way, in case you slip.

Angle grinders

These are the most recent tools to creep into the domestic
market. Ranging from small specialist tools to large unwieldy
professional (or hire shop) models—they basically consist of a
power tool which turns a circular grindstone at great speed.

**Angle grinders create an enormous amount of debris
and dust—guards should be fitted. Eye and breathing
protection is essential. Damp down the area which is to
be worked—observing ALL the rules of electrical safety!**

The basic function of an angle grinder (depending on size and
power) is to grind away material. They may be used, for instance,
for cutting through plaster on walls, when 'letting in' a new fireplace; for cutting paving slabs; for slicing through brick walls.

Larger models may have an almost 'gyroscopic effect (have
you ever noticed the tendency of toy gyroscopes to find their
own orientation when turning quickly?) and may be hard to
control. Get your balance! Be very careful if using up a ladder.
Don't overreach.

**Keep all parts of your body out of the way. Angle
grinders are very dangerous and messy. They require a
lot of concentration and, sometimes, a lot of strength.
Bad accidents on record make 'gory' reading.**

This is VITAL information for anyone who works with wood. Wood dusts, surprisingly, pose severe hazards. They can cause, depending on the type of wood, amount of exposure and the sensitivity of the individual: dermatitis, asthma, bronchitis, chronic lung diseases, lung fibrosis, eye irritation, nasal and sinus irritation, and they have been linked with cancer.

There may be specific dangers from some woods. African boxwood in your system can cause you to feel tired and lose the ability to concentrate—which could be very dangerous if operating machinery.

Wood oils and sap may be poisonous. Further hazards arise if the wood being worked on has been treated with preservatives, fungicides, insecticides, waxes, oils, polishes, glues, shellac and varnishes. Even new timber may have been treated with the first three at the timber yard.

Fungicides, for example, may be very toxic and have been linked—in addition—to anaemia and disorders of the central nervous system and liver.

Heating of the wood while working may be a factor—sanding, sawing and planing all generate heat, which may alter the chemical nature of wood dust or treatments that have been used on the wood.

Ventilation and dust extraction are VITAL. Breathing protection should NOT be taken lightly, nor should eye protection, barrier cream and—when possible—gloves. Always wash or shower well after exposure. Avoid dry sweeping or using a vacuum cleaner, unless the latter is specially designed for the job or the exhaust of a domestic model is not venting back into the work area. Small particles travel straight through domestic vacuum cleaners. Damp down dust or use a damp mop instead.

Avoid any dust from (most) plywoods, chipboards, blackboards and particle boards. The resin/glue used to hold them together may release FORMALDEHYDE—associated with headaches, 'persistent' colds and coughs, asthma, depression, insomnia, dermatitis, and possible links with cancers.

Nearly all woods have thus far been identified as requiring caution. The amount of exposure required to cause some of the more serious effects is not clear. It should therefore be considered that there is no 'safe' exposure level for inhalation. Even the most common woods (as the following, by no means definitive, list shows) are known to have caused health problems.

Beech, birch, African boxwood, South American boxwood, red cedar, dogwood, ebony, greenheart, African mahogany, American mahogany, maple, obeche, pine, ramin, redwood, Brazilian rosewood, Indian rosewood, satin wood, teak

Because of the devastation of tropical forests, more and more 'exotic woods' are becoming readily available. Treat all with caution.

Fire/explosion risks should also always be considered with the presence of dense wood dust. It is very unwise to smoke, have naked flames or create a lot of sparks in a heavily-dusty atmosphere. DON'T take ventilation lightly. It could save your life.

DIY/CRAFT HAZARDS POWER TOOLS ■ OTHER OPTIONS

HIRE TOOLS

The really determined do-it-yourself enthusiast may have occasional need for tools which are too expensive or too specialized to be worth buying and trying to find storage space for. 'Hire' shops will offer almost any equipment, from wallpaper strippers to spanners. Apart from the dangers of unfamiliarity with larger tools, there is also the danger that—when hiring by the day or by the hour—you may be tempted to hurry a job.

Plan the work and choose the tool very carefully. Try not to work alone. If an accident happens, you may need someone to raise the alarm.

ALL PREVIOUSLY STATED RULES FOR TOOLS APPLY.

When hiring equipment

○ Hire from a reputable dealer.
○ DON'T accept a heavier or more unwieldy machine than you feel you could cope with.
○ CHECK the machine is complete, with all attachments.
○ CHECK to see if special tools are required for making adjustments/fixing attachments.
○ DON'T accept machines which show any signs of slipshod repair work.
○ Be wary of machines which have just been returned—and have not been 'checked over'.
○ Ask for clear instructions, safety advice AND a demonstration, if necessary.
○ Make sure you have the correct protective gear.

When using equipment

○ Use the equipment exactly as instructed.
○ Use the equipment only for the specific type of job it was intended to do.
○ Don't over-exert yourself to get the job done quickly, because of the cost of hiring.
○ Stop and check the equipment over for leaks, worn cables or loose connections from time to time.
○ If the equipment sounds 'unhealthy' or appears to be having difficulty, STOP. Contact the supplier.
○ Don't be afraid to contact the supplier if you are the least bit unsure of the performance of the machine. Your safety is far more important than what the person in the shop might think of you.
○ Keep children and animals well away from the equipment— in fact, from the entire work area.
○ Do not leave the equipment unattended for any length of time whatsoever.

CHAINSAWS

A large petrol-driven chainsaw is possibly one of the most dangerous tools. In inexperienced hands, records prove that all manner of horrific accidents are possible—but accidents can happen to anyone. DON'T work alone. Expect the unexpected. DON'T take chances. Avoid a chainsaw 'massacre'!

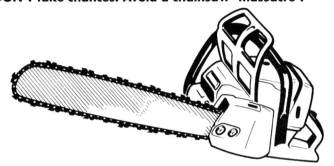

Before you start

Ensure that children and animals are kept well away from the work area. You should have thick, not loose, clothing. Long hair should be tied back. Steel-toe-capped boots are essential, as are thick well-fitting gloves. There are specially made reinforced gloves, which should be safer. There are also limb and torso guards—some are designed to 'burst' on impact, fouling the blade and stopping it. These may not be readily available, but are worth investigating.

WARNING

Under no circumstances consider using a chainsaw if you are tired, drunk or feeling ill. If the weather is very blustery or wet, forget it. DON'T work alone.

Further protection

Both you and your assistant should have impact-resistant safety goggles and hearing protectors. Debris could be thrown a considerable distance. Your assistant should be close enough to 'assist' but not to get in the way. Try to keep them 'behind the blade'—where you are.

Using the saw

Before starting the saw, check there is fuel in the tank and oil in the reservoir. If the weather is cold, you may need to use the 'choke'. Clear the area. Your assistant should stand well away.

When the engine is running, the blade should NOT run at idling speed. If it does—STOP. Consult the supplier, unless you know how to make the adjustment. Check the chain brake. Apply it and open the throttle. The chain should not move. If it does—STOP. This might not be a simple adjustment. Consult the supplier.

Fuel safety

Before refilling the tank, you MUST allow the engine to cool down. It could very easily be hot enough to ignite petrol spilled onto it. Open the cap slowly to release any 'pressure'. A funnel should help you avoid spilling fuel. DO NOT SMOKE!

After refilling, ensure the cap is tight—vibration while working can loosen the fit and fuel may spill onto you unnoticed. Sparks may come out of the exhaust and ignite the fuel.

Store spare fuel well away from the work area.

When cutting

❍ Use the part of the blade nearer the engine, to give you more control. Working close to the tip is tiring, and you are risking KICKBACK (see below).

❍ If you are sawing a long horizontal piece of wood, do so from above. Support on either side of the cut line, fairly close to it. If supports are too far apart, the wood will 'droop' as the cut is made—closing the cut and gripping the blade. This is extremely dangerous.

❍ ALWAYS cut from above—not from below or sideways.

WARNING

KICKBACK: Using the tip of the saw is very dangerous indeed. This is when some of the most serious accidents have happened. It is impossible to control the blade, which may 'jump away' from the wood violently. You or your assistant may be seriously injured. The chain may even break and fly off. NEVER work near the tip of the blade. Allow free space around it at all times. DON'T bring the tip down to touch the ground. DON'T attempt to 'pierce' with the blade.

Tree felling

It is impossible to give directions for coping with every possible variation of tree work. Accidents are numerous—both from the saw and from heavy branches falling unexpectedly. Tree

- NEVER walk about with the blade moving—or even with the motor running. If you trip, the brake may disengage
- DON'T strain when using the saw. DON'T do anything that feels awkward
- DON'T cut above your waist height. NEVER at shoulder height. NEVER above your head
- DON'T work up a ladder—even on trees. Professional tree cutters know what they're doing and use harnesses, ropes and winches—they don't just cut off branches and 'see what happens'
- DON'T set a hot saw down where a fire could start
- DON'T work in a tight space—have room for movement
- Plan your 'escape route'. If an accident happens, you may have to jettison the saw. Keep a space in mind for this, and keep your assistant out of the 'line of fire'
- Only cut wood with the saw
- NEVER cut into the ground
- BEWARE of nails in timber!
- ALWAYS disconnect the spark-plug lead if attempting to clean or adjust the blade
- The saw should do the work. Don't force it. If you have to use excessive pressure, the blade may be blunt. STOP
- Don't work in amongst brush and twigs which may prevent you using the saw safely—and being able to operate safety devices
- The chain should not be loose. If it is—STOP

felling/pruning may look easy and, to some people it seems to be quite exciting—but do NOT attempt it unless you know what you are doing.

If pruning, you may kill a tree by leaving gaping wounds for diseases and rot to set in. The tree may become dangerous in later years.

If the tree is within 10 m (33 ft) of a house, it may be affecting the foundations. Removal of the tree can also cause damage, although the effects may not be apparent for a few years. Consult a specialist.

SOLDERING/BRAZING

Most domestic soldering consists of making small joins in electrical work, using small soldering irons, or making soldered (capillary) joints for plumbing. A larger soldering iron or a propane torch may be hired for joining sheet metal. There are obvious dangers from the heat—burns and fires are a constant risk—but

DIY/CRAFT HAZARDS ■ HIRE TOOLS ■ SOLDERING/BARZING

the materials involved are also highly toxic. Good ventilation is essential to avoid breathing fumes. When judging the fire hazard, typical temperatures involved in ordinary soldering with tin-and-lead solder are 175–250°C (347–482°F), while brazing (which tends to use copper-and-zinc solder) requires temperatures of 700–900°C (1292–1652°F).

Solders

Most solders contain a combination of potentially toxic metals. Brass, copper, lead, zinc, tin and pewter are common. Do not handle excessively. NEVER inhale the fumes of molten solder.

Fluxes

Fluxes, which allow the solder to flow, are intended to 'cut through' oxides on the metal to make a good firm join. They are poisonous and corrosive. Do NOT breathe the fumes they produce. Wear barrier cream or disposable gloves. If any flux does get on your skin—use a good skin cleanser. The flux for brazing is usually a borax paste—which is often the main constituent of ant-killer! Enough said.

WELDING

Welding is only required if a strong join is needed. It's really a job for professionals, but many people hire welding equipment for repairs to cars, gates and grilles. It isn't something you do well the first time you try and great risks are involved.

Solders are not usually employed—instead, the point of contact between two metal surfaces (usually steel-to-steel or aluminium-to-aluminium) is heated until they 'melt together'. Extra metal—a filler—may be melted into or onto the joint.

You need a lot of concentration and a lot of control. Welds go wrong in lots of ways. DON'T rely on a welded joint (for example, on a car or a child's swing) unless you know what you are doing. If the temperature is too high, you may melt too much metal or make a hole. Too low and the join might break. Take a class in welding to give you an idea of what's involved.

Gas welding

Gas welding involves cylinders of oxygen and acetylene, which are mixed within a torch. In some cases, butane or propane are mixed with oxygen instead. Great care must be taken with all

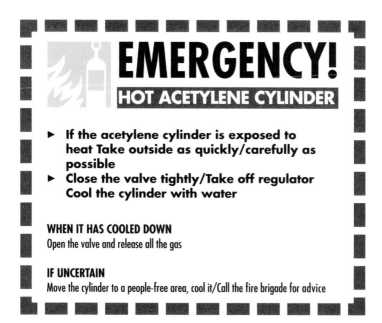

EMERGENCY!
HOT ACETYLENE CYLINDER

▶ **If the acetylene cylinder is exposed to heat Take outside as quickly/carefully as possible**
▶ **Close the valve tightly/Take off regulator Cool the cylinder with water**

WHEN IT HAS COOLED DOWN
Open the valve and release all the gas

IF UNCERTAIN
Move the cylinder to a people-free area, cool it/Call the fire brigade for advice

the cylinders. They should NEVER be handled or stored carelessly. DON'T knock or tamper with the valves or gauges.

In practice, acetylene is always lit BEFORE the oxygen is turned on. And don't forget:

❍ Do not use too great a flame.
❍ Don't touch the flame, it is extremely hot.
❍ Don't point the flame at anyone or anything apart from the workpiece itself.
❍ Clamp all work firmly in position.
❍ Wear safety gear—this is VITAL.
❍ Try to avoid having any part of your body beneath the weld site—molten metal may fall on you.
❍ Keep the hoses AWAY from the heat.

☠ WARNING

Oxygen: Fires will start and burn more easily in oxygen-rich air. Never smoke, make sparks or use a naked flame near stored cylinders.
Acetylene: It is very flammable. It can react with some metals to produce an explosion. Never replace any of the fittings on the cylinder or improvise connections on hoses or torch. If an acetylene cylinder is exposed to heat follow the emergency steps above.

- Handle cylinders with great care
- Store cylinders away from dust and debris, which might prevent a valve from closing properly
- Clean all threads—but do NOT grease them or use a jointing compound before making connections
- Don't allow children anywhere near the cylinders
- Avoid long hoses—there is more of them to be damaged
- Blue hoses should be used for oxygen, red for acetylene
- NEVER attempt to mix two or more gases in a cylinder—or to transfer gas from one cylinder to another
- Acetylene must be stored UPRIGHT
- Don't subject cylinders to extreme temperatures—hot or cold. Keep out of sunlight. DON'T smoke/have naked flames near cylinder

Arc welding

Although 'arc' welders commonly plug into a 13-amp socket, a transformer reduces the voltages BUT steps up the rate of flow of the current (amps) to a point when an intense spark or 'arc' may be formed. The immense local heat causes two metal surfaces to fuse together.

The transformer has two leads (apart from the mains lead)—one is attached to the electrode holder which you use to make the weld, the other is the earth connection which completes the circuit. The latter must be attached to the workpiece.

An arc is formed when you touch the electrode holder to the workpiece, great heat is generated and the tip of the electrode melts and flows into the joint. You MUST observe all instructions on earthing equipment and insulation.

Safety gear

Welding produces intense light and you will need a whole-face visor which is filtered to protect your skin—AND YOUR SIGHT. Apart from visible light, infra-red and ultra-violet are produced. These can cause severe eye burn, deep sunburn, skin 'cancers' and headaches.

You should cover as much skin as possible—with clothing which does not let light through. Wear several layers. Professional welders use leather aprons and gauntlets. Greasy clothes increase the risk of fire. Don't wear man-made fibres. They may melt. Heavy boots are also needed to avoid burns from droplets of molten metal.

Ventilate the area well and use breathing protection. The fumes which are produced during welding are poisonous—often extremely so.

PAINT

 One of the simplest DIY jobs that many people undertake is decorating their own home. At least, it always LOOKS simple when other people do it! Like any job you do around the home, it won't look any good if you don't prepare well beforehand and address any problems before you start. You need to plan the work, estimate the cost of and buy materials—and be sure you have a good strong stepladder before starting. NEVER balance on chairs, stools or boxes.

WARNING

Many of the chemicals with which you will come into close contact may be toxic. Stripping paint is a hazardous process—however you do it. Rubbing down old paint may be particularly hazardous—many old paints contained lead. Lead was still used as an additive in paints into the 1980s. In some countries, lead may still be used. Protect your skin and wear a facemask. Wash well after exposure—lead is very toxic.

NEVER rub down any surface you suspect is made of or contains asbestos. Wallboards and insulating/fireproof panels may be asbestos sheet. Even textured wall and ceiling coatings used to contain it (see SAFE AS HOUSES?: **Asbestos**).

Emulsion paint

Most (water-based) emulsion paints for domestic use have fairly low toxicity. Lead is no longer used as an additive (in most countries). DON'T use any paint you suspect of containing lead, especially where children may come into contact with the painted surface. DON'T buy 'cheap' paints that do not have a recognized brand name. Be wary of bright-coloured paints—they may contain all sorts of additives. See if the instructions give a hint as to their toxicity.

All paints you use should either state that they are NON-TOXIC, or clearly state any recommended precautions that you must take.

Ensure good ventilation while painting, while the paint is drying—and for a couple of days afterwards. Don't sleep in a newly-painted room for a night or so. In warmer months, paint dries a lot faster—and ventilation is easier.

Some people are more sensitive to paint fumes (even from emulsion paint), and should take extra care to ensure good ventilation. Skin contact should be avoided.

Paints containing fungicide are toxic. It may NOT be essential for you to use such a paint, unless you have a problem with damp or mould.

'Gloss' paints

Gloss and other oil-based paints contain solvents, which are poisonous. Gloves should be worn—and a mask. You should not smoke or use naked flames around solvents, nor around these paints until they are dry. Ventilation is vital. Children or pets may be much more susceptible to the effects of the fumes.

If you feel dizzy or nauseous, you are not ensuring a good supply of clean air. Open doors and windows to the room and go outside until you feel better.

Paint accidents

If you get any sort of paint at all on your skin, wash it off at once. This is easy with emulsion, if you don't wait. With gloss, or other oil/solvent-based paints, wipe away any excess at once—without spreading it further. Use a household detergent (washing-up liquid would do), without water at first. Rub this into the paint to dissolve it and rinse well with warm water. Hot water will open the pores of the skin, and may force the paint to dry faster and become sticky. If possible, buy a good skin cleaner before you start painting. **DON'T** use paint thinners, white spirit, turps, petrol or any 'solvent' on your skin.

Wear gloves and cover up other parts of your body as protection from splashes. Safety glasses should also be considered when painting ceilings or using a roller.

If you get ANY paint in your eyes, wash them with cool running water for at least ten minutes, holding the lids open. Either hold your head under a fairly low-pressure flow from a shower hose or put your head under a tap. If neither of these is possible, splash the eye for several minutes with water from a bowl. If there is any soreness, seek medical attention.

Paint spraying

There are two main types of spray equipment, which you may not choose to own, but which are readily available from hire shops. One involves an air compressor, which forces out paint droplets in a jet of air. The other forces the paint itself out under great pressure.

AVOID breathing paint mist or fumes—wear goggles and a very efficient face mask. Cover as much skin as possible. Ensure very good ventilation. Do NOT spray indoors, if possible. If you must, open doors and windows to let the air circulate. An extractor fan would be a good idea.

There is a high risk of fire and explosion—do NOT smoke or use naked flames PARTICULARLY if using oil/solvent-based paints.

⚠ WARNING

NEVER leave equipment where a child could get at it. NEVER aim the gun at any part of your body—or at anyone else. The paint in 'airless' guns is sprayed at such a pressure that it can penetrate skin and surface tissues. NEVER touch the nozzle while paint is being sprayed—on either sort of equipment. If there is a safety shield, do NOT remove it.

If you inject paint under the skin, seek medical attention at once. Explain the injury and supply all known details about the paint involved. Disconnect the machine from the mains before attempting to clear a blocked nozzle.

Aerosol cans of paint are also to be used with extreme caution—even when just doing small touch-up jobs on your car. Always wear gloves and a mask—and make sure that you don't work in a confined space.

PRIMERS/SPECIAL PAINTS

The only exception to the 'no-lead-in-paint' rule is calcium plumbate primer—no efficient alternative has been found for galvanized metal. Remember that lead is highly toxic and take great precautions to avoid exposing yourself to this risk. Never use calcium plumbate primer indoors. Avoid hand-to-mouth contact (don't smoke, eat or drink while using). Wear a mask and cover all skin. Don't breathe the fumes. If a large area of painting must be done, seek professional advice.

Many special primers contain aluminium or zinc—both of which are toxic in concentration—but are safer than any primer containing lead.

Bitumen paints for roofs and external metalwork, and creosote (commonly used as a preservative on fences and outbuildings) are poisonous. Skin contact and inhalation of the fumes should be avoided. Dermatitis (slight to severe skin irritation) is fairly certain and links are being investigated between coal tar derivatives and cancer. Two constituents are known carcinogens—chrysene and benzo-a-pyrene.

⚠ WARNING

If a child swallows any paint, do not induce vomiting—especially with oil/solvent-based paints. If the child has swallowed emulsion paint (without fungicide) give them a glass of water to drink slowly. See how the child feels after half an hour. If the child has swallowed a small amount of paint containing fungicide or a large amount of any emulsion paint, seek medical attention. If any amount of oil/solvent-based paints is swallowed, seek medical attention.

Stripping paint

Stripping paint used to be a chore. Now it is a booming industry and even a major hobby—although the 'stripper' tends to find out pretty quickly that it is basically still a chore! Large jobs tend to get messier and messier and more and more boring—and people get careless, trying to get the job finished.

If you have decided to become a 'stripper', you have two main alternatives: to heat the paint with a blow-torch (or blow-lamp) or hot-air stripper and scrape it away, or to use a paint-dissolving chemical stripper. Both have many hazards.

The likelihood, if you want a plain wooden finish, is that—having used heat—you will need a chemical stripper to clean the remains of the paint from the wood. If you are repainting, sanding would probably be adequate.

A hot-air stripper is more controllable than a blow-torch but don't underestimate the heat it can produce.

Hot-air strippers

Heating only really works with 'gloss' or oil-based paints—not on cellulose (car) paints or emulsion paints. It is only really suitable for wood—not metal or plastic. You have to heat the paint until it bubbles (not burns) and scrape it off fairly rapidly before it goes hard and 'cooks'.

A hot-air stripper is fairly easy to use. It works (basically) like a super-powered hairdryer—but is also quite dangerous. No hot stripping is easy near glass—the glass will break—and special shields must be used. A variable heat control on a hot-air stripper would give you more control. Whichever method you use, keep the heat source moving to prevent burning.

Keep a special clear area for 'resting' the hot lamp/torch/stripper when you are not using it. Keep some water handy in case a fire starts. Work in a well-ventilated area (see WARNING above right) and wear a mask and goggles. Don't wear rubber or plastic gloves which may melt onto the skin.

If working outdoors, stop if conditions become windy—the heat would not be directable. Avoid cracks and crevices around door and window frames and working under the eaves of your house—a fire may start very easily. It may be very difficult to reach it to put it out. As the debris falls it is smouldering—damp it down every few minutes with a water spray.

Clearing up

DON'T run a bare hand over a partially-stripped area or collect debris without gloves. Some of the fragments, when cooled and hardened, can be razor sharp. Wear breathing protection and gloves. Dispose of all debris carefully. Don't burn it—more fumes will be given off.

⚠ WARNING

Older paints are very likely to contain lead, which is extremely toxic. Do not breathe fumes. Wear eye and breathing protection and keep skin covered up.

Old textured paint—even only a few years old—may contain asbestos, which is highly dangerous if inhaled. Special chemical strippers claim to be safe for the job, but seek professional advice before tackling removal.

Chemical strippers

There are several kinds of chemical stripper—most are messy, all are poisonous and will burn the skin quite badly. Most produce fumes which are toxic, and make you feel dizzy or nauseous.

It's basically a question of applying the stripper, leaving it a while to do its job and scraping it off again. This often has to be done several times to completely remove all paint. If you wait too long, the solvent (or moisture, in some types) evaporates and the job becomes much harder.

Good ventilation is ESSENTIAL—as are eye/breathing/hand protection. Thick plastic gloves are fine, but will gradually dissolve. If they start to leak, replace them. Cover up as much skin as possible and try to work in a very controlled way—DON'T slosh the stuff about.

If any chemical touches your skin, wash it off with water at once. If you don't, you will very quickly feel what the chemical can do. Very painful burns are possible. If ANY of the chemical goes in your eyes, flush with cool running water for at least ten minutes—with the lids held open. Seek medical attention.

Paste strippers, which you mix with water, often come in powder form. DO NOT INHALE! Wear a mask when mixing. These strippers are usually based on caustic soda (sodium hydroxide) which will dissolve skin very efficiently.

The grain of the wood may 'lift' with water-based paste strippers and 'water-washable' liquid strippers if water is used as a final rinse. When sanding, avoid the dust which will contain residual chemicals. Cover your skin, and wear gloves, a mask and goggles.

There are aerosol liquid strippers which—even if discontinued—may still be found on some dealers' shelves. Somehow, the idea of spraying a highly toxic, highly caustic substance does not seem safe or intelligent.

If you don't have a lot of energy or patience—DON'T attempt the large-scale removal of paint. Call a specialist!

DIY/CRAFT HAZARDS PAINT

CRAFTS

There are a number of crafts that are enjoyed at home, sometimes as business ventures—many involve dangerous tools, processes, materials and chemicals that require careful handling and storage. Businesses must meet legal standards of workplace safety. The same precautions are VITAL at home.

CHILDPROOFING

Instruct children in the dangers of crafts processes and materials. Exclude them from the work area, if possible securing and locking it when not in use, lest they attempt to 'have a go' themselves.

Photography

Keen photographers will probably want to do their own processing and printing. If you use the bathroom as a darkroom make sure that you always clear everything away afterwards. If you have a proper darkroom make it lockable. The deeper you get into processing the more chemicals you will need.

Experiment with different types of tongs and plastic gloves to find the best way of avoiding contact with chemicals. Barrier creams may leave fingerprints on film or paper.

◐ NEVER be casual about mixing or storing chemicals.
◐ AVOID skin contact with any chemical—even if you think it is harmless. Once you start putting your fingers in the baths for developer, stop and fix, you will find it very difficult to get out of the habit. You may suffer from dermatitis straight away or develop sensitivity later on.
◐ AVOID inhaling chemical powder, which is easily carried in the air. This is when chemicals are at their strongest. Wear a face mask. Fit an extractor fan.
◐ AVOID hand-to-mouth contact. Don't smoke, eat or drink in the darkroom.
◐ NEVER raise bottles to your face to sniff the contents. The smell of some chemicals will knock you over—especially concentrated acetic acid 'stop' solution.
◐ DON'T empty any chemicals down the sink. Contact the manufacturer or dealer regarding safe disposal.
◐ REMEMBER you are working with a potentially fatal combination of electricity and water.

Ceramics

Whether in a pottery class or a home studio, it is very easy for the person involved in their craft to ignore the dangers from chemical glazes and from the heat of the kiln in which pots are fired. Most potters seem very casual about clay dust where they work. The dust is loaded with silica which can lodge in the lungs causing silicosis, a debilitating lung disease.

Many glazes are toxic, some very toxic, with lead and cadmium compounds common along with other dangerous metals. Some release poisonous fumes when fired.

Regularly inspect the equipment you use. Check electric wheels and kilns for safety and the flues of gas kilns. Is the pugmill safe? When pushing clay into it can you reach the screw? One day you may find out the hard way that you should not have removed the guards.

❍ NEVER breathe in particles of clay or glazes. Wear a mask—especially when spraying glazes.

❍ Instal an efficient extractor fan and ensure good ventilation in the workplace.

❍ DAMP DOWN DUST and clear up with a mop. DON'T use a dry brush (it stirs up the dust), or a vacuum cleaner—hazardous particles pass straight through!

❍ Clearly label ALL materials.

> **! WARNING**
> Some glazes are unsuitable for tableware since they are never completely stable. Poisonous particles may be swallowed.

Other crafts

Hobbyists are taking up a wide range of home crafts. At one time, if not getting poisoned by substances such as oxalic acid in chemistry sets, the main danger seemed to be the inhalation of polystyrene cement while making up plastic model kits. Today both the range of activities and our knowledge of their risks has increased.

Accidents and illnesses can be avoided if you know the risks involved before you take up a craft. Most manufacturers and dealers should be helpful in warning of the dangers of equipment and materials—but also seek expert tuition and advice in craft processes.

Casting miniature figures in 'white metal' is one of the new popular hobbies which carries severe risks from fumes and carbon monoxide—as well as the danger of serious deep burns from molten metal. Some reusable mould-making compounds release formaldehyde vapour, which is highly toxic (see POISONS: **Formaldehyde**). Silicon is a very common ingredient of mould-making 'rubber' and mould-release agents.

Enamelling carries high risk of burns and exposure to toxic fumes as powdered mixtures of ground glass and chemicals are deposited on to pieces of copper and fired at a high temperature. There are grave dangers of inhalation and ingestion of powder 'enamels'. Small kilns may have been insulated with asbestos—itself very dangerous. Be aware of the risks (see SAFETY FIRST: **Asbestos**).

Stained-glass work uses lead, which is a dangerous, cumulative poison, and glass which can splinter unexpectedly and cut. Cuts will aid the entry of lead into the blood system. Flash-coloured glass usually has a thin layer of colour applied to one side—the side you cut. As the cutter runs over the surface, tiny fragments of glass fly up. As you lean in to concentrate on the work, eye protection is essential—as are thick gloves when straightening and stretching the lead cames.

Large amounts of solder and flux are used to join the work together—avoid inhaling the fumes they give off when heated. Individual sections of glass may be etched with acids or treated with chemicals, glass powders and oxides which are applied in solution and fired on. All need careful handling.

ARTISTS' PAINTS

Paints used by artists, designers and illustrators are NOT for use by children, who should be given non-toxic play paints instead. Check that these play points ARE non-toxic.

Lead, mercury, arsenic, barium, manganese, cobalt, chromium, cadmium, titanium and antimony all frequently occur as compounds in artists' paints. Never lick your brushes—even to 'put a point on them'. Avoid excessive skin contact. If mixing dry pigment, also avoid inhalation.

Using 'airbrushes' or spray guns increases the risk greatly. Wear a dust mask and ensure good ventilation. There is also a fire/explosion risk if the air becomes heavily dust-laden—increased immensely if a solvent is involved.

Avoid hand-to-mouth contact. Don't mix eating or smoking with painting. The danger is great—after years of using artists' materials, the user may become casual about the dangers. Several of the toxic substances are cumulative poisons (in other words they build up in your system), many are carcinogenic (cancer causing), and by the time any serious effects come to light, it may be too late.

REMEMBER

If you currently enjoy or are thinking of taking up a craft, read all you can about it to discover the associated hazards—there are many other potentially dangerous crafts processes in addition to the examples given here. Enrol in classes before launching into a new activity. It will give you a chance to see whether it really interests you—and to learn about safety rules before taking unknown risks.

LADDERS

Most accidents in the home are caused by falls. Whenever you have to work above a comfortable height, you need something to stand on. DON'T use chairs, stools or boxes. They could tip over—and there is nothing to hold on to. NEVER make a pile of furniture or boxes to stand on.

Every home should have a simple set of steps—a small stool/steps combination will probably be enough for a kitchen. Make sure it has non-slip feet and is strong enough to take your weight. Never stand on the top step with both feet, unless there is a rail to hold onto. Never stand the stool/steps on anything to increase the working height.

Stepladders

You should have a small, lightweight, aluminium step-ladder, with a safety bar/handrail at the top to hold onto and good locks—to stop the ladder folding up or splaying its legs when you are on it. The treads should be wide enough to be comfortable.

Don't go for the biggest set of steps if you don't really need them. Carrying them through the house could be difficult and dangerous. There are other points to consider:

◑ Are the feet of the stepladder non-slip?
◑ Try the ladder in the shop. Feel the weight—can you carry it easily? Can you put it up and collapse it easily? Does it have locks? Does it look/feel strong enough? Are there sharp edges or places where fingers could be pinched?

If you need to get in and out of a loft, there are extendible (two-way) 'stepladders' which could double as loft ladders. The best course of action would be to fit a special loft ladder which is always there, but folds up out of sight.

There are various two- and three-way convertible alumin-ium ladders which hinge in various directions. Before conver-sion, it may be a stepladder. Undoing locks and raising one side of the ladder will form a straight ladder.

Extension ladders

Most ladders are made of wood or aluminium. Aluminium is lighter, won't rot and is generally less of a problem. The length of the ladder you need depends on how high you need to work and how much room you have to store the ladder when not in use. Ladders can be hired quite economically, which widens the range of choice considerably.

Working up ladders is neither as comfortable, nor as safe as working on a platform tower—but you may have no choice if there is no room to erect a tower.

DIY/CRAFT HAZARDS LADDERS

153

Most extension ladders are in two separate sections (sometimes three) and extend to a 'safe' working length of about 6–7 m (19–22 ft)—'safe' because there must always be an overlap between the sections when working. You will need a ladder to be at least 1 m (over 3 ft) longer than the highest point you need to reach. Always anchor it at the top for extra safety.

Using an extension ladder

If you feel nervous working off the ground, and this feeling does not pass, STOP. Not everyone is comfortable with heights. There is nothing 'sissy' about this. Working at a height is a very risky business—there are many deaths and injuries associated with using ladders.

NEVER rest the ladder on very uneven or very soft ground. If the feet seem likely to slip, DON'T climb the ladder. On uneven ground you may have to block up the feet, unless they are adjustable, and anchor the ladder top and bottom before climbing. On soft ground, rest the feet on a strong board to spread the weight. Bang pegs into the ground to stop the board slipping and anchor the ladder top and bottom. ALWAYS have someone working with you, in order to:

◗ Hold onto the ladder, with one foot on the bottom rung to steady it
◗ Get help if an accident occurs

NEVER lean sideways off a ladder—move the ladder so you can reach. NEVER climb higher than the fifth rung from the top—you need some ladder to hold onto.

REMEMBER

Use the rule of threes: Of your two hands and two feet, three points of contact should always be made. Either both feet and one hand should be in contact—or both hands and one foot. Professional climbers always use this rule to avoid risk-taking and accidents.

While working

◗ Wipe mud off your feet BEFORE you climb the ladder.
◗ When working, have BOTH feet on one rung and spread your weight evenly.
◗ Lean against the ladder with your legs, keeping your free hand behind the ladder, holding a RUNG.
◗ DON'T hold onto the side of the ladder, if you slip, your grip won't hold.
◗ DON'T lean the ladder against guttering—this is dangerous. If necessary, use a stand-off. This is a small frame which holds the ladder away from the wall at the top.

- If working in one location, tie the ladder to pegs driven into the ground.
- ALWAYS anchor the top of the ladder to stop it slipping. Tie it to a fixed/stable part of the building, through a window to an immovable object or to a batten across the inside of the window frame.
- If you need to rest the top of the ladder exactly where a window is, tie a strong batten or stout plank across the top of the ladder to rest the load either side of the window.

REMEMBER

- Check ladders regularly for signs of wear and tear
- Never use a ladder as a bridge, laid down to walk on
- Never use a wooden ladder if you suspect rot
- Never leave a ladder unattended outside. Children and burglars are quite fond of them
- Lock the ladder away when not in use — or chain to a stout tree or fixed post
- Never work up a ladder in high winds
- Never allow more than one person on a ladder at a time
- Never ascend or descend a ladder with your back to it
- Don't climb if you are drunk, tired or feeling ill
- NEVER position a ladder in a doorway

CARRYING TOOLS

DON'T carry tools in your hand, your pockets or slung about your body on hooks or straps. They could cause a serious injury if you fall. Instead, carry tools up in a shoulder bag, which you could jettison, and hang it on a hook when you reach working height. DON'T drop tools to the ground when you've finished with them!

! WARNING

NEVER use a power tool with the lock-on button pressed. If you fall, or drop the tool, it may carry on working — and cause a more serious accident.

Carrying a ladder

Everyone tells you to carry a ladder vertically. That is easier said than done with a long ladder. It takes a lot of nerve, control and practice. **If you can't manage this circus trick, don't despair. Two people can wrestle a long ladder into almost any position, better than one.**

⚠ WARNING

Always use a ladder at an angle of 75°. The distance between the base of the ladder and the wall should be one quarter the height of the uppermost point of contact.

Accessories

Look for ladder accessories which can make working at a height easier and safer. There are bolt-on or clip-on platforms which can make the rung you stand on into a wider 'perch'. Stand-offs hold the top of the ladder away from the wall so you can avoid resting the ladder on guttering. A clip-on tray would make a good area for resting tools or paint while you work. A pot of paint could be easier to manage if slung from a rung on a large S-shaped hook.

When NOT to use a ladder

DON'T use a ladder if you feel unsafe or at all worried about the height. DON'T use a ladder if you will need both hands to do a job. DON'T use a ladder for climbing onto a roof—nearly all roof coverings are unsafe and liable to be damaged in the process. DON'T use a ladder for major jobs like taking down a chimney stack.

In all these cases, call a professional—who may choose to erect proper scaffolding—or consider using a platform tower.

PLATFORM TOWERS

Platform towers are safer, by far, than using extension ladders. You need space at ground level for one to stand safely. They are probably too costly and too bulky in storage to make it worth buying one—but they may be hired.

The towers are, basically, in 'kit form'—consisting of H-frames which slot together. Each level is added at right angles to the previous one—with occasional diagonal braces dropped in for stability. At the top are placed standing boards and—very important—a hand rail.

Platform towers can be built single-handed, with a rope to haul the sections up, but really require at least two people.

When built, the towers resemble some children's climbing frames. NEVER let a child play on or near the tower. NEVER leave a tower unattended.

Working safely

The top of the tower should be at a safe working height for the job you need to do. NEVER use a ladder on top of a tower.

The height of the tower should not be greater than three times the base dimension. Square towers are safer than oblong ones, where the height should not exceed the SMALLER of the base dimensions. External supports—outriggers—increase the working height to three times the resulting base.

Towers MUST be built on level ground—choose the feet to suit the quality of the support the ground offers. On soft ground, you need base plates to spread the load and prevent sinking. On very level ground, 'castors' (which MUST be locked before the tower is climbed) allow the tower to be moved from one location to another—avoiding dismantling.

There are 'adjustable' feet which allow you to compensate for slightly uneven ground—or if it is absolutely essential to build the tower on steps. **ALWAYS level the first section, before commencing to build the tower.**

REMEMBER

Always climb *inside* the frame. Haul a bag of tools up separately, with a rope—you must have both hands free for climbing. Make sure you wipe any mud off your feet BEFORE you climb.

Safety first

- ◑ ALWAYS secure the tower at the top, to prevent toppling.
- ◑ NEVER rest a ladder against a tower—or only do so **IF** the top of the tower is secured and **IF** the ladder is leaning towards the building. **ALWAYS** secure the ladder top and bottom, before attempting to climb it.
- ◑ Fix the top boards, or remove them if you expect high winds. It is fairly possible that they may be dislodged by the wind—and do considerable damage.

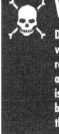 **WARNING**

Do NOT attempt roof work if you are not qualified to do so. NEVER walk on a roof! Quite apart from the damage you will do to the roof—no roof covering is intended for walking on—most roofs are not strong enough. The use of crawlboards and roof 'ladders' is ESSENTIAL. Innumerable deaths and accidents have been caused by people thinking that it was safe to walk about on roofs, because they've seen 'professionals' doing so. You must seek specialist advice.

PROTECTIVE CLOTHING

With almost all DIY processes, some form of physical protection is required. Skin may heal if superficially burned by chemicals—but body functions such as sight, hearing and breathing should not be viewed as renewable resources.

Safety equipment on sale may NOT be safe enough—safety goggles may not be shatterproof, breathing masks may be hopelessly inefficient. Always choose, not by price, but by evidence that the items have passed tests for efficiency and durability.

There is no reason why protection during amateur work should be any less effective than that used by professionals or in industry.

The basics

Tough overalls are essential for most kinds of work—not to stop you ruining your clothes, but so that dust and contamination is confined to one set of clothes and not carried around with you. A lab coat or apron, or a separate set of 'work' clothes may suffice for cleaner safer jobs.

Overalls should fit OVER another set of clothes and be roomy enough to allow for movement—such as bending down or reaching your arms above your head. They should not be loose enough to catch in machinery, particularly the sleeves. A buttoned cuff is probably safer than a plain one.

The fabric should be tightly woven and flameproof. Some man-made fibres may melt onto the skin when heat is applied.

Wash overalls separately from general laundry.

Investigate overalls with removable limb or torso guards—some of which are designed to protect you (to some degree) from the impact of a revolving cutting blade.

Gloves

The gloves you wear must not prevent you from working safely—they should fit but should not be so restrictive that you do not have full use of your fingers.

How strong they must be depends on the job you are doing. If you are stripping paint with chemicals or working with solvents, oils or acids, the gloves should be impervious to such chemicals. If handling concrete or sawing, the gloves will need to withstand a lot of wear. There are even reinforced gloves for use when operating cutting machinery—which should resist a saw blade (to some degree) turning at speed.

If working with dust or installing loft insulation, you should wear gloves which fit to the wrist with an elasticated band to prevent particles from falling inside.

Gloves should not make your hands sweat—leather and canvas seem the best materials, although there are numerous alternatives. For lighter work, consider disposable paper/plastic alternatives.

If gloves become heavily soiled, or contaminated inside, they should be discarded—unless specified as washable.

If you really CAN'T wear gloves to do the job—not even 'surgeon's gloves'—or if you cannot wear both gloves, protect the skin with barrier cream.

Footwear

Good footwear should have thick, non-slip, rubber soles—which can resist piercing, acids and grease. It should also provide you with safety from electrocution.

The uppers should ideally be steel-reinforced to prevent: crushing of toes when a heavy object falls on your feet; cutting or stabbing feet with gardening tools, chainsaws or axes; burning of the feet from (for instance) droplets of molten metal, while welding.

The shoes/boots should be comfortable enough to wear and stand in for long periods of time. Thick socks may make them more comfortable—remember to wear these socks when you try the footwear on.

Protecting eyes

To be any good at all, eye protectors must be fairly comfortable and allow near-perfect vision. If scratched or worn, they should be replaced. They should not 'steam up'. Depending on the task in hand, they should resist impact from flying debris, dust penetration (around the sides) and splashes with chemicals. It is ESSENTIAL that they be shatterproof.

There are various types available. For some jobs, where the main danger is flying debris or splashed chemicals, 'spectacle'-style eyeshields may suffice. They are easy to put on and remove. Don't use them if there is any danger of them falling off, or being knocked off. An adjustable band which goes right round the head would make them safer.

Whole-face screens, which hinge up and down—so they can quickly be lifted out of the way—are very good for flying debris and chemical splashes. They protect the whole face, and in some cases, the front of the neck as well. Drifting dust will find its way in, though. This form of protection might be your only choice if you wear glasses.

Goggles with an elastic strap which goes round your head, like those worn by skiers, may be essential to keep dust out of your eyes. They may be prone to steaming up.

If welding, you will need eye protection in the form of a whole-face shield with a clear (tinted) portion which filters out the harmful intense light—if your sight is to be protected AND your skin (see **Welding**).

Hearing protectors

Think of them as 'hearing protectors'—not EAR protectors. They should be worn with almost any power tool or noisy process. Most people don't realise that their hearing is at risk from loud noise. It's common to feel a temporary effect, which ranges from muffled hearing to a 'ringing' in the ears. Eventually, this temporary hearing loss can become permanent.

Hearing protectors (which look like headphones) are easy to wear. The headband should be movable so that you could also wear a safety helmet. You should still be able to hear a little—in case an accident happens, something falls, or someone shouts for help.

Wax earplugs are efficient, for most people, but not very convenient. Cotton wool (cotton) is useless.

Breathing protection

Inhalation of dust or fumes may present an extremely serious long-term health hazard. Breathing protection is VITAL. Smog masks, made of cloth or paper, are not sufficient—except for protection from larger debris. Even so, you should replace them frequently—as soon as you can see any build-up of dust on them.

An improvized mask of several layers of thin cotton fabric, anchored in place with elastic around the head, is more efficient than a smog mask or many simple disposable masks—providing it is 'moulded' well around the nose and mouth. If you have a thick beard, this may be your only choice, unless you go for something very elaborate.

Most face masks are called 'nuisance masks' and will cope with non-toxic larger-particle hazards like glass-fibre, loft insulation and (less so) sanding dust. They need replacing frequently. If they have replaceable filters, it may be possible to insert more than one.

Fumes or poisonous dust (even suspected poisons) require
a tougher mask altogether. It may even be necessary to go for
a gas-mask type, which covers the whole face. There are even
versions with their own air supply.

Head protection

**Look for proper helmets and accept only those guaran-
teed against impact and penetration. The helmet MUST
fit and be comfortable enough so that you don't avoid
wearing it.**

It must also stay on. A chin strap is a good idea (if comfortable) to prevent the helmet falling off. Try a helmet before you
buy it. Lean forward and see if it stays on.

There should be some air circulation, to avoid excessive
sweating—sweat might run into your eyes. If you need to wear
hearing protectors, check that this is still possible.

The cranium protectors worn by cyclists and climbers
allow free movement and afford quite a lot of protection. These
may be suitable where slips and falls are the main hazard—but
many have open sections which won't protect the head from
falling objects or flying debris.

Sun protection

Never underestimate the power of the sun when working outside. Even a ten-minute job can turn into a couple of hours, and
before you know it you could be burnt to a frazzle—with the
added risks of sun/heatstroke and skin cancer. Cover exposed
skin with high-protection sun screen, particularly vulnerable
areas such as the back of the neck, and wear a hat (especially if
you're a 'bit thin' on top!). In hotter, humid climates the recommended fluid intake for someone doing manual labour is four
litres a day—be aware of the danger of dehydration. Properly-
filtered sunglasses are a must—fashion 'shades' do not offer
adequate protection.

4

Danger lies in assuming that all household 'chemicals' are safe. Many preparations, from DIY products to over-the-counter medicines, are far from safe if not used with great care. Even aspirin may have toxic effects.

Poisons

CHEMICAL HAZARDS

 No one can avoid coming into contact with potentially-dangerous chemicals in the home. Most people have enough sense to treat all chemical preparations with proper respect—but there are many cases of poisoning which could not have been foreseen. Most dangerous or potentially-dangerous substances are labelled, but always remember: research into the harmful effects of all substances which are used in the home, in industry and the environment is constantly providing new information. Once, substances such as nicotine, asbestos, lead, aluminium, arsenic, opium and even mercury were thought to be harmless—or even beneficial to health!

Know the dangers

You may think that you never use dangerous chemicals, but the chances are that you do—everyday—and will continue to do so as part of normal life.

It's true that most household products have a fairly low toxicity (depending on the level of exposure and the sensitivity of the individual). In some cases, you would need to ingest large quantities to be at risk. But exposure over a long period of time may be as serious in the end result.

Quite apart from these dangers, always remember that you are under threat, it is now clear, from hazards in your water supply, and from pollution in the environment. Some areas carry particular dangers from lead, cadmium or aluminium. There are also the domestic dangers from asbestos, formaldehyde and carbon monoxide—even mains and cylinder gases. These may be aggravated by exposure to chemicals while at work (see WORK & PLAY).

PLEASE NOTE

Main references to some of the hazards may be found as follows:
Water supply see SAFETY FIRST: Water and ESSENTIALS
Environmental pollution see ESSENTIALS
Lead see SAFETY FIRST: **Water**
Aluminium see SAFETY FIRST: **Water**
Asbestos see SAFETY FIRST: **Safe as houses?**
Formaldehyde see POISONS
Carbon monoxide see SAFETY FIRST: **Gas**
For references to drugs and substance abuse, see **HEALTH**

Look around your home

 Look at a few of the most obvious products—not just polishes and detergents, but bleaches and toilet cleaners, air-fresheners and drain clearers, matches and firelighters, flysprays and disinfectants

 If you do DIY or enjoy gardening, you probably have paints, brush cleaners and paint strippers, thinners and solvents, pesticides and weed killers, rust removers, adhesives and sealants, creosote and bitumen

 If you own a car you may be likely to have antifreeze and de-icer, petrol and oil, batteries containing acid

 Check the 'over-the-counter' medicinal preparations you keep for emergencies: aspirins and painkillers, skin creams and depilatories, cough medicines and 'cold cures', travel sickness remedies, vitamin and mineral supplements, cornplasters, products to regulate bowel movements

 Possibly you have received or regularly require prescribed medication: for a heart condition, for migraine, for infections, for stress or sleeplessness, for birth control. If these medicines are old or unwanted, should they still be in your home?

 Most people use some sort of 'cosmetic': deodorants, perfumes, aftershaves, talcum powders, soaps, bubble baths, hair sprays and gels, nail varnishes, shaving creams (and razors!), even cosmetic eye drops and 'suntanning' tablets to name a few.

There is some form of alcohol in every home—either in the 'drinks' cupboard or in perfumes, aftershaves, surgical spirit. In small amounts, it may be present in many aerosols, window and glass cleaners and cosmetics.

As a drink, in most countries, it is socially acceptable and pleasant—in moderation. Alcohol is NOT, as many people think, a stimulant—it actually depresses the nervous system. Alcohol IS a poison. It damages the body. In particular, it affects the brain, liver, heart and stomach. It can be very 'addictive' or lead to dependency. Externally, contact with alcohol in various products can lead to dermatitis.

! WARNING

Milk tends to delay the effects of alcohol (especially the drowsiness) but, contrary to popular belief, coffee (caffeine) does NOT reverse the effects of alcohol. Drinking coffee will NOT make it safer to drive or operate machinery.

WHAT ALCOHOL DOES

Alcohol causes different effects in different individuals, but most of the following (to some degree) in everyone—when taken to excess: reduction in coordination and control of muscles, and problems with reasoning, judgement and vision. It's immediately easy to see why alcohol is related to so many accidents. The 'drinker' is probably the worst judge of what is good for him or her once affected.

The heart rate is increased. There may be dehydration, depression or 'manic' episodes. Although you may look 'flushed' and warm, there is a significant decrease in body temperature—which could be very serious (when swimming for example).

When alcohol is taken to excess, the poisoning can lead to unconsciousness, kidney and liver damage, heart disease or failure, gastric ulcers, impotence, impaired brain and nerve functioning and death. Other side effects include raised blood pressure, obesity and severe depression.

LOOK AT THE FACTS

Alcohol is a very major cause of death by poisoning in adults, not just by 'drinking and driving'. Poisoning does not always lead to death, but is more likely to in conjunction with drugs. Alcohol ranks in the highest category of

REMEMBER

It is not compulsory to drink. It is definitely not compulsory to drink to excess. If you don't want a drink, when offered, say 'no'. Social pressure often makes refusal difficult—but it is YOUR decision.

risks, with toxic fumes and exposure to gases, such as carbon monoxide. A large proportion of violent crimes are alcohol related.

One study shows that 50% of deaths in people under 25 are alcohol related. Teenagers regularly over-consume—leading to poisoning.

In children, alcohol ranks amongst the top causes of accidental poisoning, too—with turps/white spirit, disinfectants, pesticides (rat poison, for instance) and bleaches. Nearly ALL children over ten, especially boys, will have tried alcohol—knowingly. Young children will help themselves (more innocently) given the chance.

BAD COMBINATIONS

Mixing alcohol with some other substances can increase dangers. Great care should be taken with all drugs, and medications such as: anti-histamines, sedatives, anti-depressants, tranquillizers—even aspirin and codeine. The effects of these combinations may not be immediate—they may not show for several drinks.

MAJOR RISKS

- Never combine alcohol with drugs of ANY sort
- NEVER mix alcohol with breathing the fumes of a solvent
- NEVER (do you NEED to be told?) mix alcohol with driving or the operation of machinery—at home or work
- NEVER mix alcohol with swimming—quite apart from the risk of drowning, both lower your temperature. This heat loss may be fatal
- Pregnant women run an increased risk of foetal abnormality—especially if drugs are also taken

! WARNING

When a child swallows alcohol, or products containing alcohol, the level of sugar in the blood may drop abruptly— risking hypoglycaemia. First-aid treatment should always include giving the child a spoonful of sugar or a sweet drink.

PRECAUTIONS

Store all products containing alcohol where children cannot reach them. DON'T allow children to play with perfumes, aftershaves or any household product. NEVER store alcoholic drinks where children or anyone else who may be at risk may be able to get at them (see **Safe storage**).

▶ IF YOU THINK THAT YOU, OR SOMEONE YOU KNOW, MAY HAVE A PROBLEM WITH ALCOHOL—SEE HEALTH

TYPES OF HAZARD

Chemicals may be hazardous in a wide variety of ways. Obviously the toxicity of the chemical/substance is an important factor in judging risks—but it should be possible to avoid exposure to many unnecessary dangers. For SUBSTANCE ABUSE see HEALTH.

Inhalation

Chemicals in powder form, in smoke, in fumes and vapours and sprays all enter the system by the nasal membranes and the lungs—both are routes for rapid absorption into the blood and tissues. Proper ventilation and breathing protection are obviously essential, especially during DIY of any sort. Inhalation may also lead to 'aspiration pneumonia'—inflammation of or fluid in the lungs—and various forms of cancer.

Ingestion

Although ingestion may be deliberate, it usually occurs in adults from mislabelled containers and confusion. Children seem to put everything in their mouths! Care must always be taken when handling chemicals to avoid hand-to-mouth contact—NO eating, drinking or smoking.

Hygiene is important—scrub hands, especially under the fingernails. NEVER use chemicals of any sort where food is prepared or stored.

Don't store chemicals and preparations in a 'shambles', increasing the risk that the wrong substance may be drunk—or given to a sick child in the middle of the night.

Absorption

Absorption through the skin (or broken skin) is a very real danger. The chemical does not need to be liquid to be absorbed. Often, warnings on products are inadequate or unclear regarding the irritant properties of chemicals. Rubbing and scratching the area should be avoided. Always wear gloves (which should be either washable or disposable) and cover up as much skin as possible. Washing is important after contact.

Fire/explosion

If the label says 'no naked flames'—TAKE NOTICE! Fire and explosion are a risk with many chemicals, especially alcohols and solvents, and also with fumes and dusts produced when working. Fire and explosion may lead to some of the other forms of poisoning. 'Safe' plastics in the home—the foam in some soft furnishings or the PVC covering on flexes and cables, for instance, produce deadly fumes when burnt. Great care should be taken when having bonfires.

Store flammable substances away from all sources of heat—even sunlight. This does not necessarily mean that they may burst into flames, but chemical changes may take place which could be dangerous. Most very flammable substances may have a 'flashpoint'—a temperature above which they become particularly unstable. Store and use BELOW the flash-point temperature.

Caustic burns

Many chemicals cause burns to the skin, soreness or damage to the eyes and internal burns. Chemicals do not have to be in liquid form to be dangerous. Burning of the skin, mouth, throat and stomach may lead to rapid absorption into the body.

REMEMBER

Chemicals in the home represent other dangers too. They may be misused in a variety of ways—many of which could be avoided with safe storage or not having chemicals in a home where there is particular risk. Keeping some chemicals may be foolhardy where there are children, old people, blind or poor-sighted people—or when there is a risk of substance abuse (see HEALTH), suicide, arson and vandalism.

Types of poisoning

Poisoning may be **acute** or **chronic**. It may be accidental—through inadequate warnings on the product, misunderstanding warnings or just not comprehending how dangerous a product is. Self-inflicted poisoning may occur through long-term use of over-the-counter medicines. It may also be deliberate—all manner of household products and drugs are used in attempts at suicide.

Acute poisoning occurs in a fairly short time, usually by ingestion (swallowing). With most household products the risks are low, unless a fair amount of a substance is ingested—children/infants and elderly people are obviously at greater risk. But all cases of acute poisoning should be treated as medical emergencies.

Chronic poisoning is the gradual accumulation of chemicals in the body. The body may have no way of eliminating these chemicals—particularly in the case of babies and small children, whose systems are not fully developed.

Some poisons will immediately be dangerous to everyone. Other, subtler poisons will cause different effects in each individual. People susceptible to allergies, eczema, migraine or asthma may be particularly sensitive—and should take extra precautions.

Sensitivity may be 'acquired'. For instance, you might suffer no ill effects from contact with a product which has a warning that it may cause dermatitis. After using for a long period of time, you may develop a sensitivity.

People who take tranquillizers or sleeping tablets should avoid all contact with solvents, or fumes or vapours of solvents (see **Solvents in the home**). Other drugs may also affect the way the body deals with toxic substances. If you are receiving any medication, ask the doctor/pharmacist if a bad reaction or weakened resistance to any substance is a possible side effect.

High risk groups

Children under four are most at risk from accidental poisoning. They don't understand danger, can't read labels, put a lot of things in their mouths. They will often sample substances which don't look or smell at all pleasant. The younger the child, the less their physical ability to cope with poisons.

Pregnant (and breast-feeding) women should take extra care. The foetus or baby can be harmed in many ways by drugs, alcohol—almost any chemical or over-the-counter medicine. Some, such as solvents, find their way into breast milk.

Haemophiliacs may need to pay extra heed to 'caustic' warnings on products. Even aspirin can cause internal bleeding in some people. **The blind** and poor-sighted should be given help to label, store and identify products.

The elderly may be unable to read warning labels, may forget having taken medicine and take an extra dose, may be unable to count or hold small pills, may get confused and accidentally mix dangerous products. Dosages and maximum periods of over-the-counter medicines should be reduced, since their bodies will not process chemicals so efficiently.

REMEMBER

DIY enthusiasts, gardeners and anyone who looks after a home may believe that the chemicals they use—such as paints, paint strippers, drain cleaners, solvents, pesticides—are in special weak 'domestic' concentrations. Because these products are sold for home use, surely they must be 'safe'? This is often quite WRONG. Many 'industrial strength' products are common in the home.

SAFE STORAGE

The first objective is to store EVERYTHING out of the reach of young children and infants. That means all really nasty things like rat poison, paints and DIY chemicals, solvents, drain clearers should be locked away—out of the home if possible, in a garage/shed.

Children, left alone for long enough, show a truly amazing amount of ingenuity in order to reach unreachable places. The classic is to pull out drawers and use them as steps to climb up.

Most accidents with poisons have been found to occur when products weren't locked away and were less than four feet from ground level.

Alcohol should NEVER be stored with children's drinks. Remember: Young people are at risk from alcohol, as well as toddlers. Limit the amount of alcohol in the home and lock it away. Children and young people WILL want to sample drinks which adults appear to enjoy. Quite apart from poisoning, there are the risks of accidents and unconsciousness, once the casualty is intoxicated.

REMEMBER

If you store products in locked cupboards, rooms or outbuildings, KEEP THE KEY HIDDEN. Otherwise the lock presents no barrier at all.

Child-proof containers? It is accepted that up to 15 per cent of children will break through a child-proof top in five minutes. DON'T rely on these containers to keep children out. If a bottle is glass or plastic, it could easily be broken or crushed underfoot—and the contents released.

Keep chemicals in groups—wherever you store them—to reduce the risk of using the wrong product for the job. In the dark, in a hurry, when sleepy, you may take the wrong medicine, give the wrong drink to a child or spray caustic substances onto your skin.

Keep anything toxic away from food and areas where food is prepared. Keep anything flammable away from sources of heat (including sunlight) and away from power points, light fittings and inflammable materials.

Herbicides and pesticides can be very poisonous. Children, in particular, have been known to drink paraquat. Comparatively small amounts can be fatal.

Decanting

NEVER NEVER NEVER 'decant' chemicals or medicines into different containers. A high number of accidents occur through adults and children mistaking the contents of containers. Relabelling is NOT sufficient. Labels may fade or fall off— and not everyone can see clearly or is able to read.

A recognisable container such as a fizzy-drink bottle is still a fizzy-drink bottle to a small child—no matter what you write on the label. And it's just as likely that a thirsty adult could take a swig from a bottle without bothering to read the label. In some countries (including the UK) it is now illegal to decant poisons into other containers. So it should be!

Medicines

 Medicines which you are currently using—from cough syrups to birth control pills—are likely to be left lying around. NEVER do this where children are about. The elderly may leave bottles open or pills within easy reach. Take care when visiting them with children.

Every home should have a lockable medicine cabinet. Some homes have a lot of 'medicines'—so the cupboard may have to be quite large! It should be strongly wall-mounted at a height which allows adult access, but deters small children.

REMEMBER

When storing caustic, flammable and highly toxic substances — even in the garage/shed — inspect the packages and shelves for signs of rot, corrosion and decay. Some chemicals don't age well. Containers may rust, inflate, crumble — or even explode. DON'T buy huge quantities of chemicals. By keeping different sorts of chemicals separate from one another, you can reduce the risk of bad reactions occurring when two or more of them mix — which could occur from natural seepage or spillage from old stored containers. Make sure the stored chemicals are not subjected to heat or frost. Heat may alter the chemical nature of a product. Frost may crack a container.

Disposal

Disposal of unwanted chemicals can be very tricky indeed. Medicines should be returned to a pharmacist/hospital, where they can be disposed of properly. You may need to seek the advice of local authorities or environmental protection groups if you wish to dispose of really nasty substances. Do NOT pour chemicals down sinks, drains or toilets. Contact manufacturers/suppliers for advice.

ACUTE POISONING

Some poisons attack the central nervous system—causing problems with or even stopping the heart, breathing and other vital functions. Others replace oxygen in the blood or prevent the ability of the blood to absorb oxygen.

Symptoms of acute poisoning

Some or all of these symptoms may be present. You may have to act quickly—so keep calm and try to be observant:

◗ Look for evidence—a packet/container/bottle/poisonous plant which could tell you what has happened.

◗ Or has the casualty been overcome by fumes or smoke?

◗ There may be delirium or convulsions.

◗ If the poison was carbon monoxide or smoke, the casualty may have a blueness or darkening of the face, lips and fingernails.

◗ If the poison was swallowed, there may be vomiting, stomach and abdominal pains, diarrhoea.

◗ Burns round the mouth indicate that a caustic substance has been swallowed.

◗ If this is the case, severe pain may result—all the way down to the stomach.

◗ You may be able to smell the poison on the breath—petrol, solvents, bleaches, disinfectants, shoe polish?

You must act quickly

If there are symptoms of poisoning and discomfort or distress, call an ambulance and then begin first aid.

If the patient stops breathing, give mouth-to-mouth or mouth-to-nose artificial respiration. Mouth-to-nose may help you avoid traces of poison on the casualty's mouth.

If they have been covered in poison—such as a dense spray of insecticide—cover the face with a cloth and give artificial respiration through a hole in it (see **Pesticides**).

If they are breathing, conscious or not, place them on their front with the head turned to one side and the chin jutting out. On that side, draw the arm and leg away from the body. Even an unconscious person may vomit and block the airway (see HEALTH: **Recovery position**).

If there are burns and pain in the stomach and the casualty is fully conscious, up to a pint (half a pint for children) of milk or water may be slowly sipped—but only if absolutely necessary. It may dilute the stomach contents and slow down the poisoning process.

Loosen restrictive clothing. Remove clothing contaminated by caustic substances, solvents and pesticides. If there is time, wash affected skin (avoiding exposing yourself to chemical dangers) BUT don't allow the casualty to get cold.

Inducing vomiting

☠ WARNING

NEVER induce vomiting with salty water—for adults or children. Deaths have occurred by the dramatic raising of sodium levels in the body.

NEVER induce vomiting if a caustic/corrosive substance has been swallowed. What burns 'on the way down' will burn again on the way back 'up'.

NEVER induce vomiting if petrol or solvents have been swallowed. There is a danger of aspiration pneumonia if the substance enters the lungs.

Inducing vomiting is generally considered a waste of time—and may do more harm than good. In the US, the use of ipecacuanha is recommended as an emetic—or vomit inducer. In concentrated form it may poison a small child. It is not readily available in many countries, including the UK where it is considered better to spend the time summoning help and applying first aid.

Inducing vomiting may be necessary, if you are unable to get help. It is probably only suitable in cases where children have **very** recently swallowed tablets or poisonous berries. Putting your fingers very gently to the back of the casualty's throat until the natural reflex takes over, is the only way. **Causing a casualty to vomit is very risky.**

SAVE A LIFE!
ACUTE POISONING

WHEN THERE IS OBVIOUS DISTRESS, DISCOMFORT, VOMITING OR LOSS OF CONSCIOUSNESS

► **Do NOT induce vomiting**

► **Try to judge what has happened. How much poison is involved?**

► **Phone for an ambulance**

CAUSTIC?
If the poison has burnt the lips or mouth — moisten them with milk or water. Milk or water may be sipped.

KEEP AIRWAY OPEN!
Even if the casualty is conscious, place them in the recovery position.

BE PREPARED FOR CONVULSIONS, FITS OR VOMITING

IF BREATHING STOPS
Give artificial respiration, mouth-to-mouth or mouth-to-nose. Avoid poison round the casualty's mouth.
Try to collect a sample of the poison and any vomit, for medical inspection

CHRONIC POISONING

Poison that has built up in the system over a period of time may be harder to detect. Much depends on the general health of the individual and the poison involved. You should know better than anyone else if you are regularly exposed to dangers at work, like lead, solvents or pesticides. In the short term, you may notice that you feel 'off colour' when using substances, but after a few days away from them you feel better again. This may indicate sensitivity or an allergy—but it is also telling you that you may be at risk.

Chronic poisoning may lead to a large number of varying disorders such as: dermatitis, asthma, regular headaches, respiratory complaints, cancers, circulatory problems, gastrointestinal illnesses.

A thorough medical check-up should be sought if you think you are at risk. You should examine yourself regularly and decide whether you are taking the proper precautions when dealing with chemicals. Even at work, safety procedures may be inadequate—or you may be especially sensitive to a particular substance.

MAJOR RISKS

Look at the labels of products that you buy for the home. Read the warnings and obey all instructions. Exceed the precautions if you can.

Labelling of products for home use is strictly regulated. You know that, if a specific chemical is named, you are dealing with a potentially-dangerous preparation. The absence of specific chemical names should not be taken to mean that the product is safe. Danger lies in thinking that things you can buy for home use—especially medicines—must be safe.

You may be so familiar with a product, that you don't bother to read the label at all. There ARE warnings on most toilet cleaners, for instance, NOT to mix them with any other cleaning agents. A violent reaction might take place, producing choking toxic fumes.

Symbols have been developed to help you recognize hazardous substances at a glance.

PLEASE NOTE

The exclusion of any type of household chemical or medicine from the following lists should not be taken as an indication that such products are safe. Almost any chemical can be dangerous, if used without caution. The following represents some of the most common dangers in the home. Use the EMERGENCY! panels for first-aid information and guidance.

▶ Use these EMERGENCY! panels (particularly for skin and eye contact) as a guidance throughout this section.

EMERGENCY!

CHEMICAL HAZARDS

▶ SKIN AND EYE CONTACT

Skin contact 1: Wash off as soon as possible with soap and warm water. Don't scrub.

Skin contact 2: As 1, but rinse the skin for several minutes to be sure all traces of chemical are removed or diluted. If area of skin affected is large or becomes sore or a rash develops, seek medical attention.

Skin contact 3: As 2, but rinse area for at least 20 minutes under running water. There might be 'burns' — seek medical attention.

IF CLOTHING IS CONTAMINATED with caustic chemicals, solvents or pesticides, remove it and wash affected skin but don't allow the casualty to get cold.

Eye contact 1: Rinse out the eye (with the lid held open) with cool running water for a few minutes. As soon as the eye irritation settles down, you may need to seek medical attention.

Eye contact 2: As 1, but don't waste any time. Rinse out the eye for at least 20 minutes. Hold a soft cotton pad over the eye gently and seek medical attention.

IRRIGATING THE EYES must be done under a tap. Cold water will do but may be uncomfortable. Hot water will NOT do. Try to get to a mixer tap, to get a gentle flow at about blood temperature — this is less unpleasant. A shower attachment may make the job easier. Even a bowl of clean water splashed into the eye for a while will help. The process may be very traumatic for children or the elderly. Do the best you can and seek medical assistance as soon as possible.

NOTE: In industry where goggles may not be possible, but eyes may be at risk, it is common to have a bottle of eye-washing solution fitted with its own eyebath top. If you can find something like this to keep in the home for emergencies, it would be sensible.

EMERGENCY!

CHEMICAL HAZARDS

▶ INHALATION: GUIDELINES

Take dizzy or nauseous casualty to fresh air. Loosen restrictive clothing. If symptoms don't subside in a few minutes, seek medical attention. Otherwise, keep casualty calm and warm.

IF CASUALTY IS VERY DIZZY, HAS SLURRED SPEECH, OR VOMITS

Place in recovery position, to keep airway open.
Cover to keep warm and call an ambulance.

▶ INGESTION: GUIDELINES

Small amounts of simple household chemicals may do no harm. Observe casualty for some time, give milk or water to sip slowly and seek medical attention if there is any discomfort or distress. DO NOT INDUCE VOMITING.

IF CHEMICAL IS HIGHLY TOXIC

Even small amounts may be dangerous. Wipe away chemical from skin with water — especially if substance is caustic.
Seek medical attention as soon as possible. If obviously serious, call an ambulance.

IF A LARGE AMOUNT OF CHEMICAL IS SWALLOWED

If larger amounts have been swallowed, almost any chemical may be dangerous. Keep casualty calm. Call an ambulance.

IF CASUALTY IS UNCONSCIOUS, BUT BREATHING

Place in recovery position, to keep airway open. Cover casualty to keep warm and call an ambulance.

BE PREPARED FOR CONVULSIONS OR VOMITING

If fits or convulsions take place, don't let the casualty hurt themselves — particularly protect the head. If vomiting occurs in a semi-conscious or conscious casualty, they may choke.

IF HEARTBEAT/BREATHING STOPS

Be prepared to give cardiac compression and artificial respiration. With a small child, place your mouth over the mouth and nose of the casualty. Avoid the risk of poison around the casualty's mouth.

Acids in the home

Acids are common as constituents of many domestic cleaning products such as de-scalers (limescale removers), toilet cleaners, drain clearers—even car batteries contain fairly strong acid.

Plumbers are fond of 'spirits of salt'—it cleans metal very quickly. Some people use it to kill tree stumps or clean stone- and quarry-tile floors. It is actually a 32 per cent solution of hydrochloric acid, which produces toxic choking fumes.

If a product contains any named acid, follow directions very carefully, avoid skin contact and DON'T mix with other preparations. Most acids are very caustic and cause burns to human tissue and often choking acrid fumes. **See EMERGENCY! panel: Skin contact 3/Eye contact 2. DO NOT INDUCE VOMITING**

Alkalis in the home

These include sodium hydroxide (caustic soda), sodium hypochlorite, ammonia, potassium hydroxide, calcium oxide—in fact, any caustic or heavy-duty cleaning product which does not contain a named acid is likely to contain a powerful alkali. Like acids, most produce burns on human tissue and sometimes choking fumes. They are common in bleaches, toilet cleaning products, paint strippers, oven cleaners. **See EMERGENCY! panel: Skin contact 3/Eye contact 2. DO NOT INDUCE VOMITING**

Solvents in the home

Solvents are used in numerous products, from paints to glues, spot/grease removers to dry cleaning chemicals, aerosols to paint thinners, brush cleaners to nail varnish. They evaporate rapidly, even at low temperatures. The vapours are toxic and can quickly make you dizzy or nauseous.

It is not safe to drive or operate machinery if affected. They are readily absorbed into the body where they melt fats and damage tissues.

Long-term exposure risks are liver and kidney damage. Even short-term exposure may cause memory loss and disturbed concentration. Overexposure may cause headaches, vomiting, stupor or even hallucinations. Emergency treatment is required. **See EMERGENCY! panel: Skin contact 1 or 2/Eye contact 1. DO NOT INDUCE VOMITING**

REMEMBER

Ventilation when working is top priority. If your job involves the use of solvents, you should be properly protected and have regular medical check-ups to ensure you are not suffering long-term effects.

WARNING

Some drugs such as sedatives, tranquillizers, sleeping tablets — even anti-histamines and alcohol — do NOT combine well with exposure to solvents. The liver cannot process the chemicals and there is an increase of the narcotic effects (the dizziness, loss of concentration, stupor). DON'T risk exposure if you are on medication. DON'T drive or operate machinery if you are affected.

FORMALDEHYDE

There is much debate on the safety of cavity-wall insulating foam — used in many countries. The danger is that (depending on conditions such as moisture and humidity) it may release formaldehyde vapour. It is probably reasonably safe if totally contained — between two brick or block walls — less so if used behind a flimsy partition wall. The official view varies from country to country — but links are being investigated between formaldehyde and various forms of cancer.

Levels registered in all cases are very low — but it may be wise, if you are considering foam insulation, to make sure the contractor is a registered user and his/her methods are covered by national seals of approval.

Formaldehyde is known to be an irritant of the upper respiratory passages. Symptoms may include: headaches, depression, insomnia, recurrent coughs and colds, asthma and gastrointestinal problems. Old people and young children may be more at risk.

Several weeks after the foam has 'cured', the release of formaldehyde decreases. There is a background level of formaldehyde in most homes — from processed woods like plywood and chipboard (which may offer problems to the DIY enthusiast or joiner — see DIY/CRAFT HAZARDS: **Wood dust**), resin glues for wood, cyanoacrylate (instant) glues, resinous plastic and even foam-backed carpets.

The use of aerosols is a very sensitive issue in most countries—especially aerosols linked to damage caused to the ozone layer by propellant gases. Safe alternatives should be sought, also, because there are immediate and long-term health risks to the user.

The mist of chemicals produced during spraying can enter the lungs. Quite common products may cause inflammation of the nasal and throat membranes, asthma and chest irritations. There are plausible links to more serious chest complaints after long-term use.

Care should obviously be taken with hair sprays, throat sprays, deodorants and perfumes—any products which are applied directly to the body, near the face.

KITCHEN/LIVING AREAS

Washing-up liquid/washing powders

These have fairly low toxicity. They could cause vomiting or gastrointestinal upsets. Inhaling bubbles could cause aspiration pneumonia. **Skin contact 2 or 3/Eye contact 2.** DON'T hand wash with machine powders, particularly if they contain enzymes, as chronic dermatitis is a common result. Always rinse off skin. The enzymes used not only break down proteins such as those found in egg or gravy but also body proteins such as blood, skin and mucous membrane. Inhalation MUST be avoided. **Skin contact 3/Eye contact 2**

▶REMEMBER TO REFER TO EMERGENCY! PANEL FOR FIRST-AID GUIDANCE THROUGHOUT THIS SECTION

Oven cleaners

Usually highly corrosive—whether as sprays or impregnated into foam pads. They (commonly) contain caustic soda (sodium hydroxide). If swallowed or inhaled seek urgent medical attention. **DO NOT INDUCE VOMITING. Skin contact 3/Eye contact 2**

Shoe polish

The wax involved may cause gastrointestinal upset if swallowed, and diarrhoea. The solvents may include ethanol, toluene or xylene—which are more serious. If swallowed by a child, seek medical attention. **Skin contact 1**

Mothballs/mothproofing

May contain naphthalene, which is highly toxic, or paradichlorobenzene—which is less so. May also contain permethrin (see **Flying insect killer**). If swallowed by a child and there is any doubt about the amount swallowed and the nature of the product, seek urgent medical attention. **DO NOT GIVE MILK** to the casualty (or any liquid). Milk will speed the absorption of the naphthalene into the body. Mothballs (especially those containing naphthalene) should be stored in the shed/garage. **Skin contact 1 or 2**

Bleaches

May contain caustic soda (sodium hydroxide), chlorine or ammonia. If swallowed, particularly by a child, give milk or water to sip and seek medical attention. Fumes may be severely irritating to the eyes, nose and throat—avoid breathing them. NEVER mix bleaches with any other cleaner—a violent reaction producing choking fumes may result. **Skin contact 3/Eye contact 2**

Liquid ammonia

A highly-unpleasant irritant that produces choking fumes. DON'T mix with other chemicals. If swallowed, seek medical attention, or give water to sip. Judge by discomfort/distress/amount swallowed. **Skin contact 3/Eye contact 2.** Seek medical attention if skin or eyes are 'burnt'.

Glass cleaner

Products for mirror- and window-cleaning may contain petroleum-related products or alcohol and ammonia. If sprayed, avoid breathing mist as lung damage may occur. If swallowed, medical attention may be urgent. **Skin contact 2/Eye contact 2**

Felt-tip pens

Always make sure coloured pens given to children are NON-TOXIC. Some contain solvents. Some may contain aniline dyes which could cause severe poisoning if the pens are sucked. Breathing difficulties may occur and the skin would probably take on a bluish tone. Seek medical attention.

Matches

Apart from the obvious fire dangers, matches are also poisonous. The phosphorus in 'strike anywhere' matches may cause liver damage if chewed. So-called 'safety' matches are less harmful—but could cause gastrointestinal problems.

Glues

Most **water-based glues** for paper are non-toxic. Look for ones so labelled—particularly if children are to use them. Glues containing solvents should only be used in the home under well-ventilated conditions.

Instant glues (cyanoacrylates) produce toxic irritant fumes which should not be inhaled. The toxic effects of these glues has not been fully understood yet. The fumes may be linked to immediate or delayed asthma, but in domestic use the quantities used are very small. If the fingers become stuck, trying to free them from an object or from each other might tear the skin. Tiny areas of adhesion may be gently eased apart in warm soapy water with the aid of a blunt spatula or the handle of a spoon. Larger areas need MEDICAL ATTENTION. If any glue goes in the eyes or mouth, URGENT medical attention should be sought.

'Airplane' or modelmaking glues and certain **spray glues** contain toluene—a toxic solvent. Avoid breathing the fumes. NEVER breathe in the spray (a mask is advisable) as long-term use may well result in lung damage. If overexposed to the fumes of these glues there may be vomiting, cramp, impaired breathing, loss of body heat and a reduction in blood pressure—even unconsciousness and death in some cases (usually only when the substance has been severely abused). **Skin contact 2/Eye contact 2**

Butane

In many homes this is used to fuel: cigarette and gas-oven lighters, portable gas heaters, cordless curling tongs and some DIY blow torches. Avoid breathing butane gas—and keep away from children, especially the small lighter-fuel-size cannisters. If inhaled to the level where vomiting, headaches or slurred speech are occurring—**get to fresh air at once** and **seek medical attention.** Skin contact with the liquid gas may cause 'burns' similar to frostbite. **Skin contact 2 or 3/Eye contact 2 (using WARM water). Seek medical attention.**

Furniture polish

Solid kinds may contain white spirit (see **Turpentine substitute**), petroleum-related substances or other solvents.

Avoid breathing the fumes—or even excessive skin contact. **Skin contact 1.** Don't breathe in the spray mist from aerosol polishes—they may contain alcohol, or petroleum-related substances or silicone wax. The effects on the lungs could be severe. Avoid ingesting any kind of polish. Small amounts may not do any harm, but **urgent** medical attention should be sought if a child consumes a large amount. **Skin contact 1/Eye contact 1**

Disinfectants
NEVER use for medicinal purposes unless directed—and then ONLY as directed. If ingested, it could be very serious. The central nervous system may be affected. There may be fainting or unconsciousness. The lips and all tissues down the throat may be very sore. Seek urgent medical attention. **Skin contact 2 or 3/Eye contact 2**

Metal polishes
Most contain a solvent and ammonia. AVOID breathing fumes. Wear gloves during use and dispose of polishing cloths, which may also be contaminated with toxic metal particles. **Skin contact 2/Eye contact 1**

Washing soda
Sodium carbonate is fairly caustic. If swallowed give water, followed by milk/egg mixture with a little vinegar. Seek medical attention. **Skin contact 2 or 3/Eye contact 2**

BATHROOM/LAVATORY

Surgical spirit
Dangerous if swallowed—especially by children. Risks include lowering of blood sugar levels (hypoglycaemia) and possible convulsions or unconsciousness. Medical attention may be urgently required—permanent damage to the eyes may occur. Give a child a spoonful of sugar or glucose to boost blood sugar levels. **Eye contact 1**

Nail varnish/varnish remover
Contain solvents—often acetone. There may be sensitivity reactions to the nails and skin, with headaches and nausea from the fumes. Always apply in a well-ventilated area. Ingestion is unlikely—but does occur. Monitor condition and seek medical assistance if necessary. **Eye contact 2**

Deodorants

Some people may be sensitive to the perfumes or alcohol (in some kinds). See the directions, which usually tell you NOT to use on sensitive or broken skin. If armpits are shaved, the risk of a local reaction is dramatically increased. Anti-perspirants may actually block pores—and lead to a severe infection. Never spray anti-perspirants all over yourself. You would interfere with your body's cooling process. If in aerosol form, do NOT breathe the mist. Many deodorants contain aluminium compounds. Aluminium is a potentially-toxic metal. Vaginal deodorants (and bubble baths) have been linked with urethritis and cystitis. If a child swallows deodorant, seek medical attention. **Eye contact 1**

Perfumes/toilet water/aftershave

The danger arises mainly from the alcohol content. If enough has been swallowed, there may be a dramatic drop in blood sugar (hypoglycaemia) and possibly convulsions or unconsciousness. A child should be given a spoonful of sugar or glucose to boost sugar levels—and urgent medical attention sought. **Eye contact 1**

REMEMBER

Many cosmetics may cause adverse reactions—from rashes and itching, to breathing irritation and headaches. Acute sensitivity to perfumes and other ingredients may produce recurrent cold symptoms and regular headaches. If testing a new product, ALWAYS dab a little on a soft area of skin—such as the inner forearm—and wait 48 hours before further application.

Hair colourants and bleaches

ALWAYS follow the manufacturers' instructions—particularly regarding sensitivity tests. DON'T take this lightly. A whole-head application could be very serious and temporarily disfiguring if you have a bad reaction. Allergic/dermatitic reactions are common. Sensitivity may develop with prolonged use. Never use on a child without expert supervision. If swallowed, particularly by a child, give water to sip and seek medical advice. **Skin contact 1, 2 or 3/Eye contact 2**

Talcum powder

Allergic skin reactions are common—especially in areas such as the armpit (particularly if it has been shaved). AVOID all cheap talc if not by a known manufacturer. VERY strange contaminants have been found—even serious enough to cause death.

Talc may contain potentially-toxic compounds of magnesium and aluminium. Do NOT allow children to play with talc. Avoid ingestion or inhalation of airborne talc. If any rash develops, discontinue use at once. **Eye contact 1**

Lavatory cleaners
May contain acids or alkalis, both strong enough to cause burns to the skin and eyes. May also contain ammonia, caustic soda or chlorine. If ingested, lavatory cleaners may cause burns to the mouth and lips and throat, with severe damage to the stomach. DO NOT INDUCE VOMITING. Give water to sip. Seek URGENT medical attention. **Skin contact 3/Eye contact 2**

Hydrogen peroxide
The stronger the concentration, the more serious the effects. If swallowed, give water to drink and seek medical attention. **Skin contact 3/Eye contact 2.** If skin or eyes are 'burnt', seek medical attention.

Alum
A powerful astringent. If swallowed, may cause gastrointestinal problems. **Skin contact 2/Eye contact 2**

THE MEDICINE CUPBOARD

 Being able to buy 'drugs' over the counter at pharmacies—for some people—implies that the substances are 'safe'. Sometimes this is far from the truth. No product is intended to be used excessively or for a long time. Some may be completely unsuitable for children or should only be used with care by the elderly.

In some cases, where these products are used to treat only 'symptoms', a more important medical problem may be masked—or new illnesses/conditions may develop.

The information given here is very general. ALWAYS consult your doctor or medical adviser before using a product regularly. ALWAYS read manufacturers' instructions.

❗ WARNING
Suddenly ceasing prescribed medication may produce harmful effects. Seek medical advice before doing so.

MEDICINE SAFETY

 It should go without saying that anything which belongs in a medicine cupboard should NEVER be left where children could reach it. Even adult-strength iron tablets (which may look exactly like sweets) can be extremely dangerous for a child. LOCK THE CUPBOARD AND HIDE THE KEY! Also, avoid mix-ups by the elderly or poor-sighted by clear labelling and NEVER swapping drugs from one container to another.

If you are taking prescribed medicines of any kind, ALWAYS consult a doctor/pharmacist if you regularly use or intend to use any over-the-counter preparations. Some chemicals may produce undesirable combinations in the body which could prove harmful.

Pregnant women and women who are breast-feeding babies under six months old should be wary of ALL drugs—even aspirin and paracetamol—unless they are assured by a doctor that it is safe to take them. Many drugs will pass from mother to baby.

Aspirin

It is now agreed (at least in the UK and US) that aspirin and products containing aspirin should NOT be given to children under twelve. Many adults are sensitive and may experience gastrointestinal pain or problems, rashes or even asthma. Over use may damage the wall of the stomach—possibly causing internal bleeding. Never take aspirin if you have a stomach ulcer. Taking aspirin regularly may cause you to develop sensitivity. Aspirin (salicyclic acid) is a constituent of many other preparations including wart removers and corn 'cures'—both lead to absorption of the chemical into the system.

Paracetamol

Never take paracetamol if you have a history of liver or kidney problems. The main danger of long-term use is damage to the liver. Doses should be very much reduced for children AND the elderly. Do NOT take paracetamol, even at the stated maximum dosage, for more than two or three days. If it hasn't 'worked', you should see a doctor in case you are taking the wrong medicine. Paracetamol is also a constituent of many other preparations—always read labels!

Codeine

Not so commonly available these days, but should definitely not be over used. It is common as an ingredient in cough medicines and pain-killing preparations. Dependency may occur—because codeine is an opium derivative. It may react badly with other drugs. Main side-effects include constipation and drowsiness. In higher doses, there may be a clamminess of the skin or even loss of consciousness. The presence of codeine in any 'medicine' is usually clearly marked.

Indigestion remedies

If you need to take indigestion remedies frequently, you may have dietary problems. See a doctor. Common ingredients include: aluminium compounds (aluminium is a potentially toxic metal), magnesium compounds and sodium bicarbonate. Constipation may result from aluminium-based preparations in the short term, but links are being investigated with bone development problems and possible kidney damage. Magnesium is very likely to cause diarrhoea, and may be toxic if taken to excess. The kidneys may suffer from long-term use. Even bicarbonate of soda should not be used on a regular basis. It actually produces gas (CO_2) in your stomach.

Diarrhoea remedies

Most are based on opium or belladonna derivatives and NOT designed for prolonged use. You may experience a dry mouth and giddiness. Rashes have been known to be caused in some people. Treat with great caution and consult a doctor if you feel you need to take a medicine like this on a regular basis.

Cough medicines

Codeine-based products should be used with extreme caution. They can lead to dependency. Antihistamines in these preparations will make you drowsy—possibly very drowsy. By inhibiting coughing—codeine especially may prevent the lungs being relieved of mucous. Ephedrine is a stimulant, with an adrenalin-like tendency to make you tense and nervous. It can react badly with other drugs. Ipecacuanha, seen in some medicines, is an emetic (vomit inducer). It can only be presumed that high doses of medicine containing it may make you feel sick.

Hay fever/allergy remedies

Be VERY careful of the sedative effects of antihistamines—sleep may be unavoidable. Do NOT mix with alcohol or attempt to drive or operate machinery. There may be gastrointestinal or urinary problems with long-term use.

Cold 'cures'

Very high doses of vitamin C are not good for you over a long period. Problems may involve kidney stones, diarrhoea, stomach problems and heartburn. There may even be a kind of dependence developed. Aspirin in combination with vitamin C may lead to further stomach irritation.

Try to avoid gargles and lozenges with iodine or borate in them—especially for children and the elderly. There may be toxic effects from absorption.

Nasal decongestants should never be used for more than seven days. The nasal passages may be damaged—or even toughened up, and congestion may become worse. Camphor-based products are NOT suitable for babies or infants.

Travel-sickness remedies

Most are based on antihistamines (see **Hay fever/allergy remedies**). Nearly all can cause severe drowsiness. Since they are taken when travelling, there is a great danger that you may attempt to drive. You may fall asleep and miss destinations or not hear warning announcements. Regular use should be avoided, except under medical supervision. Take great care with dosages for children and the elderly.

Laxatives

Taking laxatives is a rather old-fashioned idea. It is still very common and people should try to regulate bowel movements by dietary means. Laxatives do NOT cure the cause of constipation. Users may become dependent, and not deal with the actual cause of their condition.

Vitamin/mineral supplements

These are NOT cures. They are NOT sweets. They are NOT intended to be taken in concentrated form over long periods. They are a dietary supplement—NOT an alternative to eating properly. They are good as support when health is low or in times of illness. Problems of excessive or very long-term use could be serious. If you feel you 'need' vitamins (in normal doses) for more than a couple of weeks, see a doctor. There are already well-recognized side effects of many vitamins and minerals. Vitamins A and D, for instance, can be very toxic if over used. Do NOT allow children to eat vitamins like sweets.

Germanium has been recently banned by law (in the UK, at least) and there is some doubt about ingesting selenium. Niacin has been shown to be harmful in quite small overdose quantities. Always bear in mind that 'food supplements' are often not vetted/tested as drugs.

PRESCRIBED MEDICINES

 There has been a lot of debate about most prescribed medicines. Some undoubtedly 'work'—but many produce side effects or bad combinations with other drugs and may lead to addiction or dependency. Mild or serious overdosing may occur. Most people trust medicines and doctors enough to believe that a course of medicines prescribed for them MUST be good for them. They will often carry on taking a medicine which disagrees with them. If in doubt, go back to your doctor—there may be an alternative that could be prescribed.

More and more people are looking outside conventional medicine for 'kinder' cures and therapy. It may be that a 'cure' is more frightening than the condition, in some cases.

Safety points

- NEVER share medicines with anyone else. Just because they are suffering a 'similar' condition, does not mean that the medicine is appropriate.
- ALWAYS ask the doctor or pharmacist about the known side effects of a medicine. It is their duty to tell you.
- ALWAYS ask how serious it would be if someone else—perhaps a child—took the medicine by mistake.
- ALWAYS ask if the medicine might disagree with alcohol, coffee or over-the-counter medicines. Make sure you explain what you do for a living if you are exposed to chemicals in the workplace as well.
- ALWAYS ask if addiction or dependence are possible.
- NEVER keep unwanted medicines. They may have a limited shelf-life and you shouldn't attempt to start your own pharmacy. Take them back to a pharmacist for proper disposal. Do NOT throw them down the sink or toilet.

DEPENDENCE

Dependence on medication, especially tranquillizers, sleeping tablets and painkillers, is a very unpleasant predicament to be in. You may feel unable to cope without the drug—and terrified of symptoms returning. You may even be physically addicted to a drug—your body may demand it, while your mind is trying to tell you that you don't need it. There are help groups and most doctors will be sympathetic. Quite often there may be alternative safer medicines. You must have help to escape dependence on medication.

GARAGE/SHED

All the 'nasty' chemicals for household use should be kept in the garage or shed—and locked away.

Glues

Solvent-based glues present a risk of inhaling the fumes. This can lead to dizziness and loss of judgement while working. Longterm effects may be more serious (see **Solvents in the home**). If you are badly affected by fumes, do not overexert yourself—your heart may be at risk. Get to fresh air (see EMERGENCY! panel). Only use in a well-ventilated area. These glues are also highly flammable. Continue ventilation until the glue is fully dry. Avoid skin contact. Do NOT use solvents to clean the skin. Extensive contact may require medical attention. Soaking the affected area in soapy water may allow you to peel the glue away gently. (See also HEALTH: **Substance abuse**). **Eye contact 2 or emergency medical attention**

Two-part resin-based glues have been linked to asthma. Wash off before the glue 'sets' with warm soapy water or a skin cleanser. Try to AVOID skin contact—chemicals may be absorbed into the skin. **Skin contact 3/Eye contact 2 or emergency treatment**

'Hot melt' adhesives can involve temperatures of 200°C and produce deep burns. Wear gloves and protective clothing. Cool areas of contact immediately with cold water—but don't attempt to remove the glue. Don't burn the glue—a toxic smoke is produced. These glues do not have a long shelf life—old glue may ignite, in use. **See EMERGENCY! panel: Skin contact 3 (but cool first and leave fixed patches of glue)/Eye contact 2 or emergency treatment**

Water-based (PVA) wood glues are known to cause dermatitis. Avoid skin contact. Wash from the skin before it dries. Wash/rinse immediately from mouth or nose. If a large amount is swallowed, give casualty water to sip and seek medical attention. **Skin contact 2/Eye contact 2**

Tar/grease-spot removers

Most contain fairly heavy-duty toxic solvents such as tetrachloromethane, xylene or carbon tetrachloride. If swallowed may lead to gastrointestinal problems and loss of consciousness. Avoid breathing vapours, which are also toxic—use ONLY in a well-ventilated area. If affected, get to fresh air and seek medical attention if you don't feel better in a few minutes. Tetrachlorethylene may be a constituent. You MUST avoid inhalation or swallowing—or even prolonged skin contact. There have been many serious reactions. If affected, beyond a little dizziness, seek medical attention. Regular exposure may lead to liver damage. **Skin contact 2/Eye contact 2**

Turpentine

Very poisonous. **If swallowed, do not induce vomiting**—seek emergency medical attention. May cause severe gastro-intestinal irritation, breathing difficulties and kidney damage. Urine may be dark and smell of turpentine. Use in a well-ventilated space. **Skin contact 2/Eye contact 2**

Turps substitute/white spirit

Similar to turpentine. Flammable. If enough is swallowed there may be lung irritation, fits, 'drunkenness' and headaches. In severe cases there may be lung problems—even coma. **Do not induce vomiting**—seek urgent medical attention. Use only in a well-ventilated area. **Skin contact 2/Eye contact 2**

Petrol

If swallowed, severe gastrointestinal irritation and pain is likely. The vapour alone may cause damage to the lungs or pneumonia-like disorders. Swallowing large amounts can lead to convulsions and unconsciousness. **NEVER induce vomiting**—the risk to the lungs is severe. NEVER syphon petrol by sucking on a tube. Highly flammable/explosive. Use in a well-ventilated area. **Skin contact 2/Eye contact 2**

Caustic soda

Sodium hydroxide. Used as a super-strength cleaner and drain clearer. Also the main constituent of paste paint strippers. It can react violently to acids or bleach. NEVER use in the toilet with other chemicals. Caustic soda produces a violent exothermic (heat-producing) reaction with water—the right concentrations will reach 'boiling point' very quickly. This substance is HIGHLY CORROSIVE. Do NOT breathe in the powder. **If swallowed, do NOT induce vomiting.** Give milk or water to sip—1/4 litre (1/2 pint) maximum for children and seek urgent medical attention. **Skin contact 3 (urgent)/Eye contact 2 (urgent—permanent damage is possible)**

> **! WARNING**
>
> When mixing caustic soda, always add caustic soda crystals to water—NOT water to crystals. A violent reaction may occur, with droplets of hot caustic liquid ejected.

Paste paint stripper

See Caustic soda

Liquid paint stripper/brush cleaners

Liquid strippers may contain methanol (methyl alcohol). Lowering of the blood pressure, headaches and nausea are common. If swallowed, damage to the eyesight may occur. Caustic effects may be very serious—**do not induce vomiting.** Give water to sip. Seek urgent medical help. May also contain dichloromethane, which affects the central nervous system and may be absorbed readily by the skin or by inhalation. Skin contact with all paint strippers is painful (see DIY/CRAFT HAZARDS). Serious dizziness may occur, making it unsafe to work—with long-term use the heart and lungs are at risk. **Skin contact 3 (NOT hot water!)/Eye contact 2 (urgent)**

Antifreeze

May contain methanol but more likely ethylene glycol, which is highly poisonous. Unfortunately, it has a sweetish taste and a bright colour—so children may be very interested. It is extremely dangerous if swallowed. Seek urgent medical attention. Observation is recommended for at least 48 hours. All manner of serious effects may occur from oxalic acid formed in the system—affecting soft tissues, kidneys, brain and heart. Giving a small amount of whisky or brandy to drink (these contain ethyl alcohol) should slow the progress of the ethylene glycol into the blood. Seek medical help first! **Skin contact 2/ Eye contact 2**

Methylated spirits

Contains ethyl alcohol, methanol (which can cause permanent eye damage) and pyridine. See **Surgical spirit. Skin contact 2/Eye contact 1**

Rust removers

These are highly poisonous and highly corrosive. Avoid all skin and eye contact. Swallowing could be very nasty indeed, with burning pains and vomiting. Do NOT induce vomiting. **Skin contact 3/Eye contact 2 (urgent)**

Slug and snail killers

May contain metaldehyde. Highly toxic. It is unlikely to be swallowed by humans in sufficient quantities to do harm, but may cause gastrointestinal problems, fainting, nausea or breathing difficulties. Medical attention is more likely to be needed for small children. Dogs may be at risk.

Rat/mouse poison

This rates high on the lists of dangers to children. Poisoning in adults is rare. The main constituents are anti-coagulants—agents which prevent blood clotting, such as warfarin. Symptoms are blood in the urine and faeces or bruising. NEVER have these poisons in the home if there are children about. A serious infestation of rats or mice should be dealt with by a local authority or specialists.

Firelighters

May contain metaldehyde (see **Slug and snail killers**), waxes and petroleum spirit, turpentine, paraffin or solvent to ensure burning. Store in a cool well-ventilated space. May be serious if swallowed by a child, causing gastrointestinal problems, pain and vomiting. Do not induce vomiting—seek urgent medical attention. **Skin contact 1 and 2**

Ant killer

May contain borax or pyrethrum. There may be an allergic reaction with skin irritation. Avoid skin contact and breathing dust. Pyrethrum may cause (in children) excitability or trembling—but usually only if large amounts have been swallowed. Seek medical attention. **Skin contact 2/Eye contact 1**

Flying insect killer

Usually contains pyrethrum, permethrin or other similar insecticides. See **Ant killer.** Avoid breathing mist from aerosol types. **Skin contact 2/Eye contact 2**

De-icer

May contain isopropanol or methanol (see **Liquid paint strippers**). You may experience dizziness. If affected, do NOT drive. Avoid skin contact. Seek urgent medical attention. **Skin contact 2 or 3/Eye contact 2**

Cement

Avoid skin contact. Cement is very alkaline when wet and can cause painful burns. Always wear waterproof gloves and avoid breathing the dust. **Skin contact 3/Eye contact 2**

Plaster and fillers

Plasterers often have severe dermatitis, which is common from constant handling of alkaline materials (see **Cement**). In addition, as plaster dries on the skin, it may 'pull' moisture from it. Always wear gloves or a good barrier cream. **Skin contact 2 (before it**

dries hard)/Eye contact 1 or 2 (at once). Interior decorating fillers may contain fungicides and 'resins' which will further aggravate dermatitis.

Lime (calcium oxide)

Avoid skin and eye contact, inhalation and ingestion. Lime is VERY alkaline and will burn the skin. When combined with water, it produces heat and calcium hydroxide—an irritant gas. **IF LIME GETS IN YOUR EYES, treat as an emergency. Skin contact 2 or 3/Eye contact 2 (urgent)**

PESTICIDES

Pesticides and herbicides may be some of the most unpleasant and dangerous chemicals we 'use'. There are 'organic' or natural alternatives, but a whole industry has been built around these toxins—for domestic as well as commercial use.

Pesticides include all products which control or eradicate unwanted insects/small animals. Herbicides are used to kill weeds (or unwanted plants) and fungi—as in timber treatments or the control of algae growths on stone and masonry.

DON'T be casual about the use or storage of these chemicals. You should always consider:

◑ Is this the right product for the job?
◑ Do I need something this powerful? Is there a weaker alternative?
◑ Is this the smallest amount I can buy?

WARNING

Never experiment with poisons—only use as instructed. NEVER mix products together to 'kill two birds with one stone'—you may kill a lot more than two birds. NEVER NEVER NEVER decant pesticides into other containers. In some countries, including the UK, this is prohibited by law.

Poisoning

Many pesticides involve a solvent as a 'carrier', confusing the symptoms of poisoning (see **Solvents in the home**). Immediate risks are skin and eye irritation. All the main routes to the system are open to pesticides, though. **Inhalation** is common, of the mist when spraying, or of dust in dry chemicals. Tastes in the mouth and nasal irritation are early warning signs that you SHOULD be wearing a mask.

Ingestion may be accidental or deliberate—or even a gradual process involving hand-to-mouth contact (eating/drinking/smoking while working) or incomplete washing of the hands and fingernails. Wear gloves! Pesticides may also enter the system if they have been deposited on or assimilated into fruits and vegetables.

Absorption is also highly possible. Don't wait for skin irritation to signal danger. You may not feel skin discomfort until you have been using a product for some time. Cover all skin—especially the hands and arms.

The legacy

Some pesticides are known to persist—but research may take many years to reach definite conclusions. It is certain there may be long-term effects to the planet, to animals and to man, in terms of disease and other toxic effects. 2,4,5-T has now been withdrawn or banned in many countries, because of links with birth defects and cancer. Even a new-born baby has traces (in the fatty tissues) of the now-withdrawn and banned DDT.

Disposal

Never pour chemicals, particularly pesticides, down any drain (sink, lavatory or street). They should NOT be introduced into rivers, streams or the sea. If they enter your septic tank, they may kill off the necessary bacteria there.

If you have more than a 'little' to dispose of or are unsure about disposing of unfinished packets, cans and bottles, get in contact with your local authority—or environmental concern groups—for their advice.

Using safely

◗ Don't mix pesticides.
◗ Don't store in the wrong containers.
◗ Keep children and animals away from treated areas.
◗ Only spray **what** you need to, **where** you need to and **when** you need to.
◗ Keep the spray aimed where you need it—ALWAYS using a coarse spray, NOT a fine mist.
◗ Spray gardens early morning or late evening when few pollinating insects (or the birds that eat them) are about.
◗ Don't spray in a breeze or wind. Spray drift may do damage. Protect yourself.
◗ Always wash all exposed skin after using.
◗ Don't eat, drink or smoke until you have washed.

- Gloves should be in good condition and washable or disposable. Aborbent and damaged gloves may present a long-term health hazard.
- Always wash out spray equipment and watering can to avoid contamination. If you can keep separate equipment for pesticide, it would be sensible.
- Always wear goggles and a breathing mask when spraying, or mixing dry ingredients.

First aid

Acute poisoning is rare—but if chemicals are breathed in or swallowed, the casualty should be observed. If there are any effects—particularly in a child—medical attention should be sought. **Refer back to EMERGENCY! panel for treatment details. Skin contact 1, 2 or 3** (Don't forget to wash your scalp/hair) **Eye contact 2**

REMEMBER

If you use pesticides regularly—perhaps as part of your work—you MUST protect yourself from constant exposure to the toxins. A regular medical check-up would be sensible to ensure you are not suffering long-term effects. Some pesticides may contain mercury or arsenic compounds or other cumulative poisons (which collect in you system).

☠ WARNING

Sodium chlorate and potassium chlorate weedkillers are explosive—particularly in the dry state. They may explode or cause fire if dropped or shaken violently. Both are highly toxic—swallowing either of these may be followed by abdominal pain, confusion and convulsions. A surprisingly small amount may kill. Call an ambulance immediately, if a child appears to have swallowed some.

Paraquat is a very common weedkiller. It is often used very casually—but is highly toxic if swallowed.

FERTILIZERS

Many fertilizers for plants should be mixed, applied and stored with the same caution as for pesticides. Many are poisonous. Particular caution should be used with nitrate fertilizers, which are considered to be a major hazard to the environment. They are a danger in domestic water and 'over-fertilized' fruits and vegetables. They become toxic in the body (as nitrites) and may be linked to some cancers.

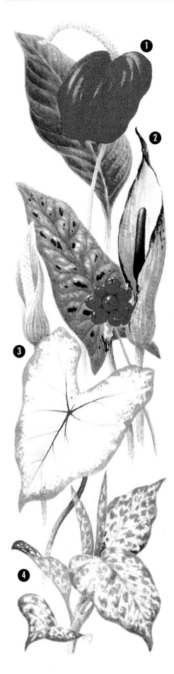

You may come across poisonous plants in domestic gardes, parks and growing 'wild' on verges—in some cases you might choose one as an 'indoor' plant. Toxic effects range from skin irritation to death. Many of these plants maybe attractive to children.

The amount of physical harm caused by these plants will depend largely on the age and general health of the casualty. It also depends on the toxicity of the plant and the sensitivity of the individual. Always seek medical attention if there is any doubt—especially if you think a child has swallowed a harmful substance. Take a sample of the 'poison' with you.

The following group of plants not only illustrates the amazing diversity of poisonous plants, but also represents some of the most common hazards.

ARUM FAMILY *Araceae*
Irritant poisons, causing burning pains in the throat and stomach, thirst, nausea, vomiting, fatigue and shock.

1►Flamingo flower (Pigtail plant) *Anthurium scherzerianum* Native to Central America, also a houseplant. Perennial, with dark-green lance-shaped leaves and scarlet spathes with golden-yellow spires at their centres. Height: 30–45 cm (12–18 in). **All parts are poisonous.** Other anthurium species are also poisonous.

2►Cuckoo-pint (Wake robin/Lords and ladies/Adam and Eve) *Arum maculatum* Native to temperate regions, in hedges, woods and ditches. Tuberous-rooted and herbaceous, with hooded spathes and bright red fruits, which are very attractive to children. Height: 30–38 cm (12–15 in). **All parts are poisonous, especially the berries. Can be FATAL.** Other arum species also poisonous.

3►Angel's wings (Elephant's ears) *Caladium x hortulanum* Hybrid developed from tropical American species, now a common houseplant. A tuberous-rooted tropical perennial with large wing-like leaves in many rich colours. Height: 23–38 cm (9–15 in). **All parts poisonous.**

4▶Dumb cane (Leopard lily/Spotted dumb cane) *Dieffenbachia maculata* Native to tropical South America and common as a garden ornamental and a houseplant. An evergreen perennial with large colourful leaves in various colours and patterns. Height: 45–120 cm (18–47 in). **All parts poisonous, especially the sap. In contact with mouth, it can render people speechless for several hours. Also avoid contact with the eyes.** Other dieffenbachia species also poisonous.

5▶Dasheen (Eddo/Taro) *Colocasia esculenta* Native to Java, a garden ornamental and a food plant. It is a cormous-rooted perennial, with large arum-like leaves. Height: 90–180 cm (35–71 in). **All parts poisonous if eaten raw, especially corms**. Tubers are edible if properly cooked.

6▶Malanga (Tannia/Tanyah/Yautia) *Xanthosoma saggittifolium* Native to the West Indies and an ornamental in southern US. Also a food plant. Tuberous-rooted perennial with large arum-like leaves. Height 75–120 cm (30–48 in). **All parts poisonous, especially uncooked tubers.**

7▶Swiss cheese plant (Breadfruit vine/Cheese plant/Fruit salad plant/Window plant) *Monstera deliciosa* Native to Mexico and Central America and a houseplant. This evergreen climber has large glossy leaves with holes, creamy-yellow arum-like flowers and greenish-white club-shaped fruits. Height: 1.8–6 m (6–20 ft). **All parts poisonous, although the fruits are edible if overripe.**

ARALIA FAMILY *Araliaceae*
Irritant poisons, causing burning pains in the throat and stomach, thirst, nausea, vomiting, fatigue and shock.

8▶Common ivy (English ivy) *Hedera helix* Native to Europe, western Asia and North Africa, as well as being naturalized in North America. Common garden ornamental and a houseplant. Woody evergreen climber with dark-green lobed leaves and green flowers. Height: up to 30 m (98 ft). **All parts are poisonous, especially the berries.**

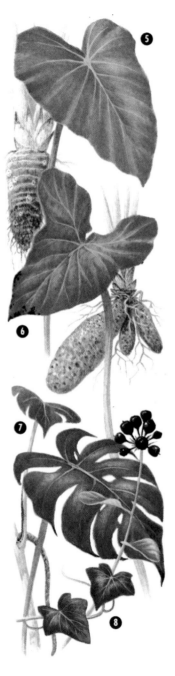

CASHEW FAMILY *Anacardiaceae*

Irritant poisons, causing burning pains in the throat and stomach, thirst, nausea, vomiting, fatigue and shock. Causes severe irritation and rashes on skin contact. Wash affected parts immediately. **WARNING:** When burning these plants on a bonfire, do NOT inhale the smoke. This has been the cause of many DEATHS.

1►Poison sumac (Poison dogwood/Poison elder) *Toxicodendron vernix (Rhus vernix)* **Native** to swamp areas of the US. Deciduous shrub/ small tree bearing many leaflets and purplish stalks. It develops greenish-yellow flowers and yellowish-white, peppercorn-like fruits. Height: 120–180 cm (47–71 in). **All parts are poisonous, including the sap.**

2►Poison oak (Poison ivy/Mercury/Cowitch) *Toxicodendron radicans (Rhus radicans)* **Native** to an area from Canada to Guatemala. Occasionally grown as an ornamental! Trailing or climbing vine or shrub with toothed or lobed leaves, velvety and hairy beneath. It develops greenish-white flowers and whitish, waxy, berry-like fruits. Height: up to 3.6 m (12 ft). **All parts poisonous, even pollen and sap.** Several plants in the rhus genus are also poisonous.

3►Poison ivy (Poison oak/Hiedra) *Toxicodendron quercifolium (Rhus toxicodendron)* **Native** to the eastern US. Deciduous sparsely-branched shrub that spreads by suckers. Its leaves are formed of several leaflets. It develops greenish flowers and whitish fruits. Height: 60–120 cm (24–47 in). **All parts are poisonous, including the sap.**

OLEANDER FAMILY *Apocynaceae*

Poisons causing numbness, tingling in the mouth, abdominal pain and vertigo and affecting the heart. Can be FATAL.

4►Oleander (Rosebay) *Nerium oleander* **Native** to a large area, from the Mediterranean to Japan. Also a garden ornamental and a houseplant. Evergreen shrub with leathery narrow leaves and clusters of white flowers. Height: 1.8–4.5 m (6–15 ft). **All parts are poisonous, especially sap and wood. Can be FATAL.**

BERBERIS FAMILY *Berberidaceae*
Irritant poisons, causing burning pains in the throat and stomach, thirst, nausea, vomiting, fatigue and shock.
5▶Barberry (Salmon barberry) *Berberis aggregata* Native to western China and a common garden ornamental. Deciduous shrub, with pale- to mid-green leaves and yellow flowers. These are followed by round coral-red berries. Height: 120–180 cm (47–71 in). **The berries are poisonous.**

6▶American mandrake (May apple/Raccoon berry) *Podophyllum peltatum* Native to the US and a common garden ornamental. Rhizomatous herbaceous perennial. The yellow fruits have edible pulp. Height: 90–120 cm (35–47 in). **The rhizomes and unripe berries are poisonous.**

BORAGE FAMILY *Boraginaceae*
Toxic effects as the berberis family.
7▶Comfrey (Russian comfrey) *Symphytum x uplandicum* Hybrid, originally introduced from Russia as animal food, a very common ornamental. Herbaceous perennial with mid-green lance-shaped leaves and blue-purple flowers. Height: 75–100 cm (30–39 in). **The leaves are poisonous.** Some herbalists recommend small quantities for medicinal purposes. Common Comfrey *Symphytum officinale* is edible.

BOX FAMILY *Buxaceae*
Irritant poisons causing abdominal pains, vomiting, nerve symptoms such as dilated pupils, headaches. Can cause convulsions.
8▶Common box *Buxus sempervirens* Native to Europe, western Asia and North Africa, and a very common ornamental. Evergreen shrub/tree with small green leaves. Height: 1.8–3.6 m (6–12 ft). **All parts poisonous.**

BELL FLOWER FAMILY *Campanulaceae*
Poisons causing numbness, tingling in the mouth, abdominal pains and vertigo and affecting the heart. Can cause paralysis.
9▶Lobelia (Edging lobelia) *Lobelia erinus* Native to South Africa and a garden ornamental. Height: 10–23 cm (4–9 in). **All parts are poisonous.**

HONEYSUCKLE FAMILY *Caprifoliaceae*
Irritant poisons, causing burning pains in the throat and stomach, thirst, nausea, vomiting, fatigue and shock.

1►Honeysuckle (Woodbine) *Lonicera periclymenum* Native to Europe, North Africa and western Asia and a common ornamental. A perennial deciduous climber with pale-yellow flowers (flushed with red) and bright-red berries. Height: 4.5–6 m (15–20 ft). **The berries are poisonous.** Other plants in this genus also poisonous.

2►Guelder rose (Whitten tree) *Viburnum opulus* Native to Europe, North Africa and northern Asia and an ornamental. Deciduous shrub with scented white flowers and translucent red berries. Height: 1.8–3 m (6–10 ft). **The berries are poisonous.** Other viburnum species also poisonous.

3►Snowberry (Waxberry) *Symphoricarpos albus* Native to the eastern US, and an extremely common ornamental. Deciduous shrub with pink flowers and snow-white fruits from autumn to late winter. Height: 1.5–2.1 m (5–7 ft). **Poisonous berries.**

4►Elder (Elderberry/European elderberry) *Sambucus nigra* Native to Europe, North Africa and western Asia, and an ornamental. Deciduous shrub with leaves formed of mid-green sharply-toothed leaflets. Yellow-white flowers and black berries in late summer. Height: 3.6–4.5 m (12–15 ft). **Poisonous roots and sap.** Berries are also poisonous in large quantities, although may be made safely into wine. Flowers are edible.

STAFF TREE FAMILY *Celastraceae*
Irritant poisons, causing burning pains in the throat and stomach, thirst, nausea, vomiting and shock. May cause DEATH.

5►Common spindle tree (Skewer wood/European spindle tree/Prick wood/Louse berry tree/Ananbeam) *Euonymus europaeus* Native to Europe and western Asia, and common as an ornamental. Deciduous shrub/small tree, with slender-pointed mid-green leaves and inconspicuous green-white flowers. These are followed by rosy-red capsules with orange seeds. Height: 1.8–3 m (6–10 ft). **All parts are poisonous. May be FATAL.** Other euonymus species are also poisonous.

SPIDERWORT FAMILY *Commelinaceae*
Irritant poisons, causing burning pains in the throat and stomach, thirst, nausea, vomiting, fatigue and shock.
6►Purple queen *Setcreasea purpurea (Tradescantia purpurea)* **Native to Mexico, an ornamental and a houseplant. Perennial with tufted, lance-like, purple leaves and pairs of boat-shaped purple bracts. Height: 30–38 cm (12–15 in). Poisonous sap.**

DAISY FAMILY *Compositae*
Irritant poisons, causing burning pains in the throat and stomach, thirst, nausea, vomiting, fatigue and shock.
7►Ragwort *Senecia jacobaea* **Native to Europe, North Africa and western Asia, introduced to the US and New Zealand. Common on waste-land. Perennial with deeply-lobed light-green leaves and yellow daisy-like flowers. Height: 90–120 cm (35–47 in). All parts are poison-ous.** Many other plants in this genus are also likely to cause poisoning, including groundsel (*S. vulgaris*).

BINDWEED FAMILY *Convolvulaceae*
Irritant poisons of a purgative nature, causing abdominal pains, vomiting and purging. These are often accompanied by drowsiness and slight nervous symptoms.
8►Bindweed (Field bindweed/Cornbine) *Convolvulus arvensis* **Native to Europe and Asia, and extensively naturalized in both temperature zones. A common weed. Climbing perennial with counter-clockwise-twining stems and pink or white scented flowers. Height: 20–75 cm (8–30 in). Poisonous seeds.**

9►Morning glory *Ipomoea purpurea (Convolvulus major)* **Native to tropical America, and a common ornamental. Perennial or annual climber, with purple, blue or pink flowers. Height: 1.8–3.6 m (6–12 ft). Seeds are poisonous.**

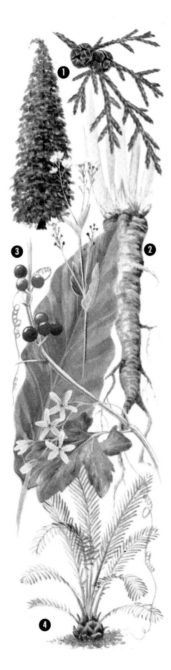

CYPRESS FAMILY *Cupressaceae*
Irritant poisons, causing burning pains in the throat and stomach, thirst, nausea, vomiting, fatigue and shock.
1►Lawson cypress (Port Orford cedar) *Chamaecyparis lawsoniana* Native to the US and a very common ornamental. Evergreen conifer with parsley-scented dark-green/grey foliage. Height: 9–15 m (30–49 ft). **The resin is poisonous.**

MUSTARD FAMILY *Cruciferae*
Irritant poisons, causing burning pains in the throat and stomach, thirst, nausea, vomiting, fatigue and shock.
2►Horseradish (Red cole) *Armoracia rusticana (Cochlearia armoracia)* Native to Europe and naturalized in the US. Herbaceous perennial cultivated for its edible roots. Height: 45–60 cm (18–24 in). **Roots are poisonous when eaten excessively.** Many other plants in the mustard family are poisonous, including wild mustard (*Sinapis arvensis*), wild radish (*Raphanus raphanistrum*) and rape (*Brassica napus*).

GOURD FAMILY *Cucurbitaceae*
Irritant poisons, causing burning pains in the throat and stomach, thirst, nausea, vomiting, drowsiness and slight nervous symptoms.
3►White bryany (Devil's turnip) *Bryania alba* Native to a large area, from Europe to northern Iran, common in hedgerows. Climbing perennial with greenish-white flowers and red berries containing several flat seeds. Height: 1.8–3 m (6–10 ft). **Roots and berries are poisonous. It is estimated that 15 berries may kill a child.**

CYCAD FAMILY *Cycadaceae*
4►Cycad (Sago palm/Japanese fern) *Cycas revoluta* Native to southern Japan and Ryukyu Islands, common in tropical gardens and as a houseplant. Evergreen palm-like tree. Starch-like edible sago extracted from trunk. Height: 1.5–3 m (5–10 ft). **Poisonous seeds.** Pith edible if washed.

YAM FAMILY *Dioscoreaceae*
Irritant poisons, causing burning pains in the throat and stomach, thirst, nausea, vomiting, fatigue and shock.

5►White yam *Dioscorea rotundata* Native to Sierra Leone and the Congo, and a common food crop. Tuberous-rooted, sprawling and climbing, with somewhat heart-shaped leaves. Height: 60–150 cm (24–59 in). **Poisonous tubers, unless boiled or peeled.**

6►Black bryony *Tamus communis* Native to south and west Europe, North Africa, Anatolia, Palestine and Syria, and common in hedgerows. Perennial climber with yellow-green flowers. Berries are bright-scarlet when ripe. Height: 2–4 m (7–13 ft). **All parts poisonous. Can be FATAL to children.**

CUP FERN FAMILY *Dennstaedtiaceae*
Irritant poisons, causing burning pains in the throat and stomach, thirst, nausea, vomiting, fatigue and shock.

7►Bracken (Brake/Hog pasture bracken) *Pteridium aquilinum* Native to and widespread in many countries, except temperate South America and Arctic. Also an ornamental. Perennial with fern-like leaves. Height: 30–150 cm (12–59 in). **Leaves are poisonous. 'Spores' thought to be carcinogenic.** Young 'fiddleheads' may be boiled and eaten.

HEATH FAMILY *Ericaceae*
Irritant poisons, causing burning pains in the throat and stomach, thirst, nausea, vomiting, fatigue and shock.

8►Calico bush (Mountain laurel/Ivybush/Spoonwood) *Kalmia latifolia* Native to the eastern US and an extremely common ornamental. Evergreen shrub with glossy, lance-shaped, mid- to dark-green leaves and bright-pink flowers. Height: 1.5–2.1 m (5–7 ft). **Poisonous leaves and sap.**

9►Common rhododendron (Pontic rhododendron) *Rhododendron ponticum* Native to Portugal, Spain, Anatolia and the Lebanon, and naturalized in Britain. Evergreen shrub with purple brown-spotted flowers. Height: 3–4.5 m (10–15 ft). **Poisonous flowers, leaves and sap.**

SPURGE FAMILY *Euphorbiaceae*
Irritant poisons inducing abdominal pains, vomiting, nerve symptoms such as dilated pupils, headaches and, occasionally, convulsions. Sap causes blisters. Some plants may cause DEATH.

1►Jacob's coat (Copper leaf/Fire dragon) *Acalypha wilkesiana* Native to Pacific Islands, a common ornamental in warm regions and a houseplant in temperate zones. Evergreen shrub with colourful elephant-ear-like leaves in many colours. Height: 1.5–3.6 m (5–12 ft). **All parts poisonous.** Other acalypha species also poisonous.

2►Croton (Variegated laurel) *Codiaeum variegatum pictum* Native to Pacific Islands and the Malay Peninsula. Also an ornamental in warm regions and a houseplant in temperate zones. Evergreen shrub with brightly-coloured leaves. Height: 1.2–2.1 m (4–7 ft). **All parts poisonous. Purgative action.** Other codiaeum species also poisonous.

3►Sun spurge *Euphorbia helioscopia* Native to Europe, Mediterranean region and central Asia. Usually seen in cultivated land. Annual with finely-toothed oval leaves and yellowish bracts. Height: 20–50 cm (8–20 in). **All parts are poisonous. Produces intense irritation of mouth and lips. Violent purgative action. Can be FATAL.**

4►Poinsettia (Christmas star/Painted leaf/Mexican flameleaf) *Euphorbia pulcherrima* Native to Central America and tropical Mexico, and a common ornamental and house-plant. Evergreen shrub with colourful bracts. Height: 1.2–2.4 m (4–8 ft). **All parts are poisonous. Sap causes skin irritation.** Many other euphorbia species are also poisonous. All should be treated with respect.

5►Manchineal tree *Hippomane mancinella* Native to Florida, West Indies and tropical South America. Tropical tree with egg-shaped saw-edged leaves and fleshy, sweet-smelling, yellowish-green flowers. Height: 3–6 m (10–20 ft). **All parts are poisonous.** It is said that not even grass will grow underneath it!

6►Coral plant (Physic nut) *Jatropha multifida* Native to tropical America and a common houseplant. Evergreen tree/shrub with large lobed leaves and scarlet flowers. Height: 3.6–6 m (12–20 ft). **All parts are poisonous.**

7►Castor oil plant (Palma christi/Wonder tree) *Ricinus communis* Native to tropical Africa, an ornamental and a houseplant. Annual with large deeply-lobed leaves and petalless, insignificant, green flowers, followed by large, round, spiny seedpods. Height: 1.5–3.6 m (5–12 ft). **Highly toxic. Up to four seeds may induce severe poisoning in an adult. Eight may be FATAL!**

BEECH FAMILY *Fagaceae*
Irritant poisons, causing acute abdominal pain and often delirium.
8►Copper beech *Fagus sylvaticus* 'Purpurea' Native to Europe. Large deciduous tree with purple leaves, inconspicuous green flowers and triangular brown nuts in autumn. Height: 9–12 m (30–39 ft). **Poisonous nuts.**

FOOD PLANTS
Many of the plants that form part of our regular diet are potentially toxic, especially when eaten raw. The potato and tomato are both members of the nightshade family (*Solanaceae*), which produces many extremely poisonous plants. Potato tubers become green and highly toxic when sufficiently exposed to light — killing many people (especially children) each year. The fruits of tomato plants are not harmful when ripe, but the stems and leaves ore poisonous. Rhubarb leaves are extremely poisonous, but the stems are quite safe to eat after they have been cooked.

1►White oak *Quercus alba* Native to southeast Canada and the eastern US, and an ornamental. Large deciduous tree with lobed dark-glossy-green leaves. Height: 18–25 m (59–82 ft). **Poisonous acorns and leaves.**

FLACOURTIA FAMILY *Flacourtiaceae*
Irritant poisons, causing burning pains in the throat and stomach, thirst, nausea, vomiting, fatigue and shock.
2►Pangi (Kapayang) *Pangium edule* Native to the Malay Peninsula. Quick-growing, spreading tropical tree, with large heart-shaped leaves and reddish-brown edible fruits. Height: 6–7.5 m (20–25 ft). **Poisonous seeds,** edible after cooking.

FUMITORY FAMILY *Fumariaceae*
Irritant poisons, causing burning pains in the throat and stomach, thirst, nausea, vomiting, fatigue and shock.
3►Bleeding heart *Dicentra spectabilis* Native to Japan and a common ornamental. Herbaceous perennial with fern-like grey-green leaves and rose-red heart-shaped flowers. Height: 45–75 cm (18–30 in). **All parts are poisonous.**

HORSE CHESTNUT FAMILY
Hippocastanaceae
Irritant poisons, causing burning pains in the throat and stomach, thirst, nausea, vomiting, fatigue and shock.
4►Horse chestnut (European horse chestnut) *Aesculus hippocastanum* Native to the Balkans and an ornamental. Large deciduous tree with leaves formed of several leaflets. White red-blotched flowers and prickly green fruits containing mahogany-brown 'conkers' (seeds). Height: 7.5–9 m (25–30 ft). **All parts are poisonous.** Not to be confused with sweet chestnuts, which are edible.

IRIS FAMILY *Iridaceae*
Irritant poisons, causing gastroenteritis.
5►Yellow Iris (Yellow flag/Water flag) *Iris pseudacorus* Native to Europe and Asia and a common ornamental. Deciduous perennial with sword-like leaves and yellow flowers. The fruit capsules contain bright orange or scarlet seeds. Height: 75–90 cm (30–35 in). **All parts are poisonous.** Sweet chestnut (*I. foetidissima*) is also poisonous.

LAUREL FAMILY *Lauraceae*
Poisons causing numbness, tingling in the mouth, abdominal pains, vertigo, vomiting and purging and affecting the heart. Can cause paralysis.
6►Headache tree (Californian laurel/Californian bay/Pepperwood) *Umbellularia california* Native to California and Oregon, and an ornamental. Evergreen aromatic tree with large, glossy, aromatic leaves, yellowish-green flowers, and dark-purple fruits. Height: 15–25 m (49–82 ft). **The leaves are poisonous.**

REMEMBER
Some plants are extremely poisonous to everyone. With many plants, however, the toxic effects depend greatly on the sensitivity of the individual. Some people may be highly allergic to some species — even apples, tomatoes or strawberries may provoke a violent reaction. Unfortunately, in these cases, the casualty only discovers the allergy when the (possibly dangerous) symptoms develop.

Many of the potentially-deadly plants, such as foxglove, have very unpleasant tastes or odours — which is a deterrent to most people. Animals are nearly always aware of the warnings given by taste or smell, but children may be attracted by colourful or unusual seeds, leaves or berries. Very small children are highly inquisitive and likely to ignore a nasty taste or smell.

PEA FAMILY *Leguminosae*
Causing a range of effects from loss of coordination and muscular movements, to double vision and deep sleep. Some have irritant poisons and may affect the nervous system, sometimes leading to coma. May cause DEATH.

1►Rosary pea (Indian licorice/Weather plant) *Abrus precatorius* Native to the tropics and a very common ornamental in warm regions. Woody climber with rose to purple flowers and oblong fruits enclosing black-and-scarlet glossy seeds. Height: 1.5–3 m (5–10 ft). **Poisonous seeds. May be FATAL.**

2►Laburnum (Golden chain/Golden rain tree) *Laburnum anogyroides* Native to south and central Europe and a common ornamental. Deciduous tree with pendulous bunches of yellow pea-like flowers. Height: 6–9 m (20–30 ft). **All parts are poisonous, including the roots, bark and wood. May be FATAL Many children die each year. Beware, too, of other laburnums.**

3►Sweet pea *Lathyrus odoratus* Native to Italy and an extremely common ornamental. Annual climbers, with pairs of mid-green leaves and scented flowers, similar to those of the pea plant, in a wide colour range. Height: 1.5–3.6 m (5–12 ft). **All parts are poisonous, especially the seeds.**

4►Lupin (Lupine) *Lupinus polyphyllus* Native to the western US and a common ornamental. Herbaceous perennial with tall spires of blue or red flowers. Wide colour range and many varieties. Height: 120–180 cm (47–71 in). **All parts are poisonous, especially** seeds. Causes respiratory difficulties. Other lupins are also poisonous.

5►Kidney bean (Haricot bean/Wax bean) *Phaseolus vulgaris* Native of tropical America and a very common food crop, yielding beans in pods. Height: 120–180 cm (47–71 in). **Poisonous seeds when uncooked. Kidney beans must ALWAYS be boiled for at least 15 minutes to destroy toxins. May be FATAL, otherwise, especially to children or the elderly.**

6▶Locoweed (Crazyweed) *Oxytropis campestris* Native to northern Asia and the US. Only occasionally grown as an ornamental. Herbaceous perennial with small, narrow leaves and flowers in a wide colour range, white to bright purple. Height: 45–75 cm (18–30 in). **All parts are poisonous.**

7▶Milk vetch *Astragalus canadensis* Native to the US. Herbaceous perennial with leaves formed of many leaflets. It has greenish-cream flowers, then fruits. Height: 90–150 cm (35–59 in). **All parts are poisonous.**

8▶False acacia (Black locust/Yellow locust) *Robinia pseudoacacia* Native to the eastern and central US, and an ornamental. Deciduous tree with leaves formed of light-green leaflets. Pods contain up to ten seeds. Height: 7.5–15 m (25–49 ft). **All parts are poisonous.**

9▶Japanese wisteria (Wisteria) *Wisteria floribunda* Native to Japan and a common ornamental. Deciduous climber with light- to mid-green leaves formed of many leaflets. It has violet-blue or white flower clusters. Height: 7.5–15 m (25–49 ft). **All parts are poisonous.** Other wisterias also poisonous.

10▶Gorse (Furze/Whin) *Ulex europaeus* Native to Western Europe, naturalized in many areas and a very common ornamental. Evergreen spiny shrub with honey-scented yellow flowers. Height: 120–180 cm (47–67 in). **Poisonous seeds.**

1►Common broom (Scotch broom) *Cytisus scoparius* Native to Western Europe and an extremely common ornamental. Deciduous shrub with small mid-green leaves and rich-yellow flowers. Height: 1.8–2.4 m (6–8 ft). **Poisonous seeds and leaves.** Spanish broom (*Spartium juncium*) is also poisonous.

2►Horse-eye bean (Cowhage/Cowitch) *Mucuna pruriens (Stizolobium pruriens)* Native to the tropics. Climber, with pods covered with irritant hairs. The large mottled seeds resemble a horse's eye. Height: 6–9 m (20–30 ft). **Poisonous pods and seeds.**

LILY FAMILY *Liliaceae*
Causing excitement of cerebral functions, loss of coordination and muscular movements and affecting the heart. Some cause numbness, tingling in the mouth, abdominal pain, vertigo and paralysis. May cause DEATH.

3►African lily *Agapanthus campanulatus* Native to Natal and a very common ornamental. Deciduous perennial with long, arching, sword-like leaves and umbrella-like heads of pale-blue flowers. Height: 60–75 cm (24–30 in). **Poisonous fruits.**

4►Asparagus *Asparagus officinalis* Native to the seacoasts of Europe, North Africa and Asia. Common herbaceous perennial with edible young shoots. Tall stems bearing fern-like foliage and red berries. Height: 90–120 cm (35–47 in). **Poisonous berries.**

5►Asparagus fern (Emerald fern) *Asparagus densiflorus* 'Sprengeri' *(Asparagus sprengeri)* Native to Natal and a very common houseplant. It has feather-like light-green foliage borne on wiry stems. Star-like green-white flowers and clusters of red berries. Height: 60–90 cm (24–35 in). **The berries are poisonous.**

6►Autumn crocus (Fall crocus/Meadow saffron) *Colchicum autumnale* Native to Europe and an extremely common ornamental. Cormous, with lance-like green leaves and lilac-coloured flowers. Height: 20–25 cm (8–10 in). **Poisonous corms, seeds and leaves. May be FATAL.**

7►Kaffir lily *Clivia miniata* Native to Natal and a very common ornamental and house-plant. Fleshy-rooted perennial with strap-like leaves, orange-to-red flowers and bright-red berries. Height: 38–45 cm (15–18 in). **Poisonous berries.**

8►Lily-of-the-valley *Convallaria majalis* Native to Europe, naturalized in the US, and an extremely common ornamental. Herbaceous perennial with pairs of elliptic mid-green leaves and white, waxy, bell-shaped, sweetly-scented flowers. It also bears red berries. Height: 15–23 cm (6–9 in). **All parts poisonous, especially the berries. May be FATAL.**

9►Snowdrop *Galanthus nivalis* Native to Europe and a very common ornamental. Bulbous, with strap-like leaves and white flowers with green markings. Also develops berries. Height: 13–20 cm (5–8 in). **Poisonous berries.**

10►Gloriosa lily (Glory lily/Climbing lily) *Gloriosa superba* Native to tropical Africa and Asia, an ornamental and a houseplant. Tuberous-rooted perennial climber with yellow, red and orange flowers. Bright-red seeds. Height: 120–180 cm (47–71 in). **Poisonous seeds and tubers. May be FATAL.**

1▶Bluebell (Wild hyacinth/Harebell) *Endymion non-scriptus (Hyacinthoides non-scripta)* Native to Europe, in woods and shady places, and an ornamental. Bulbous, with narrow mid-green leaves and violet-blue flowers. Height: 25–30 cm (10–12 in). **Poisonous bulbs and seeds.**

2▶Hyacinth *Hyacinthus orientalis* Native to Eastern Europe and western Asia, and widely grown as a garden ornamental and house-plant. Bulbous, with upright, strap-like, mid-green leaves and spires of waxy-flowers in a wide colour range. Height: 20–25 cm (8–10 in). **Poisonous bulbs and fruits.**

3▶Tiger lily *Lilium tigrinum* Native to China, Korea and Japan, an ornamental and a houseplant. Bulbous, with upright stems bearing orange-red flowers. Height: 90–150 cm (35–59 in). **Both the bulbs and fruit are poisonous.**

4▶Wild daffodil (Lent lily) *Narcissus pseudonarcissus* Native to Europe and a very common garden ornamental. Bulbous, with long strap-like leaves and lemon-yellow trumpet-like flowers. Height: 20–30 cm (8–12 in). **Poisonous bulbs and fruits.**

5▶Star of Bethlehem (Nap-at-noon/Summer snowflake) *Ornithogalum umbellatum* Native to Europe and North Africa, naturalized in eastern US and a very common ornamental. Bulbous, with stout stems bearing white flowers. Height: 30 cm (12 in). **The fruits are poisonous.**

6►Solomon's Seal (David's harp) *Polygonatum x hybridum* Extremely common ornamental. Herbaceous perennial, with mid-green stem-clasping leaves and arching stems bearing white flowers. Later, red or bluish-black berries appear. Height: 60–100 cm (24–39 in). **Poisonous berries.**

7►Tulip *Tulipa* genus. Native to Europe and extremely common as an ornamental and houseplant. Bulbous, with flowers in many colours at the tops of stiff upright stems. Height: 25–38 cm (10–15 in). **Poisonous fruits and bulbs.**

8►False hellebore (White hellebore/Itchweed) *Veratrum viride* Native to the US and an ornamental. Rhizomatous-rooted herbaceous perennial with basal and stem-clasping mid-green leaves. It also bears branching sprays of yellow-green flowers. Height: 1.5–2.1 m (5–7 ft). **Poisonous rhizomes and fruits. May be FATAL.**

FLAX FAMILY *Linaceae*
Poisons that act on the brain, causing hallucinations, delirium, thirst and dryness in the mouth.
9►Common flax (Linseed) *Linum usitatissimum* Native to southwest Asia, naturalized in many countries and an ornamental. Annual, with narrow leaves and pale-blue flowers. Height: 50–60 cm (20–24 in). **All parts are poisonous.**

REMEMBER
If you suspect that you or someone else has been poisoned by a plant, seek urgent medical assistance. Do NOT wait to see if the problem 'gets better'. If possible, take along a sample of the plant believed to have caused the poisoning.

Use gloves to place the sample in a plastic bag. Don't restrict the sample to a few leaves. If there are flowers or berries, take those along as well. This will enable the plant to be IDENTIFIED quickly. Make a note of **where** the plant was found — this may help speed up its identification.

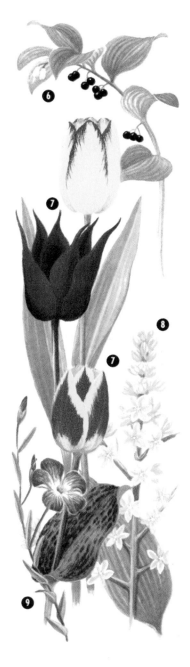

POISONOUS PLANTS

LOGANIA FAMILY *Loganaceae*
Poisons that act on the central nervous system, causing spasms, spasmodic swallowing and death.

1▶Carolina jasmine (Evening trumpet flower) *Gelsemium sempervirens* Native to the US and Central America, and an ornamental. Twining shrub with glossy lance-shaped leaves and fragrant bright-yellow flowers. Height: 1.8–2.4 m (6–8 ft). **All parts are poisonous. May be FATAL.**

2▶Strychnine (Nux-vomica tree) *Strychnos nux-vomica* Native to southern Asia and widely grown commercially. Evergreen tree. Berries produce seeds yielding the drug *nux vomica* (from which the poison strychnine is prepared). Height: 6–12 m (20–39 ft). **Poisonous seeds. May be FATAL.**

MOONSEED FAMILY *Menispermaceae*
Poisons that act on the brain, causing problems with vision, delirium, dilated pupils, thirst and dryness of the mouth. May also cause paralysis.

3▶Moonseed (Yellow parilla) *Menispermum canadense* Native to the eastern US and eastern Asia, grown commercially and an ornamental. Rhizomatous, with woody twining stems and white or yellowish flowers. Height: 1.8–3.6 m (6–12 ft). **Poisonous seeds.** The dried rhizomes are used for medicinal purposes.

MYRTLE FAMILY *Myrtaceae*
Irritant poisons, causing burning pains in the throat and stomach, thirst, nausea, vomiting, fatigue and shock.

4▶Paperbark Tree (Punk tree/Swamp tree) *Melaleuca quinquenervia* Native to eastern Australia, Papua New Guinea and New Caledonia. Large evergreen tree with white peeling bark and white flowers. Height: 4.5–7.5 m (15–25 ft). **All parts are poisonous.**

FOUR O'CLOCK FAMILY *Nyctaginaceae*
Irritant poisons, causing burning pains in the throat and stomach, thirst, nausea, vomiting, fatigue and shock.

5▶Four o'clock plant (Marvel of Peru/Beauty of the night) *Mirabilis jalapa* Native to tropical America and an ornamental. Herbaceous perennial with fragrant flowers in many colours.

Height: 45–60 cm (18–24 in). **Poisonous seeds and roots.**

OLIVE FAMILY *Oleaceae*
Toxic effects as the myrtle family.
6►Winter-flowering jasmine *Jasminum nudiflorum* Native to China and a very common ornamental. Deciduous wall shrub with bright-yellow winter flowers. Height: 1.8–3 m (6–10 ft). **All parts are poisonous.**

7►Privet (Californian privet) *Ligustrum ovalifolium* Native to Japan and extremely common as an ornamental, especially for hedges. Evergreen shrub with glossy green leaves and cream flowers. Height: 1.5–2.1 m (5–7 ft). **All parts are poisonous.**

WOOD SORREL FAMILY *Oxalidaceae*
Toxic effects as the myrtle family.
8►Pink oxalis *Oxalis articulata* Native to eastern South America and naturalized in many countries. Fleshy perennial with bright-pink flowers. Height: 5–10 cm (2–4 in). **Poisonous leaves.**

EVENING PRIMROSE FAMILY *Onagraceae*
Toxic effects as the myrtle family.
9►Enchanter's nightshade *Circaea lutetiana* Native to Europe, western Asia and North Africa. Herbaceous perennial with white flowers. Height: 45–60 cm (18–24 in). **All parts poisonous.**

PAEONY FAMILY *Paeoniaceae*
Toxic effects as the myrtle family.
10►Paeony (Peony/Chinese paeony) *Paeonia lactiflora* Native to Siberia and Mongolia and a very common ornamental. Herbaceous perennial bearing white flowers. Height: 60 cm (24 in). **All parts are poisonous.**

PALM FAMILY *Palmae*
Toxic effects as the myrtle family.
11►Burmese fishtail palm (Clustered fishtail palm) *Caryota mitis* Native to wide area, from Burma to the Malay Peninsula, Java and the Philippine Islands. Palm-like, with fishtail leaves. Height: 3.6–12 m (12–39 ft). **Poisonous fruits and sap.**

POPPY FAMILY *Papaveraceae*
Narcotics that act on the brain, causing giddiness, dimness of sight, contracted pupils, headache, noises in the ears and drowsiness passing into insensibility. May cause DEATH.

1▶Opium poppy *Papaver somniferum* Native to southeastern Europe and western Asia, grown commercially and an ornamental. Annual with deeply-lobed leaves, white, pink, red or purple flowers. Height: 60–75 cm (24–30 in). Unripe fruits produce a milky sap from which opium is derived. **May be FATAL.** Other poppies also poisonous.

POLYGONUM FAMILY *Polygonaceae*
Irritant poisons, causing burning pains in the throat and stomach, thirst, nausea, vomiting, fatigue and shock.

2▶Rhubarb *Rheum rhoponticum* Native to Bulgaria and widely cultivated for its edible stems. Herbaceous perennial with long stems that bear large leaves. Height: 45–90 cm (18–36 in). **Leaves are poisonous, stems may be too if eaten raw.**

POKE FAMILY *Phytolaccaceae*
Irritant poisons, causing burning pains in the throat and stomach, thirst, nausea, vomiting, fatigue and shock.

3▶Pokeweed (Red-ink plant/Pigeon berry) *Phytolacca americana* Native of the eastern US, and widely grown as a garden ornamental. Herbaceous perennial with green leaves. It bears white flowers and purple berries containing crimson juice. Height: up to 2.4 m (8 ft). **All parts are poisonous.**

PRIMULA FAMILY *Primulaceae*
Irritant poisons, causing burning pains in the throat and stomach, thirst, nausea, vomiting, fatigue and shock. Other plants may cause rashes when touched, especially primulas.

4▶Cyclamen (Sowbread) *Cyclamen purpurascens (Cyclamen europaeum)* Native to central and southern Europe and a very common ornamental. Cormous, with silver-mottled kidney-shaped leaves and fragrant carmine flowers. Height: 10 cm (39 in). **All parts are poisonous.**

☠ POISONOUS PLANTS

BUTTERCUP FAMILY *Ranunculaceae*
Poisons causing various effects, depending on species. Some act on the heart, causing numbness, tingling in the mouth, abdominal pains, vertigo and vomiting. Others have irritant poisons causing abdominal pains, vomiting and purging, cramps, drowsiness and slight nervous symptoms. May cause DEATH.

5►Monkshood (Helmet flower/Garden wolfsbane) *Aconitum napellus* Native to Europe and Asia and a very common ornamental. Herbaceous perennial with deeply-cut leaves and hooded deep-blue flowers. Height: 90–100 cm (35–39 in). All parts are poisonous. **May be FATAL.**

6►Baneberry (Herb Christopher/Necklaceweed) *Actaea spicata* Native to Europe and Asia as far as China. Herbaceous perennial with white flowers and purplish-black berries. Height: 30–60 cm (12–24 in). **All parts are poisonous, especially the berries. Can be FATAL.**

7►Wood anemone *Anemone nemorosa* Native to Europe and western Asia and an ornamental. Herbaceous perennial with deeply-lobed leaves and white flowers. Height: 15–20 cm (6–8 in). **All parts are poisonous.**

8►Columbine (Granny's bonnet/European crowfoot) *Aquilegia vulgaris* Native to Europe, North Africa and Asia as far as China. It is naturalized in the US and a very common ornamental. Herbaceous perennial with violet, pink or white flowers. Height: 45–60 cm (18–24 in). **All parts are poisonous.**

9►Marsh marigold (Meadow bright/Kingcup) *Caltha palustris* Native to Europe, Asia and the US, in marshes, fens and ditches. Also an ornamental, especially in boggy areas. Herbaceous perennial with rounded leaves and cup-shaped bright-yellow flowers. Height: 30–38 cm (12–15 in). **All parts are poisonous.**

POISONOUS PLANTS

219

1►Clematis *Clematis* x 'Nellie Moser' **Hybrid** ornamental. Deciduous climber with large mauve-pink flowers. Many other varieties, in a wide colour range. Height: 1.8–2.4 m (6–8 ft). **All parts are poisonous.** The traveller's joy (*C. vitalba*) is also poisonous and can be FATAL.

2►Delphinium *Delphinium* 'Blue Tit' Ornamental. Herbaceous perennial with spires of indigo-blue flowers. Many varieties, in a wide colour range. Height: 150–180 cm (60–71 in). **All parts are poisonous.** Larkspur (*D. ajacis*) is also poisonous.

3►Winter aconite *Eranthis hyemalis* Native to Europe, naturalized in the US and an ornamental. Tuberous-rooted perennial with bright-yellow flowers surrounded by green ruffs. Height: 10 cm (4 in). **All parts are poisonous.**

4►Hellebore (Green hellebore) *Helleborus viridus* Native to Europe and very common as an ornamental. Herbaceous perennial with yellowish-green cup-shaped flowers. Height: 30 cm (12 in). **All parts are poisonous.** Other hellebores are also poisonous.

5►Greater spearwort *Ranunculus lingua* Native to Europe and Siberia. Perennial with stoloniferous roots, toothed and lance-shaped leaves and bright-yellow buttercup-like flowers. Height: 50–90 cm (20–35 in). **All parts are poisonous.** Beware of other plants in this genus.

6►Lesser celandine (Pilewort) *Ranunculus ficaria* Native to Europe and western Asia, and naturalized in the US. Found on shady grassy banks, streamsides, meadows and woods. Herbaceous perennial with bright yellow flowers. Height: 15–25 cm (6–10 in). **Sap is poisonous.** Sometimes used externally for medicinal purposes.

BUCKTHORN FAMILY *Rhamnaceae*
Irritant poisons, causing burning pains in the throat and stomach, thirst, nausea, vomiting, fatigue and shock.

7►Alder buckthorn (Black dogwood) *Frangula alnus* Native to Europe, the Urals, Siberia and North Africa, especially on damp heaths and woods. Deciduous shrub/tree with small green flowers and red berries, which turn black. Height: 4–5 m (13–16 ft). **All parts are poisonous.**

8►Common buckthorn *Rhamnus catharticus* Native to Europe, western Asia and North Africa, found in hedges and on scrub land. Deciduous shrub/tree with small green flowers and black berries. Height: 4–6 m (13–20 ft). **All parts are poisonous.**

BE SAFE
It is impossible not to come into contact with poisonous plants—they are so numerous. Get to know as many species as possible in your area. When arranging a garden, especially if children will play there unsupervised, plan it carefully and try to make it as safe as possible.

When buying new plants for your garden, check to see if they are poisonous. Make a note to position them away from the fronts of borders, where children might touch them.

Don't plant trees with poisonous seeds near garden ponds. Laburnum will poison fish.

DON'T put poisonous plants, especially members of the arum family and plants such as poison sumac (*Toxicodendron vernix*), poison ivy (*T. quercifolium*) and poison oak (*T. radicans*), on bonfires. Inhalation of the smoke can be FATAL.

Don't eat any plant that cannot be positively identified as a food. Some plants may resemble known food plants. For instance, the leaves of fool's parsley (*Aesthusa cynapium*) resemble parsley, and the roots of white bryony (*Bryonia alaba*) look very like turnips.

ALWAYS wash your hands in worm soapy water after handling plants.

ROSE FAMILY *Rosaceae*

This family includes many common fruit trees. The seed kernels contain cyanogenetic glycosides, which break down into prussic acid. Poisons cause numbness, tingling in the mouth, abdominal pains, vertigo, vomiting, purging and paralysis. Can be FATAL if large quantities are consumed.

1▶Common almond *Prunus dulcis (P. amygdalus/P. communis)* Native to western Asia and North Africa, cultivated commercially and an ornamental. Deciduous tree with pink flowers on naked branches, followed by fruits. Height: 5.4–7.5 m (18–25 ft). **Fruits poisonous when unripe.**

2▶Apricot *Prunus armeniaca* Native to China, grown commercially and an ornamental. Deciduous tree with pinkish flowers, followed by fruits. Height: 4.5–5.4 m (15–18 ft). **Poisonous seeds inside the stone.**

3▶Gean (Wild cherry/Mazzard) *Prunus avium* Native to Europe, naturalized in the US and an ornamental. Deciduous tree with white cup-shaped flowers, followed by fruits. Height: 7.5–12 m (25–39 ft). **The seeds are poisonous.**

4▶Portugal laurel *Prunus lusitanica* Native to Spain and Portugal, and a very common ornamental. Evergreen shrub with scented cream flowers and small fruits, which turn black. Height: 4.5–6 m (15–20 ft). **Poisonous seeds.** Cherry laurel (*P. laurocerasus*) is also poisonous.

5▶Peach *Prunus persica* Native to China, grown commercially and an ornamental. Deciduous tree with rose-coloured flowers, followed by fruits. Height: 4.5–7.5 m (15–25 ft). **Poisonous seeds inside the stone.**

6►Rowan (European mountain ash/Quickbeam) *Sorbus aucuparia* Native to Europe and western Asia and an ornamental. Deciduous tree with white flowers and globular orange-red berries. Height: 4.5–7.5 m (15–25 ft). **Poisonous seeds.**

7►Fish-bone cotoneaster *Cotoneaster horizontalis* Native to China and a very common ornamental. Deciduous shrub with fishbone-like stems and pink flowers. Also develops round, red berries. Height: 60–75 cm (24–30 in). **Poisonous seeds.**

8►Hawthorn (Quickthorn/May) *Crataegus monogyna* Native to Europe and a very common ornamental and hedging plant. Deciduous shrub/tree with scented white flowers and small crimson fruits. Height: 6–9 m (20–30 ft). **Poisonous seeds.**

9►Firethorn *Pyracantha coccinea* Native to southern Europe and Anatolia and an ornamental. Evergreen shrub with white flowers and bright red berries. Height: 2.4–3.6 m (8–12 ft). **Poisonous fruits.**

RISKY HOUSEPLANTS?

Many plants that are grown indoors or in conservatories are poisonous, but no one need ever come to harm, if sensible precautions are taken.

Keep young children away from them. Especially at the crawling stage, children put everything they touch into their mouths.

When propagating plants that are known to have poisonous sap, wear gloves. Wash them afterwards. DON'T allow sap to enter cuts.

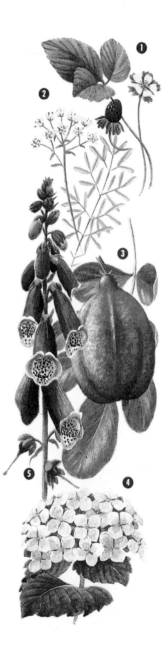

1▶Indian strawberry (Mock strawberry) *Duchesnea indica* Native to India, naturalized in North America and an ornamental, used in hanging baskets and as ground cover. Perennial with runners and bearing yellow flowers and red fruits. Height: 3–4.5 m (10–15 ft). Poisonous fruits.

RUE FAMILY *Rutaceae*
Poisons causing vertigo, vomiting, abdominal pains, confused vision, convulsions, delirium and paralysis and affecting the heart.
2▶Rue (Herb of grace) *Ruta graveolens* Native to southern Europe and very common as a culinary herb and an ornamental. Evergreen shrub with blue-green leaves and sulphur-yellow flowers. Height: 60–90 cm (24–35 in). **All parts are poisonous in large quantities.**

SOAPBERRY FAMILY *Sapindaceae*
Irritant poisons, causing burning pains in the throat and stomach, thirst, nausea, vomiting, fatigue and shock.
3▶Akee *Blighia sapida* Native to West Africa, naturalized in tropical and sub-tropical regions and grown commercially. Evergreen tree with small greenish-white flowers followed by fruits. Height: 9–12 m (30–39 ft). **Poisonous seeds. Fruits especially dangerous when under-ripe and over-ripe, or when arils are dis-coloured.**

SAXIFRAGE FAMILY *Saxifragaceae*
Toxic effects as the soapberry family.
4▶Hydrangea (French hydrangea/Horten-sia) *Hydrangea macrophylla* Native to China and Japan, an ornamental and a houseplant. Deciduous shrub with large flowerheads. Height: 120–180 cm (47–71 in). **All parts are poisonous.**

FIGWORT FAMILY *Scrophulariaceae*
Poisons causing numbness, tingling in the mouth, abdominal pains, vertigo, vomiting, purging, delirium and paralysis and affecting the heart. May cause DEATH.
5▶Foxglove *Digitalis purpurea* Native to western Europe, cultivated for the preparation of heart medicines, and an ornamental. Biennial or perennial with tall spires of bell-like, flowers.

☠ POISONOUS PLANTS

Height: 60–150 cm (24–60 in). **All parts are poisonous. May be FATAL.**

NIGHTSHADE FAMILY *Solanaceae*
Poisons that act on the brain, causing distortion of vision, delirium, dilated pupils, thirst, dryness in the mouth, and paralysis. May cause DEATH.

6►Deadly nightshade (Dwale/Belladonna) *Atropa belladona* Native to Europe, North Africa and western Asia to Iran, in scrubland, woods and rocky places. Perennial with pointed oval leaves, purple or greenish flowers and glossy black berries. Height: 60–150 cm (24–60 in). **All parts are poisonous. For children, half a berry can be FATAL.**

7►Thorn apple (Stramonium/Jimsonweed) *Datura stramonium* Native to temperate regions and the sub-tropical northern hemisphere, on waste and cultivated land. Annual with jagged-toothed, oval leaves and trumpet-shaped white flowers followed by spiny fruits. Height: 90–100 cm (35–39 in). **All parts poisonous, especially fruits and seeds. DEADLY — many child fatalities.**

8►Henbane (Black henbane/Stinking nightshade) *Hyoscyamus niger* Native to temperate Asia and naturalized in the US. Frequently found on sandy wasteground. Annual or biennial with jagged-edged, stem-clasping leaves and bell-shaped, creamy-yellow flowers. Height: 60–75 cm (24–30 in). **All parts are poisonous, especially leaves and seeds. May be FATAL.**

9►Tobacco plant (Jasmine tobacco/Flowering tobacco) *Nicotiana alata* Native to South America and a very common ornamental. Perennial with oblong leaves and white, tubular flowers. Height: 60–90 cm (24–35 in). **All parts are poisonous.**

10►Cape gooseberry (Chinese lantern/ Japanese lantern) *Physalis franchetii (P. alkekengi franchetii)* Native to southeastern USSR and China, and a very common ornamental. Herbaceous perennial with stiff stems bearing white flowers, followed by colourful lanterns that enclose orange fruits. Height: 45–60 cm (18–24 in). **All parts poisonous.**

1►Potato *Solanum tuberosum* Native to the Andes and widely grown as a food plant. Herbaceous perennial, with stems bearing white or bluish flowers, followed by yellowish or green fruits and subsequently by edible tubers. Height: 30–60 cm (12–24 in). **Leaves, green potatoes and sprouting shoots are poisonous and can be FATAL Potatoes should never be eaten raw.**

2►Woody nightshade (Bittersweet) *Solanum dulcamara* Native to Europe, Asia and North Africa, in hedges, woods and on wasteland. A scrambling, sometimes prostrate, perennial woody climber with bright-purple flowers. It also bears berries, first green, then yellow, later red. Height: 100–180 cm (39–71 in). **All parts are poisonous, especially the berries. Can be FATAL.**

3►Black nightshade (Poisonberry) *Solanum nigrum* Native to many regions and, effectively, worldwide. A weed of cultivated and wasteland. Sprawling and erect annual with white flowers and poisonous berries, first green then black. Height: 45–60 cm (18–24 in). **Poisonous berries. May be FATAL.**

4►Red·pepper (Ornamental chilli) *Capsicum annum* Native to the tropics, and a very common ornamental and houseplant. Shrubby short-lived perennial with white-to-greenish-white flowers, followed by red, yellow or green fruits. Height: 45–60 cm (18–24 in). **All parts are poisonous.**

5►Trumpet flower (Goldcup/Cup-of-gold) *Solandro nitida* Native to Mexico, a common tropical ornamental, also common in greenhouses and as a houseplant. Climbing shrub with yellow and purple flowers followed by fleshy berries. Height: 1.8–4.5 m (6–15 ft). **All parts are poisonous.**

6►Willow-leaved jessamine *Cestrum parqui* Native to Chile, an ornamental and a houseplant. Bushy, deciduous shrub with fragrant, greenish-yellow flowers. Height: 90–120 cm (35–47 in). **All parts are poisonous.**

7►Tomato (Love apple) *Lycopersicum esculentum* Native to Andean South America and widely grown as a food plant. Perennial or annual grown for its succulent fruits. Height: 1.2–1.8 m (4–6 ft). **Poisonous stems and leaves. Raw fruits are poisonous when green.** Harmless when cooked, or ripened to orange or red.

YEW FAMILY *Taxaceae*

Irritant poisons that act on the nervous system, causing abdominal pains, vomiting, purging, dilated pupils, headaches, spasms, convulsions and coma. May cause DEATH.

8►Yew (English yew) *Taxus baccata* Native to Europe and a very common ornamental. Evergreen conifer with dense dark-green foliage and fleshy, red, cup-shaped fruits. Height: up to 18 m (60 ft). **All parts are poisonous. Usually FATAL Survival after poisoning is rare.**

DAPHNE FAMILY *Thymelaeaceae*

Irritant poisons, causing burning pains in the throat and stomach, thirst, nausea, vomiting, fatigue and shock.

9►Mezereon (Mezereum/February mezereum) *Daphne mezereum* Native to Europe, Anatolia and Siberia, and a very common ornamental. Deciduous shrub with clusters of pale purple-pink to violet-red flowers on bare stems, followed by scarlet berries. Height: 90–150 cm (35–59 in). **All parts are poisonous, especially the bark and berries. Can be FATAL.** Wood laurel (*D. laureola*) is also poisonous.

POISONOUS PLANTS

227

VERBENA FAMILY *Verbenaceae*
Irritant poisons, causing burning pains in the throat and stomach, thirst, nausea, vomiting, fatigue and shock.

1▶Mangrove *Avicennia* species Native to the coastal regions of the Old and New World tropics, and Australia. Tropical tree. Height: 3–7.5 m (10–25 ft). **Poisonous sap.**

2▶Lantana (Yellow sage) *Lantana camara* Native to the West Indies, naturalized throughout the tropics, a very common tropical ornamental and a houseplant. Evergreen shrub with domed flower heads in a wide colour range, followed by black fruits. Height: 45–100 cm (18–39 in). **Dangerously poisonous fruits.**

UMBELLATE FAMILY *Umbelliferae*
Poisons that cause vertigo, vomiting, abdominal pains, confused vision, delirium and paralysis and affect the heart. May cause DEATH.

3▶Fool's parsley (Lesser hemlock) *Aethusa cynapium* Native to the British Isles and a common weed. Annual with parsley-like leaves and white flowers in umbrella-like heads, followed by egg-shaped fruits. Height: 60–120 cm (24–47 in). **All parts are poisonous. May be FATAL.**

4▶Cowbane *Cicuta virosa* Native to northern and central Europe, and Asia eastwards to Japan, in ditches, marshes and shallow water. Perennial with umbrella-like heads of white flowers. Height: 30–120 cm (12–47 in). **All parts are poisonous. May be FATAL.**

5▶Hemlock (Poison hemlock/California fern) *Conium maculatum* Native to Europe, Asia and North Africa. It has been introduced to North America, the West Indies, South America and New Zealand. A weed of damp places and woods. Biennial with umbrella-like heads of white flowers. Height: 1.5–2.1 m (5–7 ft). **All parts are poisonous. May be FATAL.** It is dangerous for children to use the hollow stems as peashooters.

6►Water dropwort *Oenanthe crocata* Native to the British Isles, France, Spain, Portugal, Italy, Sardinia and Morocco. Found in moist places. Perennial with parsley-scented and parsley-like leaves and umbrella-like heads of white flowers. Height: 60–150 cm (24–59 in). **All parts are poisonous. May be FATAL.**

7►Giant hogweed *Heracleum mantegazzianum* Native to the southeastern USSR and introduced into many countries. Found on wasteland and in moist areas, and as an ornamental. Giant biennial with parsley-like leaves and umbrella-like heads of white flowers. Height: 3–3.6 m (10–12 ft). **All parts are poisonous, even by skin contact.**

NETTLE FAMILY *Urticaceae*
Poisons causing numbness, tingling in the mouth, abdominal pains, vertigo, vomiting, purging, delirium, paralysis and affecting the heart.
8►Wood Nettle (Devil nettle/Fever nettle) *Laportea canadensis* Native to eastern US. A tropical shrub with large oval leaves. Height: 120–180 cm (47–71 in). **The whole plant is covered with small, irritant hairs. Leaves are poisonous.**

ZAMIA FAMILY *Zamiaceae*
Irritant poisons, causing burning pains in the throat and stomach, thirst, nausea, vomiting, fatigue and shock.
9►Coontie (Comptie/Seminole bread) *Zamia floridana* Native to Florida and the West Indies, a very common tropical ornamental and a houseplant. Palm-like plant with pinnae (leaves) to 15 cm (6 in) long. It has a short, tuber-like trunk. Height: 60–90 cm (24–35 in). **Poisonous tuber.**

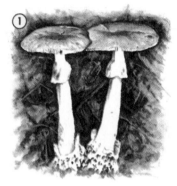

Most fungi are associated with trees or decaying wood. The mushrooms or toadstool-like form we see is only the fruiting body—a large part of the fungus consisting of fine threads, the mycelium, spreading underground.

Here are some of the more poisonous types in Europe, the US and Australia—they are NOT the only ones. Fungi from some areas are not well documented. The toxicity of even Australian species, for example, has not been extensively investigated.

Eat ONLY fungi that you can positively identify as SAFE. Some poisonous kinds look very like edible ones—DON'T GUESS.

ALWAYS avoid small or large brown mushrooms, especially if the gills below the cap are purple-brown or pink. Discourage children from eating or touching any fungi found growing in the city.

1►Death cap *Amanita phalloides* Usually found under birch, beech, conifers or dogwood in autumn. The stem is paler than the cap, with a bag-like volva at base (only visible when fungus is dug up). Cap diameter: 6.5–15 cm (2 1/2–6 in). **Symptoms: stomach pain, vomiting, diarrhoea (after 8–14 hours). After 2 days: deep coma and jaundice. DEADLY.**

2►Fly agaric *Amanita muscaria* Typically grows in clusters under pine or birch in autumn. Older specimens are very pale with loose white flecks. Cap diameter: 5–25 cm (2–10 in). **VERY POISONOUS. Symptoms: delirium, vomiting and coma.**

3►Destroying angel *Amanita virosa* Found in mixed woods and grass, with bag-like volva and shaggy stem. Smells sweet and sickly. Cap diameter: 5–12 cm (2–5 in). **DEADLY. Symptoms as death cap.**

4►Panther cap *Amanita pontherina* is common in conifer and deciduous woods, especially birch. The brown cap is spotted with white warty scales. Smells like raw potatoes. Cap diameter: 2.5–15 cm (1–6 in). **Very poisonous, can be FATAL. Symptoms: delirium, muscle spasms, coma.**

5►Deadly galerina *Galerina autumnalis* Found on decaying wood in log piles, spring–autumn. The small dark cap fades to buff. Cap diameter: 2.5–6.5 cm (1–2 1/2 in). **DEADLY. Symptoms: vomiting, diarrhoea (may not happen until after 10 hours). Kidney/liver failure may result.**

6►Deadly lawn galerina *Galerina venata* Found on lawns, often growing from buried decomposing wood. Its moist reddish-brown cap fades to buff. Tastes bitter and smells mealy. Cap diameter: 1–3.5 cm (3/8–1 3/8 in). **DEADLY.**

7►Deadly lepiota *Lepiota josserandii* Found near shrubbery and oaks. The cap has concentric bands of flattened copper-coloured scales and a musty odour. Cap diameter: 2.5–5 cm (1–2 in). **DEADLY.**

8►Deadly cort *Cortinarius gentilis* Found under conifers. Cap and stalk are a deep orange-brown. Smells of radishes. Cap diameter: 2.5–5 cm (1–2 in). **Symptoms don't appear for 2–3 weeks after ingestion. DEADLY. Risk of kidney failure.**

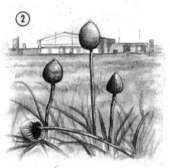

Never underestimate the deadly toxicity of poisonous fungi — you don't necessarily need to eat a large quantity to be poisoned. What's more, not all poisonous fungi have an unpleasant taste so it's easy to be fooled into believing you're eating an exible variety. You may not even realize you've been poisoned until 8 to 14 hours later with some deadly fungi, when the first symptoms start to develop (perhaps abdominal pain, vomiting and diarrhoea). For some poisonous fungi there is no known antidote.

1►Cortinarius speciosissimus
Found, singly or in small groups, near conifers, often among mosses. Red-brown cap with yellow margins, velvety texture. The closely related *C. orellanus* is slightly lighter coloured and grows in broad-leaved woods. Both smell radishy. Cap diameter: 2.5–8 cm (1–3 in). **DEADLY.**

2►Liberty cap *Psilocybe semilanceata* Found scattered by paths in fields, heaths and lawns in summer/autumn. The gills begin cream but darken almost to black with age. Cap is sticky when damp. Cap diameter: 0.5–1.5 cm (1/4– 1/2 in). **Contains LSD-like compound. May cause serious physical damage.**

3►Sweating mushroom *Clitocybe dealbata va. sudorifica* Commonly found scattered in deciduous woods. Crimson to pale pink in colour. Cap diamater: 1–4 cm (1/2–1 1/2 in). **Contains toxins which cause severe nausea, sweating and circulation disorders. Serious but seldom fatal.**

4►Red-staining inocybe *Inocybe patouillardii* Found in grass near beech and other woods. Cap is white to yellow, discolouring where bruised to pink. Gills white to brown. Smells fruity. Cap diameter: 3–7 cm (1 1/4–2 3/4 in). **DEADLY.**

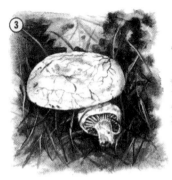

Symptoms develop rapidly: vomiting, diarrhoea, collapse.

5►Poison pie (Fairy-cake mushroom) *Hebeloma crustuliniforme* Found in autumn, growing in groups or rings in rich soil in deciduous woods and in gardens. Cap is sticky when moist, shiny when dry. Smells of radishes. Cap diameter: 3–9 cm (1 1/4–3 1/2 in). **Poisonous. Symptoms: vomiting, diarrhoea, respiratory difficulties.**

6►Emetic russula (Sickener) *Russeula emetica* Found in autumn on moist ground in woodland and parks. The cap is scarlet with shiny, moist surface. Flesh smells of coconut. Tastes very peppery. Cap diameter: 2.5–7.5 cm (1–3 in). **Not poisonous if cooked, but powerful emetic effect if eaten raw.**

7►Brown roll rim *Paxilus involutus* Common in woods, especially beech and birch. This species is commonly eaten in Western Europe. In wet weather, cap surface glutinous. Cap diameter: 4–20 cm (1½;–8 in). Very common. **Poison accumulates in the body and can eventually cause liver failure. AVOID.**

8►Alcohol inky (Common inkcap) *Coprinus atramentarius* Grows on trees, stumps and buried wood, usually in clusters. Cap diameter: 5–7.5 cm (2–3 in). Edible but, as with some other *Coprinus*, contains substance like Antabuse (drug used to treat alcoholics). Problems occur only if alcohol is consumed within 1–2 days of eating. Flushing of face and neck, tingling fingers, headache and nausea, although symptoms will pass.

5

Poisonous fumes and smoke claim many more lives than fire itself, and there are potentially lethal substances in every home and workplace. Knowing the risks, devising escape drills and understanding fire could save your life.

Fire!

FIRE FACTS Know the enemy • The fire triangle • How fire spreads • Fire in the home • Fire at work • Public buildings • Flashover • How fire kills • Explosion

EQUIPMENT Smoke detectors • Fitting • Fire extinguishers • Fire blankets • Fire escapes • Fire ladders • Fireproofing • Combustion-modified foam

DRILLS At home • Upstairs • Night patrols

FIRE! At home • Seconds count • Emergency!

ESCAPE! Moving through a burning building • Testing all doors • Staying low • Fireman's lift • Trapped • Jumping • Getting help

235

FIRE FACTS

Fire is an ever-present threat to life and property in the urban environment. In Britain alone there are about 50,000 accidental domestic fires a year—at least 70 per cent of fire deaths occur in the home. There are almost monthly reminders from around the world of the dangers of fire in places like underground stations, sports stadia and hotels.

At work the risks are just as great, no matter whether you work in known hazardous conditions or in the most up-to-date office block. However, one advantage of living in the city is the short response time of fire brigades—although you cannot rely on that as your only safeguard.

To reduce the risks, to protect yourself and possibly save lives you MUST understand more about fire—how it can start, how it behaves and how it kills.

KNOW THE ENEMY!

Fire is a chemical reaction—but it's easy to forget what can make that reaction happen. If you were starting a fire in a grate or on a barbecue, you would need kindling or firelighters on to which you would put coal, wood or charcoal. This is your **FUEL**. A lighted match is the first source of energy or **HEAT**. The fuel might not burn easily, so you blow on the firelighter or fan it. By doing so, you are providing the third vital ingredient, **OXYGEN**. If there is a good supply of oxygen the firelighter will burn freely and generate more heat and the fuel will start to burn, result—**FIRE**.

THE FIRE TRIANGLE

The three ingredients can be remembered as the fire TRIANGLE. The scientific term for the chemical reaction is oxidation or combustion. The fire will continue to burn, as long as the three ingredients—the points of the fire triangle—are present. If one of the points of the triangle is removed, it will collapse—the fire will go out.

REMEMBER

The fire triangle idea should be applied to any location or situation you find yourself in. Fire prevention means identifying, in advance, where the three points—HEAT/FUEL/OXYGEN—are going to come together with possibly deadly results.

Breaking the triangle

Whether you are snuffing a candle or putting out a house fire, your intention is the same—deprive the fire of at least one of the three points of the fire triangle. This can be done by removing:

○ **FUEL** Fire breaks (wide channels with no trees) in forests do this. A forest fire will spread only as far as the fire break—it can go no further because there is nothing to burn. In the urban environment or in your home 'fuel' is everywhere!

○ **HEAT** Water is the classic way to cool a fire—with a hose or a bucket.

○ **OXYGEN** Literally suffocating the fire. This is almost impossible outdoors, but it is the key to stopping or slowing down fires inside. On a small scale, this is how a fire blanket works.

> ☠ **WARNING**
>
> The speed at which a fire can spread is something that should never be underestimated! Using evidence from fires in homes and public buildings, fire experts have established that even a large building or structure can be totally engulfed by flames WITHIN TWO MINUTES.
>
> Man-made structures like houses are built from and contain large quantities of fuel. Most of our possessions and furniture are highly combustible. They are made either of natural products such as wood, or man-made materials—many of which contain highly-poisonous and highly-flammable chemicals. Most parts of a building are well ventilated. Large windows, doors, corridors and ventilation systems provide a ready supply of oxygen.
>
> If the third ingredient (heat) is provided under these circumstances, a fire of extreme intensity can be created in a frighteningly short space of time. You may have only two minutes to make sure all the occupants of your home are safe.

How fire spreads

When a building fire starts it spreads rapidly, heating up its surroundings by:

○ **DIRECT CONTACT** Flames passing from one object to another.

○ **CONVECTION** Heat rising, carrying gases and smoke.

○ **RADIATION** Heat rays causing flammable objects in the vicinity to burst into flames.

Convection is the most life-threatening, as it is the fastest way the smoke and flames travel—a fire spreads rapidly UP a building. This is why stairwells are so dangerous.

Almost 90 per cent of fires in the home can be traced to one of ten causes. Here they are, in order of importance, starting with the most common:

COOKERS The largest single cause of domestic fires. Fat fires, especially those caused by chip pans, frequently get out of control.

PORTABLE HEATERS Knocked over accidentally or left too close to furniture or flammable materials, the main culprits are electric, cylinder gas and paraffin (kerosene) heaters.

CIGARETTES Each cigarette is a potential fire-starting heat source. Accidentally-dropped cigarettes fall on to upholstered furniture or beds, smoulder for a while and then flare up.

MATCHES Left either too close to fires or within reach of young children.

ELECTRICAL WIRING Old or badly-installed wiring, incorrect fuses and over-loaded sockets generate heat that starts fires.

ELECTRIC BLANKETS Old blankets develop faults, often because they have been folded up too tightly. The element inside breaks or becomes exposed—when plugged in, it short circuits and catches fire.

BLOW TORCHES/LAMPS Used for 'do it yourself' and by professionals, they start fires when they are used carelessly, are left unattended or are not maintained properly.

TELEVISIONS/COMPUTER SCREENS These operate using very high voltages—far higher than other domestic appliances—and electrical faults cause fires and explosions.

CHIMNEYS Neglected for too long, soot builds up and catches fire—sometimes undetected until serious structural damage has been done.

CANDLES A simple unguarded flame—easily knocked over and capable of igniting furniture and fittings in seconds. Forgotten or neglected candles may produce disastrous results.

FIRE AT WORK

In industry, flammable materials such as chemicals, fuels and gases MUST be stored according to specific fire safety regulations. They should ALWAYS be stored away from sources of heat and heat-generating processes like machining, smelting and firing. They should also be kept apart from each other—avoiding the formation of deadly 'cocktails' in the event of fire.

Full training should be given to minimize accidents, although human error is only one factor. A faulty piece of machinery or badly-maintained plant could easily be a cause of fire.

In an office there may not seem to be as many obvious fire risks as there are in a factory. Nevertheless, offices contain all the ingredients that can lead to serious fires. Furnishings, equipment and stored paper are found in abundance. Old buildings could have faulty wiring or poorly-maintained electrical fittings. Electrical equipment and wiring could, in the event of a malfunction, be the heat source that starts a fire.

MODERN BUILDINGS

In large modern offices, particularly those which rely heavily on computers, there is a different set of fire hazards — providing architects and office designers with major problems. These offices consume a large amount of electricity and are served by complicated wiring systems. At the same time the environmental temperature must be controlled by air conditioning for higher efficiency and comfort.

Accommodating both wiring and air conditioning, but keeping them out of sight, means that many offices are designed with false ceilings and cavity floors. In most cases these meet fire safety requirements. In the event of a fault developing (modern dockside office developments in London, for example, are plagued by mice and rats which can chew through wiring) a fire could start undetected, in close proximity to good ventilation from air conditioning. Even a reinforced concrete structure can burn!

PUBLIC BUILDINGS

Fire experts have identified two categories of public building which give rise to the greatest concern: sports grounds and large shopping premises.

The safety of sports grounds depends largely on their design and construction. Football stadium fires in the past have served as tragic lessons in how not to build such structures. Wooden seats, stands or roofs are obviously dangerous. The build-up of discarded rubbish which needs only a carelessly-thrown cigarette to spark a fire greatly adds to the danger. The difficulty of evacuating large crowds in a very short space of time is often the main reason for high death tolls.

In a shop, fire could be started by any of the causes mentioned already but two factors make the outbreak of fire even more hazardous. Shops may contain large quantities of combustible materials, and they often have complex layouts (and the casual shopper is unlikely to be aware of fire exits). These factors, particularly if the shop is crowded, will make the staff's job of evacuating panicking shoppers extremely difficult.

The open plan design of many shops means that fire and smoke can spread easily. In an environment designed to encourage the free flow of shoppers, it is difficult to restrict the free flow of smoke. Large shops with an open well rising through one or more floors are, in effect, large smoke-generating chimneys to which the public have access!

REMEMBER

In public buildings, the causes of fire are essentially the same as they would be in your own home. The difference is that the responsibility for preventing fire is out of your control. Nevertheless, be aware at all times of where fire can start and know what action to take.

FLASHOVER

As fire takes hold two things happen. Firstly, the air temperature in a room increases dramatically, discolouring furniture and causing scorching which can produce smoke and poisonous gases. Secondly, fresh supplies of air are sucked into the room as hot air rises and escapes. The air temperature increases until it becomes superheated and flashover occurs. Parts of the room not in contact with the flames are set alight by the heat of the air itself—engulfing a room in flames in a matter of seconds! Once flashover has occurred the air remaining in the room will be hot enough to cook the lungs of anyone breathing it!

HOW FIRE KILLS

 Fire does not just 'burn people to death'. Understanding how fire kills could save lives. In most cases death occurs BEFORE flames have reached the victim—caused by SMOKE and TOXIC GASES. These lead to:

ASPHYXIATION Carbon monoxide and the large volume of ash and carbon particles present in smoke prevent oxygen from reaching the lungs and the bloodstream. A person will lose consciousness and suffocate very quickly (see SAFETY FIRST: **Carbon monoxide**).

POISONING When some materials burn, particularly man-made ones, not only do they produce carbon monoxide but also toxic gases. These are often highly poisonous and will kill much faster than the effects of smoke.

The stark truth is that most of the contents of our homes contain or are made from products that produce toxic fumes when burnt. Old foam-filled furniture can be a killer! The most common and potentially-lethal types of man-made or synthetic materials and the toxic fumes that they give off are:

Polyurethane foam (often used in furniture and mattresses)
► Cyanides including hydrogen cyanide
PVC (used to cover electrical wires and in plastic goods)
► Hydrogen chloride

EXPLOSION

Like fire, to which they are linked, we should have a fuller understanding of explosions in order to prevent them. Fire dramatically increases the risk of explosions — gas, cylinder gas, petrol, paraffin (kerosene) and solvents all have the potential to explode.

Flammable substances may combust rapidly, with a blast-like force. There is a risk of 'explosion' with any such substances stored in sealed containers — for example, a petrol canister, aerosol can or gas cylinder. If the contents are ignited through a rupture or defective valve or if the container is subjected to heat, the pressure built up in the container will cause an explosion. The initial force of the explosion releases the contents at high speed — in a shower or cloud. This burns as it mixes with oxygen in the air. All this happens in a fraction of a second.

Explosions can also happen if an explosive atmosphere is created. Gas, heavy fumes from solvents and liquid fuels and certain types of dust are all flammable and can build up in a confined atmosphere until a critical level is reached. A strong smell of gas, for example, is a warning of this. There are very few circumstances where this will by itself result in explosion unless the substance is highly volatile. But if the mixture of air and flammable gas or particles is ignited — even by a tiny spark — an explosion will occur.

Domestic gas causes many explosions. The source of the ignition is usually electrical. Even turning on or off a light switch, ringing a doorbell or using the telephone may provide the spark! Other causes include lighted cigarettes and matches — even static electricity, which you can produce by combing dry hair! In addition to the threat from gas leaks, many industrial and DIY processes involve substances that can create an explosive atmosphere.

FLASHPOINT
The flashpoint is the lowest temperature at which a substance can give off a mixture of vapour and air, which may be ignited by a spark or static electricity. Flammable substances MUST be kept and used at a lower temperature than their flashpoint to reduce the likelihood of ignition. The main categories are:

■ **HIGHLY FLAMMABLE** Substances with a flashpoint of 32° C (90° F) or below
■ **FLAMMABLE** Substances with a flashpoint above 32 °C (90 °F)
Some petroleum mixtures have a flashpoint below 23 °C (73° F), making them extremely flammable.

EQUIPMENT & DRILLS

With the threat of fire all around you must know how it behaves in order to be able to prepare the right sort of defence. This can be done in two ways: firstly by reducing the causes of fire, and secondly by planning what to do if a fire breaks out. The fight against fire has led to the development of a wide range of equipment that either detects the presence of fire or acts as protection against it. It ranges from the latest in high-tech sophistication to a simple bucket of water. But in order for any equipment to be truly effective, at home or anywhere else, you MUST understand how it works.

Be prepared

Your own level of preparation for fire is just as important as equipment—how will you react? What will you do? Asking the right questions is essential and, when put into the form of a fire drill, will help to eliminate panic. In a fire situation, if you know what to do, you can protect yourself and others.

SMOKE DETECTORS

The most effective early warning device for fire is a smoke detector/alarm. Smoke detectors are NOT a substitute for taking every possible precaution against fire, but they do alert the occupants in the early stages of fire while conditions may still allow for a safe escape.

There are two basic types of smoke detector, both of which are reliable. Most are the ionization type: these contain a tiny radioactive source which ionizes air inside the detector, producing electrically-charged particles (ions) and allowing a small current to flow from a battery. When smoke enters the detector, it impedes the current flow and triggers the alarm. **Ionization detectors are better at detecting hot blazing fires.**

The other type of detector uses a photoelectric device, which triggers an alarm when a light beam is interrupted by smoke. **Photoelectric detectors tend to be more sensitive to smoke from smouldering fires.**

WARNING

Even the minute amount of radioactive material in an ionization detector may be hazardous. Do NOT break open the sealed parts of the detector and always follow manufacturers' recommendations for disposal.

SMOKE ALARM FEATURES

- All smoke detectors are electrical, usually battery powered, in most cases using a small nine-volt battery which should last about a year
- They should have a bleep warning indicating that the battery is running down and a button to test the detector
- Many models have a small red light, which flashes from time to time to show that all is well
- Some models have a built-in emergency light (powered by another battery), which may help you to see where you're going if the electricity supply fails

REMEMBER

False alarms can sometimes be caused by DIY jobs or cooking. Some models give you the chance to avoid this happening with an override button, which disables or desensitizes the detector for a short time.

WARNING

More and more smoke detectors are coming on to the market. ALWAYS check for the features listed above and look for a label that shows the detector meets national safety standards. Don't be hurried into buying a smoke detector by sales people—a useless detector will give you a false sense of security—take your time and shop around.

REMEMBER

Choose a smoke alarm that is small and light to take with you when you travel. Try to sleep with the air conditioning off—it helps to prevent the rapid spread of smoke in the event of a hotel fire.

Fitting

A minimum of TWO smoke detectors are recommended for a small house. The centre of the ceiling of the downstairs hallway is a good place to fit a detector—if a fire breaks out at night in the kitchen or living room, the smoke will be detected before the actual fire reaches the upstairs bedrooms. A detector on an upper landing ceiling will alert people downstairs if a fire breaks out in a bedroom. The best precaution is to fit a detector in EVERY room, particularly if there are young children, elderly or disabled people who will need extra help to be evacuated in the event of a fire.

It is VITAL that any detector can be heard, even when the occupants of the house are asleep. Some models can be interconnected, so if one detector is triggered the others in the house will also sound.

○ **DON'T** fit detectors by walls or in corners where the free flow of smoke will be hampered

○ **DON'T** fit detectors in poorly-ventilated kitchens, where fumes and steam may trigger false alarms

○ **DON'T** position detectors in hard-to-reach places, this may make testing difficult

FIRE EXTINGUISHERS

Many fire experts are sceptical about fire extinguishers, because they may encourage untrained people to 'have a go' when they should be evacuating a burning building. You should only consider trying to put out a fire when it is small and containable (see EMERGENCY! panels). It is safer to raise the alarm and evacuate. There is an added risk that you may use the wrong type of extinguisher on the fire.

TYPES OF EXTINGUISHER

In Britain and elsewhere fire extinguishers are labelled with relevant information concerning the type of fire they are designed to be used against, their capacity and their contents. There are three main categories of fire:

A Burning materials like furniture, cloth and wood

B Burning liquids like oil

C Burning gases

A number, in front of the letter code which indicates the type of fire, shows the capacity of the extinguisher—what size of fire it can cope with. Finally the label will tell you what the contents of the extinguisher actually are (the colour of the tank should also indicate this):

■ WATER (red) Cools fire. Heavy, difficult to handle. Suitable only for A fires. Must NOT be used on liquid or electrical fires.

■ MULTI-PURPOSE FOAM (cream) Cools fire. Suitable for class A and B fires. Must NOT be used on electrical fires.

■ DRY POWDER (blue) Smothers fire. Poor for cooling. Suitable for class A and B. Safe for use on live electrical appliances.

■ CARBON DIOXIDE (CO2) (black) Good for burning liquids and electrical fires. Poor for cooling.

■ HALON (BCF) (green) Good for burning liquids, small fires, solid fuels, electrical fires. The agent BCF can cause nervous disorders if it exceeds five per cent concentration in confined spaces.

FIRE BLANKETS

Made from non-flammable cloth (usually glass fibre), fire blankets are placed over a burning object or small fire, starving it of oxygen and smothering the flames. Domestic fire blankets are usually about 1 m (over 3 ft) square which is adequate for dealing with fires in the kitchen like fat fires. They MUST be easily accessible in case of emergency. Choose one that has a wall-mounted storage tube and fix it within easy reach of your cooker, but not so that you have to reach across the hob!

☠ WARNING

Older fire blankets were made of asbestos. It is very likely that, by now, they are unsafe to use. Asbestos in this form is particularly dangerous and safe disposal should be arranged (see SAFETY FIRST: Asbestos).

FIRE ESCAPES

If you live on the first floor or higher in a multi-occupied building, your home should conform to national fire safety laws, depending on how old the building is or when a conversion was done. Doors to communal staircases MUST be fire-resisting and self-closing. A flat should be designed so that you do not need to go from the bedroom through the living room or kitchen to reach the exit.

In purpose-built office blocks, workplaces, schools, hotels and other public buildings, strict laws apply concerning fire exits and escapes. If a building has designated fire escapes then they should be safe, regularly maintained, kept clear and should not pose a security risk. It's up to YOU to know where fire escapes are and how to use them (see **Drills**).

Fire ladders
If your home does not have adequate fire escape routes, you can position a rope, rope ladder or fold-away escape ladder by a window or balcony that will allow safe escape. Do NOT leave ladders anywhere where they might be a security risk (see DIY/ CRAFT HAZARDS and SECURITY).

FIREPROOFING

Much of the risk of fire comes from flammable materials that are a part of our daily lives—in many cases there are steps we can take to reduce these risks. Some items can be treated to make them fire-resistant or flameproof, and recently furniture manufacturers have developed safer, fire-resistant foams. When buying furniture always check to see if it is fire-resistant (see **Combustion-modified foam**).

Architects and interior designers are also more aware of fire risks in public buildings and are required to incorporate safer materials and fire-resistant features. At home, however, it is up to YOU to keep to a minimum the amount of dangerous flammable materials.

Fire-retardant treatment
There are various spray treatments designed for upholstered furniture and other flammable household fittings, such as curtains and carpets. Some must be applied by specialists who visit your home—others are available as a DIY spray. All spray-on treatments are water based, which has drawbacks: they will be partly removed by washing, sponging or even when liquids are spilt on them.

DIY sprays can be difficult to use effectively. It's not always easy to judge whether your furniture is suitable for treatment and it is difficult to estimate the correct rate of application. It is advisable to get treatment done professionally, with a respray at regular intervals.

Nightwear
Most fires in the home happen at night. One precaution worth considering, particularly for children and the elderly, is to wear nightclothes with low flammability (some synthetics—including

nylon—will melt onto the skin!). These are either made from low-flammability materials or treated with fire-retardant chemicals. Take notice of any special washing instructions.

COMBUSTION-MODIFIED (CM) FOAM

 In some countries, including Britain, recent legislation means that furniture manufacturers must use only fire-resistant or flameproof foam in their products. CM foam is the most commonly used 'safer' foam and unlike normal PVC foam, which will flame and produce smoke and toxic fumes, it melts away or chars—preventing fire from spreading rapidly inside furniture.

New furniture should also be covered in fire-resistant fabrics. Depending on the age of furniture and legislation, both these features should be clearly labelled. If you are buying new furniture ask about fire-resistant features and inspect products for fire safety information.

REMEMBER

If you have had furniture for some years it is unlikely to contain CM foam or meet current safety standards. Furniture made within the last 30 years is likely to be dangerous—anything older will probably be made of traditional materials, such as horse hair, which may produce less toxic fumes but may still burn at a surprising rate.

DRILLS

No smoke- or fire-detecting device/alarm is going to help you unless you have planned and rehearsed what you would do in the event of fire—BEFORE A FIRE HAPPENS. Preplanning is essential and it's up to YOU!

Think!

No matter how clear and well-rehearsed fire drills and procedures are—you are most likely to be involved in them in workplaces and schools—they are only helpful if you think yourself into a fire situation. Imagine your surroundings under the sort of conditions a fire can create in seconds: darkness, thick choking smoke and fumes, intense heat and panic. Ask yourself questions, like: How would I get to the fire assembly point from here? What are the obstacles in my way? What if I am in an unusual part of the building?

These considerations are VITAL, but very often you will only have the vaguest knowledge of escape routes from a public building. In a hotel, department store, theatre or sports stadium, for instance, the last thing you may be thinking of is the threat of fire. But you must go through a mental fire drill WHEREVER YOU ARE. Make yourself aware of fire instructions. Look for fire exits. Take note of where staff or stewards are located.

AT HOME

YOU are responsible for thinking through fire risks and escape drills in your home. If children, elderly or disabled people live with you, then YOU are responsible for them too. YOU must plan and rehearse a fire drill with all the members of your household. Could you completely evacuate your home in the middle of the night—in the two minutes it could take for a fire to get out of control?

Walk around your home

If you live in a house with more than one floor, start on the ground floor. This is where your principal escape routes will be—the front and back doors and windows. Make a note of any ground floor windows that have bars on the outside—these will be useless for escape unless they have a release mechanism. Fires are most likely to start in the kitchen or living room—if you have to go through these rooms to get to an exit, you will need to consider which windows you can use instead. Make sure that everyone in the house—including children—is able to unlock doors and windows from the inside and knows where keys are kept.

> **REMEMBER**
>
> Laminated or wired glass may prove difficult to break through (see SAFETY FIRST: Glass). Double or treble glazing that is not designed to open fully may be hard to break. Position a hammer or special breaking tool by windows which must be used as escape routes so that they CAN be used as escape routes if fire breaks out.

Go upstairs

Escaping from upstairs bedrooms will probably have to be done at night. It will be dark. How easy are the routes? Keep all passages and hallways CLEAR and consider fitting handrails to guide and support young and elderly people—particularly where there is a steep flight of steps. If it becomes impossible to escape through the ground floor, look at all possible upstairs escape routes. Do any windows open onto flat roofs, for example? If it is not safe to jump, consider installing some kind of fire escape (see **Fire ladders**).

REMEMBER

Practise! Take your family through the fire drill regularly. DON'T just read it, DO IT. In particular, practise difficult escape routes. If these are hazardous, do this as safely as possible—some may be too risky to try except in an emergency. Someone should make sure escape routes are possible, though! If someone comes to stay, remember that they will not know these procedures. Young children could draw or paint their own copies of the fire drill—it will help them to memorize the instructions in it. If you're staying in a hotel, make sure to read the fire safety information and do a practise walk to the fire exit to familiarize yourself.

Night patrol

Devise a list of things that should be checked before you go to bed at night and get into the habit of going on patrol. Nightly checks should include the following:

- Make sure you know who is in the house and where they are.
- Switch off and unplug all electrical appliances in every room, especially televisions.
- Check that cooker rings and burners are off.
- Check that fireguards are in place.
- Turn off gas fires (except pilot lights)—especially portable cylinder gas fires.
- Unplug electric under/overblankets.

FIRE! EQUIPMENT & DRILLS ■ AT HOME

○ Recheck that paraffin stove wicks are completely out.
○ Check that exterior lights are off.
○ Make sure no cigarette ends are left burning.
○ Switch off lights and close doors.
○ Anything else specific to YOUR home.

☠ WARNING

Smoking kills: Cigarettes cause an enormous number of domestic fires every year. If you are a smoker or you live with one, follow this special drill:
- DON'T balance cigarettes precariously
- STUB OUT cigarettes properly and make sure matches are out before discarding
- PROVIDE deep ashtrays for smokers
- DON'T smoke if you are tired or drunk
- DON'T smoke in bed
- CHECK furniture—feel down the sides of cushions
- DON'T discard lighted cigarettes in wastepaper bins

REMEMBER

- Remove old newspapers and flammable objects stored under stairs—DON'T build bonfires anywhere in your home
- DON'T overload power points
- DON'T let curtains hang down round the back of television sets
- Get your household wiring checked by an electrician and check all plugs and appliance wiring
- Replace foam-filled furniture and mattresses if you can or treat them with a fire-retardant spray
- DON'T use a time switch on unsuitable appliances like electric fires
- Keep aerosols and flammable substances away from sources of heat and out of direct sunlight—they can explode
- Keep matches and lighters away from children
- Don't leave mirrors and bottles in direct sunlight—reflected and concentrated sunlight can start fires
- Don't use open fires in children's play areas—use a fireguard and do not leave them unattended
- Don't plug appliances into light fittings
- Never overfill pans with fat for frying
- Keep clutter out of halls, off staircases and keep escape routes CLEAR
- Do not place candles/hot drinks/vases or anything else on top of the television—you could start a fire
- Always use correctly-rated fuses in appliances
- In the event of a power cut, switch off electrical heaters. It's easy to forget about them and you may go to bed before the power returns

FIRE!

Unfortunately there is no way the threat of fire can be ruled out completely. In the urban environment the human error factor—perhaps a discarded cigarette or a faulty piece of wiring—will always limit the extent to which you can protect yourself. Having taken all the precautions, without totally transforming all buildings, the next step is to learn what to do if you find yourself in a fire.

PLEASE NOTE

The safest course of action in any fire situation is to follow the rules for containing it (by closing doors and windows), to evacuate the building and call the fire brigade. This applies to all types of fire which have taken hold at the flame stage—in certain cases you may have to try to put out a small or developing fire.

- **At home** Put your fire drill into action
- **At work** Follow the fire drill
- **In a public place** Seek assistance from responsible members of staff etc and make your way to fire exits as quickly and calmly as possible. Help anyone in difficulty.

At home

Initially, a fire in your home may seem containable or controllable. You may think—perhaps because you want to preserve your belongings or to save troubling the fire brigade—that you can cope with a burning sofa or a flaming television, but you are NOT a trained fire-fighter. You are severely at risk from highly toxic fumes in either of these cases. It would be almost impossible to remain detached or calm. If you tried to put out a fire in that state of mind you could help to spread it. Every decision you take COULD make the difference between life and death.

You might walk into a room, discover a fire and run out—in your shock forgetting to close the door behind you. Your attention would be completely taken by the fire—not the fire-feeding draught of air you have inadvertently created by leaving the door open!

Seconds count!

Time is the vital factor in a fire—containing it could give precious minutes to ensure all the occupants of a building reach safety (let the professionals do the rest). Rather than **fighting fire**, the two watchwords are SLOWING and CONTAINING fire. There are often some simple and fast things you can do to slow the progress of a fire. Equally there are 'safe' ways of containing a fire in a room or a section of a building.

⏱ EMERGENCY!
DOMESTIC FIRES

Try to deal with domestic fires BEFORE they get out of control. Read and commit this information to memory BEFORE the need arises. In all cases, ACT QUICKLY but stay as calm as possible.

CLOTHING FIRE

Most materials burn easily — 'synthetics' can even melt onto the skin

► YOU ARE ON FIRE
Cross your arms over your chest. Put your hands on your shoulders. This should help to protect your face from the flames
DO NOT RUN
This will fan and spread the flames, which will impede breathing
LIE DOWN AND ROLL
Roll over slowly to smother the flames
WRAP YOURSELF UP
In a carpet, wool blanket, fire blanket or non-synthetic curtains

► SOMEONE ELSE ON FIRE
Restrain them if they are panicking
LOWER THEM TO THE FLOOR
Trip them up if necessary!
SMOTHER THE FIRE
Roll them in a carpet, wool blanket, fire blanket as above
Cool the temperature of the burns soon as possible with water
Don't try to remove any clothing which has become stuck to the skin
CALL AN AMBULANCE!

ELECTRICAL FIRE

SWITCH OFF THE POWER
Try to unplug an appliance OR turn off all the power at the consumer unit/fuse box
AT NIGHT
Only switch off at the mains if you have a flashlight so that you can find your way about
SMOTHER THE FIRE
Use a fire blanket or a dry extinguisher — dry powder (blue), CO_2 (carbon dioxide/black) or halon (green)

TV/VDU FIRE

▶ **SWITCH OFF AT THE MAINS SOCKET**
OR
▶ **SWITCH OFF AT FUSE BOX/CONSUMER UNIT**

DO NOT STAND IN FRONT OF THE SCREEN—THE TUBE MAY EXPLODE

USE A DRY EXTINGUISHER AIMED INTO THE TV/VDU
OR
COVER THE TV/VDU WITH A FIRE BLANKET
OR
COVER WITH A DAMP BLANKET/TOWEL (SWITCH OFF POWER)

EVACUATE
Because of the risk of toxic fumes

IF THE FIRE GETS OUT OF CONTROL
Evacuate and call the fire brigade

ELECTRIC BLANKET FIRE

Do **NOT** open windows
Do **NOT** roll back bedding to inspect damage

SWITCH OFF AT THE SOCKET
If necessary, switch off at consumer unit/fuse box
DRENCH THE BED
A bucket of water, at least
EVACUATE
Because of the possibility of highly-toxic fumes
CALL THE FIRE BRIGADE

PARAFFIN HEATER FIRE

Do **NOT** use water or a water extinguisher (red) to put out a fire, even if there is a spillage of burning fuel

DO NOT MOVE HEATER
On no account try to carry a burning heater
SMOTHER THE FIRE
Use a fire blanket or a dry powder (blue) extinguisher. Aim it right into the appliance or at the base of flames
DON'T PEER UNDER BLANKET!
Evacuate and call the fire brigade

CYLINDER GAS FIRE

- ▶ In the event of a fire involving a gas cylinder
  Turn off the gas at the cylinder
  Move it away from the flames
  Use a fire extinguisher on the flames
- ▶ If you cannot reach the cylinder, or if the fire is spreading
  GET OUT!
  Get well clear of the house, call the fire brigade and tell them a gas cylinder is involved in the fire

FURNITURE FIRE

Synthetic fabrics and older 'non-safe' foams have a tendency to 'flare up' suddenly and burn fiercely. **Do NOT** stand too close

BEWARE TOXIC FUMES
Evacuate! Do **NOT** attempt to extinguish a fire involving synthetic materials/foams. If the fire is very small and easily smothered, you **MUST** still evacuate and call the fire brigade

NON-SYNTHETIC FURNITURE
Cool with water or water extinguisher (red). Keep going until smouldering stops. Evacuate to fresh air

NEVER

- ■ Try to take burning furniture outside
- ■ Stay in a room filled with smoke and fumes
- ■ Open a window to ventilate unless fire is extinguished

CARPET FIRE

Vulnerable to fire from sparks or a dropped cigarette. Most carpets won't burn rapidly

IF YOU'RE QUICK
If the spark or cigarette has just fallen, stamp out fire (or smother). Douse with water to cool

IF FIRE HAS STARTED
Use fire blanket to smother, or water/water extinguisher (red) to cool

SYNTHETIC CARPET/FOAM UNDERLAY
Evacuate. There is a risk of highly-toxic fumes. Call the fire brigade

FAT FIRE

▶ **TURN OFF THE HEAT**

IF YOU HAVE A LID FOR THE PAN
Replace it immediately to smother the flames

IF YOU HAVE A FIRE BLANKET
Approach with the blanket held up to protect your face and smother the flames

IF YOU HAVE NO LID OR BLANKET
Cover the pan with a damp towel or chopping board

DO NOT
Move the pan until the fat has cooled

REMEMBER

Burning fat/oil and water are a very bad combination. Throwing water on burning fat (or using a water extinguisher) will splash burning fat and spread fire. In some circumstances water will make a fat fire burn extremely fiercely

CHIMNEY FIRE

A chimney fire could be very serious. You may not know you have a chimney fire at first. If you see sparks or flames coming out of anyone's chimney, you should tell them

CALL THE FIRE BRIGADE
Before you do anything else

CLEAR THE AREA
Move carpets, furniture and any flammable objects away from the fuel burning appliance/fireplace. Burning debris is likely to fall down the chimney

CLOSE DOORS AND WINDOWS
This won't smother the fire, but it will reduce its air intake

USE EARTH
Use garden earth to smother the fire in the fireplace

USE SOAPY WATER
Detergent helps water to smother burning embers. Steam will be produced — this will help dampen the fire further up the chimney

IF IN DOUBT
If the room is filled with smoke — or the fire seems to have spread — do **NOT** waste time! Evacuate and wait for the fire brigade

ESCAPE!

In all but the most MINOR fire situation, you should implement your fire drill immediately. Everyone in the home or workplace will know what to do and how to escape. Nevertheless, it is impossible to predict how a fire will behave—even in your own home you might find yourself in a situation that you could not have prepared for.

Outside the home or workplace—in a building you are not familiar with—the range of pre-fire precautions that you can take are even more limited. Both these factors make it essential that you know the rules for escaping from a burning building and what to do if you become trapped. Mistakes in these crucial seconds and minutes could cost you your life.

Your priority at all times is to get out of a building as safely as possible. This means finding a safe way around a blaze and through a ground floor door or window.

Moving through a burning building
There are three rules for moving through a burning building:
◐ Test all doors for fire on the other side
◐ Close doors/windows as you go
◐ Stay as low to the floor as possible

WARNING
If a door to a room fits well and is closed it can contain fire for some time. Wood in the structure of a building is surprisingly fire resistant. Often after a house fire timbers, though charred, are still in place after the rest of the house has been gutted. Opening a door without caution could mean that you are suddenly engulfed by flames.

Testing all doors
This is important, because there may be little sign that there is a fire on the other side. In the case of a door which opens towards you, opening the door of a room that contains fire is VERY dangerous—the sudden influx of oxygen can cause flames to flare up and blast anyone in the doorway with ultra-hot air. The door knob (especially if made of metal) is the best conductor of heat. Put the back of your hand to the knob, if it is HOT your hand will jerk away immediately. The back of your hand is usually more responsive to heat. You might burn the skin and you need your fingers! Do NOT open the door.

If the door knob is not hot you may proceed, but before opening the door brace your foot against the bottom of it. Only then open the door by a couple of centimetres. Look into the room to check. If it is safe, open the door and enter. Had the room contained fire your foot would have stopped the door being blown open by a flare-up.

Close all doors

As you proceed, close all doors and windows, behind you. Closed doors hold back fire and closed windows cut down the amount of oxygen reaching a fire.

Stay low

Smoke fills a room from the ceiling down. At all times keep your head near the floor—crawling if necessary—where the air will be cleaner and safer. Staying low also avoids tripping or stumbling over objects. Hold a handkerchief or cloth (wet if possible) over your mouth/nose.

HIGH-RISE FIRE

 All the same rules as above apply if you live or work in a high-rise building, however there is a greater risk that you will not be able to reach ground level safely. Follow these rules:

- NEVER use a lift during a fire—you could be trapped if power fails or doors might open automatically onto a blazing floor.
- NEVER leave your flat or room (in a hotel) without your keys—you may need to retrace your steps so you do not want to be locked out.
- Set off the fire alarm and bang on other people's doors to alert them.
- NEVER try to descend a stairway that is blocked by fire. Climb upwards away from the fire and onto the roof or a balcony (see Trapped!).
- In very large buildings, if you know which side the fire is burning, make for the other side and use staircases there.

 **WARNING**

Over 50 per cent of fire deaths are due to toxic fumes and smoke. Inhaled smoke can irritate the throat causing it to contract in a sudden spasm—closing the airway. Someone found in a smoke-filled room may be unconscious and their breathing may have stopped. You should:

- Drag victim away from smoke, preferably to safety outside
- If victim is breathing but unconscious put the victim in the recovery position
- If breathing has stopped or is difficult, give artificial respiration until help arrives

FIRE! ■ FIRE! ■ ESCAPE!

FIREMAN'S LIFT

If you need to rescue an unconscious person, the safest and easiest way of moving them is the fireman's lift. If you cannot manage, and the danger is great from smoke/fumes or flames, drag the person out of the building any way you possibly can!

1 Lean forward slightly and at the same time lift casualty's arm over your shoulder and behind your head. Bend over until your right shoulder is at the same level as the casualty's stomach.

2 Pull the casualty across your back and shoulders—they will take the weight. If it helps, you can rest on your right knee. Your right arm around or between the casualty's legs will add support.

3 Taking the weight on your right shoulder, reach around the casualty's legs to hold his/her wrist, and lift. Push yourself up by pressing down with your left hand on your left knee.

4 Standing up, you can now proceed to take the casualty to safety. The position of the casualty's head is not ideal if there are head or neck injuries, but there may be no choice.

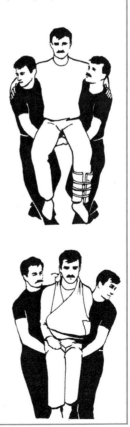

Lifting an inconscious casualty If necessary, turn the casualty face down. Kneeling at the head, slip hands under the shoulders to the armpits. Firstly, lift the casualty into kneeling position. Then, standing yourself, lift the casualty upright. Proceed with fireman's lift, 1–4.

TWO-PERSON SEATS

Two people can easily carry even a heavy casualty to safety by making a 'chair'. Holding your left wrist with your right hand, reach under the casualty and grip the other carrier's right wrist with your left hand (shown). The casualty holds both carriers' shoulders for support.

If the casualty has suffered arm injuries, carriers will have to provide extra support. Carriers should clasp each other's forward hands under the casualty's thighs, using padding in a hook-like grip (shown). At the same time, reach behind the casualty's back and either grip clothing or the other carrier's upper arm.

TRAPPED!

If your escape route to the ground floor is blocked and there is no way that you can escape from the building safely, you must try to **raise the alarm** and **signal for help**—position yourself where you can be rescued, as far from the fire as possible. Protect yourself from the actual fire and smoke and fumes.

◑ Go to a room (which is not on fire) with a window. Close doors, windows and fanlights.

◑ If you cannot get out onto a balcony or roof, slow the approach of the fire. If there is water in the room, douse the walls and door—this will slow the flames. Block any gaps or holes that will let smoke into the room. Use rolled-up cloth or blankets (wet, if possible) to block cracks and openings round doors and windows.

◑ Open a window and attract attention by shouting or waving.

◑ In an office building, throw paper out of the window—it should catch the eye of someone below.

◑ If a window is sealed and cannot be opened use a chair or your feet to knock out the glass. If you have to use your hands wrap them in cloth or a garment for protection. Use your elbows if you are wearing long sleeves.

◑ Break double- or treble-glazed windows with a hammer.

◑ Laminated glass is difficult to break, but keep trying. Once all the glass is broken, it may push out of the frame.

◑ If there is a balcony or flat roof that you can reach safely, get out on to it and close the window or door behind you.

JUMPING

■ If you are on the first floor and the ground below is soft it may be safe to drop. Throw pillows or bedding on to the ground to break your fall.

■ Hang from a window sill or ledge at full stretch. Just before you drop, hold on with one hand and with the free hand push your body away from the wall. Keep your feet and knees together with legs slightly bent, head tightly on chest.

■ When you let go, relax and try not to resist the ground. Do not jump or spring away.

■ Don't try to break a fall with your arms. Use them to protect your head.

■ If you can make a sheet or curtain rope, this will at least shorten your fall. Make sure it is tightly secured.

■ Do not jump from an upper storey. It is safer to position and protect yourself properly and await rescue.

◑ If you have to climb through a broken window, knock out jagged glass and place a blanket or clothing over the base.

> **! WARNING**
>
> Jumping to safety: With any type of fall there is a risk of injury. You should ONLY consider this as a last resort to escape fire. Risks increase greatly with every storey above the first floor. Jumping from the third floor and above must only be considered if there really is no other choice. Even if the ground is not hard, the fall could prove fatal.

> **REMEMBER**
>
> If you can make your way to the roof, move with extreme caution. If it is a pitched roof, stay on the windward side away from smoke. Attract attention by waving and shouting, NOT by throwing roof tiles or heavy objects to the ground—you could kill or injure someone below.

> **WARNING**
>
> Most roof surfaces are extremely treacherous. Slates or tiles may easily become dislodged beneath you. Sheet roofs may easily collapse—causing you to fall through. Assess the risk!

Getting help

Once you have escaped from a burning building you should still follow your pre-planned drill. This will help you account for the occupants of a building. The fire brigade must be called immediately—do NOT assume this has already been done. Delegate someone to telephone from a call box or from a neighbour's home (see **Equipment & drills**) and make sure the full address and location of the fire is given.

> **REMEMBER**
>
> Do not let ANYONE other than a trained fire-fighter enter a burning building. However brave you think you are, an untrained ill-equipped person is no match for a blaze. Resist the urge.

Outside

Get clear as soon as possible. Do NOT hamper fire-fighting or rescue operations. Beware falling debris—the whole building may collapse. If you are injured, make the fact known to medical personnel—but accept the fact that the most life-threatening injuries must be dealt with first. People may be very frightened or upset. Try to keep calm and reassure others.

6

No one wants to live or work in a fortress, but you must protect yourself from the risk of attack by intruders. Thieves and vandals seem to be more daring than ever — you'll have to do more than fit locks to keep them out.

Security

SAFE & SECURE

Intrusion, break-ins, theft and vandalism may not be life-threatening in themselves—but they cause an enormous amount of stress. Your home, where you normally feel safe, can be rendered uninhabitable in minutes. The fact that ANYONE has entered your property by force can be very traumatic. Burglars could wreck a home—or empty it! At the very worst, you may be injured or even murdered by intruders. At work, apart from the loss of valuable equipment and man-hours, priceless data may be destroyed if computer systems are sabotaged. Competitors may take extreme measures to secure highly valuable information.

Insurance policies may cover basic replacement of stolen or damaged property, but money cannot replace possessions of great sentimental value or the time spent decorating and furnishing your home. At work, it may be even more difficult to assess the extent of the damage done, including the loss of business advantages.

Many burglaries involve only petty theft. At home this may mean that 'casual' thieves are only looking for televisions and video recorders, cash and small valuables. At work 'casual' thieves may be more interested in cashboxes and employees' personal possessions. Many 'professional' thieves aim for higher stakes—they know where there are things worth stealing and times when they can be stolen.

Take action to counter the threat of intrusion and all the damage that may be done as a result. You MUST realize that there is more to safety and security than just fitting efficient locks.

REMEMBER

The number of break-ins to homes and offices every week is huge—and steadily increasing. In many cases, unless there is physical attack, arson or something very valuable has been stolen, there is little the police can do. Fingerprinting (when it takes place at all) does more to reassure you that something is being done. Tell the police if you have marked your possessions in any way, provide photographs of valuables and any information which may assist them in the recognition or recovery of your possessions.

EMERGENCY!

BURGLED!

▶ **If you arrive at your home/work and have reason to suspect that A BURGLARY IS TAKING PLACE**

DO NOT GO IN
If you think that intruders may still be inside, call the police and find a safe vantage point—outside or at a neighbour's. You may see the thieves leaving. Be ready to jot down descriptions of people/vehicles Do NOT attempt to apprehend the thieves

IF YOU ARRIVE ON FOOT
Walk on. If you have already reached the door, turn round and get away as fast as possible

IF YOU ARRIVE BY CAR
Drive on. Park where you will be safe, call the police and keep an eye on the property. If you have already pulled into a driveway, pretend you were using it to turn the car round. Back out and drive off

IF YOU DISTURB INTRUDERS
GET OUT! Do NOT call out. Do NOT block likely exit routes

▶ **If you arrive at your home/work and have reason to suspect that A BURGLARY HAS TAKEN PLACE**

DO NOT TOUCH ANYTHING
If the thieves have left, call the police. While you wait, start a visual search to ascertain what is missing. If distressed or unable to cope alone, call a friend to be with you, go to a neighbour's or wait outside for the police to arrive

▶ **If you hear intruders at night**

Do NOT leave the room. Do not attempt to apprehend the thief. Look for something to defend yourself with. If you have a bedside telephone, call the police immediately. If not, wait until the intruder has left. Move about noisily, turn lights on and hold a conversation. If you are alone, pretend to be talking to a male companion

Ground floor and basement properties are the most vulnerable. Detached houses, especially when set well back from the road and screened by walls, fences, trees and bushes, are easy targets. They may suggest a wealthier lifestyle and more valuables—it is also less likely that the burglar will be observed by neighbours or passers-by. Properties backing on to wasteland, alleyways, public parks and other routes allowing easy access (and getaway!) have a greater risk of burglary. An adjacent building site or scaffolding makes a house particularly vulnerable.

You don't have to be wealthy to be burgled! Most homes have a television, stereo, video equipment or other electrical items—all relatively easy to dispose of. Small items such as cash, cameras, watches, medals and jewellery may be slipped into pockets—an important consideration if you have to leave a house by climbing out of an upstairs window!

You can improve the security of the site where you live and also make the house itself more difficult to break into. Security is also an important consideration when looking for a new home—knowing some of the high-risk factors could be one of the influences in your choice.

Visibility

Balance your need for privacy against that for security. Fences, shrubbery and trees screening your home will all give intruders cover. Keep hedges low so that the entrances to your home can be seen over them—thorny ones may help to deter people from trying to break through them.

Garden walls and fences are less easy to climb if they have trelliswork mounted on top of them. Make it too high and too frail to be climbed. This won't keep out a determined burglar but will deter the opportunist. Such protection is only effective if there are no sections of the perimeter where a wall, roof or other solid structure allows easy access or a weak gate provides entry at ground level.

Outdoor illumination of your home should be designed to expose anyone lurking—'decorative' garden lighting may offer more hiding places in shadows. A light above or beside a door will show up anyone trying to force locks or break the door open. It will also allow you to see people at night through your security peephole.

A timer or a light-sensitive switch (which is activated when daylight fades) will turn lights on when no one is at home. A better idea for a porch light is to use an infra-red movement detector, which is activated when anyone approaches—a welcome for friendly visitors and an immediate deterrent to someone who does not wish to be seen. The impression given as you step into the field of the detector is that someone has heard you and switched a light on.

Keeping intruders out

Check ALL possible means of entry. Do they have secure locks? The smallest window (in a bathroom or over a doorway, perhaps) may be big enough for a person to wriggle through. Skylights and upper windows are vulnerable if a ladder is used or if there is access from the roof of another part of the building—or even from another building. Even a coal hole or delivery chute may be big enough for a child, if not for an adult. In old houses even chimneys may allow access. Chimneysweeps' boys used to climb inside them!

When deciding priorities, deal with the ground floor first, then any upper windows near flat roofs, beside a drainpipe or otherwise more accessible. On wood-framed windows you can improvise **temporary** security measures in a few minutes, by inserting screws to prevent latches, catches and staybars from being used.

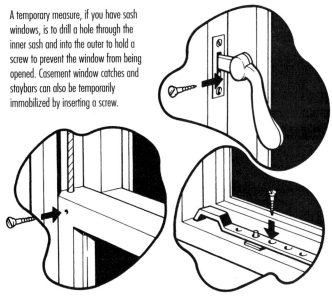

A temporary measure, if you have sash windows, is to drill a hole through the inner sash and into the outer to hold a screw to prevent the window from being opened. Casement window catches and staybars can also be temporarily immobilized by inserting a screw.

Windows could be firmly screwed shut, but this is inadvisable if the window needs to be used as a fire escape.

- Doors need to be strong, so that they cannot be smashed or kicked in. Glazed panels are a liability—use laminated glass, which is very hard to break. Fit secure locks and bolts—at least a rim lock and a mortise lock (see Door locks).
- French windows are easily kicked open if they rely on a single catch. Fit rack bolts top and bottom (see Window locks).
- Louvre window panes can be lifted from their fittings. Fix them in place with an epoxy resin glue.
- Most domestic glass can be broken or cut. Leaded panes can be removed (almost silently) by peeling back the lead. Double and treble glazing will act as a deterrent, but won't keep a burglar out. Laminated or wired glass will slow a burglar down, but the only certain protection is a metal grille or shutter outside or a sliding metal grille inside (which can be hidden by a pelmet and curtains).
- ALWAYS lock garages/outbuildings—not just to protect the contents but to prevent the use of tools and ladders. Use a toughened-steel security padlock, one with a close-fitting or shielded loop (shank) to make the use of bolt-cutters difficult. Use coachbolts (with rounded heads) instead of screws to make the fittings more difficult to prise off. If you must use screws, drill out the slots to prevent the use of a screwdriver—or use tighten-only screws.
- If ladders are kept outside, chain them to the wall, a fixed pipe or post.
- Paint downpipes with security (anti-climb) paint—it remains slippery, making pipes difficult, if not impossible, to climb. It's also VERY difficult to get off the skin and clothing, so the thief will be 'marked'.
- Don't allow ANYONE the luxury of a close 'recce' of your property. If you have a side gate, LOCK IT!

REMEMBER

Keep your glazing putty in good repair. If it is crumbling and can be picked off, the glass can be removed quite quietly. The sound of a window being smashed usually attracts attention.

Building work?

You are particularly vulnerable when building work is in progress—especially if there is scaffolding outside to give easy access. It doesn't have to be the builders, scaffolders or someone they tipped off about what's inside—it is obvious to any passer—by that the house is vulnerable.

With workpeople coming and going, it is less likely that even neighbours will realise the house is being burgled. Always INSIST that ladders are hauled up at night and chained to scaffolding, well out of reach of ground level.

Unoccupied houses

Houses left unoccupied for a long period are sometimes stripped of everything—fireplaces, curtain rods, doorknobs, floorboards, stained glass panels. There is a booming market in 'architectural salvage'! If you move out of your home to allow major work to take place, or delay fully moving in, your furniture, curtains and carpets may disappear! Few people question a removal van being loaded in broad daylight.

Moving in?

The danger is even greater when you are moving into a new house and YOU are unknown to the neighbourhood. Introduce yourself to your immediate neighbours. SOONER OR LATER YOU MAY HAVE TO RELY ON ONE ANOTHER. They will, at least, know that the person climbing through the window is not YOU. Ask them to keep an eye on the place for you, and offer to do the same for them.

How burglars get in

Most burglars enter through ground floor doors and windows. British figures suggest that nearly 50 per cent get in through a side window. About 25 per cent get in through the front door and almost as many by a rear or side door. In many cases, burglars find a door or window unsecured. Forty per cent have to force a door or window, but it's a lot less likely that glass will be broken to gain entry. Subtler, but less common methods include posing as a tradesman or using a key.

WARNING

The stereotype of the burglar with a sack of 'swag', a mask and a torch climbing down a ladder at night is misleading. About half of all domestic burglaries happen during daylight hours—when most people are at work.

Marking property

It is not very common to get back stolen goods. When things are recovered, positive identification of the objects is extremely difficult unless you have photographs and full details.

Make a list of all your valuables. This has to be done when arranging full insurance cover. It will also serve as a check list for losses. For antiques, art objects and valuable jewellery, a photographic record is essential.

Many consumer goods have a serial number—however, these may be removed or defaced. Adding your own marks will identify an object even more clearly. If they cannot be engraved

Most burglars are looking for things which are easy to carry and easy to sell. It is obviously much easier to dispose of small antiques, than to find a secret buyer for a painting by Picasso or Rembrandt! Plenty of people are only too willing to buy a watch or video recorder that 'fell off the back of a lorry'. The most common carry-away items stolen include:

- Jewellery/watches
- Video recorders
- Cash
- Hi-fis/radios
- Televisions

- Credit cards/cheque books
- Cameras
- Antiques
- Personal computers
- Silverware

in an unobtrusive place, use a special marker pen with a pigment which glows under ultra-violet light. It won't be noticeable, unless you know where to look. Use your house number (or the first letters of its name if you have no number) followed by your postal or ZIP code.

DOOR LOCKS

There are many types of lock, but the most secure are those in which the key trips several levers or tumblers. The key is therefore more complicated to duplicate without a master and the lock is more difficult to pick. Those which have a deadlock action make it impossible to force the bolt back into its casing without turning a key. Most mortise locks are deadlocks. Some rim locks may double as deadlocks, and are activated by an extra turn of the key.

From the inside a securing catch will prevent the opening of any rim lock from outside—even with a key. Most people use the catch to keep the door locked open when they're frightened of locking themselves out. **Some burglars flip the catch so that no one can open the door and catch them in the act. While you're fumbling with your key and wondering why it won't turn, the thief is escaping.**

REMEMBER

Do NOT label keys with your address, room descriptions or house numbers. Doing so only helps a thief if he/she gets hold of them.

MORTISE LOCKS

Mortise locks are fitted INTO the door, with the bolthole reinforcement (striking plate) let into the door jamb. These locks MUST be drillproof. The screws are concealed when the door is closed. The bolts MUST have reinforcing pins which prevent them being sawn through.

Mortise locks which also have handles (sash locks) are available for interior or exterior doors.

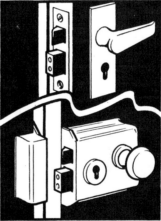

RIM LOCKS

Most are designed for surface mounting. The screws are vulnerable, except for those in the door edge and the jamb (when the door is closed).

Preferably, go for a rim cylinder lock with as many 'extras' as possible. Interior and exterior deadlocking facilities are sensible, as are pegs which won't allow a blade to be inserted from outside to force the bolt back into the lock.

On external doors ALWAYS fit a mortise deadlock. If the door is too thin to take one, get a thicker door! A mortise lock MUST have at least five levers—cheaper ones have only two. Fit a rim lock as well. Use the rule of thirds:

The rim lock **(A)** should be positioned one third of the way down from the top of the door. The mortise lock **(B)** should be positioned one third of the way up from the bottom of the door.

SLIDING DOOR LOCKS
Sliding doors are much more difficult to lock securely than conventional doors. Use a hook-and-bolt type, the nearest equivalent to a mortise deadlock. If you have metal- or plastic-framed sliding doors and need to fit a lock, the only easy ones to fit are surface-mounted.

REMEMBER

Locking internal doors, especially those of rooms with vulnerable windows, will make life difficult for a burglar. On the other hand a really determined burglar will kick them open. Unfortunately you will then have more damage to deal with.

Choosing locks

Ask your locksmith how many 'differs' (key permutations) there are for the make and type of lock you intend to buy— choose a lock which has as many as possible. If the style of lock you use is only available with a few key patterns, a burglar could easily carry all the necessary keys. Top quality locks may have only one key combination. New keys can only be cut on written request.

COMBINATION LOCKS

Electronic or mechanical push-button combination locks are useful for people who tend to lose keys, or when there have to be a large number of keyholders. You can change the combination whenever you wish—and won't have to cut new keys. Drawbacks are that potential thieves may see you punch the number in, or that you might forget the number. Do NOT write code numbers down. If you are worried about forgetting yours, include it disguised as part of a fax or phone number in your address book, so that YOU will recognize it but others will not.

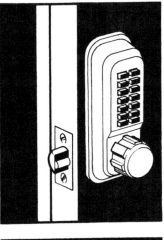

Sliding bolts on the inside of a door give added security—although these are easily opened by an intruder, once they are inside. They can quickly be rammed home if you need to prevent forced entry while you are in.

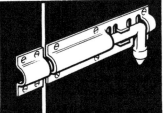

Rack bolts can be used on all doors and on wood-framed windows too. There is only one 'key' pattern, but keyholes are on the inside only—making it very difficult to detect these bolts from the outside.

SECURITY HOME SECURITY ■ DOOR LOCKS

On the hinge side, hinge bolts (dog bolts) will help prevent the hinge side of the door from being forced open. These are more difficult to fit than you may imagine—you may have to call in a professional. The metal 'bolt' locates in the reinforced hole when the door is closed.

REMEMBER

No matter how many locks, bolts and hinge bolts you fit, if the door or the frame is weak, you have a problem. Thieves can break doors and crowbar frames out of the wall. If the door faces a solid wall, a jack might even be used to force the door. Steel doors, steel-reinforced doors and steel-reinforced door frames are a last resort.

To prevent the door from being kicked in (at least), a stout steel bar fixed firmly down the hinge side of the door frame should prevent the hinges from being broken out of the wood. On the 'lock' side, the steel bar needs to retain the staple of the rim lock.

⚠ WARNING

You should always 'lock yourself in' with a mortise lock— especially when settling down for the evening or going to bed. Intruders have been known to break in and rob a house while the occupants were watching television or sleeping. DON'T FORGET: You will need a key to open the door—and may need to do so in an emergency. When you leave a multi-occupied building, check that you haven't locked someone in.

'Entryphones'

If you want to check on callers without actually having to go to the door, instal an audio or CCTV (closed-circuit television) 'entryphone'. These are becoming increasingly common, especially

A door viewer (peephole) — a small lens which gives you a good look at anyone outside the door — is a sensible addition which allows you to check all callers. DON'T forget that if you have a bright light on your side of the door you will block the light as you look through the peephole, making it obvious that you are in. Make sure the viewer has a cover which you can slide open when your head is close to it. Don't give your presence away by noisy footsteps as you come downstairs or along a hallway.

❗ WARNING

Most peepholes distort the view, even though they allow about 180° vision. If the caller is too close, they may look quite frightening.

■ If you are very short or very tall, the comfortable height for you might not give a good view of the caller
■ A peephole might not let you see someone else lurking behind the caller, or to one side
■ If in any doubt, do NOT open the door

for multi-occupied buildings. However the voice of an unknown caller would not necessarily give you any clue as to their intentions. A camera is relatively easy to put out of action and may not reveal an accomplice. Consider fitting a spyhole as a useful back-up.

Limiters/chains

Door limiters, made from toughened steel, are stronger than chains and can act as an additional bolt when the door is closed. A chain won't deter a violent intruder. Once you've opened the door, one kick will tear the chain free. Always use longer screws than those usually supplied with these devices, or they give a false sense of security.

REMEMBER

Check all callers: NEVER let strangers into your house without making sure WHO they are and WHAT their business is. Try NEVER to admit them when you are on your own. Thieves may pose as representatives of gas, water or electricity companies, social workers — even police officers — to gain entry (see SELF-DEFENCE: Attack at home).

They MUST allow you to take their identification from them and close the door. Don't just call the number on any business card they show. The number could be that of an accomplice. Use the telephone directory!

WINDOW LOCKS

It is preferable to fit locks on ALL windows. Different types are made for different forms of window. It's something a DIY person could handle, though you may need a locksmith's help if the frames are plastic or metal, instead of wood.

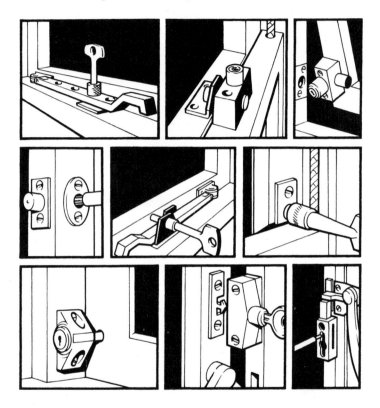

Most window 'locks' use very simple screw mechanisms. There are only a few key shapes—the average burglar is bound to know them all. For particularly vulnerable windows extra precautions must be taken, fitting REAL locks or locking catches. This may be an expensive option—or a pointless one if all the burglar needs to do is break the glass. Fit a grille! Don't forget that YOU may have difficulty opening the window if it is an emergency escape route.

Accessible windows should be fitted with net curtains, venetian blinds or louvre shutters—anything which prevents an easy view of the contents of the room.

DOGS

Almost all dogs will act as burglar deterrents—their barking will attract attention. Even a small dog can be aggressive in defending its home against a stranger and can sound 'bigger' than it is! Many dogs are friendly to almost anyone who shows interest in them and will be no actual defence at all—but their presence may be enough to make the opportunist thief look elsewhere.

Large dogs and breeds originally bred as fighting or guard dogs may be both psychological deterrents and practical defenders. All dogs need training, but with powerful breeds it is particularly important for you to be able to control and trust them. Dogs are pack animals and need a leader, or they may take over this role and become uncontrollable. Some have a strong sense of territory and will even deny access to your friends.

Domestic pets (especially when there are children about) should never be encouraged to be overly aggressive. Professional training classes are advisable for both dog and owners.

ALARM SYSTEMS

Just the sight of an alarm box may discourage a casual burglar from attempting to break in. Professional burglars know how to deal with alarms and will do so when they think the job is really worth the effort. Many people feel that the sight of an alarm box on the front of the house alerts thieves to the fact that you have possessions that you think are worth protection!

If the burglar is not aware that there is an alarm system, the surprise of bells, sirens and flashing lights may scare them off—at least, they will try to get away as fast as possible. Alarms will alert you (if you are in), and everyone in the neighbourhood, that a robbery is taking place.

When choosing an alarm system, remember that you will have to set it each time you leave the house or go to bed. If the setting process is fiddly or has to be done in a hurry to avoid sounding the alarm, you might get very fed up with the system. There is no point in installing an alarm if you do not use it. Domestic systems are not usually very complicated, but try to ensure that yours offers some flexibility and can be adjusted to suit your needs.

If you think the main value of a burglar alarm is its deterrent effect, fit a dummy alarm box on an outside wall. It's cheaper and much less hassle than having the whole works!

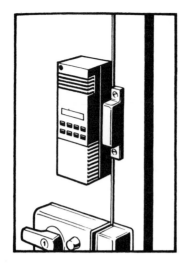

A simple battery-operated alarm will alert you if someone opens a door or window. To set the alarm you either turn a key or punch a digital code. If you live above ground in an apartment block with only one entry door to your property and you want to be alerted if someone enters a garage or outbuilding, or if you want to restrict access to part of your home (a workshop, perhaps), an alarm like this could be ideal.

Types of system

Installing an alarm system provides a greater feeling of security but is no substitute for strong locks/grilles and other security measures. Alarm systems are constantly being updated—take your time to choose one that meets your requirements. Most alarm systems are made up of the same elements, although the way in which these are linked varies considerably. The three main options are:

- ◑ Open circuit. When the circuit is completed by a detector/ triggering device the alarm sounds. To prevent deactivation by the cutting of a wire, multicore wires are used which contain a back-up circuit that reacts to tampering.
- ◑ Closed circuit. When the circuit is broken by a detector/triggering device, the alarm sounds. Back-up circuits prevent deactivation by attempting to bridge circuits with additional wires.
- ◑ Wireless systems. Each sensor/triggering device is a small short-range transmitter—if the control panel receives a signal from any of them, the alarm sounds.

Control unit

This is the 'brain' of the system. Options may include:

- ◑ Being able to switch the whole system on/off (with a key or digital keypad).
- ◑ Being able to deactivate sections of the system, which would allow you, for instance, to protect the lower floor of a home while the occupants are using the upper floor.
- ◑ Adjustments of the time allowed to leave the building and close the main door.

○ Fail-safe devices and tamper alarms, if any attempt is made to deactivate the control panel.
○ A back-up power source, which cuts in if there is a power failure for any reason.

Detectors/sensors

There are numerous triggering devices available, allowing you to choose a combination to protect your home in many ways.

Magnetic switches will alert you if doors or windows have been opened. Some are surface mounted, but there are 'hidden' versions which fit into the door or window and frame. There are ways to 'fool' them. False alarms are unlikely.

Vibration detectors will alert you if a door or window is subjected to slight movement—a door may be kicked or someone may attempt to break a window. They need to be adjusted to allow for 'normal' vibrations to avoid false alarms caused by sudden gusts of wind or heavy traffic.

Foil strips on windows (in closed-circuit systems) break contact if the glass is broken.

Pressure mats are activated by the weight of a person treading on them. They are usually placed under doormats and carpets in places that no one can avoid. Frequent traffic over them, in places such as the bottom of the stairs, may wear them out or cause their shape to show up through the carpet.

Movement sensors which emit infra-red, radio or ultrasonic waves are triggered by anything which comes within their field. False alarms may be caused by children, pets—anything of any size which moves—a tumbling cushion, a large moth near the sensor. Most movement sensors are adjustable—but take care not to render them too insensitive.

Movement sensors are advisable for external use. On your doorstep, a light can be arranged to come on as anyone approaches the door. Powerful floodlights can be activated to expose anyone who attempts to approach your home from the front, sides or rear.

Alarms/false alarms

Bells, sirens, flashing lights (inside and out) are common. The intention should be to attract attention and (hopefully) to scare off an intruder. Do NOT invest in a system which cannot be programmed to switch off after a selected period (20–30 minutes or as dictated by local legislation)—otherwise false alarms which sound for several hours will make neighbours very angry indeed. Try to arrange for a trusted neighbour to have access to deactivate the alarm. Too many false alarms in a neighbourhood and people may stop paying attention to them.

Inform the police that you have had an alarm fitted and supply the names of all keyholders who can enter your home and switch off the alarm. More recent innovations are alarms which cease to sound after a predetermined period—and then re-arm themselves.

In an area where alarms are a nuisance or there is no one to hear them, they should be linked directly to the police or a security company. There need be no indication to the intruder that the alarm has been triggered and that the police are on their way. This arrangement can be very expensive, especially since there are penalties for false alarms.

To protect security system wires from malicious damage, ask telephone and security companies NOT to run them on outer walls where they can be cut or tampered with. If this is unavoidable, feed them through sturdy conduits.

PANIC BUTTONS

Most systems can incorporate panic buttons. These are particularly sensible for the elderly and people living alone. Have as many as you feel you need—one beside the bed, one in each room. If you are attacked in your home, or have an accident that requires urgent medical attention, hit the button! Panic alarms for wireless systems may be powerful enough to work outside the house—if you were sitting in the garden, for instance.

Personal alarms

If you are elderly, or live alone, you may need to rely on your neighbours in an emergency. A simple wired 'doorbell' from your house to theirs may suffice, but the problem is that neighbours are not always in.

Systems are now available which involve you carrying a small transmitter or wearing one like a wristwatch. When triggered within an area that would easily include house and garden, it will automatically transmit a signal down a telephone line if help is suddenly needed because of illness, accident or threat. This alerts a 24-hour monitoring centre, which calls back immediately to check for accidental false alarms and will send help if there is no reply.

If such a system is used, a codeword or phrase should be arranged from the outset. A victim of attack may be forced to say that nothing is wrong, but if the codeword is used the monitoring centre will know they are under duress.

If you need advice

Advice on all security matters—from locks to full security systems—is available from several sources:

- Local police, who will also have a good knowledge of any particular local problems
- Insurance companies
- Lock and alarm manufacturers and installers

KEEP WATCH

Neighbourhood Watch, Crimewatch and comparable vigilance schemes, in which neighbours form a group to beat local crime, have been started in many cities all over the world. The idea is for everyone to keep an eye out for any peculiar or suspicious activity—strangers paying unusual attention to a house, for instance—and report it to the police. Such groups also serve to develop members' awareness of security.

Times have changed—small close-knit communities could usually spot a stranger, but it is impossible nowadays to know the pattern of other people's lives and to recognize their friends and visitors in most large cities. On the other hand, in densely-populated areas there may be more eyes to do the watching! Even where there has been no positive reduction in the local crime rate, such schemes seem to have led to better relations within the community and to a reduction in stress from fear of crime.

If you want to start or join such a group, your local police station will be able to tell you who to contact—or how to start one.

LOOK OUT FOR . . .
- Strangers knocking on front doors and peering through windows
- Anyone loitering suspiciously
- Cars cruising slowly
- Strangers hanging around schools and playing fields
- Strangers approaching children
- 'Casual' window cleaners
- Strangers trying car doors
- People you do not recognize disappearing round the back of properties

LEAVING THE HOUSE

- Always lock all doors, close and lock all windows. Check that keys have not been left in locks.
- Leave lights on if you are out at night—not just a hallway or porch light. Use a timeswitch so that they come on at dusk (random ones switch lights on and off 'realistically').
- At night, ALWAYS draw curtains or close shutters so that no one can see in.
- Leave a radio on.
- Burglars tend to be wary of a dog—it might attack them or attract attention by its barking. If you have no desire for a

dog, you could try a 'dog alarm'. Some of the more recent ones have quite a realistic 'bark'. You could set it up with a vibration or movement detector.

❍ Make sure valuables are not visible through windows.

❍ Always close a garage door—an empty garage is an indication that you are not at home.

❍ Always lock garage/outbuilding doors to prevent access to tools or ladders.

❍ Do not leave a key in ANY hiding place—however 'clever' you think it is. Leave a key with a friend or trusted neighbour in case you lose yours.

❍ If you let workmen have keys to the house, change the main rim locks afterwards (and consider changing the mortise locks too). Keep the old rim lock cylinder and key—you can replace them if you need to change the keys again. This is where a digital combination lock can be an advantage! You only have to change the combination.

REMEMBER

Random timers, which switch the lights on and off erratically, can help create an impression that the house is occupied. Regular timers with one or more on/off phases allow you to create your own pattern of lights. A very observant burglar, who is keeping an eye on YOUR home over several nights, may notice repeating patterns. It will certainly be noticed if no lights ever come on!

PHONE SENSE

■ NEVER leave an answering machine message for callers to tell them that you are out. A message such as 'We can't come to the phone at the moment' leaves a certain doubt as to whether the house is occupied or not.

■ If you regularly need to give callers numbers at which you can be contacted, NEVER do so on your prerecorded answering machine message. Arrange with the telephone company for calls to be automatically transferred. You will have to pay for the service, but callers will be unaware that they have not got through on the number they dialled.

Going away

❍ Ask trusted neighbours to be observant and keep an eye open for any suspicious activity.

❍ Cancel all deliveries—but do NOT leave notes outside the door to advertise the fact.

❍ Ask a neighbour to ensure that nothing is left outside and that mail/circulars are not hanging out of the letterbox.

- Leave curtains OPEN—it looks less unusual to have them open at night than to have them closed during the day.
- If possible get neighbours or friends to come in and switch lights on and off and close the curtains during the evening. Their activity will help to make the house look occupied.
- If you have a driveway, ask a neighbour or friend to park there occasionally instead of outside their own house.
- If away for a long period and you have a lawn, get a neighbour to trim it for you or it will be another sign that you are away.
- Take valuables to the bank safe deposit.
- Best of all—arrange for a reliable friend to move in while you are away. Cities are full of people sharing accommodation or living with relations. They may welcome a break on their own—they solve the problem of looking after pets and plants as well!

Returning home

By day: Ensure everything is as you expect it to be. Give the house a quick scan from as many sides as possible for obvious signs—a forced door or broken windows. If there are ANY signs of disturbance or anything that makes you suspicious, go to a neighbour's house and telephone the police. DO NOT GO INTO YOUR HOME. THE INTRUDERS MAY STILL BE THERE.

By night: Having an outside light left on, or one activated by a movement detector, makes it easier to check for signs of entry. Are curtains disturbed? Are the correct lights on/off? If you arrive by car and see any sign of trouble, drive on. Go to a neighbour's or a callbox to call the police. Even if all seems OK, DON'T put the car away. Park/lock the car and go closer to check that everything is normal. If you still feel something is wrong, do NOT go in.

INTRUDERS AT NIGHT

If you are woken at night by the sound of intruders, do **NOT** attempt to confront them! Don't pretend to be asleep either! Switch on lights and make plenty of noise as though you are unaware someone else is in the house. If you are on your own, call out and pretend to be talking to someone else (it may help if the imaginary companion is male). Telephone the police from the bedroom or as soon as you can safely reach your phone.

▶ SEE EMERGENCY! PANEL AT THE BEGINNING OF THIS CHAPTER FOR FURTHER DETAILS. SEE ALSO RELEVANT HOME SAFETY INFORMATION IN THE **SAFETY FIRST** CHAPTER IF YOU'RE GOING AWAY.

SECURITY AT WORK

A large commercial or industrial establishment with secret processes and highly 'sensitive' information MUST be as secure as a prison—only in reverse. Several rings of security may be needed to keep people OUT. Devices of the types used for domestic security, often in more sophisticated forms, can be used to alert people to a break-in outside work hours. During work hours, thorough security procedures will also be necessary.

Obviously, the level of secrecy of the work in hand dictates the level of security needed. Most businesses have some sort of information which may be useful to competitors—yet many businesses have VERY low security standards! In extreme cases an outer fence, patrolled at night or protected by guard dogs, should be considered. The number of gates should be kept to a minimum and strictly controlled.

There should be only one main entrance, monitored by a security guard or team. Final entry to the premises should be through a controlled door—to avoid people 'slipping' past. Inner layers of even greater security may be required, with only one route leading to areas where secret work is undertaken or strong rooms/safes and 'sensitive' records are kept. Access to such areas should be limited to authorized personnel.

Visitors must NOT be allowed to wander unescorted around a building at their leisure. It is amazing how many companies DO have 'secrets' of one sort or another, yet report that 'casual' thieves have wandered in and stolen employees' coats, bags and possessions! A 'professional' thief should find easy pickings in such a place.

When work hours are over, all unnecessary access routes should be sealed. Staff in 'sensitive' areas should be disciplined to leave NOTHING on show, to lock computer disks away and to be VERY careful what they throw in waste bins. Spies are not above 'raiding the garbage'.

Reception desks

High-security measures are inappropriate for many businesses, but some form of control must be exercised to ensure that unauthorised visitors do not gain entrance. A voice communication link—an 'entryphone'—will only allow nominal identification. The drawbacks are enormous—the telephonist cannot see if the person is alone or if they are carrying a weapon, for instance. If all expected visitors' names and the purposes of their visits are kept in a log book, a receptionist could (at least) tell if the caller has an appointment.

A closed-circuit television (CCTV) entry system would give more chance to vet a visitor, who could be asked to show credentials to the camera. The camera should be linked to a video recorder—this alone might deter a dishonest person who does not want to be recognized.

A receptionist is the most common way of forestalling visitors, and an effective one if access beyond his/her position is physically controlled. A 'panic' button should be fitted below the reception desk, or somewhere else within easy reach, so that assistance can be rapidly summoned. An extreme example might be that a receptionist could be attacked or be taken hostage for bargaining by hostile intruders.

> ## ❗ WARNING
>
> There must ALWAYS be someone at a reception or security desk. Depending on the level of security required, no responsible member of staff should leave their post for ANY reason (least of all a reason suggested by the visitor) without someone 'standing in' for them.

WALK-IN THEFT

Walk-in theft is common. Thieves may be quite daring—although when most people are busy at their job, it is easy to steal things without being noticed. DON'T MAKE IT EASY FOR THIEVES. If you have a one-person office or workshop, lock the door whenever you leave—even if only for a few minutes. NEVER leave wallets or clothing unattended or on your desk. Bags, in particular, should never be left hanging over a chair or on the floor.

Portable expensive equipment, such as computers, should be protected in some way. Companies often 'brand' their names into the plastic casings, instal 'theft alert' tags which sound an alarm if the equipment is carried past a 'theft detector' at a prescribed point, or fix the equipment to desks. There are also cables which, when unplugged from equipment, will trigger alarms.

Escort all visitors

Breaches of security are more usual at night, but many thieves and spies just walk in and walk out unchallenged. ALL members of staff should be instructed to question EVERY unknown person not accompanied by known personnel. This can be done by offering help and escorting the visitor to the appropriate person/department, or waiting with them until someone arrives to deal with them. NEVER let a non staff-member move around the premises unaccompanied. AVOID escorting unknown visitors alone if you have to pass through unoccupied areas or corridors. NEVER walk in front of unknown visitors—keep an eye on them at all times.

SCREENS/CHECKPOINTS

Identity cards

Each ring of security must be controlled by security card-operated or digital keypad locks or manned by security staff. Staff at reception/checkpoints should NOT be expected to remember all faces and names of all personnel. Identity cards with photographs are commonly used and can be worn on duty, giving continual authorization throughout the building. These may also incorporate magnetic stripes which carry information regarding the bearer's identity and security status.

In companies where there may be a high turnover of staff, or where a large number of staff need access to many areas, ordinary keys would be totally impractical. Magnetic and keypad codes could be constantly updated, whereas ordinary locks would have to be changed and new keys cut.

PIN systems

Greater levels of security can be obtained by computer-controlled systems, where each employee not only has a card with a magnetic stripe, but also has to use a personal identity number (PIN)—just as you do when using a bank's self-service cash machine.

Transmitter/infra-red

An alternative to consider is a system which uses a small radio transmitter (similar to ones used to open automatic garage doors) or an infra-red 'remote control' (similar to those which operate car alarms and domestic electronic equipment such as televisions and hi-fis). It may be possible to 'personalize' these to individual employees, but this system should not be expected to work on its own. A transmitter might fall into the wrong hands.

PFI systems

In extreme cases, security can be provided by systems which use recognition of personal features to screen individuals. PFI (personal feature identification) can be based upon voice, hand or fingerprints or, amazingly, the pattern on the retina at the back of the eye.

PFI systems used to be relatively slow in operation, but the development of sophisticated microchips has now made them an effective but expensive way of controlling access. Their only real drawback (apart from the cost) is that they have a small error rate—between one and five per cent for voice prints. Although a system like this may exclude (and annoy) some people who are entitled to access, it will NOT admit anyone it does not recognize.

Timer locks

Timing devices can be used with both electronic and conventional locks to refuse entry. Even authorized personnel will be denied access. Not only will this prevent members of staff from moving around the premises when they are not entitled to, but it will also deny them access if they are placed under threat in any way. All safes and vaults should ALWAYS incorporate a timer.

Double-door entry systems

For an extremely high-security reception checkpoint at a main entrance or between security levels, a 'trap' between two sets of locked doors could be considered. The 'trap' could be monitored by CCTV, a video recorder and a microphone for further communication with a visitor. A built-in metal detector could be used to detect weapons, metal objects and tape recorders. A theft detector could sound an alarm if an attempt is made to carry out security-tagged portable material or equipment.

Such systems would slow down escape from high-security areas in the event of fire or other emergency. Safety may have to be balanced against security. There is always a risk of a 'fake' emergency being created or a door held open after the passage of a legitimate user—though this would become especially difficult with a 'trap' system.

GUARDS/GUARD PATROLS

Stationary guards or watchmen may require electronic alarms and CCTV surveillance equipment to enable them to know what is going on around a building, supplemented by guard patrols. Dogs can be used to announce the presence of intruders as well as deterring them.

Patrol times should not follow obvious patterns or else intruders would know when to expect them. Intruders may be able to calculate time taken between one point and another—routes should be varied. Professional thieves might even use their own temporary electronic sensors/detectors to let them know when someone is approaching.

Finding reliable guards is difficult. Do NOT seek to save money—good security pays for itself! Use an established and reputable company or train your own guards, if your organization is big enough to do so.

Limit the number of tasks a security guard has to do—he/she cannot be in several places at once! Monitor the guard's work to ensure that he/she does not become too bored. They must be alert at all times.

Never forget that it is lonely, slightly-risky work sitting at a security desk. Sooner or later a friendly 'visitor' is bound to try to start a conversation with the guard—who may be glad of the companionship. If the 'visitor' is clever, he/she may be able to gather all kinds of information.

BASIC BRIEF FOR GUARDS

- ALWAYS go by the rule book
- NEVER allow access without full authentication by the person/department expecting the visitor
- CONFIRM identification for any repairmen, service engineers, 'official' visitors and representatives
- DON'T confirm by telephone at numbers or using names supplied by visitors. Use your own sources. Make prior arrangements for regular visitors—these may include ringing a contact number to confirm arrival of a visitor and an appropriate security password

External vulnerability

You may not be in a position to organize a high perimeter wall, rolls of barbed wire, electric fences and guards in watchtowers but you CAN apply the same basic principles to any situation. It is very unlikely in the centre of a city that a site could be surrounded with a protective zone. It's likely that there may be public access around the building, which of course makes doors/windows highly vulnerable.

Really determined intruders may make approaches from underground, through adjoining buildings or from the roof. Ideally all possible routes of entry or weak areas require protection, although most 'professional' thieves find they don't have too much trouble. They KNOW the obvious steps which may be taken to secure premises.

Concentric rings of defence can compensate for external vulnerability. Penetrating the outer rings will not necessarily allow access to the real target of the intruder. A concentric system also makes it much more difficult to use even sophisticated surveillance techniques from outside the building.

Specialist advice should be called in when very high level protection is required or when the site is vulnerable to a break in security.

INDELIBLE EVIDENCE

Video recorders linked to a CCTV surveillance system permit close scrutiny of intruders and their methods. Dye-smoke devices are sometimes used to spray dye at intruders, marking clothes and skin with a harmless but long-lasting colour. Some of these dyes are invisible except in special lighting conditions.

⚠ WARNING

Can entry be made to your premises:
- Via ventilation and other ducts?
- Via sewers or other underground service channels?
- From the roof—especially if there is access from adjoining or neighbouring buildings within roping distance?
- Through the walls of adjoining properties, including cellars and roofspaces?

GUARD DOGS

 Dogs for guarding commercial and industrial premises are usually chosen from breeds which have been developed for their strong territorial sense and fighting qualities. German shepherd dogs (Alsatians), Rottweilers and Dobermans are frequently used. They may develop close alliances with their handlers, but should not be treated as domestic pets. To treat a guard dog as a family pet is to confuse both roles, and you risk undermining the dog's effectiveness. More seriously, the dog could pose a threat to other family members.

Guard dogs MUST be reared to follow a strict discipline and this must be maintained. They MUST be trained to control and restrain intruders, not to kill them, but must be tenacious up to the point of inflicting serious injury. Very careful training is required for both dogs and handlers.

An intruder may try to make friends with dogs by bringing food several times. Dogs must be taught to resist such temptations. Drugging/poisoning dogs is an old technique which should be a thing of the past. Dog 'whistles' and 'deterrents' which emit a sound that is beyond the range of human hearing should have no effect either. Real guard dogs should be impervious to all distractions—even the sound of gunshots.

SAFES/VAULTS

In a concentric security system—consisting of rings of tight security—the innermost ring is the safe, strongroom or vault. It should protect its contents from the actions of thieves, from fire, explosion and flood.

A safe should be totally immovable but, just in case, should withstand being dropped from a great height. Skilled safe-crackers use a variety of means, from acids and electronic listening devices to crowbars, high-powered burning equipment and explosive charges to open a safe. Safes are actually graded according to the time it takes a 'top' safe-cracker to open them. To be efficient a safe cannot rely on locks, but must withstand all forms of attack.

Vaults must protect their contents, even if the building around them is destroyed, and must also withstand the effects of fire hoses being played upon their heated outer casings. Doors and walls of a vault are usually rated according to the number of hours for which they can withstand fire.

The degree of protection you must seek to achieve should be matched against the importance of the 'valuables'. Find out exactly what specifications mean on any safe/vault. If there was an explosion, would the contents be totally unaffected? Would documents still be legible?

DOCUMENT SECURITY

Restricting access to documents and keeping them in a secure safe is only the beginning of document security. There are other ways of discovering their content. Rough notes, unwanted research material and all 'sensitive' waste MUST be destroyed. Carbon papers carry an impression of everything typed using them, as do typewriter and printer ribbons. They must also be destroyed.

Computer print-outs, however obscure, can provide information—facts and figures can be interpreted or passwords, codes and program structures may be revealed to a skilled interpreter. Sensitive waste should be placed through a slot into locked waste bins and shredded at the end of the day within the secure area.

COUNTERING ESPIONAGE

'Spying' isn't always about selling vital national secrets—it goes on in all areas of business, especially where there is great competition. ANYONE can buy bugging equipment and other 'toys' from specialist shops or by mail order.

Modern lenses make it possible to view and photograph images at considerable distances. Listening devices can monitor conversations. Interception of electronic data and even the electromagnetic radiation from computer terminals can provide a source of information. Sensitive material should be protected from ALL such forms of spying.

> ## ⚠ WARNING
>
> Phone tapping is illegal in most countries. In Britain, you must have a warrant from the Secretary of State or the consent of the person being tapped! You could face a fine or imprisonment. Mere possession of bugging equipment is not illegal, but use could constitute a prosecutable offence.

Cameras

Cameras should obviously be forbidden in areas where secrets are kept, but it is very difficult to identify a camera built into a wristwatch, a briefcase or a cigarette lighter—these are a cliché, but ARE available and DO get used!

Unless you make all staff and visitors change into special clothing and remove all jewellery how can you be sure they do not have a camera? Vigilance for any suspicious activity on the part of all members of staff is necessary. Ask yourself why the visitor keeps looking at his/her wristwatch, or why he/she seems to place his/her briefcase with such care?

Cameras can now be equipped with endoscope attachments which can be inserted through keyholes, though any small aperture drilled through a wall, through ventilation grilles and cabling ducts.

Long-range lens systems allow photographs to be taken from considerable distances. For maximum security never handle documents or place typewriters, printers, or computer screens where they can be seen from a window—at ANY level in the building. Even with fairly amateur equipment, a readable photograph of a document can be obtained at a distance of 90 m (300 ft).

Telephone tapping

Governments may employ telephone tapping as part of surveillance of suspected terrorists, and to provide information about political activists and criminals. SUCH TAPPING HAS TO BE OFFICIALLY AUTHORIZED AND ANY OTHER TAPPING IS ILLEGAL. It does not necessarily involve linkage directly into the main telephone system—bugging individual telephones, for instance, is far too easy and far too common.

Radio telephones, car phones, and cordless phones are even easier to 'tune in' on. They should never be used for conversations which need to be kept secret. Even the use of a baby alarm or similar domestic listening device makes it possible for outsiders to eavesdrop more easily.

Dealing with tapping

With a standard telephone system an efficient telephone tap is undetectable to the subject, although many people who suspect tapping of their phones have reported clicks, echoes and an unusually high number of service problems on the line. However, special equipment is available which will indicate the presence of a tape recorder, a telephone transmitter or a direct wire tap.

If you suspect tapping, or wish to protect information, fit a scrambling device. Systems are available on the open market which can offer a choice of 145 billion different codes. If these are continually changed, only chance (or other forms of espionage) will enable the eavesdropper to identify the code. Since the recipient of a call MUST be using a scrambler with the same code, this is only a practical solution for a limited number of regular calls. Portable, less-versatile versions enable scrambled calls to be made from any phone.

A device which works with all calls on the line is a telephone observation neutraliser. This sends a powerful current down the line which will melt transmitter wires—it could also damage the telephone line so its use is NOT permitted by most telephone companies!

Bugs

Microphones with transmitters can be fixed to a telephone or hidden ANYWHERE in a room, allowing others to eavesdrop and/or record everything that is said.

Several kinds of bug detector are available, disguised in many ways, from packets of cigarettes to standard desk accessories and office equipment—so that they can be used in circumstances where a certain amount of subtlety is required.

To defeat any undiscovered bugs, a transmitter can be used to 'jam' their transmissions. To avoid having to use elaborate equipment to identify the exact transmitting frequency of the bug, the jamming equipment should constantly sweep through a wide range of frequencies. This will have the effect of swamping the spy signal.

Long-distance listening devices

Highly directional microphones can focus on sounds at very long distances. There are mass-market devices which claim to detect a whisper at 30 metres (100 feet). The advertisements always have phrases such as 'a boon to the hard-of-hearing'—whereas they are actually a boon to anyone who wants to invade people's privacy. If sufficiently sensitive, long-range listening devices can 'read' a conversation from the vibration of the windows of a room! This is how laser 'microphones' work!

Such equipment can be confused by using another sound source—loud music, a noisy computer printer or other machinery close to your conversation point.

More thorough protection, which will interfere with all audio-reception devices, is given by using a device to generate white and pink noise at frequencies which are inaudible to the human ear.

Avoiding bugs

If your security needs are not great enough to warrant investing in special equipment, the best way of avoiding eavesdropping equipment is to carry on sensitive conversations in the open and in very public places where there is plenty of background noise—traffic, fountains, music.

Keep moving so that eavesdroppers continually have to change range and focus. Indoors go into a bathroom and run a shower with a radio playing. Don't forget the possibility of lip-reading or that either you or the person you are speaking to might be 'wearing' a bug.

Strange but true

If you think cameras concealed in lighters and briefcases with secret compartments belong in James Bond movies, think again—the range of security devices currently available is phenomenal, and in many cases, ingenious. Micro technology means that bugs, transmitters, even tape recorders can be reduced to minute proportions and, when concealed in 'ordinary' items, are undetectable to the casual observer. A microphone can be reduced to the size of a match and a whole room-monitoring system, complete with camera, can be concealed within a 'normal' book.

Entire offices can be kitted out with desk accessories and equipment where nothing is quite what it seems. A desk light can double up as an observation system, with a miniature video camera and recorder installed into the base. This means that the user can secretly see and record the activities in a room without being present. Alternatively, how about a framed print or picture for a conference room wall—again, complete with its own observation system. Even a tropical plant placed decoratively in a corner can house a wireless transmission/reception device, artfully concealed in a bamboo cane!

On the move, an anti-kidnap device is equipped with a quartz pulse-transmitter to help in locating the victim. A jacket button can double up as a microphone, and a handy can of 'X-ray spray' makes the contents of sealed envelopes visible. Watches (and lighters!) with secret built-in cameras complete the picture.

COMPUTERS

Computer eavesdropping

Although modern computer terminals emit much less electromagnetic radiation than in the past, it is still possible to use sensitive equipment to pick up radiation from computer terminals. Radiation travels through water pipes and power lines as well as through the air.

The proximity from which this radiation can be detected varies considerably from system to system—the cost of such eavesdropping, especially at any real distance, means that it is only likely to be used if the stakes are high. Tests can be carried out to discover the radiation levels your system emits. It has been reported that certain national military control centres are enclosed in copper-lined vaults to avoid such 'eavesdropping'.

Computer hacking

Any computer linked to a large network or outside sources, such as the telephone line, is vulnerable to 'hacking'—unauthorized access to gain information or to alter or corrupt data and programs. Whether the hacker's aim is to steal secrets, sabotage a competitor's files or simply to play a malicious joke, the means will be the same.

> ## ⚠ WARNING
> Computer hacking is ILLEGAL. Spreading computer viruses is a 'joke' that could land you in jail!

To avoid casual tampering or snooping at your terminal, arrange sensitive files carefully. Name them in ways YOU can recognize but others would not. It is possible on most good computers to arrange 'partitions' to hide really sensitive information. If done correctly, no casual 'spy' will even spot the partition, let alone be able to breach it without a password. Anyone who can open up the program and interpret it might detect the partition.

Coded access

Codes may be used to prevent unauthorized access to files, parts of files or whole systems. As with all codes, it is possible that other people may be able to decipher them. The more complicated they are and the more stages there are to discover the more difficult this will be, but also the more time it will take for those with authorization to gain entry. Too many codes and the procedure may be too difficult to remember. There would be a temptation to write codes down—which immediately makes them vulnerable.

Many people, quite naturally, choose codes with personal associations—the names of their children in birth order, the name of a favourite holiday resort, their favourite food. However, a saboteur could research such personal information to try to arrive at the correct codes. Using a series of figures linked with telephone numbers, social security numbers or birth dates is risky for the same reason.

In selecting codes, mixtures of letters and figures are preferable—to produce a wider range of variables. The longer the code the more numerous the variables become.

> **REMEMBER**
>
> When unattended, secret and sensitive disks should be locked in a safe or vault. Simple storage boxes could be smashed open or taken away and broken into later. Information on disks could not only be read, but also altered. Fortunately the time taken and expense involved in breaking codes and operating surveillance can be prohibitive.

Computer viruses

Viruses that infect computers or computer networks behave, in some ways, like real diseases. A computer network could suddenly become very 'sick'. Most are stupid, but often devastating, practical jokes—others may constitute serious sabotage. Only when a program begins to go wrong does the 'sickness' become apparent.

Viruses are 'caught' when a program containing the virus is fed into the system. Once it has entered, it can affect other programs, disks, terminals and networks. Viruses can also be transmitted down telephone lines.

Screening for viruses

Fortunately, screening programs are now available which can check every new disk for viruses. These screening programs should be used at the beginning and end of every work period (at least) on every terminal and before and after ANY new disk is inserted.

Checks on disks from outside sources should be made at a terminal which is not linked to others and not used for other work. In this way a virus can only be caught by that terminal and can be 'cured' or erased on that terminal. Modems (phone links) should NOT allow access to a whole network—a terminal should be set aside to receive incoming transmissions so that outside 'infection' can be contained.

Anti-viral 'disinfectant' programs are available to 'cure' most known viruses.

Funny but deadly

Most viruses are created by fairly clever people with 'nothing better to do', although there may be cases of revenge by unhappy employees. One of the first known was a 'time bomb'—viruses are often designed to do their work at a prescribed time, or to be triggered when certain conditions occur. The time bomb involved a 'Happy Birthday' greeting to one of the largest computer corporations on a special anniversary. The fullscreen greeting popped up on VDUs all over the world.

Viruses can cause extremely odd problems—and may not be spotted until it's too late. Some only 'mess up' programs or systems—others are more 'imaginative'. It's impossible to predict what the next joker may dream up but—to give you an idea—here are some examples of real viruses:

❍ The cursor dies and falls to the bottom of the screen or becomes 'bouncy' and impossible to control with a mouse.
❍ One figure in a huge column of figures spontaneously changes—every nine becomes a four, for example.
❍ A picture appears across the screen—one famous example was a very rude drawing of a naked woman. While the operator stares in surprise the entire memory of the hard disk in his/her terminal is erased.
❍ Obscene words appear in the middle of innocent documents, often triggered by common word combinations.
❍ Obscene insults to the user flash up on the screen.

Perhaps the weirdest aspect of viruses is that some have been known to lie dormant for long periods of time, to mutate or even combine with other viruses to produce new ones. Strict screening procedures have helped to reduce the spread—but enormous damage has been done.

VEHICLES

Cars should always be securely locked when parked. Choose a fairly busy location—under a streetlamp at night, if you can't park in a garage or off the road. Make sure that all windows and sunroofs are secured. In Britain alone, a car is stolen every two minutes!

A variety of devices is available to prevent the car from being driven, even if a thief has a key that fits the ignition or bypasses the ignition. There are simple restraining bars to lock the steering wheel to a pedal, or immobilize the gear stick. Increasingly, locking devices may be built into the car itself.

Car alarms, like those for buildings, operate in a variety of ways. Some use infra-red beams to detect any movement within the car. Others are vibration detectors which respond to movement of the car itself—some are so sensitive that even touching the car can set them off. Careful adjustment is necessary if there are not to be repeated false alarms.

Radios

NEVER leave valuables in a car—or if there is no option then make sure that they are locked out of sight—don't tempt thieves! Car stereo systems are a favourite of thieves. Some car manufacturers are coding digitally-tuned radios to match the individual car. An experienced thief should recognize this type and leave it alone. One drawback that was found with this system is that they have to be returned to the manufacturer to be reset, if your battery ever runs down.

Other radios are made to slide into and out of the fascia board. They can be removed easily so that you take them with you. If you forget or don't bother, it is just as easy for a thief to slip the radio out! DON'T hide it under the seat—that is the first place a thief will look! Take it with you. The drawback is that you may simply lose the radio!

Motorcycles/bicycles

Motorcycles have steering locks to prevent them being ridden away by thieves but that will not deter professionals who may load them on a lorry. Chain them to railings, a lamppost or other fixture. A cable alarm system can be used which will be activated if it is severed.

Bicycles are particularly vunerable to theft. The whole bike needs securing—thieves commonly steal individual bicycle parts. An expensive front wheel should have a quick-release mechanism. Detach it, run the security cable or chain through it, through the back wheel and the frame and around a fixed post. Even this, however, will not deter a determined thief.

7

Many jobs involve hazardous processes or chemicals, but even working in an office may pose severe risks. Leisure activities, too, carry a level of danger — this may be part of the thrill of many sports, but serious injuries are common.

Work & play

THE WORKPLACE Hazardous chemicals • Diseases/disorders • Physical damage • Noise • Radiation risks • Work posture • The unhealthy office • 'Sick' buildings • Legionnaire's disease • Humidifier fever • The public • Working outside

LEISURE Television • Music • Barbecues • Spectators • Angling • Playgrounds • Radio-controlled models • Fireworks • On the town

SPORT Strength • Stamina • Suppleness • Warming up/cooling down • Women and sport • Types of injury • Clothing • Sport by sport

THE WORKPLACE

Although factories, shops and business premises may be regulated by laws concerning work conditions and safety measures, it is the workforce itself which must see that these laws are implemented. The people who actually do the jobs should be most able to see when laws are ignored, and to recognize when safety measures are inadequate. Discover, through your union or staff organization, what regulations apply and see that they are followed. An occupation should NOT involve you risking your life or your long-term health.

PLEASE NOTE

It is not possible to give full details of EVERY type of work hazard for EVERY occupation. The information given here is designed to highlight types of hazard, some of which may be quite unusual. Think about YOUR job and its risks. Are you safe?

Fire

The provision of appropriate smoke detectors, alarms, sprinkler systems, fire-fighting equipment, fire exits and escape routes is VITAL. Use fire-fighting equipment only if you have been trained and you know what kind of fire it is appropriate for. Know the location of fire exits. Fire drills should be held regularly and taken seriously. Responsibility for helping any disabled or injured colleagues, or anyone else who might need special help, should be assigned and such assistance practised (see FIRE!: **Equipment & drills**).

! WARNING

Don't increase fire risks! NEVER let piles of inflammable waste or scrap build up in the workplace. ALWAYS clear up spills of oil, fuels and solvents.

HAZARDOUS CHEMICALS

Where hazardous chemicals and other potentially-dangerous materials are stored or are used, 'hazchem' symbols should be displayed. These indicate the nature of the danger to the emergency services and give them guidelines

HAZCHEM CODES

CODE	DANGER OF VIOLENT REACTION/EXPLOSION	PROTECTIVE CLOTHING	BREATHING APPARATUS f = when on fire	ACTION
P	YES	YES	NO	
R	NO	YES	NO	
S	YES	NO	YES	**DILUTE**
S	YES	NO	YES **f**	
T	NO	NO	YES	
T	NO	NO	YES **f**	
W	YES	YES	YES	
X	NO	YES	YES	
Y	YES	NO	YES **f**	**CONTAIN**
Y	YES	NO	YES	
Z	NO	NO	YES	
Z	NO	NO	YES **f**	

REMEMBER

Even if you don't know the meaning of the codes yourself, ALWAYS give them to the emergency services when alerting them to a fire or accident. Doing so may save seconds — or lives!

for dealing with it. Unfortunately the codes are not internationally identical.

Most hazchem codes are made up of two or three characters. The first is usually a number which denotes the type of fire-fighting or controlling agent to be used:

1 = Water   **2** = Water fog   **3** = Foam   **4** = Dry agent

The first letter indicates the type of dangerous substance which is involved, what precautions must be taken and whether or not the toxicity of the substance means that all spillages must be contained. Where **dilute** is indicated, the residue may be washed away to drains. The instruction to **contain** implies that it is important to prevent chemicals reaching drains or watercourses.

Where **E** is shown, people should be evacuated from the building or the vicinity of the incident. There may be other better-known symbols or whole words included—**CYLINDERS** indicates the risk of a serious explosion.

Care at work

Chemicals such as mercury compounds were once used in the millinery trade. These have been known for some time to damage the brain, to produce 'mad hatters'. In many cases, such as this, the dangers were recognized too late.

The effects of chemicals are frequently cumulative—they build up in the blood and may cause disability or even cancer later in life. If your work involves handling any potentially hazardous substances or inhaling fumes or airborne particles, you should ensure that proper provision is made for your protection.

Protective clothing is only one element to consider. Methods which control and/or contain the hazard at source are the first requirement and proper ventilation is essential (see DIY/ CRAFT HAZARDS: **Protective clothing**).

The storage and handling of flammable, explosive, caustic and toxic materials, especially liquids and gases, is another obvious area where strict safety measures are required.

Graphic designers/illustrators

Spray glues are commonly used—although steps have been taken to discourage such substances. The main dangers are from the solvents employed (see POISONS: **Solvents**). Although these aerosol glues allow for fast working methods, the breathing of solvent vapours and glue particles has been proven to cause respiratory disorders. Other symptoms which may indicate inadequate ventilation are: headaches, nausea, short-term memory loss and dermatitis. Long-term effects may be far more serious—involving heart and liver damage and cancers.

Airbrushes—devices which atomize paints, inks and dyes—are also common. Many pigments are HIGHLY dangerous, especially when inhaled or ingested (see DIY/CRAFT HAZARDS: **Artists' paints**).

☠ **WARNING**

NEVER put paintbrushes in your mouth to 'put a point on them'. Depending on the types of pigment involved, the cumulative effects could be extremely harmful—even fatal. The most dangerous pigments include almost all metal/metal oxide colours and lamp black. Aniline dyes are also poisonous.

Adequate and specialized ventilation systems or special spray booths should be used for better spray glues and airbrushes. Comfortable breathing protection would be sensible. The alternatives to spray glues include special adhesive waxes which are relatively safe for long-term use.

SOLVENT RISKS

Many shoe repair and dry-cleaning shops may have insufficient ventilation. Shoe repairs commonly involve the use of powerful solvent-based adhesives. Dermatitis is an extremely common side effect, also caused by the leather dyes and polishes, but there are far more serious risks from constant inhalation of solvent vapours. Dry-cleaning solvents are equally toxic. Short-term effects may include headaches, nausea and memory loss. Long-term effects may involve heart and liver problems and cancers (see DIY/CRAFT HAZARDS and POISONS).

REMEMBER

You MUST understand the dangers of dealing with chemicals on a long-term basis (see also SAFETY FIRST and DIY/CRAFT HAZARDS):
- MAKE SURE you know what to do in the event of any accident with chemicals — spillages may combine together to produce toxic or dangerous effects
- NEVER transfer chemicals from original marked containers to improvised bottles and jars
- NEVER leave containers open
- MAKE SURE all containers are labelled correctly and clearly
- NEVER use chemicals from indistinctly-labelled containers
- BE TIDY! BE SAFE! Clear up spills as advised and dispose of as directed. Disposal information MUST be available to you

DISEASES/DISORDERS

For centuries, it has been known that workers in certain industries and occupations were prone to particular diseases and disorders. Coal miners and quarrymen were often stricken by the lung diseases pneumoconiosis and silicosis.

There are many other trades in which the risk of illness, injury or early death was the price for being in work. Today, an increasing number of hazards are now being recognized which were previously ignored or not understood.

► Additional information on safety in the work area and protective clothing can be found in DIY/CRAFT HAZARDS.

Dusts of many kinds, including metals, woods, minerals and even flour, may seriously damage health—affecting lungs, nose, throat, eyes, skin, digestive and urinary systems, the brain and nervous systems (see, for instance, DIY/CRAFT HAZARDS: Wood dust). Wear the correct level of breathing protection.

☠ WARNING

Mineral oils!: All kinds of mineral oil, from bitumen and petroleum to paraffin and lubricants, present health risks. Most common problems are acne and dermatitis, but there are also risks of cancer and of respiratory problems from inhalation of fumes and particles. Cutting oils—usually formed by mixing a lubricating oil with animal or vegetable fats and other additives, present a particular risk—especially when unrefined. Some machine operators may be continually exposed to or splashed by such oils. A man straddling or rubbing against a machine may become oil-soaked. The scrotum is particularly susceptible to contamination and there is a great possibility of cancer. Women can also develop genital cancer from contact with mineral oils. NEVER keep oily rags in a trouser pocket. ALWAYS wash your hands free of oil BEFORE going to the lavatory.

REMEMBER

The warning above is just one example of a serious work hazard—do you know all the risks associated with your occupation?

It's up to you!

Take a look at your colleagues at work—especially older ones who have been doing a job for many years. Do many of them show similar health problems? Those who do may have been heavy smokers or have other similarities that make them more prone to diseases—but could their problems be work related? Will the same things happen to you if you continue to work in the same way?

When people have 'always done it this way and it's never done me any harm', they may be being blind to the obvious.

► Additional information on safety in the work area and protective clothing can be found in DIY/CRAFT HAZARDS.

If there is a chance your job puts you at risk, are there ways in which you could change working methods or conditions to make it safer? If you feel the risk is unavoidable, is it really worth it? Would it be better to look for something safer? You may not have many choices of occupation but, if people refuse to do unhealthy jobs, eventually the employers may be forced to make the jobs safer.

PHYSICAL DAMAGE

Many jobs, especially those involving manual work, place physical stress on the body and can cause damage to it, quite apart from accidental injuries. Those at risk range from construction workers to teachers, nurses to hairdressers, shop assistants to computer operators. Common causes of injury are:
- Bad posture
- Physical strain
- Vibration
- Repetitive operations

! WARNING

You don't have to go out to work to suffer from work-induced illnesses and disorders. Domestic chores can involve the same strains and cause the same problems and stresses.

Home workers and freelances should be particularly aware of the dangers of their jobs and the stresses induced by poor work environments. They may develop bad work practices or take undue risks because they have no one else to alert them.

Accidents

Many jobs involve working with dangerous tools, fast-moving machinery or in perilous locations. Where appropriate, helmets, goggles or face masks, gloves and other protective clothing or safety harnesses s hould always be worn. Hair should NOT be worn loose and trailing parts of garments must be covered, so that they cannot catch in machinery. Safety guards on tools and machinery must be used, even if it is possible to work more quickly without them. A slight increase in productivity is not worth a damaged limb or a lost life.

WARNING

NEVER operate a machine unless authorized and trained to do so.
NEVER attempt to clean or clear a blockage on a machine which
is in motion or connected to a power source. **SWITCH IT OFF AND
DISCONNECT IT FROM THE POWER SOURCE.**
NEVER wear loose clothing, long hair, dangling chains or other
jewellery which could get caught up in moving parts. NEVER
distract other people who are operating machinery.

Scaffolding and other high work sites must have barriers to prevent anyone accidentally stepping off.

SAFE MACHINE OPERATION

When operating any machinery—industrial or otherwise—never start the
machinery until you can answer 'yes' to ALL the following questions:
- Do you know how to stop the machine?
- Are all fixed guards fitted properly and all mechanical guards in working order?
- Is the area round the machine clean, tidy and free from obstruction?
- Are you wearing appropriate protective clothing, safety glasses, shoes,
  gloves etc?

IF you have ANY doubt about whether a machine is working properly, inform
your supervisor immediately.

Machine guards

Fixed guards prevent contact by any part of the body with dangerous parts of the machine. Interlocking guards ensure openings are automatically closed before the machine can be operated and stay closed until the operation is ended.

Automatic guards may be operated by photoelectric cells, contact with a physical barrier or pressure-sensitive mats. These stop the machine or prevent it starting if someone is too close to the machine.

REMEMBER

Check that guards are in place and automatic devices are working properly
BEFORE starting any machine. NEVER TAKE RISKS! By the time a machine is
stopped you may have lost a limb or your life.

It's up to you!

Sometimes an employer or manager may fail to implement proper safety measures, to save money or speed up output. The individual worker may make the decision to work without protection—especially when safety equipment is cumbersome or uncomfortable. It is the responsibility of EVERY individual to know and understand risks and to see that adequate provision is made to ensure work safety.

A study by the British government's Health and Safety Executive showed that carelessness and human error were the cause of only a small proportion of accidents at work—the main cause was the employer's failure to provide a safe system of work. Make SURE that this is not the case at your workplace.

Vibration

The long-term use of vibrating tools, such as pneumatic drills, hammers or chisels, chain saws and electric grinders, can cause hand/arm problems leading to permanent damage.

Vibration damages the tendons, the nerves and the blood capillaries and may even lead to decalcification of the small bones in the hand. This damage frequently shows itself as vibration white finger (VBI, also known as dead finger or Reynaud's phenomenon)—an untreatable condition which has been known as a risk to stone cutters, riveters and road workers for many years. Although there is no 'cure', the conditions may improve if exposure to vibration is avoided.

The first noticeable symptom of white finger may be an intermittent tingling in the fingers. Next is intermittent numbness. Even when there is a blanching of one or more fingertips (with or without tingling or numbness) this may not interfere with the use of the hand. The condition can progress to affect more of the finger.

In cold winter weather, when the reduction of the blood supply is more noticeable, the condition is likely to be worse and the victim may find it interferes with the use of the hands.

When blanching is extensive and begins to occur in warmer months, the condition is usually bad enough to affect all finger use. If you think you have the early symptoms, see a doctor to be sure, and try to change the kind of work you do before the condition develops further.

Whole body vibration

This occurs when the whole body is shaken by a machine or vehicle. The effects vary according to the frequency and strength of the vibration and the length of exposure to it. A frequency of less than once per second can induce motion sickness; from one to 80 vibrations per second can lead to tiredness, disturbed sleep, blurred vision, piles, hernias and damage to internal organs. Bus and heavy goods vehicle drivers may be subjected to constant engine juddering.

Suspended cabs in vehicles and improved mounting and maintenance of machinery can reduce the problem by eliminating vibration or helping to isolate workers from it. If vibration does occur, the periods of exposure should be kept as short as possible to minimize the health risks.

NOISE

A great many jobs are noisy by necessity. Noise-induced hearing loss is among the most common work-related injuries and disorders. It is usually due to temporary threshold shift (TTS)—the persistent noise level causes the hair cells of the inner ear to shift their field of sensitivity upwards and they no longer respond to soft sounds.

Temporary hearing loss ('auditory fatigue') can occur after only a few minutes' exposure to intense noise, but is usually reversible after a period of time away from noise. When exposure occurs over months or years, only partial recovery may be possible. Deafness to certain frequencies or even total hearing loss may be experienced. Where those exposed to noise are never away from work for more than two days, temporary hearing loss is likely to become permanent hearing loss.

Workers in manufacturing industries exposed to high levels of machine sound are the most likely to develop noise-induced hearing loss. The woodworking industry creates some of the highest noise levels:

❍ 100 dB (bandsaws/panel planers)
❍ 105 dB (edge banders/multi-cutter moulding machines)
❍ 107 dB (bench saws/high-speed routers)

Road and rail workers, airport ground staff and nightclub workers are among many others exposed to high noise levels.

If noise cannot be eliminated or reduced by redesigning the process or the machinery, some alleviation can be obtained by enclosing the machinery or blocking the noise transmission path, using acoustic chambers, hoods or guards.

Enclosing workers employed in quieter areas is a parallel measure if they are not working on or do not have to be in direct contact with the noise-producing machine. The risk can be further reduced by reducing shift times in the noisy areas.

Individual hearing protection

Ear muffs, plugs or canal caps on a headband can all be used to block sound from reaching the ears—but they do not remove the hazard. Personal hearing protection must be worn BEFORE entering the work area or switching on noise-producing machinery.

It must **not** be removed while still exposed to high noise levels—not even for a couple of minutes. In a jet engine room, for instance, where noise levels reach 117dB, removing ear protection for as little as one minute would give a 'dose' equivalent to the recommended maximum of 90dB for eight hours.

CHECK: Hearing protection should be compatible with other safety equipment—especially when worn with helmets (see DIY/CRAFT HAZARDS).

! **WARNING**

Wearing ear protectors makes verbal communication difficult (unless radio headphone systems are used) and also muffles any warning alarms or disguises the source of sounds, which could lead to accidents.

PROTECT YOUR SIGHT

Does your work put you at risk from any of the following?
■ Flying particles of any material
■ Splashing of liquids
■ Dust
■ Irritation by fumes or vapours
■ Molten metal splash
■ Exposure to infra-red or ultra-violet light, lasers or strong visible light
YOUR SIGHT MUST BE PROTECTED. Whether plain or filtering goggles, head covers or separate fixed or movable shields are suitable depends upon the process (see DIY/CRAFT HAZARDS: Protecting eyes).

REMEMBER

It is not only the operative who is at risk. Bystanders where oxyacetylene burners and similar tools are in use are particularly liable to 'arc eye', acute eye inflammation (and possible permanent damage) caused by exposure to ultra-violet light (see DIY/CRAFT HAZARDS: Welding).

►Additional information on safety in the work area and protective clothing can be found in DIY/CRAFT HAZARDS.

RADIATION RISKS

 Everyone must now be aware of the need for protection from radioactivity, whether in the nuclear industry, medicine or the laboratory.

Any form of radiation can be harmful above certain levels and radiographers, dental nurses and others who operate x-ray machines must ALWAYS be screened.

Light is a form of radiation and, although visible levels of light are not usually harmful, exposure to ultra-violet light, emitted by many photocopiers, can damage the eye. The lid on the machine should cut out most of this radiation, but some blue light often gets through and can damage the retina of the receptor cells at the back of the eye.

Fluorescent lighting also emits ultra-violet light, but at lower levels. Many people find that working under fluorescent lighting causes sore eyes, headaches—even nausea.

Lasers can cause retinal damage—NEVER stare at them.

Workers in jobs where they routinely look at very hot surfaces, such as molten metal and glass, may run the risk of developing opacities in the lens of the eye. Filtered goggles should be worn. Radiant heat can also cause overheating of the body and heatstroke.

> # ❗ WARNING
>
> **Photocopiers: In addition to ultra-violet radiation photocopiers produce both hydrocarbon and ozone emissions. Nitrous oxide is also sometimes produced and toner dust is suspected of being carcinogenic. Liquid toners can also give off fumes which can produce dizziness and nausea if inhaled. Site copiers where ventilation is good. Limit time spent near a copier. Do NOT make copies with the machine uncovered, no matter how laborious or time-consuming the job.**

VDUs

Visual Display Units—the screens of computers—are like televisions (see **Leisure**). They emit some radiation, but are usually manufactured to higher standards than televisions. Working at a VDU does demand that the operator sits at much closer proximity than the viewer of a television would choose for comfort or safety.

Japanese studies (unconfirmed) suggest exposure to radiation from VDUs could influence miscarriages or deformity of the foetus during pregnancy. It has also been suggested that sperm may be affected.

Measurable levels of very and extra-low frequency radiation do occur, though there is no firm evidence that these have any biological effect on humans. These emissions do not come from the screen, but from the transformer windings and would be present in many kinds of domestic apparatus.

It has also been suggested that using a VDU damages the eyesight. It seems probable that eye problems among VDU users have been revealed by the use of the screen rather than caused by the VDUs themselves, but any prolonged and concentrated viewing can cause eyestrain.

REMEMBER

Anyone with uncorrected, or imperfectly-corrected eye defects is likely to suffer greater eyestrain when using VDUs than those with properly-corrected vision. A thorough eye test by an ophthalmologist is recommended before anyone begins work with VDUs, and annually thereafter. Restrict working time to about an hour, before taking a break for a few minutes to rest the eyes.

The flickering of the screen, though usually imperceptible, may be a risk to people already subject to forms of epilepsy affected by flickering light or striped patterns.

The build-up of an electrical charge on the screen may be linked to forms of dermatitis on the face and hands of some computer operators with particularly sensitive skin.

Most VDU problems are caused by bad work postures, poor lighting conditions or concentrated work for long periods without break or exercise.

RADIOGRAPHERS

Radiographers, dental nurses and others who operate x-ray equipment are continually at risk from radiation, if they do not take proper precautions. Always retreat behind the appropriate shields before operating the x-ray (and make sure that the subject is also protected, except for the area which is being investigated). There may still be some slight risk from cumulative exposure, leading to cancer, though this may not become apparent until ten or more years after you have given up the job.

WORK POSTURE

Badly-designed seats, having to stand for long periods, poor lighting that forces people to get too close to their work, uncomfortable bench or desk heights and reaching too far, are all causes of bad work posture. They place parts of the body under undue strain.

Back problems and knee strain occur from lifting heavy weights, whether office furniture, sacks of cement or hospital patients. Sitting all day in a bad posture will also cause back problems, whether driving a bus or working at a desk. Piles and varicose veins can be caused/aggravated by the same conditions, and also by standing all day.

A brisk walk, or lying with the legs raised, will help to counter a build-up of circulatory back pressure and improve the flow of blood to the heart. During a break, 'put your feet up'!

Optimum bench/desk heights

❍ Work surfaces should be at a level which allows the arms to be bent at the elbow at right angles to the body.

❍ It should be possible to reach work by extending the arms while still keeping the back straight.

❍ Computer and typewriter keyboards should be at a level which allows the forearms to be at right angles to the body or slightly lower so that the wrists are not bent upwards. This is lower than a convenient writing height.

❍ When sitting, thighs should be parallel to the ground, with the legs bent at a right angle at the knee. Feet should touch the floor or a footrest used to prevent pressure on the back of the thighs.

 Can you avoid lights or images reflected in the screen?

 Screens should NOT face windows. If direct or reflected glare cannot be avoided, it must be shaded and direct sunlight into the room should be diffused. Anti-glare screen filters also help to cut down on reflections.

Is normal office lighting suitable for working at computer stations?

No. Ideal room illumination for operating a VDU is 300 lux, for general officework 500 lux or above. If you do other work at your computer station you should have separate desk illumination.

 Should you be able to read the screen without leaning forward?

You should be able to read the screen by looking down slightly, without bending your neck.

How long should you work at your computer?

 You should take frequent pauses, long enough for eyes to recover from the effort of reading the screen and to interrupt the repetitive strain of keyboarding. Breaks should be at least 15 minutes per hour of intensive work or every two hours of less intensive work. Mixing VDU work with other activities is the most efficient way of providing these breaks.

On your feet all day?

There are many jobs which involve standing for long periods which can lead to circulatory problems producing varicose veins and aggravating conditions such as haemorrhoids (piles). Resting when possible, preferably with the feet raised, and taking a brisk walk whenever there is an opportunity will help minimize these effects. Other aspects of the job will cause further problems. These are just some examples:

Hairdressers Staying at your work station while an assistant runs your errands is not a good idea. By all means let your assistant wash a client's hair, but collect the client from reception yourself and take the need to fetch towels or equipment as an opportunity for exercise. Assistants shouldn't be reluctant to fetch a client's coffee—it all helps to counter the static nature of your job!

Nurses Are subject to many of the problems which doctors have to face. They frequently have to lift heavy patients in awkward positions, placing great strain on the back. Follow all the advice given in training to make lifting easier and get help with heavy patients. Back strain is a major cause of nurses being off duty and of them leaving the profession. Undue strain not only damages you—it puts patients at risk as well!

Aircraft cabin crew Stewards and air hostesses are not only on their feet, they have to ferry trays and push trolleys, often when the plane is flying at an angle. However, their constant activity tends to counter some of the effects of standing. Longhaul routes, with consequent 'jet lag' problems, not only affect sleep patterns but interfere with menstrual cycles. Dehydration caused by flight conditions dries skin and hair and can also cause or aggravate sinus and other conditions. Drinking plenty of water during the flight is even more important for cabin crew than for their passengers!

Shop assistants If stuck behind a counter, make sure you have a chair or stool to relieve pressure on your legs, ankles and feet. If selling on an open floor take the opportunity to move around—not just to greet customers and make a sale but also for the sake of preserving your own health.

Petrol station and car park staff Forecourt attendants do a lot of standing, but they and people who work in car parks are also continually exposed to petroleum fumes, contact with oils and exhaust fumes. Ensure that enclosed spaces are well ventilated and position yourself where you can get as much fresh air as possible, when not attending to clients—this may be difficult at roadside stations as passing traffic provides even more dust and pollution.

REPETITIVE STRAINS

These are exactly what the name suggests, injuries caused by repeatedly making the same demand on tendons, muscles and joints—such as using a hand screwdriver all day or pounding on a keyboard. They include a number of conditions, from ganglions (small round swellings which are usually painless) to inflammation of tissue, of tendons and of tendon sheaths, and can lead to osteoarthritis.

Repetitive strain injuries have been around for a long time, especially in jobs which require the same small movements to be carried out at speed as, for example, in a production line. However, the victims, usually unorganized labour (so that no action was taken to change the system) often gave up their jobs and were replaced by others.

It was not until the introduction of word processors and computer keyboarding and the appearance of repetitive strain injuries in white-collar workers (especially journalists and others motivated to draw attention to the problem) that it was widely accepted—although there are still some people who deny that the condition exists!

Proper ergonomic working conditions in offices which prevent undue pressure on wrists, elbows and fingers make it less likely that the condition will develop, but the length of time spent at a keyboard should be limited and broken up by periods doing other types of work.

The problem does not usually occur with old-fashioned manual typewriters, partly because the physical actions are more demanding so that muscles are more generally used. Speeds are usually slower and typists stop to change paper, to make corrections and to carry out other office activities.

THE UNHEALTHY OFFICE

Offices might be expected to be much safer than industrial premises—no dangerous machinery, risky processes or toxic chemicals, no working at heights or in difficult conditions.

In fact, all these hazards CAN be present in the office environment—from guillotines to ozone emissions, and from cramped conditions to electrical dangers.

Inadequate ventilation, poor lighting and high noise levels are common in many offices. All add to stress levels and undermine both health and work efficiency.

'Sick' buildings

The workforce in, or users of, some buildings sometimes report a higher incidence of illnesses than other similar places of work—for no identifiable reason. Offices, hotels and shops are among the types of building usually affected. Common symptoms which occur are:

- Irritation of the eyes, nose and throat.
- A feeling of dryness of the skin and mucous membranes.
- Skin rashes and itching.
- Mental fatigue.
- Headaches, nausea and dizziness.
- Recurrent coughs, colds and throat infections, hoarseness and wheezing.
- Recurrent gastrointestinal upsets.

There may be no obvious cause for some of these disorders, but there may be common features in these buildings, such as:

- Forced and closed ventilation system with air-conditioning.
- Windows cannot be opened.
- Lighting is ill-considered, or unsympathetic.
- Whole interior relatively warm, with no temperature variation from one area to another.
- Indoor surfaces often covered with textiles—carpets or fabric-covered walls and room dividers.
- Poor standard of hygiene with drinks machines or staff 'kitchens'.

It has been suggested that the causes of ill-health may include chemical pollutants, from carpets, paints, furniture, office equipment and chemicals used with it—all of which can collect if ventilation is poor. Dust and fibres from furnishings and carpets, and bacterial contaminants, are also possible airborne causes. Low humidity, inadequate air movement, the lack of negatively-charged ions in the air, overheating, poor lighting and other undesirable environmental factors could all be contributory factors.

Apart from complete replanning of the interior of the building and its maintenance systems, there may be little that can be done to put things right—precise causes are often almost impossible to identify. Individuals can improve their own circumstances by managing to get a window open or at worst using a portable fan to combat stuffiness and high temperatures. The most effective action if you feel at risk is to change your job!

Humidifier fever

Sometimes part of the 'sick building' syndrome, but a problem which can occur in otherwise satisfactory buildings, humidifier fever is a flu-like illness. It is caused by the inhalation of fine droplets of contaminated water from the reservoirs or holding tanks of humidifier systems. As well as being used in air-conditioning systems, humidifiers are used by some printing firms to stabilize paper size and condition.

Symptoms of humidifier fever vary from a mild disorder with headaches and muscle aches to an acute fever and cough with chest tightness and breathlessness on exertion. Symptoms usually appear four hours or more after the beginning of a shift at work and most frequently on the day of return to work after the weekend or a longer break. The body usually manages to overcome the symptoms after 12–16 hours, but the problem is that they recur on the return to work.

If humidifier fever is suspected, blood tests and samples from the sludge in the system can give confirmation. Preventing a recurrence, even with regular cleaning and servicing of the system, often proves difficult and it is best to replace it with one which does not produce water droplets.

REMEMBER

AVOID humidifier fever by using:
- **Steam humidifiers, which produce no droplets. Their high temperatures also kill micro-organisms**
- **Compressed air atomizers, which do not use water reservoirs**
- **Humidifiers which use evaporation instead of spray (though these can also develop contamination in their reservoirs)**

THE PUBLIC

Police officers, psychiatric nurses and social workers know that they may often have to deal with violent and difficult clients or members of the public and, like people in the armed services, must accept some physical risk as part of their job. They are usually given training to help them deal with people, but there are many other jobs that involve direct contact with members of the public that pose similar risks.

Bus and taxi drivers are now frequently protected by screens to prevent direct assault, but are still vulnerable when not behind that protection. The same often applies to staff in banks, booking halls and enquiry offices when interview or information desks are situated in the open without protection.

In many situations involving the public, feelings can run high and may lead to verbal abuse and actual physical violence—often as a release for pent-up frustration or through fear or embarrassment. Disputes, whether over tickets for a concert, in a restaurant or in a store, can result in similar physical assault. In such circumstances there may be some build-up to the situation and, if so, call for help or take some kind of defensive action (see SELF-DEFENCE).

An exposed location makes you vulnerable to psychopaths or troublemakers, who may have no particular argument with you other than a focus for their anger.

To minimize the risk, such attack locations should be contained in some way. Preferably there should be a wall to prevent approach from the rear. If this is not possible, a fixed screen should be erected. The site should be as close as possible to a door leading to a non-public area which can be easily secured against entry.

A help button should be placed below the desk so that assistance can be called if a situation seems likely to escalate or a sudden attack is made. A telephone should always be available, for direct contact with security staff or the police without having to go through a switchboard. Desks should be deeper than arm's length, so that a blow cannot be directly delivered across them.

Anyone working in such circumstances must develop considerable tact in dealing with the public and keep calm in the face of provocation. Never return violence, though you should seek to contain it. Members of the public may come to your assistance if you are attacked but, before requesting such help, consider how much support might be given to your antagonist! You might provoke a riot!

Doctors

Doctors are exposed to the many diseases of their patients. Although doctors seem to develop considerable immunity, they are particularly at risk from the viruses that produce flu-like illnesses. However, it is stress which takes the largest toll. Not only do doctors have to cope with the strain of being responsible for their patients—often in life-and-death situations—but also with their social and emotional problems.

Many family doctors, interns or hospital housemen and women work excessively long shifts and are on call at unsocial hours. This can disrupt their own personal and social lives and increases the stress load. Nicotine, alcohol and drug abuse, despite their professional knowledge of the damage these can do, are sometimes a resort as a relief from tension, which can lead to or aggravate depression.

Doctors need to give particular attention to their own stress management and to take their own health seriously—all too often it is badly neglected.

Banks/finance houses

The competitive work of dealing-room staff on the money and commodity markets places them under continual high levels of stress. Coupled with high earnings, this may lead them to live the 'high life' in their free time. When recession hits or markets collapse their jobs can collapse too.

It's not only the wheeler-dealers who come under pressure. The most staid bank manager is dealing with clients' problems—not just handling loans and financial advice—and can be perceived as being the cause of them at times of foreclosure. This can lead to all kinds of psychosomatic problems, ranging from backache to loss of libido. The traditional round of golf can be the cure-all if it is simply used for relaxation and exercise—but not if the golf club becomes yet another place for client contact or competition. Wherever you relax or exercise, be discreet about your job.

REMEMBER

Handling cash: Shopkeepers, bank staff and anyone responsible for handling cash or paying in takings to the bank are all vulnerable to holdup or assault, adding to the stress level of their jobs. Strict security procedures should be followed. Never take risks. Companies and businesses should always have insurance cover and, if life or limb are threatened, heroism is NOT an option.

WORKING OUTSIDE

Dress appropriately for the weather, making allowances for possible changes. Alternatively, take appropriate changes of clothes to your worksite. It may be relatively easy to strip off if it gets warm, but having no extra protection against the cold and wet is inviting health risks if you are exposed for long periods. Feet and hands are at risk from working continually in wet conditions. Fingers, toes, nose and chin are vulnerable to frostbite when temperatures are very low.

Long-term exposure to heat and sunlight can have serious consequences. Apart from the very real dangers of dehydration, sunburn and sunstroke/heatstroke are common. In severe cases these lead to death. The possibility of skin cancers should be considered.

Try to work in the shade and cover skin with light clothing. Sunblock creams can be used to protect the nose, ears and shoulders. The head and the back of the neck should ALWAYS be covered. Other areas of skin should be protected by creams which reduce the harmful effects of the sun's rays. Make sure you have access to cooling drinks at all times.

Never underestimate the powerful effects of the sun on the naked eye. Eye damage can be permanent and disabling, and anyone working outside is at risk. Sight should be considered a non-renewable resource—arm yourself with high-protection sunglasses which filter rays and reduce glare.

LEISURE

 There are no areas in our lives where safety can be ignored. The activities we choose for relaxation often put us at risk in some way. Sometimes the element of danger is part of the appeal of a particular pursuit, because it makes it more exciting. Some sports—jet-skiing or rock climbing, for instance—are inherently dangerous. Many others— football, rugby, ice hockey—involve considerable rough and tumble that can lead to injury (see Sport).

Do-it-yourself and crafts may carry a wide variety of health risks (see DIY/CRAFT HAZARDS).

TELEVISION

Less obvious are the dangers from passive leisure activities such as watching television! Television plays a very important role in our lives, gives a lot of pleasure and genuinely aids relaxation—but all colour televisions produce x-rays. Although shields within the appliance give protection, people who watch for many hours sitting close to the screen could be doing themselves harm.

The level of risk to adults may be small but added to natural background radiation levels and the small amounts emitted by other household equipment, it will be cumulative. Avoid too much exposure and reduce the risk of cancers by not sitting too close.

When playing video and computer games you are much closer to the screen than when watching television programmes. This increases the risk from radiation and contributes to eyestrain, so take extra care.

❗ WARNING

Do not leave babies for long periods near a television that is switched on—not just in front of the screen, but at the side or behind the appliance too. They could be affected by the radiation it produces.

Out of condition

A greater risk for the 'couch potato' is the reduction of any kind of physical activity, which is encouraged by watching excessive amounts of television.

It's easy to get out of condition. If you haven't played sport or taken exercise for some time, be wary of being talked into joining the office sports team or taking the family on a hike. You may not be up to it—as many people discover when they try to run to catch a bus!

MUSIC

More quantifiable is hearing damage from the use of 'personal stereos' played at high volume. A Swedish study showed that people regularly listen to pop music at a volume above 100 dB—the European Community has a legal limit for workplaces of 90 dB over a day and never higher than 96 dB for more than two hours.

Wearing a personal stereo in the street, or while walking, cycling or driving could be very dangerous. Pedestrians and cyclists rely on their hearing as well as their eyes to warn them of approaching vehicles and other hazards. Even while driving it is important to hear what is happening around you so that you can be prepared to take appropriate action.

> # WARNING
> Don't risk hearing loss. Turn the volume down before you switch your personal stereo on and then adjust to a comfortable level. Sudden exposure to a blast of sound is particularly damaging.

Sound levels at rock concerts are often well above the 100dB level, and in nightclubs may reach four times the European Community's workplace limit. Disc jockeys in nightclubs are particularly at risk. They should either work in a cubicle where the sound level is greatly reduced or wear earphones to hear the music at a safer level.

> # REMEMBER
> Hearing loss usually begins with the higher notes—a buzzing or ringing sensation may develop and persist after the music has stopped. At first the damage is usually temporary—but heed the warning and avoid further regular exposure to high noise levels.

Electric/electronic musical instruments
Many electric/electronic musical instruments rely on full mains voltage—treat them with respect. Do NOT attempt rewiring

or experiment with combinations of instruments/amplifiers/ accessories, unless you really understand what you are doing. Make sure that all plugs and leads are capable of taking the loads imposed on them—physical and electrical (see SAFETY FIRST: **Electricity**). BEWARE of playing in rain or wet locations. You could cause a short circuit, start a fire or—at worst—be electrocuted!

BARBECUES

Brick-built barbecues, with steel cages for the coals, are safest—if you want to build your own. NEVER be tempted to use stone or concrete slabs as a base for the coals—they could explode. Ready-made barbecues could be safer still—look for expected national seals of approval.

■ Watch out for partly-cooked meat—serious food poisoning may result
■ Don't barbecue meat or fish straight from the freezer. It may not be fully cooked inside when it looks burnt on the outside
■ Partially cook food in a microwave before barbecuing
■ Choose wood-handled tools, where possible, to avoid burning your hands
■ Keep children/pets away from the cooking area
■ Don't site the barbecue under trees, near foliage, fences or sheds which might catch fire
■ ALWAYS have a fire extinguisher or a fire bucket handy
■ NEVER throw petrol or other flammable liquids onto a sluggish fire. Use small amounts of lighting fuel BEFORE lighting
■ Store fuel and firelighters SAFELY—away from children

SPECTATORS

There have been a number of recent cases of disasters in sports stadia. Check out exits and escape routes as you enter. Choose locations where you could not be trapped in a crowd. Don't let your involvement in the game blind you to the development of an ugly atmosphere among spectators or the start of violent confrontations. Get out before you get involved (see SELF-DEFENCE: **Caught in a crowd**).

Spectator injuries

Some grounds have netting between spectators and the pitch, but at many—especially amateur matches—there is always the chance of a ball or puck being kicked or driven into the crowd. Be as alert as you want your home team to be!

At motorsports events avoid positions by the rails at dangerous corners. Such locations are not only more risky for the spectators, but increase the dangers to drivers if they try to avoid crashing into the crowd. Crash barriers may be inadequate if there is a serious collision involving flying debris, catapulted vehicles or fire.

> ## REMEMBER
> **Don't create extra dangers yourself: Don't smoke—and NEVER drop burning cigarettes. Don't drop drinks cans or other objects which could cause someone to trip or fall. NEVER throw any object—even at the referee! Cheer your own team, but DON'T abuse rival supporters. Acting like a hooligan will not win you friends!**

ANGLING

Angling means long hours spent sitting at the water side waiting for 'a bite'. Makeshift stools or tree stumps don't offer the body much support. Provide yourself with a comfortable chair—it is as important as a comfortable sitting position at work. Make sure you get up and walk around at frequent intervals.

Night anglers, especially, may be exposed to the cold and wet. Remember that temperatures drop at night, so take suitable clothing for the chilly hours. If you do carry a drink laced with alcohol, you take the extra risk of lowering your blood temperature. You may also lose your judgement and fall in, exposing yourself to the danger of hypothermia.

Avoid fishing on your own. Apart from the companionship and competition fishing with friends can provide, it also means that there is someone there who can help in an emergency.

In an attempt to get good bait, some anglers use substances like carbon tetrachloride to subdue wasps and allow access to wasp larvae. Most chemicals used for this purpose are highly toxic—even if exposure is short term.

The risks of water-related sports include the danger of drowning. Angling clubs should be committed to training their members in resuscitation techniques.

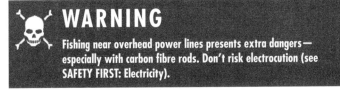

WARNING
Fishing near overhead power lines presents extra dangers— especially with carbon fibre rods. Don't risk electrocution (see SAFETY FIRST: Electricity).

Although playgrounds are far safer for children than playing on the streets, or in derelict buildings or waste dumps, there are still many hazards to be borne in mind. Playgrounds vary enormously—some have adult supervisors or are directly controlled within parks, others may go for years without proper maintenance and may be unsafe. Most parents expect older children to have some 'rough and tumble' play, but younger children may be at risk and should never be left to play unsupervised.

Apart from the dangers of vandals and bullies, who seem unable to leave the playground behind, look out for some of the following:

■ Ground surfaces should be impact absorbing—a fall onto concrete or tarmac could be very serious. Fatal head injuries have been sustained by falling less than two feet. Even grass-covered soil is too hard.

■ Litter—especially broken glass

■ There should be a sign to tell you the rules of the playground (rules that no one ever reads) and who to go to for assistance. There should be a telephone close enough to be used for any emergencies. The sign should give details of the nearest hospital casualty unit.

■ Swings, slides and climbing frames should be far enough apart for safety—also no piece of apparatus should be close to a wall or steps. If a child should fall from any piece of apparatus, what would they land on?

■ Is all the apparatus regularly maintained?

■ Swings for small children should have retaining bars to prevent them slipping off the seats. Slides and climbing frames should have good handrails and handholds.

☠ WARNING

Playgrounds are very often near or in areas where people walk their dogs. There are very real dangers to children from parasites carried in dog mess. These parasites persist in the soil and are very common. Small children, with their lack of discernment, are particularly at risk (see HEALTH: Zoonoses).

RADIO-CONTROLLED MODELS

Almost any full-size vehicle can be reproduced in model form and operated by radio control. Modellers build boats, cars, aeroplanes and helicopters which can take to the water, road or air like their full-size counterparts—with some of the same (and added) safety risks.

Model aeroplanes and boats in particular must be constructed with a great deal of care and often with the same structural integrity as the real thing. A fault in design or construction can spell disaster. At the building stage, modellers must be careful with the hazardous materials they are using like cellulose-based products and glues which give off dangerous fumes, and flammable/toxic epoxy resins and fibre glass (see POISONS).

Model aeroplanes create special risks for operators. Propellers are the main cause of injury—the average speed of a model aeroplane propeller is 17,000 rpm—and the face, chest and hands are particularly vulnerable. Flying radio-controlled models requires a lot of skill and dangerous crashes are most common among the inexperienced.

The best way to learn control skills is by joining a club. Membership is monitored so that users do not operate on the same frequency, which could lead to loss of control. Clubs enforce other safety rules—for example, model aeroplanes should always fly beneath controlled airspace to avoid the risk of interfering with REAL air-traffic control—all designed to make this popular hobby safer for everyone.

FIREWORKS

The attitude to fireworks varies enormously from country to country. Most have legislation which prevents young children from buying fireworks. Despite warnings, which are repeated again and again:

■ NEVER return to fireworks which fail to light
■ ALWAYS light fireworks at arm's length, with a taper
■ Move well back as soon as the firework has lit
■ ALWAYS aim rockets and fireworks which shower sparks or shoot out missiles AWAY from spectators
■ One person should be in charge of the fireworks during a display—or a well-organized team. Fireworks should be lit within a cordonned-off area. Children should be kept well away.
■ NEVER hold fireworks—unless specifically stated on the label that you may do so. Even so, hold at arm's length and point away from your face. Some 'safe' fireworks can cause severe injuries
■ ALWAYS lock pets indoors—some people and young children may also be frightened by the newer 'improved' bangs
■ Keep unlit fireworks well covered—away from sparks and sources of heat

REMEMBER

What goes up must come down. In most cases the remains of rockets and other airborne fireworks DO come back to earth again. Try to aim fireworks away so that they will not rain down upon people, buildings, cars or anywhere where there is a fire risk.

ON THE TOWN

When you go out for recreation you may be concerned mainly with avoiding any potential dangers on the streets—and probably give little thought to the safety of restaurants and places of entertainment.

There are laws in most countries which cover hygiene in cafés and restaurants and fire regulations for places of entertainment, shops and fairgrounds. Can you be sure that these regulations are being adhered to?

Checking restaurant hygiene is difficult. In some countries customers are welcome to walk into the kitchen, in others it might seem an unusual request—but not an unreasonable one. Any reputable establishment should be proud to show clients where food is prepared.

While most establishments may be concerned to ensure your safety as well as your continued patronage, calls from inspectors are probably infrequent and you will be dependent upon the vigilance and sense of responsibility of the staff. Attitudes to safety may sometimes be extremely casual or members of the public may often be uncooperative—objecting to not being allowed to stand where they wish at a concert or in a theatre, for example.

Be responsible yourself. Never obstruct an exit or exit route. If you see regulations being ignored—perhaps an emergency exit locked or blocked—have a word with the manager in the interests of everyone's safety.

Better safe than sorry

When you go to a restaurant, a theatre, a cinema or a nightclub, make sure you know where the exits are and work out the easiest routes to them. It can save precious moments, if there should be the need to get out quickly. It only takes a moment as you go in and settle down—once you get into the habit it will become automatic wherever you go. There have been very serious fires in many public places in recent years—involving the loss of many lives.

If possible leave coats and bags in a cloakroom. Even in normal circumstances it is easy for people to stumble over them in the dark as they pass along a row of seats or walk down an aisle. Even on a seat next to you, clothing and bags may be stolen—especially in the dark.

Don't drop drink cans, ice cream cartons or chocolate boxes on the floor. If smoking is permitted, be VERY careful to stub out cigarettes properly—in an ashtray, NOT on the floor.

Awareness of the benefits of fitness has risen greatly over the past couple of decades and increasing numbers of us regularly play some form of sport.

A fit body can be your most valuable urban survival equipment. On the other hand, participating in sport with inadequate discipline or training can easily lead to injury. A basic understanding of how your body works, together with an awareness of your limitations and a thorough knowledge of your sport, will enable you to avoid all injuries but the purely accidental.

Why be fit?

A fit body feels good and lasts longer! Regular exercise increases suppleness, strength and stamina and keeps weight down. It helps control hypertension (high blood pressure) and diminishes the risk of heart attacks, diabetes and strokes. When you're fit, you are better able to fight off the effects of poisons, or to defend yourself against physical attack. A fit person not only remains mobile longer—evidence suggests that they live longer too.

Fitness is literally being in a suitable condition to perform a given task. Nowadays, it has come to mean an improved level of efficiency in the muscles and cardiovascular system through physical exertion. Overall fitness can be broken down into

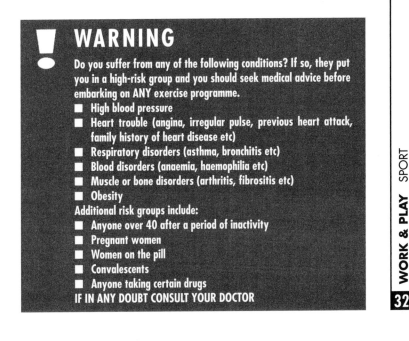

WARNING

Do you suffer from any of the following conditions? If so, they put you in a high-risk group and you should seek medical advice before embarking on ANY exercise programme.
- High blood pressure
- Heart trouble (angina, irregular pulse, previous heart attack, family history of heart disease etc)
- Respiratory disorders (asthma, bronchitis etc)
- Blood disorders (anaemia, haemophilia etc)
- Muscle or bone disorders (arthritis, fibrositis etc)
- Obesity

Additional risk groups include:
- Anyone over 40 after a period of inactivity
- Pregnant women
- Women on the pill
- Convalescents
- Anyone taking certain drugs

IF IN ANY DOUBT CONSULT YOUR DOCTOR

three categories: **strength, stamina** and **suppleness.** Different types of activity lead to improvement in each of these areas to varying degrees. Weightlifting, for instance, will not on its own improve your aerobic capacity (stamina), or flexibility (suppleness)—as squash will—but will greatly increase your muscle bulk (strength).

REMEMBER

Specific kinds of fitness are needed for specific sports, and anyone taking up a new physical activity should be aware of the potential risks. If you throw yourself too strenuously into a new activity without preparation, you are courting disaster. Take an initial fitness assessment to pinpoint areas of weakness, especially if you are over 40. However unfit you may be, there will be a sport you can undertake safely.

STRENGTH

All physical activity will increase your strength to some degree. Muscles convert energy into movement. They do this most efficiently when plenty of oxygen is present, during **aerobic** exercise. If you continue to demand work from your muscles when there is no oxygen present, they are able to produce energy anaerobically—but only for a short time.

Anaerobic exercise produces waste products, chiefly lactic acid, which leads to unpleasant 'burning', usually followed by stiffness and cramps. In building strength, you must gently teach the muscles to work more efficiently without oxygen, and with fewer undesirable side effects.

REMEMBER

A well-built musculature protects vulnerable body parts. Strong abdominals (stomach muscles), for instance, help protect the lower back: quadriceps (front of thigh) help support the weak hinge-type knee joint. Strong muscles mean you are less likely to suffer injury through everyday tasks (gardening, lifting, housework etc.). Complementary cross-training can greatly improve your performance in your chosen sport.

Training for strength involves short periods of intensive work interspersed with longer periods of gentle exercise or rest. There are NO short cuts. Increasing your workload too fast only leads to lactic acid build-up, bunched, strained and torn muscles with undue stress laid on the tendons which connect muscle to bone. **ALWAYS** warm up adequately before working out to enable your muscles to work efficiently and to reduce the likelihood of injury. **ALWAYS** stretch after weight training to help disperse lactic acid and to lengthen bunched muscles which could limit movement.

STAMINA

Stamina means endurance and relates to cardiovascular fitness. For maximum efficiency in converting energy into movement, you must be able to deliver the oxygen in the red blood cells to your muscles quickly. Aerobic fitness is essential for many sports—squash, tennis, football, rugby, for example.

At rest, an adult heart pumps five litres (about nine pints) of blood round the body. During exercise, the amount of blood pumped can increase to as much as 40 litres (70 pints) per minute in a super-fit person. The heart of an unfit person will supply perhaps 25 litres (44 pints) per minute, despite beating far faster! The more efficiently your heart works, the longer you can endure exercise, and the sooner you will recover from it.

Aerobic exercise should keep your heart working at 70–80 per cent of its full capacity. If you build up a sweat and are breathing faster, yet still able to hold a conversation, you are exercising at about the right level. Keep this up for 20 minutes, at least three times a week, and you will improve your stamina gradually and safely. As you become fitter, you will need to increase the frequency and/or duration of the exercise in order to maintain or carry on increasing stamina.

◑ ALWAYS stop exercising if you feel dizzy, nauseous, short of breath or tight in the chest. These are signs that you could be overtaxing your heart.

◑ NEVER stop abruptly after aerobic exercise. Walk around as you cool down.

REMEMBER

As your heart becomes more efficient, the risks of hypertension and coronary heart disease are reduced. More blood passing through organs such as lungs, liver, kidneys and brain increases their efficiency too. Aerobic activity burns calories, converts fat to lean tissue and speeds up the metabolism, helping you to control your weight. Improving your stamina makes you feel more energetic and more confident.

SUPPLENESS

However efficient the heart and lungs, however strong the muscles, the body is useless if it can't move freely. Flexibility is the third component of overall fitness, and is essential for good performance in most sports—you will also be less likely to suffer strains and ruptures from the sudden reaches, kicks and jumps. Gymnastics, dance and martial arts demand and encourage great flexibility. Aerobics classes should involve a stretching sequence.

To encourage suppleness and avoid bunched and torn muscles and strain on the heart, adequate **warming up** and **cooling down** periods are essential. Each individual's level of suppleness depends on the length and flexibility of muscles,

THE WARM-UP

Moderate exercise—jogging, walking, cycling, swimming—is a warm-up in itself. Most racquet sports and some team games include a 'knockabout' period, which warms the muscles and raises the pulse. IN ALL OTHER CASES, OMITTING THE WARM-UP IS THE EASIEST ROUTE TO INJURY.

DON'T stretch a cold muscle. Begin the warm-up with a period of 'cardiovascular' exercise to get the blood flowing to the muscles and to avoid sudden strain on the heart.

DON'T exercise after a meal. A large proportion of your blood will be circulating around the stomach and intestines during digestion, and trying to divert it elsewhere can lead to stitches and cramps—potentially lethal in watersports. Allow 90 minutes for digestion.

DON'T 'bounce'. Alternately stretching and contracting is more likely to tighten than lengthen the muscle, and it puts strain on the tendons.

Ease into the stretch. Get into position, relax and breathe. Concentrate on the muscle groups which will do the most work—hamstrings and quads for a footballer, for example. Be especially careful to warm up ANY muscle which is stiff, aching or has suffered a previous injury.

ligaments and tendons. Women are naturally more supple than men. You must always work within your limitations. Again, there are NO short cuts—the muscle will stretch when it's ready. If you try to force it, it will tear. Swimming is excellent for lengthening muscles.

Increase the time you spend warming up and cooling down and incorporate more gentle stretches to encourage longer, more flexible muscles. 'Little and often' is the key.

THE COOL-DOWN

Prepare the body to return to its normal state. During exercise, your muscles help the heart by pumping blood to the lungs for re-oxygenation. If you stop too suddenly, blood can pool in the muscles, starving other areas and making the heart pump faster to compensate. Dizziness, trembling and even blackout can result if you come to an abrupt halt.

Cooling down and stretching are essential to disperse lactic acid and discourage the formation of inflexible scar tissue over tiny tears in contracted muscles (which lead to stiffness). Slow down GRADUALLY and repeat the warm-up stretches after exercise.

WARNING

Pain in the chest or shooting down the arms can mean your heart is under too much strain. It doesn't necessarily mean that you're having a heart attack, but you should still get yourself checked out medically just for peace of mind.

WOMEN AND SPORT

Specific differences that may affect women's performance in sport, or the risks of particular sports, include:
- Smaller hearts and lungs than men
- Less muscle bulk and strength than men
- Better fat-burning mechanism than men (useful in endurance sports)
- Higher fat percentage (25% on average)
- Lower centre of gravity than men
- Smaller stature (useful in some sports—riding, for example)
- Wide pelvis alters angle of femur—can lead to knee problems in running
- Breast pain and jogger's nipple can be caused by running
- Joint looseness (useful in some sports—gymnastics, for example)
- Reproductive organs well protected
- Menstrual cycle can be affected by strenuous exercise
- Pregnancy hormones loosen ligaments
- The contraceptive pill increases risk of cardiovascular strain

'NO PAIN, NO GAIN' IS WAY OUT OF DATE. Yes, stretching and strenuous exercise can be uncomfortable, but there's a difference between a working bum and the pain of a torn muscle! Be sensitive to the difference! If the pain is sudden, unusual or excessive, slacken off, STOP and find out what it is. The natural 'high' — the endorphins and adrenalin stimulated by exercise, the increased blood flow and oxygen levels — can blind you to your body's distress signals. DO NOT use painkillers, stropping, sprays and creams to mask injury. Pain is your body's early warning system. LISTEN TO IT!

CHILDREN AND SPORT

Medical advice and expert tuition should be sought to determine the effect of the stresses and strains of different sports on growing bones and muscles. Exercise for young children should mainly involve learning balance, coordination and spatial awareness. As the young person reaches puberty, hormone production and growth may be affected by strenuous exercise. NEVER force a child to over-use muscles and joints — by lifting weights, for example. If a particular interest is shown in any type of sport, seek specialist advice to make it SAFE.

TYPES OF INJURY

Some of the most common injuries associated with sports, together with basic preventative measures and first aid, are dealt with here. ALL coaches should have training in the recognition and first-aid treatment of injuries associated with their sports. ALL sports centres, sports grounds and gyms should encourage their members to learn the basics, too.

Muscles

Tears, strains, sprains and **pulls** of the elastic fibres that form muscle are very common. Many of these injuries can be prevented by keeping the muscles in good condition, and by warming them up before exertion EVERY TIME. The quadriceps at the front of the thigh and hamstrings at the back of the leg are particularly at risk from sudden movement, imbalance of power between the two sets of muscles and/or insufficient warm-up. More serious is quads **haematoma** (bleeding within the muscle caused by a blow), most common in contact sports.

ACTION: Rest, ice, compress and elevate the affected part. This procedure is used for many sports injuries. The sooner it can be administered, the more effective it is. Rest means REST! Stop playing, however much you want to go back on. 'Ice' can be anything cold to hand—a cold drink can will do. Apply for about 15 minutes every two hours. Compress with a bandage, but NOT too tightly. Remove IMMEDIATELY if limb turns pale or 'blue', or if pain is severe. Elevate the affected part to reduce swelling and throbbing pain. Use a high sling for arms. A normal dose of aspirin can be taken as an immediate measure to reduce pain and inflammation. Seek medical attention—URGENTLY for haematoma, especially if the area seems to be swelling.

WARNING

Aspirin should NOT be given to children under 12. It can lead to bleeding of the stomach, and is potentially harmful. Do NOT use painkillers to enable you to continue playing with an injury! Do NOT use painkillers on a regular basis.

Tendons

Tendons bond muscles to bone. Since bone is rigid and muscle is flexible, it is the inelastic fibres of a tendon which often tear, swell and become inflamed under stress. **Tendinitis** (inflammation of the tendon) is most common in the biceps, around the patella and the Achilles tendon. These are often 'over-use' injuries, caused by repeated stress and can develop slowly. If you notice pain directly after exercise, coupled with stubborn stiffness, or a burning sensation during exercise, you could be developing tendinitis.

ACTION: Rest and ice. Aspirin, if necessary. Seek medical attention.

WARNING

Ironically, Achilles tendinitis can be caused by pressure from the high 'Achilles protector' on the back of some training shoes. If your shoes have these, CUT THEM OFF. Better still, choose training shoes without them.

The most common tendon **rupture** is of the Achilles tendon. It feels like a kick from behind accompanied by severe pain at the lower end of the calf muscle or in the tendon itself. You may be able to feel the gap where it is broken.

ACTION: Rest, ice and elevate. CALL AN AMBULANCE. Surgery is almost always needed at once.

Ligaments

These are tough bands of white, fibrous connective tissue that link two bones together at a joint. They are flexible, but not very elastic, so are vulnerable to **strain** and **rupture.** The most common injuries occur in the ligaments of the knee, caused by sudden twisting or bending, and in the cartilage ligament if it gets caught between the bones. Local pain, swelling and bruising are felt, accompanied by a popping sound if the ligament is ruptured. Strengthen the quads to prevent these nasty injuries.

A **sprained wrist** occurs when the ligaments are stretched suddenly beyond normal range. It is less serious than a **sprained ankle**, which is one of the commonest of all sports injuries. Ligament strain is a frequent cause of **back pain**, and is often due simply to bad posture. Keep your back supple through swimming, aerobics or yoga and strengthen the abdominal muscles.

ACTION: Injuries to the knee ligaments can be serious. A rupture usually requires surgery. Rupture of the cross-shaped cruciate ligaments (inside the knee joint) may not cause intense pain at first. You may hear a 'pop', and the knee swells within an hour. A well-padded splint can help, but ONLY IF YOU KNOW WHAT YOU'RE DOING. Seek urgent medical attention. For most ligament injuries, rest, ice, compress (lightly) and elevate. Seek medical attention.

WARNING

The hormones of pregnancy slacken the ligaments, making them more vulnerable to injury. Always take medical advice before embarking on a course of exercise during pregnancy.

Cartilage

This is a dense connective tissue at the joints of movable bones. The meniscus cartilage in the knee protects the joint surfaces, provides stability and acts as a shock absorber. Twisting on a bent weight-bearing knee can tear and trap some of this cartilage in the joint—one of the most common knee injuries. The knee might click/lock/give way and will swell within hours. Avoid twisting movements. Work on your quadriceps at the front of the thigh to support the knee.

ACTION: Rest, ice, compress and elevate. Seek urgent medical attention. Partial or total removal of this cartilage is a common operation—partial is preferable, since total removal increases pressure on the joint fivefold.

Bursae

A bursa is a small sac of fluid-filled fibrous tissue which helps reduce friction round joints and where ligaments and tendons pass over bones. **Bursitis** (inflammation of the bursae) has many causes, most commonly pressure caused by ill-fitting shoes, over-repetition of movements etc. It is characterized by tenderness and swelling and is rarely serious.

ACTION: Rest, ice, compress and elevate. Seek medical attention if there are signs of infection or if inflammation does not subside.

Bones

Fractures of the ribs, collarbone, radius (forearm), scaphoid (base of thumb), knee, shinbone, ankle and foot are occupational hazards if you play contact sports, ski or ride. Pain is usually intense, immediate and pretty unmistakable—although ankle and radius fractures are commonly difficult to distinguish from sprains.

ACTION: It's better not to move the casualty if you suspect something is broken. Seek urgent medical attention.

Head/neck

Eye/nose/mouth

After any **blow, scratch** or **cut** in or around the eye, if vision is clouded or obscured or if the eye looks cloudy or bloody, seek urgent medical attention. If it seems a curtain has been pulled across your eye, you may have suffered a **detached retina**, for which surgery is required. Apply an ice pack to a **black eye** as soon as possible.

If you wear glasses or sunglasses for sports, make sure they have shatterproof polycarbonate lenses. If you wear contact lenses, remember that they can be knocked into (or out of) the eye. Seek specialist advice.

Nosebleeds are caused by rupture of the blood vessels on the membrane inside the nose. Pinch nose, lean forward and apply ice pack. If bleeding continues for more than a few minutes, seek medical attention.

Cut lips and **tongue** tend to bleed profusely, which can be quite frightening. Apply pressure and do not swallow blood as you might be sick. Seek medical attention for large cuts, if bleeding doesn't stop or if a tooth goes through the lip.

Broken teeth should be saved as they can often be replaced. AVOID swallowing or inhaling teeth. ALWAYS seek medical attention if you suspect you've swallowed a tooth, or for someone who has been knocked out and lost a tooth. If you wear partial or complete dentures, it's wise to remove them before playing vigorous or contact sports.

CLOTHING

Protective clothing is VITAL for safety in many sports. New, more comfortable and safer versions are always being developed—it's worth keeping YOUR equipment up to date.

Shields/pads

You wouldn't think of playing cricket without shin pads, but many people balk at wearing them for football or field hockey. Since shin injuries are so common, it's worth overcoming the 'image' problem! Extensive body armour is worn for American football and ice hockey. Boxers always wear gumshields, which can be custom-made. Men should always wear a box (genital protector) for vigorous or contact sports.

Gloves

Wear gloves to protect against blows (cricket, ice hockey), cold (skiing, motocross), friction burn (sailing) or to improve grip (riding, cycling, golf, weightlifting etc).

Helmets/goggles

Injuries to the head and eyes are potentially EXTREMELY serious, even fatal. Headgear should always be properly fitted and up to date—spare no expense. Cricket helmets with faceguards, similar to the baseball catcher's mask, are increasingly worn, as are helmets for cycling. Do NOT consider riding a horse without a proper helmet—the old-fashioned peaked hard hats need to be the specially-reinforced type.

Skiers should wear special filtered goggles to protect eyes against glare. Cyclists' goggles will keep out grit and insects and prevent the eyes watering. Swimmers need goggles to protect the delicate lining of the eye from chlorine and salts, while squash players should wear glassless goggles to guard against the ball impacting into the eye socket.

WARNING

ALWAYS buy a new helmet once yours has protected your head from a heavy blow. It may not have the strength to protect you a second time, even if it looks undamaged.

Watersports

Buoyancy aids are essential for non-swimmers, but life preservers should be worn by EVERYONE who is at risk—canoeists, sailors, windsurfers, surfers. Never forget that you may be knocked unconscious—it's not a case of how good a swimmer you are. Wetsuits protect the body from cold and abrasion. NEVER wear earplugs for diving—the water pressure could force them into your ears.

Footwear

Buy the best you can afford and, if possible, have a pair of shoes specifically designed for every sport you play. Many sporting injuries could be avoided by wearing the correct footwear. A running shoe, for instance, has a wide base which is too stable for the rolling and twisting of the feet in racquet sports.

Buy shoes with uppers made of natural materials—leather, cotton, canvas. Try shoes on while wearing the same socks you use for sport. REMOVE plastic motifs from leather or canvas shoes—they don't expand with the shoe and can cause blisters, sores and chafing. DON'T wear somebody else's shoes. You could be asking for verruccas, athlete's foot or other fungal infections. DON'T play sport barefoot—your feet need protection and support.

SPORT BY SPORT

Here's a summary of the most common risks of some of the most common sports. Know YOUR sport and take every precaution to protect yourself from serious injury. Never forget that, apart from the dangers of permanent damage or death, quite simple injuries may prevent you from ever participating in your chosen sport again.

COMBAT SPORTS

BOXING/WRESTLING/JUDO/FENCING

Most boxing injuries are to the face and head, with concussion a constant danger. Protective headgear can be worn, but some feel this impairs reflexes, slowing movement and leading to a false sense of security. Amateur boxers are now forced to rest after a knockout to minimize the likelihood of becoming 'punch drunk' — a form of irreversible brain damage. Blows to the eye can detach the retina. Both wrestling and judo are surprisingly-safe sports, probably because they are well supervised, though shoulder separation or dislocation, and fractures of the ribs, collarbone, fingers or toes do occur in the latter. Some arm positions can also lead to tennis elbow and ligament injury. In fencing, faulty equipment is often the cause of injuries and accidents.

CONTACT SPORTS

RUGBY/AUSTRALIAN RULES FOOTBALL/ AMERICAN FOOTBALL

Injuries are common in these relatively-violent sports, though they are safer than they used to be—especially American football, thanks to the extensive padding and protection worn. Collapsing the scrum in rugby is now illegal since players have been paralysed this way. Common injuries include: shoulder separation; costochondral injuries (front or chest wall where ribs join sternum); quads haematoma; scrum pox (impetigo of the skin caught from cuts). Back injuries from twisting in the scrum/scrimmage can be minimized by conditioning the muscles; cauliflower ear by wearing a sweat band. Broken bones and spinal injuries are not uncommon.

BASKETBALL

NETBALL/HANDBALL/VOLLEYBALL

Dislocated and fractured fingers, mallet finger (see Field sports) and thumb sprain all arise from mistimed catches. Foot injuries and ankle sprains are common due to play on hard surfaces. The actions of jumping, backing and overhead throwing can lead to jumper's knee (patella tendinitis—tendon inflammation), footballer's groin and shoulder problems—typically rotator cuff (tendons on top of the shoulder joint), rupture and impingement.

FIELD SPORTS

CRICKET/BASEBALL/SOFTBALL

As in other social sports, the under-fit weekend player risks injury. Common complaints include 'pitcher's elbow' (inflammation where the muscles are attached to the bone), the children's version—'little league elbow'—and radiohumeral joint injury, caused by round-arm throwing. Fractured or dislocated fingers and mallet finger (a bruised, swollen and often sagging end-joint caused by a blow to the tip) are hazards of hard ball sports. The actions of bowling, pitching or sliding can also lead to shoulder and back injuries.

GOLF/HOCKEY

A similar set of back and hamstring injuries can arise from these sports since they're both played with a bias to one side. In golf, upper back pain is common, and golfer's or tennis elbow arises when the club is gripped too tightly or frequently hits the turf. Hockey players are vulnerable to high knee and hip pain, due to running while bending over, and to footballer's groin.

TARGET SPORTS

SHOOTING/ARCHERY

Properly supervised, these sports should not be dangerous. In archery, tennis elbow can develop in the bow arm if technique is faulty. In shooting, elbow and knee pads protect against bursitis. Earmuffs are essential to avoid damage to the tiny bones of the middle ear. Most injuries are the result of accident. ALWAYS follow the gun code:

- NEVER point a weapon at a person, even in jest
- NEVER let a child play with a gun, even if you're sure it's unloaded
- ALWAYS assume a gun IS loaded
- ALWAYS break the gun before putting it in a car or climbing an obstacle
- ALWAYS ensure barrel points to the ground — move stock to barrel
- See SELF-DEFENCE: The law

TRACK AND FIELD

RUNNING/SPRINTING/ LONG JUMP/HIGH JUMP/JAVELIN/ SHOT PUT/DISCUS

Running events carry the risk of adductor strain and hamstring sprains — particularly from explosive or sudden movements and one-sided running. Long-distance runs on hard surfaces leave runners prone to shin splints and stress fractures. Kick-type movements aggravate quads mechanism injuries (Osgood Schlatter's disease, or shinbone inflammation and jumper's knee). Hurdlers or discus throwers can suffer footballer's groin. Javelin throwers and shot putters both have their 'own' injuries — javelin thrower's elbow is caused by round-arm throws and shot putter's finger is a sprain of the three fingers used for the final acceleration. In high jumpers, Fosbury flop ankle is caused by twisting the lower leg on take-off.

SPORT BY SPORT

RACQUET SPORTS

TENNIS/SQUASH/RACQUETBALL/ BADMINTON

Tennis elbow (inflammation where the muscle is attached to the bone) is the classic injury, arising from faulty technique in the backhand (tennis), or from an over-tight grip in the other three sports. An over-tight grip can also lead to 'squash player's finger'. Foot, ankle, knee and toe injuries are further perils of squash and racquetball, which are often played by the under-fit who come unstuck twisting and turning at speed, maybe in ill-fitting shoes. Badminton players are prone to shoulder injuries owing to the frequency of overhead shots. Twisting and lunging are features of this sport too, putting ankles and knees at risk.

FOOTBALL

Again, inadequate fitness leads to injury, and adductor strain (inside thigh), caused by a kick with the inner side of the foot, and hamstring sprains are especially common in weekend players. Knees are vulnerable to ligament and cartilage strain or rupture. Shin bone fractures and sprained ankles are common. Footballer's ankle occurs when repeated minor injuries have stretched the ligaments, and footballer's groin (a loosened symphysis ligament leading to pelvic inflammation) can be caused by sidestepping or backing movements.

ICE SPORTS

SKATING/ICE HOCKEY

Fractures are more common than ligament injuries on ice—radius and scaphoid fractures are the most usual (from falling onto the outstretched hand) and stress fractures of the tibia (shin) can result from jumps in figure skating. Improved proficiency is your best guard against this. Ill-fitting boots can lead to ankle ligament strain/rupture or to skater's heel (inflammation of bursa over heelbone). Full padding/armour is essential for playing ice hockey, since most injuries result from blows of the puck or stick.

SKIING

Around 30,000 people are injured skiing each year. About half of these result from skiing in off-piste areas, which can be extremely dangerous. Avalanches and colliding with unanticipated obstacles are among the hazards. The modern ski boot protects against ankle fractures, but puts more stress on the knee, with ligament strains and ruptures caused by twisting or bending. Have your bindings set locally where they know the conditions.

Train adequately before you ski. Fatigue leads to mistakes. If you have never been skiing before, take classes to strengthen your muscles — well in advance. ALWAYS ski with a companion, in case you sustain an injury.

If you must ski alone, set your bindings looser — it's easier to put a ski back on than to cope alone with a broken leg or twisted ankle.

If you are injured: Apply snow to the injury to reduce swelling. Make a splint from ski sticks or branches. DON'T use outer garments for bandage or splint. You must keep warm. DON'T walk across snow. Lie on your stomach or skis and slide.

ON HORSEBACK

In Britain, riding is the commonest cause of death among young girls. Falling off the horse, of course, is the most dangerous hazard, expecially falling onto the head or neck. A proper helmet (see Clothing) minimizes the risk. If a fallen rider complains of neck pain, numb or weak limbs, DON'T assume the hat prevented injury, seek urgent medical attention. Other common fall injuries include strained or ruptured knee ligaments, fractured thigh, shin or collarbones and shoulder separation. Always respect the horse — it is dangerous at both ends (and uncomfortable in the middle!). Learn some 'horse sense' — horses are easily startled by sudden noises. Give all hindquarters a wide berth.

ON WHEELS

CYCLING/BMX/ROLLERSKATING/ STREET HOCKEY/SKATEBOARDING

Falls are VERY common — adequate head protection MUST be worn for all these sports. It may look 'uncool' but it could save your life! For all except cycling, knee and elbow protectors are advisable to protect against jarring blows as well as grazes and cuts. The kneecap is particularly vulnerable in 12- to 16-year-old girls — any bent-knee exercise puts extra strain on the joint. 'Street' sports also carry an increased risk of lung pollution from the carbon monoxide, lead and sulphur compounds in traffic exhaust fumes — the intake of which is increased by rapid deep breathing.

SPORT BY SPORT

GYM SPORTS

AEROBICS/GYMNASTICS/WEIGHTS

These three sports are most commonly practised by three distinct groups and cause quite specific injuries. Aerobics has contributed to a rise of heart attacks in women over 35, who launch too quickly into over-strenuous exercise and fail to recognize warning signs — exhaustion, pain, discomfort or overheating. Ensure your instructor is up to date in the latest developments of this sport and never 'go for the burn'. Ideally, gymnasts need supple joints and ligaments, short stature and nimbleness, which accounts for the sport's popularity with young girls. Flexible joints can easily be overstretched, though, and common injuries include sway back elbow (lax ligaments leading to an unstable joint) and back injuries, caused by frequent hyperextension (overstraightening). Bad technique is the usual route to injury in weightlifting/weight training. The back must always be straight — a belt helps in this by transferring more force to the abdominals. Too much pressure in the abdomen, and a hernia (rupture) is possible, or fainting due to blood not returning to the heart and brain. Biceps strain, shoulder separation and chest wall pain are all hazards of overdoing bench presses.

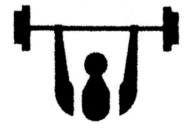

WATERSPORTS

SWIMMING/DIVING

On the whole, swimming is a very safe sport, and is often used as part of the treatment/therapy for all kinds of sports injuries. Chlorine content of public baths can lead to conjunctivitis (inflammation of the eye). Conversely, unchlorinated pools carry the risk of bacterial or fungal infections — from athlete's foot and verruccas to ear infections and impetigo. Long-distance swimming can cause 'swimmer's shoulder' (rotator cuff rupture or impingement) or 'swimmer's knee' (ligament strain) due to over use and/or bad technique. Diving should ALWAYS be properly supervised, or potentially fatal neck injuries can result. Impact velocity can lead to wrist, thumb and shoulder injuries.

WATERSKIING/WINDSURFING/SURFING

A well-fitting wetsuit protects against a high-pressure enema (rectal or vaginal) in waterskiing, hypothermia in windsurfing. Bad posture can lead to back pain and injuries in both sports — the risk decreases with competence. Life jackets MUST be worn, even by competent swimmers in case of blows to the head. Waxes used on boards and skis can cause dermatitis. There is a danger with all these sports that you may find yourself far from shore, or in waters you would not choose for swimming.

SAILING/CANOEING/ROWING

Back pain can be a problem, especially in the latter two sports—developing the abdominals guards against this. Paddler's wrist (inflammation of tendons on outer side) and tennis elbow can be caused by gripping oars/paddles too hard and twisting. Rowers are prone to blisters (rub surgical spirit on palms to toughen them or wear gloves) and haemorrhoids (seek medical advice if 'piles' persist). Sailing involves much isometric exercise—where the muscles contract, but the limb is static—and quads strain or abdominal strain are common. Sailors should also guard against dehydration and hypothermia—take drinking water, wear sun screen and protect the eyes.

WARNING

All watersports should be considered potentially DANGEROUS. Death by drowning is particularly associated with people taking unnecessary risks or with alcohol/drug use. NEVER drink and swim! Alcohol is thought to be responsible for up to 50% of drowning deaths each year—in Britain alone, one person under 25 dies each week. Apart from 'errors of judgement', alcohol lowers the blood temperature—as does swimming. This combination alone can be fatal.

NEVER swim within 90 minutes of eating. Blood around the muscles is in short supply and cramps can result. IF you get cramp: stop swimming, turn onto your back and float. Stretch the cramped muscle. When the cramp eases off, swim to the side using a different stroke.

NEVER dive into very cold water as the shock results in immediate hyperventilation, increased blood pressure and pulse. The shock itself may kill you. Loss of body heat weakens movement and reduces coordination. This is especially dangerous if you are elderly, convalescent or unfit.

NEVER swim in flooded sand or gravel pits or quarries. Submerged objects are dangerous and steeply-shelving sides may make climbing out impossible.

All motorists know when they're taking risks. City driving, in particular, has become aggressive and competitive. Here are some tips on self-defence for motorists. Going on holiday? Don't let your brain take a holiday too!

In transit

ON THE MOVE Motoring • Defensive driving • Normal conditions • Adverse conditions • Breakdowns • Basic car maintenance • Spare-parts kit • Accidents • After an accident • Car fire • Motorbikes • Cycling

PUBLIC TRANSPORT Underground • Bus • Train • Water transport • Man overboard • You overboard

LIFTS/ESCALATORS Types of lift • Hatches • Safety devices • Using lifts • Trapped in a lift

TRAVELLING ABROAD Research your destination • Before you go • Immunization • Be prepared • Packing • Air travel • Plane crashes • Restricted and prohibited items • Sea travel • First-aid kit • When you arrive • Food and drink • Accidents • Assess the risks • Acclimatization • Heat • Sun • Sensible tanning • Humidity • Altitude • Cold • Time zones • Jetlag

ON THE MOVE

Many people are constantly on the move, so this chapter deals with common situations that arise when you are away from home—on your way from A to B. Whether you're driving, cycling, riding public transport, or catching an aeroplane to another continent, a little forethought can make your journey safer. Accidents happen, even on foot. In fact, the chance of you being injured or killed when out walking is around 1 in 900. Motoring accidents, however, account for around 6000 deaths per year, in Britain alone.

MOTORING

If it is true that you drive as you would like to live, there are an awful lot of frustrated jet-setters on our roads. However careful a driver YOU might be, there's a maniac round every corner looking for an accident. Traffic accidents can be due to many factors, but these narrow down to three overall headings: the driver, the car and the road.

A responsible and careful driver of a well-maintained car on a road where visibility is good should be relatively safe. You owe it to yourself, your passengers, to pedestrians and other road users to do everything in your power to minimize the risks of motoring. But it's essential to know what to do when the worst comes to the worst.

DEFENSIVE DRIVING

In a nutshell, this means being alert to possible dangers BEFORE they become dangerous. It doesn't mean crawling along at a snail's pace with your eyes glued to the rear-view mirror, but simply adopting a kind of sensible road safety which should become second nature to every driver.

Normal conditions: TOWN

Around 95 per cent of pedestrian accidents occur on roads in built-up areas. If you don't want to contribute to this depressing statistic, you should increase your awareness of the hazards on every high street—try to anticipate what is likely to happen and be ready to react. In general, be alert, observant, and USE YOUR COMMON SENSE.

◑ If there's a bus ahead, watch out for passengers running to catch it or jumping off, even between stops.

- Taxis are a law unto themselves, especially in large cities. Be ready for them to slow down or swerve with little warning and watch out for people hailing one from the kerb.
- If you hear a siren, slow down and be ready to move to the side. DON'T stop, swerve or accelerate suddenly.
- The people in the car in front are gazing at the shops, pointing or gesticulating. They may be looking for a parking space or a street number—or having a row—but they're not aware you're on their tail. KEEP BACK, or bip the horn.
- Approaching a parked car: is there someone inside? Is the indicator flashing? Is there smoke coming out of the exhaust pipe? It could be about to pull out.
- Check for signs of movement behind or between parked cars, and especially around ice cream vans—a child could run out suddenly.
- In moving traffic, if brake lights go on on the car in front, touch your own brakes to warn the car behind.
- Use reflections in shop windows at junctions and on blind bends to help you 'see round corners'.

Normal conditions: TOWN/COUNTRY

- Don't assume the car in front is about to turn left because its indicator is signalling left—the driver may have forgotten to cancel it, or may change his/her mind.
- Give cyclists as much room as you would a car. They may swerve to avoid a pothole. They may want to pull out.
- Turning left at a junction? Make your intention obvious—signal, take up a dominating position on the road, pull out onto the crown to allow traffic to pass inside.
- Be patient with learners—remember you were one once. They are unpredictable. If stopped behind one on a hill, beware of them rolling back before moving off.
- Animals are also unpredictable (especially cats, which ALWAYS run the wrong way). Brake hard and keep in a straight line if one should run in front of you—swerving could cause a worse accident.
- Headlight flashing should only be used to alert the other driver to your presence or as a danger warning. Never assume it means 'please go ahead'.
- Garage and pub forecourts are invitations to careless drivers. Watch out for cars braking and pulling in with little or no warning.

Normal conditions: OUT OF TOWN

- Drive EXTRA carefully if there's mud on the road, especially in the rain.

- Drive VERY slowly past horses and avoid any sudden noise.
- It's best to wait for a flock of sheep or herd of cattle to pass.
- Hay and mud can warn of a slow moving tractor ahead; droppings can warn of animals.
- Farm produce shops and roadside stalls mean the car in front may brake or turn off suddenly. Be prepared for carelessly parked cars on the verges ahead.

Night

- Ensure your windscreen is spotless—reflected lights at night can reduce your vision dramatically.
- If a car approaches with headlamps on full beam, DON'T look directly at them, but follow the nearside kerb until it passes. Retaliating by switching your own lights to full beam is **OFF**ensive driving—in other words, likely to cause an accident.
- In the country, be prepared for changes in street lighting and road surfaces at county boundaries.

Adverse conditions: FOG

Fog is one of the most hazardous of all driving conditions. Not only is visibility severely reduced, but sound is muffled, the road becomes more slippery than during rain and it's very difficult to judge distance or speed. Here's what to do when you run into fog.

- Day or night, drive with dipped headlights. Ensure the lenses are clean, since dust and dirt can reduce beam intensity by half.
- Use fog lamps if you have them—they are low mounted and give a wide arc of light. If you have a single fog lamp, switch on headlights too to avoid being mistaken for a motorbike.
- Drive SLOWLY. Since you should be able to stop within your range of vision, you might have to limit your speed to as little as 8 kph (5 mph).
- Stay back from the car in front. Braking distance is some 20 m (about 70 feet) more than in normal, dry conditions, and restricted vision slows reaction time.
- Beware of driving in the middle of the road and following the Catseyes—you don't want to meet someone doing the same in the other direction.
- Following the tail lights of the car in front could mean following an idiot into a ditch or a head-on crash. Keep to the side of the road and follow the kerb instead.
- Keep your window open. The unnatural silence is disorientating and you might pick up aural clues to the whereabouts of invisible things.

- Use windscreen wipers and keep the inside demister on to clear the windscreen.
- If you have to turn left, first stick your head out of the window and listen especially hard.
- NEVER NEVER overtake—unless somebody is about to give birth on the back seat.
- Leaning forward to peer through the screen makes you tense and tired. You'll see—and drive—better if you try sitting normally.
- Try not to park on the road. If you must, switch on hazard lights and set up a warning triangle 90 m (about 100 yds) behind—135 m (about 150 yds) on a motorway.

Adverse conditions: SNOW/ICE

- Service your vehicle so it doesn't let you down.
- Stock up with anti-freeze, windscreen wash, de-icer and scraper BEFORE cold weather sets in!
- Carry a small emergency kit including food, flashlight, matches and first-aid kit.
- Emergency triangles will alert other motorists if you get into difficulty.
- Two candles can create sufficient heat to warm a car interior and save running the engine.
- A shovel and a bag of sand (or even cat litter, which doesn't freeze) could help get you out of deep snow.

STUCK IN SNOW

To get out of this slippery situation:
- **DON'T accelerate hard. This will compact the snow and probably force it into the tyre treads, too**
- **Straighten the wheels and start in second gear**
- **Place something under the driving wheels (is your car front-, rear- or four-wheel drive?) to provide traction. Sacking, cardboard, sand, grit, twigs, a rug, blanket or clothing will do. Dig/scrape the snow away as much as possible first**
- **Get a push from your passengers, but . . .**
- **DON'T let them stand behind the car—it could roll or slide back**
- **DON'T stop for them until you're on firmer ground**

DEEP SNOW
- **Try 'rocking'. Drive forward as far as possible (in first gear, gentle revs), then quickly engage reverse and go back slowly a few inches. Repeat until you've built up a track**
- **If that fails, dig the snow from around all four wheels and proceed with above techniques**

- In sub-zero temperatures, metal parts of the car may 'burn' the skin of your hands.
- Heavy items in the boot will increase traction in a front-wheel drive car.
- Check tyre pressures—if they're low, grip is reduced. Inflate wide-grooved radial tyres slightly more than normal.
- Pump the brakes to avoid wheel lock (this also warns the car behind) and don't steer at the same time. Brake GENTLY.

EMERGENCY!
SNOWBOUND

If you find yourself trapped in a blizzard or snowdrift, you must above all KEEP WARM and KEEP AWAKE. DON'T try to go for help—there are plenty of cases of people dying of exposure a few yards from civilization.

Take your **survival pack** from the boot, CLEAR an area around the exhaust pipe so you can run the engine and heater without risk of **carbon monoxide poisoning,** and stay in the car. **DON'T** run the engine constantly—too much heat makes you **drowsy,** you'll create too much exhaust and you could run out of **fuel**—ten minutes an hour is sufficient.

DON'T keep the radio on—you'll flatten the **battery.** Wrap up in clothing, blankets, car carpets, newspapers—whatever is to hand—and move limbs, fingers and toes from time to time so that you keep blood circulating.

Falling asleep is **dangerous** because your body temperature lowers and you could succumb to **hypothermia.** You could also **suffocate,** if the car gets buried.

DON'T drink alcohol as this also lowers the body temperature, as well as clouding judgement and encouraging drowsiness. Open a window (away from drifting snow) occasionally and, if the car is getting buried, poke a wheel jack or umbrella up through the snow to make an **air channel.**

If you're stranded with others, huddling together in one vehicle creates warmth and boosts morale. Take it in turns to nap. If another vehicle comes into range, possibly a rescue service, flash your lights and bip your horn. They may not be able to see your car, or may not realize there is anyone inside.

- DON'T drive peering through a small clear patch on a misty screen. Liquid soap or washing-up liquid rubbed on the glass will help prevent misting up.
- Drive at low speed in as high a gear as possible. Stopping distance in ice can be ten times greater than normal and a high gear reduces the chance of wheel spin.
- Uphill: try to ascend without stopping or changing gear. Wait at the bottom until you have an uninterrupted climb.
- Downhill: descend very slowly in a very low gear and try not to brake.
- Travel by daylight and use major highways.
- Tell someone where you're going—if something happens, you'll be easier to find.

WINTER MOTORING SURVIVAL KIT

To be prepared for the worst, always carry the following: A shovel, sacking or grit, blankets, extra clothing, sleeping bag, wellington boots, a flashlight and spare batteries.

For longer journeys, especially out of town, add: Food, a flask of hot drink, a book or a battery-operated radio/tape player.

Adverse conditions: HOT SUN

What's adverse about that? Apart from the problems of visibility due to glare, sunny skies tend to make for lapses in concentration—two thirds of all road accidents happen in good weather. Heat also frays tempers, and accidents are accompanied by unnecessary extra injuries when angry drivers come to blows. If your car overheats, stop to allow the engine to cool. If you are driving in unfamiliar territory and stopping is out of the question, switch on the heater. This will give greater volume to the cooling water and, although the inside of the car will get even hotter, the engine will cool.

When convenient stop and open up the bonnet. **Do not undo the radiator cap until the temperature drops.** Check the radiator and all hoses for leaks. If the radiator is leaking, adding the white of an egg will seal small holes.

If there is a large hole, squeeze that section of the copper piping flat to seal it off. It will reduce the size of the cooling area but, if you drive very steadily, you will be able to keep going.

Metal gets HOT. Be careful! All metal parts of a car can become hot enough to cause blisters.

When adding fuel in sandy conditions, sand and dust can get into the tank. Rig a filter over or just inside the inlet to the tank to prevent contamination.

If you're involved in an accident, keep your cool. If necessary, stay in the car until the police arrive. Certain road surfaces soften in hot sun—check tyre treads since they can collect sticky grit this way.

REMEMBER

NEVER leave children or animals in the car in hot weather unless you definitely know it will only be for a couple of minutes. If you must, open the window as far as you safely can and return soon. Invest in shades for front and back windscreens to guard against overheating the interior, when you can't park in the shade. Perforated metal roll-down blinds are available to stick on side windows too.

Adverse conditions: FLOOD

○ Try to assess the depth of a 'sheet' of water before entering, using other cars, trees, lampposts or depth marker posts as gauges or by throwing in a heavy rock.

○ NEVER drive fast through water. You could send a wave over the engine and stall it, you will obscure your vision and may skid.

○ Wait for other cars to go through first.

○ Drive on the crown of the road where the water is at its most shallow.

○ Drive slowly in first or second gear, keeping the revs high, and slip the clutch.

○ NEVER change gear in water and try not to stall. This can suck water through the exhaust and into the cylinders, damaging pistons, connecting rods and crankshaft.

○ Water higher than the fan blades—not even knee deep—will make them send a spray over the engine and can do a lot of damage.

○ Once through the water, TEST YOUR BRAKES—they're probably soaked and useless. To dry them out, drive slowly with light pressure on the brake pedal, then pump them gently. Don't speed up until they're OK.

Skid control

When the tyres lose their grip and slide along the road surface, you're in a skid. It happens principally when the car is being driven too fast for the road conditions. DON'T PANIC! You must first identify which wheels are skidding and then act swiftly and calmly.

FRONT-WHEEL SKID: You've turned the steering wheel, but the car keeps going straight. Usual cause: accelerating too hard into a bend.

In a **rear-wheel-drive car**, take feet off ALL pedals and gradually turn the wheel in the direction the car is moving. When the front wheels grip, accelerate gently and steer out. In a **front-wheel-drive car**, ease off the accelerator, retaining contact, and continue steering smoothly in the direction you want to go. When you're back on course, straighten the wheels and accelerate gently.

REAR-WHEEL SKID: The back of the car swings round. Usual cause: taking corner too fast in a front-wheel-drive car.

Take feet off ALL pedals and gradually turn the wheel in the same direction the back of the car is sliding. When all wheels are aligned, accelerate gently.

FOUR-WHEEL SKID: The wheels lock and the car carries on travelling forward without losing speed. Usual cause: hard braking.

Ease off the brake until the wheels start to turn. Once you can steer, straighten the wheels and pump the brakes to avoid locking again.

WARNING

NEVER slam on the brakes as you start to skid. This 'natural' reaction is the worst thing you can do.

BREAKDOWNS

Not only inconvenient, but dangerous if you're alone in a quiet area (see SELF-DEFENCE) or you've driven out of town in severe weather. If you carry a spare parts kit at all times, AND learn how to use it, you'll be prepared for most situations.

Basic car maintenance

To minimize the likelihood of breakdown or accident (and reduce garage bills!), do these simple checks on your car once a week:

◑ Keep the oil (the lifeblood of the engine) topped up (and don't fail to replace it along with the filter at the recommended mileage).

◑ Check for leaks (of ANY fluids).

◑ Tighten loose spark plugs—check their leads are secure and look at the distributor to check the points.

◑ Check/top up levels of brake and battery fluid, washer bottle and radiator.

- Add anti-freeze to the radiator in winter.
- Tighten exhaust manifold bolts.
- Check alternator drive/fan belt.
- Clean and grease battery terminals.

EMERGENCY!
BREAKDOWN

Despite ALL preparations and all your good intentions, something WILL go wrong. These temporary measures will help you to get to the garage or a callbox:

► FAN BELT
SYMPTOM: IGNITION LIGHT COMES ON WHILST DRIVING
DON'T use tights or string for a temporary belt, unless your car is an old model. Modern fan belts work under very high tension and most improvised belts won't last a minute. Many modern engines are cooled by motor, but the ignition light might indicate a malfunctioning or broken alternator.

Carry a replacement belt. Failing this, let the engine cool (for about 30 minutes) and pick out broken bits of belt. Turn off unnecessary electrics (the battery won't be getting topped up), then drive slowly in high gear for no more than 5 km (about 3 miles).

► FUSES
SYMPTOM: One part of electrical system fails
There should be spare fuses in the fuse box. If not, borrow one of equal rating from a less essential circuit, for example, the rear screen heater. **NEVER** use silver foil, ordinary wire or a paper clip to bridge a fuse—it could start a fire. 13-amp household fuse wire is OK.

► FROZEN LOCK
SYMPTOM: Obvious! You can't insert or turn the key
Heat the key with a match and insert key until it turns. **NEVER** breathe on the lock—your mouth could stick to it and condensed moisture will make the problem worse.

► FLOODED ENGINE
SYMPTOM: Engine won't fire after too much choke
Push manual choke in, slowly depress accelerator fully, hold down and turn starter for a few seconds. This should blast plug ends dry.

► CRACKED DISTRIBUTOR CAP
SYMPTOM: Engine misfires or won't start

Lift off cap, scrape burnt plastic with nailfile/screwdriver and seal crack with nail varnish or drop of oil from dipstick.

► JAMMED ACCELERATOR
SYMPTOM: Pedal won't rise, revs stay high

Hook your foot under the pedal to bring it up, signal left, look for safe place to stop. Braking will be hard if revs are still high. Make temporary spring of strong rubber bands with a wire hook. Drive no faster than 50 kph (about 30 mph) to garage and DON'T jab accelerator.

► SHATTERED WINDSCREEN
SYMPTOM: Obvious! Toughened glass 'crazes', obscuring vision

DON'T punch a hole as you drive — you'll cut yourself badly and could get glass splinters in your eyes. Lean forward to see, slow down and STOP as soon as possible. Cover heating vents, spread newspaper over bonnet and front seats, wear gloves or wrap hand, push ALL the glass out onto paper and wrap it up.

Fix tape around edge of screen or remove rubber flange and fit temporary screen if you have one. Otherwise, drive slowly to garage, wearing glasses or sunglasses to protect your eyes.

REMEMBER

When driving on a road with loose chippings, place your hand on the inside of the screen. This helps dissipate the shock if a stone hits and may prevent shattering.

SPARE PARTS KIT

- Spare wheel, wheel brace and jack (check the tyre pressure regularly)
- Fan belt, emergency windscreen, headlamp bulbs
- Fuses (at least one of each type for your car)
- Length of HT lead (between coil, distributor and spark plugs)
- Can of fuel (replaced every few months, as it deteriorates)
- Can of water (replaced frequently), can of oil
- Spark plug (only one necessary — they rarely all fail at once)
- Strong elastic bands, rolls of insulating/plumbers' tape
- Spare keys in a concealed box outside the car
- Screwdrivers (slot and crosshead types)
- Spanners (including one for spark plugs)
- Jump leads
- Baling wire (for temporary repairs to, say, exhaust pipe)

CAR SURVIVAL KIT

As well as the spare parts kit, cars should ALWAYS carry the following:
- Warning triangle
- Fire extinguisher—BCF or dry powder
- First-aid kit
- Powerful flashlight with red filter and 'flashing' mode
- Maps

ACCIDENTS

More than 20 times as many accidents happen in built-up areas as on motorways, and most are avoidable. Often they're caused by sheer carelessness, as is proven by the most 'popular' times that they occur:

- In summer and autumn
- In good weather
- On Sunday afternoons
- Late in the evening
- Closing time for public houses and nightclubs
- After work

WARNING

Fatigue creeps up slowly, especially on a monotonous ring road. Do you have full recollection of the last few miles? Are you yawning frequently? Are you imagining shadows are obstacles? You're dangerously tired. Open a window, listen to a talk show (but only if you're interested, otherwise it's a soporific drone), chew a sweet, lick a finger and dampen the inner corners of your eyes. Stop for a hot drink.

REMEMBER

Obvious as it may seem, ALWAYS try to avoid any head-on collision with another vehicle or a fixed object such as a wall or tree—these are the most dangerous accidents of all. Aim to scrape along a wall. Go for dense bushes, not a solid tree. If a car is heading straight for you, brake (not violently, or you'll skid) and scan the road for an escape. Remember the other driver might try to pull over in the same place as you, so only turn when you see which way he's swerving. Turn away from him, even if you have to side-scrape an 'innocent' car.

TOP TEN ACCIDENT CAUSES

Statistics tell us everything we need to know about our driving 'mistakes'. Unfortunately, the resultant death toll is enormous—and increasing. If you are honest with yourself, you KNOW if you are taking unnecessary risks when driving. Look through this list and see if you can make using the roads safer. Starting with the most common cause of accidents:

- Careless, dangerous or drunk-driving on a straight road or ignoring/ misjudging traffic signals
- Misjudging a right turn or crashing into someone turning
- Overtaking dangerously
- Bad parking or getting out of a parked car
- Impatient driving—for example, overtaking on the inside or U-turns
- Turning left and swinging out too wide
- Stopping suddenly or unexpectedly
- Pulling out carelessly or bumping the car in front in a slow queue
- Reversing into something or someone

After an accident

If you are involved in any accident—major OR minor—the law requires you to take specific steps. Your insurance firm will need to know certain details. You may be injured or in shock—you will definitely be shaken up—but it's in everyone's interests to KEEP A COOL HEAD.

WARNING

If there is serious injury, the first priority MUST be to apply first aid and call an ambulance. If fuel is leaking, and there is ANY danger of fire or explosion, DON'T smoke. Move everyone back to a safe distance.

1 STOP! Even if your own vehicle is OK (for example, if someone else swerved to avoid you and crashed).

2 If you've caused an obstruction, ensure there are no further accidents by getting someone to stop or direct oncoming traffic.

3 DON'T argue about what happened, and don't admit liability.

4 If anyone saw the accident, take their name and address so they can make a statement to your insurance company.

5 Call the police (and an ambulance) if there is serious injury, traffic obstruction or damage to vehicles or property. Inform them later if you damaged an unoccupied car—it's an offence not to.

EMERGENCY!

CAR FIRE

Car fires are usually started by faulty wiring and/or leaking petrol. A fire may start as a result of a collision or a dropped cigarette on a garage forecourt or similar.

► **You must act quickly, to put out the fire**
Petrol is extremely volatile—if the vapour ignites, it's likely that the tank may explode

► **Switch off the ignition**
In most cases, this will stop the fuel pump

FIRE ON THE MOVE
Coast to the side of the road, STOP and get everyone out of the car

FIRE UNDER BONNET
DON'T raise bonnet—you'll fan the flames higher. Open it just enough to insert fire extinguisher nozzle. Direct at base of flames, from edge to middle, side to side—leave no patch unsmothered

IF YOU HAVE NO EXTINGUISHER
Smother flames with any thick material to hand—car rug (wool, NOT man-made fibre), wool coat. Call fire brigade. Get all passengers back to a safe distance

FIRE NOT UNDER BONNET
Disconnect battery. Use fire extinguisher, if site of fire is accessible. Back off and call fire brigade

REMEMBER

You MUST carry a fire extinguisher in your car. Keep it mounted on the dashboard or by your feet—NOT in the boot!

6 Make a sketch of the scene before anything is moved: road names, position of trees and lampposts, road markings, position of vehicles in relation to each other, length and position of skid marks.

7 Make a note of damage sustained by ALL vehicles involved. If you've got a camera with you, even better—you need as accurate a record of the damage as possible.

8 Collect names, addresses and registration numbers and give your own. If anyone refuses, note their car registration number — it can be traced.

9 If someone is injured, you MUST produce your insurance certificate there and then, or report to the police within 24 hours.

10 Contact your insurance company within 24 hours and ask for accident report and claim forms. Supply them with full, accurate details and a written estimate of repair costs from a garage, any letters or bills from the other driver(s), plus any police 'Notice of Intended Prosecution'.

> # REMEMBER
> **People may be expecting you to arrive. Telephone ahead to say that you are going to be late.**

MOTORBIKES

Why does mother despair when Johnny gets his first motorbike? Because a motorcyclist is eight times more likely to crash than a car driver. Why does Johnny want a motorbike? Because it combines the speed of a car with the freedom of a bike—and enhances his 'image'. These also being the reasons for the worrying statistics.

The commonest cause of motorbike accidents is a driver failing to see a rider at a junction, and injuries are often far more serious than they need to be because of insufficient body protection. The manoeuvrability of a motorbike encourages recklessness, especially in experienced, (over) confident riders. Keeping out of danger on a 'mean machine' entails being VISIBLE, being PROTECTED and riding SAFELY.

Visibility
ALWAYS keep your headlight on, even in daylight. Remember that the lamp—especially on machines under 250 cc—may NOT be powerful enough to make you visible on its own. **ALWAYS** wear something bright coloured or fluorescent, especially at night and in poor weather. **ALWAYS** flash your headlight before overtaking and be aware of the driver's blind spot. **NEVER** overtake on the inside.

Protection
ALWAYS wear a safety helmet (required by law in Britain) and replace it if the finish becomes crazed or cracked or once it's

saved you in a fall. It may look undamaged, but DON'T trust it. **ALWAYS** wear goggles or a visor to protect your eyes—against grit, wind and insects, as well as injury.

ALWAYS cover your body when riding, since even a minor fall easily results in nasty cuts and grazes. Leathers with reinforced knee and elbow patches give the best protection against abrasions (looks 'cool' too). Add a one-piece storm suit for cold or wet weather. **NEVER** leave off the gloves—your hands are very vulnerable. **NEVER** ride in any footwear other than sturdy boots with thick soles.

Safety

ALWAYS be aware of the vehicle in front. In dry weather, the safe braking distance is 1m (about 3 ft) for every 1.5 kph (about 1 mph). Try this test using a road marker. It takes two seconds to say 'only a fool forgets the two-second rule' (test yourself with a watch to get the timing right). If you can fit it in between the car in front and yourself passing the marker, you're at a safe distance behind. Pull back a little, to be sure. Remember to DOUBLE the braking distance on wet roads.

- Pull more firmly on the front than the rear brake when riding upright on a dry road.
- Use even pressure on both brakes on wet roads.
- DON'T use the front brake when you're leaning, turning or riding on a surface with loose grit.
- DON'T line up or form a pack when you're riding with others. Make a 'V' formation for maximum visibility.

CYCLING

Cyclists in the city are vulnerable—roughly 90 per cent of accidents involving bicycles occur in built-up areas. Around a third of these casualties are teenagers. Since no licence is needed to ride a bike, many cyclists lack proficiency or disobey (or are ignorant of) road signs and markings. But bikes ARE traffic and should no more ignore a red light, for instance, than should a car.

As with motorbikes, many accidents happen due to the cyclist not being seen, and—as with motorbikes—visibility, in conjunction with adequate body protection, good road sense and proficiency, makes cycling safer.

Visibility

ALWAYS use front and rear lamps at night. In Britain, lamps and rear reflectors are required by law. A bike must not be sold without rear, front, pedal and wheel reflectors. Wear something made of retro-reflective material at night. Arm and ankle bands or Sam Browne belts are lightweight and effective. Wear bright clothing in the day and/or a reflective belt.

Protection

Wear a lightweight cycle helmet and consider using a filter mask to protect against exhaust fumes. It may not look 'cool', but surely everyone understands these days about the dangers of cycling (and breathing hard) on a regular basis in heavy traffic. Your long-term health is at risk. Investigate masks thoroughly. Some offer no real protection at all.

Safety

ALWAYS slow down as you approach a junction and keep a sharp eye out for cars turning out or crossing your path unexpectedly. **ALWAYS** use the correct arm signals.

DON'T hug the kerb—you need room to pull in if a car scrapes by. **DON'T** pass too close to parked cars—a door might open. **DON'T** help give cyclists a bad name. Drivers tend to think of bicycles as the mosquitoes of the road, so make sure YOU signal clearly, don't wobble and don't crawl along in the middle of the lane.

CYCLISTS' SURVIVAL KIT

Though you'll want to keep down the weight you carry, it's advisable to take the following on EVERY trip:
- Puncture repair kit
- Pump
- Spare valve
- Adjustable or barbell spanner
- Screwdriver

PUBLIC TRANSPORT

The generally more-or-less efficient, more-or-less up-to-date networks of buses, railways, underground trains, coaches and—in some countries—trams and boats, are the principal means of transport for millions of city dwellers, and many thousands rely on them entirely for mobility. Most of us assume we'll be safe on public transport, but sadly this is no longer necessarily the case.

Muggers and bag snatchers, pickpockets and rapists DO operate on public transport, and fire, crashes and breakdowns are ever-present threats. Wherever you're heading, and especially if you're going into unfamiliar territory, research the route beforehand. Not only will you find your way more quickly, you'll avoid appearing lost and vulnerable by stopping to peer at maps and signs.

In many cities, you can ring a 24-hour information service for help with the route and an information line for details of how the services are running.

REMEMBER

If you possibly can, tell someone WHERE you're going and HOW and WHEN you intend to get there. That way, you'll have raised the alarm in advance should something happen to you.

Underground train

London's tube—along with other cities' underground train networks—is not getting safer. Trains, escalators and lifts are deteriorating faster than the modernization programme can cope with. An ever-increasing volume of passengers puts extra stress on an already-overloaded system.

BE SAFE—NOT SORRY

Avoiding danger every time you travel underground is largely a matter of common sense. If you're travelling late at night, get in a compartment with other people—or preferably find another means of transport. If a rowdy or drunken group of 'hooligans' gets on, avoid eye-to-eye contact until the next station, then change compartments.

Especially if you're a woman, don't be shy of allying yourself with a fellow, trustworthy passenger if there's somebody frightening in the carriage—chatting will normalise the situation and reduce your sense of fear. Also, two people are less likely to attract an attacker than a lone passenger.

Follow the New Yorker's example and always read (or pretend!) on the tube so you avoid eye contact that could be interpreted as provocative. DON'T get deeply involved in what you're reading, though. Keep alert and concentrate on what's happening around you.

Keep valuables in a secure inside pocket or in a zipped compartment on the inside of your bag to foil pickpockets—especially during rush hours and at stations in the city.

In general, make yourself inconspicuous and remain aware of your situation. To prepare for the worst (and you MUST), see SELF-DEFENCE for advice on avoiding danger and dealing with attack.

The terrible and tragic fire at King's Cross in 1987 provided proof—if proof were needed—that disaster can strike even in the most mundane of situations. Since that catastrophe, of course, smoking has been banned on most public transport, in London at least. This has gone some way towards reducing the risk of fire.

Recorded sex crimes on the tubes (from 'flashing' to rape), on the other hand, increased by 40 per cent in 1989, and the incidence and seriousness of random violence is on the increase. Controversy surrounded the London debut of the Guardian Angels—New Yorker Curtis Sliwa's voluntary, uniformed crime-fighting organization—since some people felt, at first, that the Angels were more likely to incite violence than avert it (see SELF-DEFENCE: Guardian Angels). The signs are, however, that London's tube network is beginning to catch up with New York's beleaguered subway system—and EVERY passenger should be alert to danger.

Bus/coach

Apply the same degree of common sense to bus, coach and tram journeys, especially at night. If you're a woman on your own, sit near to the driver or ask the conductor to keep an eye out, and take care at bus stops in deserted or ill-lit streets.

Again, research the journey—look at a street map of the route you're taking so you always have a rough idea of where you are. Make sure you're familiar with the area around your destination (see SELF-DEFENCE).

Train

Again, common sense should keep you out of trouble on a train journey—many of the same rules apply. The 'it is dangerous to lean out' signs are NOT a joke. Trains may pass each other a few centimetres apart and poles, signs and other obstacles are often close to the tracks. You can imagine the consequences and, yes, people have been beheaded! Also, if the train jolts, you could be thrown out. **NEVER** try to close an open door on a moving train—move away from it and call the guard instead.

Although there is (statistically) far less chance of your being involved in a train crash than in a car accident, it's a common fear—possibly because you have no control over what happens. If the train you're on crashes:

◗ You won't have much time to think, but if you feel the emergency brakes come on, throw yourself onto the floor (clear of windows and doors), brace yourself against something fixed and pull your chin down to your chest to protect against whiplash. You MUST be supported when the impact comes.

○ NEVER try to throw yourself out of a moving train—your body would receive the full impact, instead of being partially protected by the carriage. There might be live tracks or pools of battery acid to fall into.

Water transport

You might think twice before stepping onto a riverboat, since the horrific night in August 1989 when 51 people perished in the middle of London's Thames. Nobody could have predicted the Marchioness disaster—it was a freak accident, which is highly unlikely to be repeated. However, it is as well to be prepared when you're travelling over water, even in the middle of a city.

Many aspects of water safety are taken out of your hands on a ferry, waterbus or pleasure cruiser. The crew MUST be responsible for passenger welfare—and trained to cope in any emergency. Certain situations, however, may call for an instant reaction.

Man overboard!

Your priority as a passenger is, of course, to alert the crew— IMMEDIATELY. The greater part of the following procedure will be their responsibility:

○ Start turning the boat to approach him/her from downwind and shout 'man overboard' to alert all crew to the situation. Throw life belts (BESIDE the person) and KEEP WATCHING. As you approach, stop the engine before hoisting them in (over the stern, in calm water). If the person appears in difficulty or is losing consciousness, someone (who MUST be wearing a life jacket) must jump in and help them.

YOU overboard!

How you cope depends on whether the water is cold or warm. In either case, your first action is to **SHOUT** and inflate your life jacket, if you have one.

COLD WATER: If you have a life jacket, curl up in the foetal 'HELP' (heat-escape-lessening) position to keep warm. If you have no life jacket, remove heavy footwear, but keep all your clothes on—even when soaked, they help you conserve heat. Try to float on your back, moving as little as possible and breathing slowly.

WARM WATER: If you have no life jacket, you can make a float from a pair of trousers by knotting each ankle, holding the waistband open behind you and whipping them over your head, trapping air in the legs. Failing life jacket and trousers, adopt the 'drownproofing' position—lying relaxed face downwards in the water, with arms forward. **DON'T** try to swim to shore when at sea or in a wide river, nor after the boat—you'll exhaust yourself quickly swimming in clothes.

FIRE ON BOARD

Fire on board a boat can spread rapidly, especially if the boat is constructed from wood/fibreglass. It is doubly dangerous, because of the proximity of fuel tanks. Again, the crew should deal with any fire, but if you discover it first:

■ SHOUT! Turn off the engine. Throw any burning equipment or fittings overboard. Get everyone out if the fire is below deck, close doors and hatches and fight the fire from above, so that you're able to retreat to safety if necessary.

■ NEVER throw water on blazing petrol or gas. Use an appropriate extinguisher (see FIRE!).

LIFTS/ESCALATORS

In many buildings there is little option but to use a lift to reach upper floors. Few people have the stamina to climb regularly to higher than four floors let alone the average 14 or 15 in a large block of flats! Many people may not be able to manage that sort of climb if, for instance, they are elderly, carrying heavy loads or have a heart or breathing problem. Lifts have been a part of life for over a hundred years. In principle, most are quite simple, but they can still suffer from frequent operating problems.

Types of lift

There are two main types of lift: hydraulic and traction. Almost all lifts operate electrically and are obviously to be avoided during known periods of power failures.

Hydraulic lifts, mainly in low-rise buildings, are driven by a pump that raises and lowers a ram (piston) and—with lots of variations—usually conform to one of the following principles. The ram may be situated directly beneath the car, pushing it up and down the shaft on guide rails OR the ram may be situated to one side of the shaft and use steel suspension ropes to pull the car up and down the shaft on the guide rails.

Traction lifts employ a motor, gearbox and brake to raise and lower the car on guide rails by a system of pulleys, steel suspension ropes and counterweights. High-speed lifts usually do not have a gearbox—they are driven directly by the motor.

REMEMBER

In Britain, at least, it is a safety requirement that the steel suspension ropes in a lift are capable of taking twelve times the load imposed on them — so they are extremely unlikely to break. There is a minimum of two ropes (usually four or five), any ONE of which would support the weight of a full lift car.

Hatches

Lifts used to have hatches in the ceiling, towards the back of the car, for engineers' access. They were designed for engineers to be able to climb INTO the car—NOT for passengers to climb out. Most are locked outside the car. In Britain, hatches are being phased out. There have been very serious accidents when vandals or show-offs have used these hatches—probably because they saw someone do so in a movie.

Safety devices

The worst thing that could go wrong with a direct hydraulic lift (where the ram is situated directly below the car) is that—for one reason or another—there may be a sudden loss of hydraulic pressure. The car would sink to the bottom of the shaft, but would do so slowly.

If the steel suspension ropes were to break (in some hydraulic and all traction lifts) there is a non-electrical speed monitoring device. This recognizes that the car has begun to descend too rapidly—a ten per cent increase in speed will trigger it—and immediately engages wedges beneath the car to arrest the downward movement.

At the bottom of the lift shaft there are buffers which will absorb the impact of a falling lift car.

Lift doors must be closed before a lift will operate. If it is possible to operate the lift with ANY door open, do NOT use the lift. It must be serviced.

Some lifts (especially modern ones) recognize when the load capacity has been exceeded, and will not operate until someone steps out and lightens the load.

Automatic doors should always reopen if a person or object prevents them from closing. The car should not be able to move if the doors are not fully closed—someone could be caught in the doors. This is obviously a VITAL safety precaution.

Using lifts

When entering or leaving the lift car, check that the lift has stopped exactly level with the floor. This is a surprisingly

common problem and causes many accidents, as people trip up or down the uneven levels.

The maximum number of persons the lift can carry will be clearly displayed within the car. Do NOT exceed it—some modern lifts won't allow you to—although the greatest risk is that the motor will be electrically overloaded and break down. It is also possible that the lift will descend more rapidly than usual and trigger the speed-monitoring device—stopping the car anywhere in the shaft.

Lattice-gate (cage) lifts should ALWAYS incorporate a 'stop' button—in case someone or something gets caught in the lattice. Solid-door lifts may also have 'stop' buttons.

All lifts MUST have alarm buttons or emergency telephones. Alarm buttons simply sound a bell or buzzer to draw attention to the fact that there is a problem with the lift. If you can't hear the alarm from the car, it may be sounding in a security office, the caretaker's room—even in the street. Press the bell even if you cannot hear the alarm—in a power failure, a back-up battery will operate it.

Some emergency telephones are 'hot lines' to a rescue service. Some require you to dial an emergency number. If that doesn't ring or no one answers, dial the fire brigade.

❗ WARNING

Do NOT hold automatic doors open by holding back the doors. Most have an override function—after several attempts to close, the doors will attempt to push any obstruction out of the way so that the car can move. A warning buzzer may sound at the same time. Use the 'door open' button instead.

REGULAR MAINTENANCE

Whoever owns the lift is responsible for any damage or injury caused by it. Whoever manages the lift MUST keep a record of reported operating problems and MUST ensure that regular maintenance is carried out. Every day someone must check:

- The alarm button
- The telephone
- The 'door open' button
- The 'stop' button
- The lights
- The floor indicators

It is also vital to report the failure of the car to stop exactly level with each set of doors. There should be no step up or down into the car.

TRAPPED IN A LIFT

Stay calm. Keep others calm. Most people have seen 'disaster' movies and many will immediately assume that the car will plummet to the bottom of the shaft. Some will be angry as well as anxious, if they have an appointment that they will miss. Take charge. Understanding how lifts work will help you reassure others. Discourage anyone who starts to tell 'lift disaster' stories.

EMERGENCY!
LIFTS

1 Press the lowest and highest floor buttons, NOT all the buttons. If the lift car does not start moving again ...

2 Press the 'door open' button. You might be at a floor. It's quite common for the doors to close, but for the car to stay put.

3 Press the alarm button. You may not be able to hear it sounding. Hold your finger on it for as long as you like.

4 There may be a phone. Some contact emergency services directly, whereas others work as 'outside lines'. Call the number indicated on the lift instructions or call the fire brigade. If there is an outside line, don't let anyone use it for personal calls until rescue has been arranged.

5 If there is no alarm/phone, bang loudly and rhythmically on the doors. Several people should bang in unison using keys, shoes or coins to make as much noise as possible. Pause every few beats to see if you can hear anyone coming to help.

6 If you have made contact with the outside, try to wait as calmly as possible.

Will the lift plummet down the shaft?

Extremely unlikely. Even when power is cut to the lift, fail-safe brakes prevent the car from moving.

Could someone climb out of the hatch?

Climbing out is possibly the WORST thing you could do. If you could get on top of a lift car, you would find it very greasy, dirty and slippery. A fall could easily prove fatal. Once out there, it is very unlikely that you could reach or open the door to the floor above. The hatch may fall shut and the lift may start moving. Opening the hatch breaks a circuit and immobilizes the lift car. DON'T TRY IT.

 What if there is no engineer available?

Rescues may be handled by the fire brigade, who are trained for lift emergencies.

 What happens if there is no power?

The lift can be winched up and down the shaft. Back-up batteries ensure that lights and alarms still function.

Is it possible to force the lift doors open?

Sometimes—but you would probably not be able to reach or open the doors onto any of the floors. You might be between floors. When the car is halfway above a floor, the gap beneath the car would be extremely dangerous.

 Is it possible to suffocate from lack of air?

The car may become stuffy, but all lift cars have vents which will allow a flow of air. Some are concealed above false ceilings or in gaps around the doors.

 What if there is no response from the outside?

This usually only happens in office buildings which are closed for the night or the weekend. In multi-occupied domestic buildings, there is a much better chance of attracting rescue, day or night. Even in office buildings there is usually a caretaker, janitor, security officer—or cleaners in the evening or early morning.

Q What if the lights go out?

A The back-up battery will give at least an hour's light once power has failed.

Q What if it's an office building and it's the weekend?

A The worst that can happen is that you will have to wait until the building opens after the weekend.

ESCALATORS/WALKWAYS

❍ ALWAYS hold the handrail. If there was a power failure or someone pressed the emergency stop button, you and everyone else may fall. This is particularly dangerous on escalators. See **Reading the signs** (colour pages).

❍ DON'T stand so that your feet rub against the sides. The heat generated by the friction could soften rubber and plastic soles and drag them into the machinery.

❍ STEP off the escalator—DON'T drag your feet over the comb. If edges are worn you risk a nasty accident.

❍ GET CLEAR of the escalator/walkway as soon as you reach the end. If there are people behind you, they may have no choice but to shove you out of the way to prevent a 'log jam'.

❍ DON'T allow children to play on escalators/walkways.

❍ ALWAYS carry pets, small children, pushchairs and prams, soft luggage or shopping. It may get caught as the steps 'open' and 'close' on escalators—or at the sides of escalators/walkways.

❍ ALWAYS stand on and keep luggage to one side of the escalator to allow people to walk up or down.

❍ When escalators are stationary, walk up and down with extreme care. The irregular depth of the steps could easily cause you to trip and fall.

TRAVELLING ABROAD

 The world is getting smaller! Once mysterious and inaccessible foreign lands are now equipped with airports, hotels and at least one branch of a well-known hamburger restaurant. Tourism is BIG business; business is international. Whether it's for a holiday or to strike a deal, most of us now leave hearth and home at least once a year, taking journeys across continents, time zones, cultural and language barriers—often quite unprepared.

The pitfalls of a foreign trip are legion. There are any amount of disasters that can—and do—befall even the most seasoned traveller. As we travel further afield, there are more local customs and laws to take into account.

This section addresses all aspects of our behaviour abroad—from practical considerations (such as staying healthy and safe) to matters of courtesy and fitting in with the local customs and culture.

Research your destination

You may think, if you're only nipping a few hundred miles across a continent to lie on a beach for a fortnight, there's nothing you need to know (and to some extent the travel industry has encouraged you to think that). But even in a place you THINK you know well, there's ALWAYS plenty to learn—you've probably even been startled by an unfamiliar street in your own home town.

As soon as you enter a foreign country, you face many differences—language, diet, climate, dress, religion, customs and laws which may vary widely from those you take for granted, but aren't necessarily obvious straight away.

It is a matter of courtesy and common sense to know what to expect in any situation. Even if you intend to do nothing but sunbathe, you might want to visit the local church or hire a car—do you need to cover your head/shoulders/legs? Is today a local holy day? Do you need an international driver's licence/special insurance? What are the local roads like? What the hell does THAT road sign mean?

On business trips it's especially important to know how not to offend your host/contact/potential client, how to ensure they'll be eager to welcome you back.

> # REMEMBER
> No matter what the purpose of your trip, when you're abroad, you're a GUEST in somebody else's home.

BEFORE YOU GO

Immunization
Up-to-date information regarding which vaccinations are mandatory or advisable for your destination is obtainable from the

A reliable guidebook to your destination is a worthwhile investment. Choose one for its depth of information rather than its lyrical descriptions—those you can write yourself! You need specific details on ALL aspects of the country, which could be condensed under the following headings:

LANGUAGE English is the native language of 330 million people—after Mandarin Chinese (748 million speakers), the commonest language in the world. Your hosts are more likely to have learned English than any other tongue, but take the trouble to acquire a few basic phrases in their language. It's only courteous (at least) to ATTEMPT to communicate in the vernacular, and it's often appreciated. In remote areas, it's entirely possible that NO English is spoken at all, making a phrase book an absolutely essential piece of equipment.

CLIMATE Air travel means you can be freezing in an English February one day and sweltering in the tropics the next. Obviously you will need to know whether the heat you're heading for is humid, as in Singapore, or fierce and dry, as in desert regions such as Egypt. In other words, extreme variation is possible, even within a single country—the coast of Senegal, for instance, is humid whilst the interior is extremely hot and dry. Forewarned, you will be able to pack appropriate clothing and avoid the most extreme seasons.

CURRENCY You should know not only that 1 kwacha = 100 ngwee in Zambia, whereas in Papua New Guinea you get 100 toeas for your kina, but also how many toeas a cold drink costs, how many kinas—if any—you're allowed to import. Will dollars or pounds turn out to be more useful? Are there two rates of exchange—government and black market? If so, what are the risks? Are traveller's cheques and credit cards going to be any use to you, and if so, which? Some countries have an inadequate supply of small denomination coins in circulation, and shops might give sweets or a box of matches in place of the odd penny.

DRESS Climate and purpose of trip will be the main factors in deciding what to pack, but you should find out what the locals wear. Do the women cover up—and do visitors have to do so, too? Are you going to offend people if you wear shorts? Is topless/nude bathing tolerated/restricted/illegal? Is a suit and tie right for meetings?

RELIGION A potential can of worms for the uninformed visitor. Eating in public during Ramadan or asking for an alcoholic drink in a Muslim country, women wearing shorts in Italian churches, entering a Burmese temple without a temple sash—all these will SERIOUSLY offend. Whatever your religion or attitude, you should be respectful of others' beliefs if only by taking the trouble to find out what they are.

CULTURE You DON'T criticize the monarchy in Thailand (it's a SERIOUS offence) or touch anyone on the head; you DON'T use soap in a Japanese bath; you DON'T show anger in Bali ... Other countries have different concepts of what constitutes good behaviour. To avoid offending someone (or risking prosecution) with your rude unclean habits, learn what you can about your destination and respect its ways.

DIET These days, of course, you can find the same bland 'international' menu the world over, but you invariably eat far better at local restaurants. Knowing something about the cuisine of the land will enhance your understanding of its people, not to mention your enjoyment of their hospitality. Rudimentary menu translation ability will enable you to avoid any foods you dislike or are allergic to, and steer away from the more unusual offerings—sheeps' eyeballs, rat, insects and live monkeys' brains are not as common as some people would like to think.

HEALTH You MUST ensure you are innoculated against endemic diseases (see Immunization), that your travel insurance is sufficient to cover ALL eventualities, and that you take an adequate supply of any prescribed medication. It's as well to know something about the standard of health care in the area you're visiting—on the whole, it's wise to avoid medical or dental treatment in Third World countries. Many countries have a favoured panacea for minor ailments—as anyone who's been advised by a Greek doctor to swab themselves with their own urine (which has antiseptic properties) to the affected part can affirm.

RED TAPE Do you need a visa? What are the customs restrictions? What would you do if you lost your passport? Is hitchhiking allowed? What if you're arrested? As with health care, it's as well to be prepared for the worst—collect phone numbers/addresses of the consulate, embassy and travellers' cheque/credit card 'lost or stolen' departments in one place and keep them with you. Ensure your travel insurance covers legal costs.

airline—they have a vested interest, since they are obliged to fly you home at their expense if you don't have the necessary jabs. Additional vaccinations are advisable for visits to the tropics and subtropics, since the natural immunity of people from a temperate climate might not be strong enough to withstand the barrage of unfamiliar diseases. Your GP should be able to advise you, and administer the innoculations. Allow plenty of time before you're due to travel.

The following list of the major vaccinations, when to have them, how long they last and for which destinations they're required or advised is a general guide. ALWAYS check your vaccinations are up to date before you travel. NEVER assume you'll be OK without them.

SMALLPOX On 1 January 1980, the WHO (World Health Organization) declared the entire world free from this disease, and innoculation is no longer necessary.

CHOLERA An acute infection of the small intestine, cholera is spread by the *Vibrio cholerae* bacterium in food or water contaminated by the faeces of someone suffering from the disease. Symptoms are severe fluid loss—profuse sweating, vomiting and diarrhoea ('ricewater stools'). Over 50 per cent mortality rate, if untreated.

Vaccination: At least six days before travel; revaccination necessary after six months—if within that time, no waiting period is required. Some (Middle Eastern) countries may require two injections. **Required:** Pakistan, India, Burma (renamed Myanmar). Many Middle Eastern and Far Eastern countries and parts of African continent. Gives limited protection. Avoid local water.

YELLOW FEVER Spread mainly by the *Aedes aegypti* mosquito, yellow fever—or yellow jack—affects the liver and kidneys. Symptoms include chills, headache, fever, black vomit and jaundice. It can be fatal.

Vaccination: Only given in registered centres, at least ten days before travel and not within 14 days of any other live virus vaccine (except polio). Valid ten years and 100 per cent effective. **Required:** Central Africa (15°N of equator to 10°S), Central America (northern border of Panama state to 15°S of equator—except Bolivia and part of eastern Brazil).

TYPHOID The digestive system is infected by the *Salmonella typhi* bacterium, causing weakness, high fever, chills, sweats, a red rash on chest and abdomen and—in severe cases—inflamation of spleen and bones, delirium and haemorrhageing. *Salmonella paratyphi* A, B or C cause the milder paratyphoid fever.

Vaccination: A course of two monovalent typhoid injections two–six weeks apart gives protection for up to three years. If time is short, a single injection gives some protection for six–eight weeks. **Advised:** Worldwide except northwest Europe, Canada and the USA.

TETANUS Affects the nervous system after contamination of a wound with the bacterium *Clostridium tetani.* Symptoms are muscle stiffness and spasm followed by rigidity, starting in the neck and jaw (hence 'lockjaw'), spreading to the back, chest, abdomen, limbs and possibly the whole body, causing it to arch backwards. Accompanied by high fever, convulsions and extreme pain.

Vaccination: An initial course of three injections of the tetanus taxoid, spaced out over six—twelve months, then booster every three–five years. A patient receiving an open wound, especially if dirty, should have a booster if their last was more than a year before. **Advised:** Worldwide.

POLIOMYELITIS Still a problem in the Third World, polio or infantile paralysis is spread by the faeces of people who have the disease and affects the central nervous system. Paralytic poliomyelitis is the most extreme, less common form, causing muscle weakness and eventual paralysis. All forms are more severe when contracted in adulthood, and there is no cure.

Vaccination: The preferred form is the Sabin vaccine—three drops on a sugar lump, taken three times at monthly intervals. Pregnant women are given the Salk vaccine by injection. Booster every five years. **Advised:** Worldwide.

MALARIA Spread by the *Anopheles* mosquito, malaria is thought to have killed Oliver Cromwell, who refused quinine—then known as Jesuit's Bark—on religious grounds. Nowadays the disease has left Europe (though it can theoretically be present anywhere warm where there is water). There are four varieties, of which Falciparum Malaria is the worst. Symptoms are high fever with alternate shivering and sweating, intense headache, nausea and vomiting.

Prevention: There is no vaccine, but preventative measures are usually effective. Cover up and use insect repellant after dark when this type of mosquito bites (only the female bites humans). Take prophylactic Proguanil, Chloroquine, Pyrimethamine, Maloprim or Fansidar tablets, the first of which is available without prescription. ALWAYS consult your GP about malaria pills, however, as the map of resistant strains is always changing. You may have to take more than one type. **Advised:** Practically worldwide except Europe, Australasia, the South Pacific and North America.

PACKING

From the life-threatening to the purely practical ... Packing is the inescapable chore you must perform every time you travel, and it's worth becoming an expert at it. Successful packing means you're equipped for every eventuality, from a 14-hour delay at Charles de Gaulle Airport to sunstroke in Bombay.

The aim is to cut down the time you spend and the baggage you end up having to lug around. Most people 'overpack' shamelessly, which they come to regret when they can't find a taxi at some remote stopover and are faced with a two-mile walk in the midday sun.

On a flight, for short trips, it's worth trying to avoid using the hold for your luggage and packing everything into your carry-on bag. That way you avoid the often-interminable wait for your luggage at the carousel. If you're driving, the less you take, the less fuel you use.

Air travel

Other than jetlag, most people should experience no difficulties in modern civil aircraft—generally, the longer the flight the more uncomfortable, though! However, people suffering certain medical conditions should inform the airline when they make reservations.

The decreased oxygen in a pressurized cabin can cause problems for passengers with any form of heart disease or recent thrombosis, anyone suffering severe respiratory disease and elderly people with hardened arteries.

People with sinus trouble might experience sinus pain and earache owing to the slightly-rarefied atmosphere. This can also cause expansion of intestinal gases—uncomfortable for anyone with a recent gastric or intestinal lesion, operation or haemorrhage. If in doubt, ask your doctor, or even the airline—all the major ones have medical departments.

Plane crashes

Actually, there is a common problem with flying. It is FEAR. Since flying is far safer than motoring, this fear is irrational, if understandable. Doctors think it's more about the loss of control or a mild claustrophobia than a belief that the plane will go down. REMEMBER! You are more likely to die in your armchair (from coronary diseases) than in a plane crash!

You will, of course, have listened and watched CAREFULLY during the cabin crew's demonstration of emergency procedure—if the unthinkable happens, you will know roughly what to do.

- Ensure each garment can be worn with several others (it used to be called 'mix and match'!). That way you get more mileage out of fewer clothes. Favour crease-resistant fabrics.
- Fit the sleeves of one shirt/blouse over the sleeves of the next, button it up, and continue until you've 'put them all on' the first, then roll the whole thing up. You end up with a bundle that takes up only a little more space than a single shirt, and you prevent creasing. Do the same with skirts, jackets and trousers.
- Roll up socks, tights, underwear and stuff them into shoes.
- Shoes are the heaviest items, so take only one spare pair.
- If you're going to need an overcoat and heavy jacket, wear them.
- If you make frequent business trips, refine your packing technique until everything fits in a briefcase—it is possible!

HOLIDAY

- DON'T be tempted to cram the suitcase full. You're bound to need space for souvenirs/duty-free goods etc.
- Take only as much shampoo, sunscreen, soap and the like as you can use in the time. There ARE shops in other countries!
- Take a leaf out of the backpacker's book—tear up your books! Buy paperbacks and discard sections as you read them. Tear out the relevant pages of guidebooks and leave the rest at home.
- Pack essential overnight requirements in your carry-on bag, so that you can cope with long airport delays or if you lose your luggage.
- Bring film for your camera—it invariably costs more abroad.
- Wear a moneybelt for your valuables.

REMEMBER

ALWAYS attach name/address labels to your cases to differentiate them from similar ones and to identify lost luggage. Mark your cases inside too.

ALWAYS attach something to your luggage to make it instantly recognizable on the carousel—a bright strap, bright tape on the strap or handle, a sticker. It is not vital for you to spot your luggage quickly, but it may be vital to stop someone with similar baggage from taking yours.

ALWAYS ensure your bags are strong and securely fastened—tie a belt round dodgy cases. Baggage handlers invariably hurl the luggage about and it's inconvenient, not to mention embarrassing, to find your dirty laundry scattered over the conveyor.

NEVER pack cash, valuable jewellery, important papers, fragile items, liquids or vital medications.

Sixty per cent of plane crashes actually happen on take-off or landing, so you may as well relax for the most part of the flight. The cabin crew are highly trained in all aspects of emergency drill and will guide you, if anything does happen. DON'T add to the horror by screaming and panicking—be the one that helps the crew. Keep up morale and comfort others.

Sea travel

Since the Herald of Free Enterprise went down yards from Zebrugge harbour in 1987, some people might have thought twice about the risks involved in travelling by sea. That disaster created such a stir, however, partly because it was so unprecedented. Nowadays ferry operators have been obliged to check into their standards of safety. On a modern ferry, ship or cruise liner, the worst problem you're likely to encounter is seasickness.

Just about all ships these days are fitted with stabilizers which cut down the severity of the rolling motion. Often, once you become accustomed to the motion, you find your 'sea legs' and can enjoy the crossing. Otherwise, travel-sickness pills are very effective. Beware of taking these on a short ferry crossing, when you intend to drive afterwards—they can cause severe drowsiness. AVOID mixing such tablets with alcohol.

As on aeroplanes, the crew is highly trained in emergency procedures, so if anything serious does go wrong, you will be advised what to do. Long cruises always begin with a muster drill, for which you will wear your life jackets. Note the positions of muster stations on shorter crossings, if you're feeling nervous. Bon voyage!

First-aid kit

Obviously, if you're going to Paris or New York for the weekend, items such as antimalarial tablets will be unnecessary. Know your destination! What will you be likely to need? Make your own check list.

Insect repellent
Antihistamine ointment
Sunscreen lotion/sunblock
Calamine lotion
Antiseptic cream
Diarrhoea remedy
Aperient (mild laxative)
Antimalarial tablets
Indigestion tablets

Travel sickness tablets
Painkillers
Water purification tablets
Dressings/plasters
Cotton wool
Scissors
Tweezers
Safety pins
Thermometer

REMEMBER

If you are prone to a recurring condition (haemorrhoids, mouth ulcers, gingivitis, cystitis), don't forget to include your usual medication—it might be unavailable or very expensive. If you are receiving prescribed medication, make sure you take enough to last for the trip.

ANTI-AIDS KIT

If you're travelling to a Third World country, it has been suggested you could take this kit to protect against contaminated blood or equipment should you need emergency treatment (to be administered by a doctor):

- 2 X 5 ml syringes
- 1 X 10 ml syringe
- 4 X red needles 23 g X 1¼"
- 4 X green needles 21 g X 1½"
- 1 X scalpel handle, plus blades
- 2 X curved needles with fixed silk sutures 26–30 mm

If you're staying a long time, consider including:
- 2 X Macrodex saline plasma expander 500 ml
- 2 X Dextrose saline 500 ml
- 3 X intravenous infusion sets

> # ! WARNING
>
> If you have any medical condition that could affect your ability to travel safely, or your comfort abroad, seek medical advice before you leave home. You need to be aware of any special risks, how much medication to take with you and who to contact if you have a problem. It really would be worth learning some likely phrases such as: 'I am a diabetic. My bag has been stolen. I am in desperate need of insulin.' Among the groups who should seek advice are anaemics, asthmatics, arthritics, diabetics, epileptics, haemophiliacs and those who have recently suffered a heart attack. Pregnant women should also check before travel.

WHEN YOU ARRIVE

Apart from having all the necessary vaccinations before you leave home, there are certain precautions you can take to minimize the risk of falling ill or having an accident.

Food and drink

Funny tummy, Delhi belly, Montezuma's revenge, Rangoon runs, Aztec two-step, Tokyo trots—whatever you call it, traveller's diarrhoea is practically an occupational hazard of leaving home—or it is for those who don't take care.

This holiday blight has been shown to be more prevalent among the under-30s than in older people. Possibly older people tend to have travelled more and developed immunity to foreign bugs, or they may be more careful and sensible eaters.

Still, a change of climate and diet alone can upset the system enough to make you prone to infection. Nobody seems quite certain whether traveller's diarrhoea can be entirely avoided. Generally it lasts no more than three days.

> # ! WARNING
>
> If diarrhoea persists for a week or more, you MUST seek medical attention—it could well be a symptom of something more serious. Typhoid and the paratyphoids, cholera, amoebic or the bacillary dysenteries, giardiasis and various worm infestations all start with these symptoms.

Many cases of holiday diarrhoea are due to germs finding an ideal habitat in poorly-cooked or unhygienically-kept food or in the local water supply, from which they find another ideal habitat in your body. Don't give them a chance! Avoid:

- Unsterilized water
- Unboiled milk
- Bottled water with a broken seal on the cap
- Ice cubes—they usually ARE local water
- Unpeeled fruit—including tomatoes and cucumber
- Salad, unless washed in safe water
- Watercress
- Tepid cooked food or food from a display
- Shellfish, unless very hot and fresh
- Food from street traders
- Locally-made ice cream—stick to proprietary brands
- Anything from a fly-infested restaurant

If you do succumb, you could take a diarrhoea remedy, but it may be better to let the infection take its course. Do what you have to do. Drink plenty of non-alcoholic fluids, preferably bottled water (not tea, coffee OR fruit juice—they'll irritate the stomach further). Eat nothing—or a little bland food—and rest until the storm has passed.

REMEMBER

The contraceptive pill and other medicines are likely to be expelled from your body before they've taken effect.

Accidents

Being involved in an accident away from home can be a miserable, or even a fatal, experience. Of course, misfortune can befall you anywhere, but don't imagine a holiday means a holiday from being sensible.

The first priority is to make sure your insurance is adequate to cover medical costs at your destination. It's difficult to be overinsured if you're heading for the USA or Canada—go for the maximum cover. You can buy insurance from your travel agent along with the tickets, or direct from an insurance firm. It's usually included in a travel package which covers cancellation and lost or stolen property too.

Take the original cover note with you, and make a copy to leave at home. Check whether you need to inform the insurance company straight away if you make a claim, and get a receipt if you need to pay upfront for any treatment.

As always, keeping safe is a matter of common sense—a virtue that sometimes deserts otherwise quite sane people in the excitement of being on holiday. Driving abroad—though you may be on the 'wrong' side of the road—is just the same as driving at home. That's to say, even if the police aren't so vigilant, you NEVER drink and drive. If everyone else seems to drive like a maniac, YOU must observe all the safe practices of defensive driving you use at home.

Assess the risks

Be careful with mopeds and motorbikes in holiday resorts. They're often very old and badly maintained. If you're not used to riding one, it might be better not to learn on a pitted road with loose stones, a sheer drop on either side, hairpin bends and local drivers driving towards you at enormous speeds.

In fact, if you try any activity for the first time—windsurfing, parascending, skiing, waterskiing—make sure you get some lessons, and that your teacher is competent/qualified. If you don't speak the local language, look for a teacher who speaks enough English for you to be able to understand the instructions.

If you do hire a car, try to patronize a reliable firm. If you're unsure of the quality of the vehicles, don't be shy—use the following check list to vet the car:

HIRE CAR CHECK LIST

Before you drive:
- Does the ENGINE start easily first time?
- Check TYRE treads and pressures. Is there a jack and tools for changing the wheel? Check the SPARE TYRE too
- Is the tank full of FUEL? Is OIL at the correct level?
- Is there enough WATER in the cooling system? Washer bottle?
- Do SEATBELTS fasten, unfasten and retract?
- Do all ELECTRICS, especially lights and wipers, work?
- Do all DOORS open and close? Do the KEYS work?

Now check you have all the necessary documentation, and preferably a handbook too. Note any dents, scratches and chips in case you're charged for them. Drive round the block before accepting the car
- Is the CLUTCH smooth?
- Are the GEARS easy to find? Do they crunch?
- Are the BRAKES smooth? Are they even? Do they screech?
- Do all the INSTRUMENTS work?

REMEMBER

DON'T walk alone in unfamiliar territory—especially in extreme cold or heat. If you must, then tell someone where you are heading and when you expect to be back. All sorts of unforeseen hazards await the unwary rambler—from sudden mist and storms, wild animals, falls, all forms of attack to simply getting hopelessly lost.

Acclimatization

When you transfer yourself suddenly from a temperate climate to tropical heat, your body does not rise in temperature. Conversely, your temperature doesn't fall in extreme cold. Instead, the 'heat regulator'—the hypothalamus—situated in the brain, springs into action to acclimatize your body. It is thought to take up to six weeks for full acclimatization to be complete, but most of the changes occur within the first two weeks.

What are the changes? The resting temperature of your body actually falls, encouraging your glands to sweat more copiously for longer periods, starting at a lower temperature. It is the evaporation of the sweat which increases the heat lost by the body.

Heat is also transferred from inside the body to the surface at a higher rate, in order that heat can be further lost via convection. That happens through an increased flow of blood to the skin, and dilation of surface blood vessels starting at a lower temperature.

Factors that slow the rate of acclimatization are age, fatigue and obesity—and a low testosterone level (the male hormone that stimulates sweat). In other words, an overweight female pensioner, tired from a long journey, is going to feel uncomfortable! Here's how to help YOUR body adjust:

Heat

The MOST important step is to increase your intake of fluid. It is essential that your body sweats sufficiently—you MUST constantly replenish the water (and salt) lost in this way. Drink two litres, plus one, per 10°C (one pint of water per 10°F) EVERY 24 HOURS. Yellow-coloured urine is a sign you are not drinking enough.

It isn't often that you're advised to eat more salt, but your body requires around double the normal amount in tropical heat. NEVER raise your salt intake without increasing your fluid intake accordingly. NEVER rely on any alcoholic drink to replenish your fluid levels.

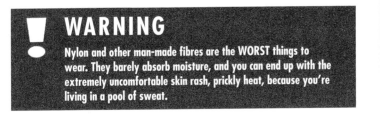

WARNING

Nylon and other man-made fibres are the WORST things to wear. They barely absorb moisture, and you can end up with the extremely uncomfortable skin rash, prickly heat, because you're living in a pool of sweat.

Since it is necessary for the sweat to evaporate to keep the body cool, what you wear is important. Loose (to trap a layer of air) light clothing made of absorbent fabric is the rule—white cotton is by far the best choice. Cotton absorbs 50 per cent of its weight in water. White clothing reflects the light, reducing the solar heat load by up to a half.

Sun

These days, much publicity is given to the dangers of excessive sunbathing—to the extent that one is less likely to feel envy for a white-skinned person burnt to that once-fashionable nut-brown colour. The link between sun exposure and skin cancer has now been proven, and the incidence of malignant melanomas in temperate countries has been rising at a rate roughly equivalent to the spread of mass tourism.

WARNING

If you WANT skin cancer, statistics indicate that all you need to do is work in an office for 50 weeks a year, then lie on a beach for the other two—preferably using tanning oil that offers insufficient protection and acquiring a nasty case of sunburn in the first few days. Anyone with red or blonde hair and/or a freckled skin will be particularly susceptible.

Sensible tanning

❍ Use a high SPF (sun protection factor) cream or lotion on ALL exposed parts for the first few days, changing to a lower SPF only once your skin has darkened.

❍ REMEMBER! The SPF levels of sun preparations are not standardized. One company's factor 15 might be the same as another's factor 32.

❍ Especially if you're fair skinned or unused to the sun, consider a total sunblock for vulnerable areas (bridge of nose, knees, shoulders, breasts).

❍ REMEMBER! Burning rays reflect off water, snow and white sand, intensifying their effect. Always use waterproof lotion for swimming, since the rays penetrate several feet below the surface.

❍ The sun can burn even through glass, clothing and in shade. A cloudy sky is not a protection against sunburn.

❍ ALWAYS build up your sun exposure time gradually. 15 minutes is quite sufficient for the first day, especially if it's winter back home. In general, you can double your exposure time daily, but use your common sense! YOU know how your skin reacts to sun.

❍ NEVER go out in the midday sun! In fact, avoid it between 11 am and 2 pm.

WARNING

SUNSTROKE/HEATSTROKE: Normally this unpleasant condition is heat syncope, or heat collapse, brought on by sudden, prolonged exposure to heat and is not serious. Symptoms of the mild form are dizziness, fatigue, nausea and fainting, possibly accompanied by blurred vision and yawning. Treatment is rest in a cool room and plenty of fluids.

Heat hyperpyrexia, or heatstroke, however, is far more serious and potentially life-threatening if untreated. The body temperature rises to 40–41° C (105–106° F) due to the failure of the heat regulation mechanisms, sweating stops and the patient loses consciousness.

The moment you suspect hyperpyrexia, SEEK URGENT MEDICAL ATTENTION. Cool the casualty, preferably by spraying with cold water in a stream of dry air, or by wrapping in a wet sheet or immersing in a cool bath. It is VITAL to take the casualty's temperature (rectally) every five minutes or so and STOP the cooling procedure as soon as it drops to 39°C (102°F).

The casualty MUST drink plenty of water and take salt. This can be given intravenously if unconscious. This condition tends to be rare in the ordinary traveller—expatriate manual workers in hot climates are most at risk.

❍ NEVER fall asleep in the sun. Sunburn is dangerous, especially if the skin blisters. The medical names for sunburn are acute actinic dermatitis and solar erythema—let that put you off!

Humidity

You sweat just as much in dry heat, but the sweat evaporates. In a humid climate, the atmosphere is all but saturated with moisture already and can't absorb much extra. This means your sweat stays with you, and you may believe you're sweating more copiously. You're not.

A relative humidity of 40–70 per cent is generally agreed to be comfortable for humans, with an air temperature of 15.5–27°C (60–80°F). Mogadishu (Somalia) has a maximum relative humidity of 81 per cent; Bahrain, 89 per cent; Mauritania, 91 per cent; and Oman, 94 per cent! Top of the hell-on-earth league, though, is the United Arab Emirates, where it is possible for the air to become saturated to the maximum—100 per cent relative humidity.

Altitude

Several important cities are at or higher than 1830 metres (about 6000 feet) above sea level—for instance, Nairobi,

Johannesburg, Bogota and Mexico City (the most populous city in the world, incidentally, with some 19 million inhabitants). On arrival, you will undoubtedly find you are shorter of breath than usual, owing to the rarified atmosphere—there is less oxygen in the air.

Tolerance for altitude varies considerably from one person to another, but the symptoms of 'mountain sickness'—headache, nausea and shortness of breath—are very unlikely to appear in anyone below 3050 metres (10,000 feet).

The body copes with altitude by making new blood cells to enable you to take in more of what little oxygen there is—this process takes up to three weeks. So, even if you feel quite well, you should NOT exert yourself when you arrive. The British athletes at the 1980 Olympics were brought to Mexico City—which is 2255 metres (7400 feet) above sea level—four weeks before the Games and were forbidden to train at all for the first four days to allow their systems to adjust.

Cold

In a very cold climate, the small arteries (arterioles) become restricted, reducing the flow of blood to the surface in order to maintain warmth. This increases the blood pressure, meaning that the heart works harder and needs more oxygen. Angina sufferers feel worse in the cold, and those with arteriosclerosis (hardened arteries) will suffer freezing cold feet and hands, since the blood flow is impeded.

In extreme heat, loose clothing traps a layer of air and encourages perspiration. In the cold, you should also wear loose clothes—this time to trap warm air. The rule is, wear several layers—you stay much warmer than with one thick jumper. Clothing should be close fitting, but not tight enough to restrict movement or prevent that insulating layer of air.

Don't wear TOO much. If you sweat profusely, the moisture can freeze. Equally, ensure that your outer clothing always stays dry. Extremities (toes, fingers and especially ears) are prone to frostbite.

Time zones

Aside from climate changes, the traveller has to cope with readjusting to different time zones. It's a peculiar fact of modern life that you can now get around the world so fast that you can live through a few hours twice, or never have them at all.

The internationally-agreed time change line is drawn along the 180° meridian, zig-zagging around lands in the Pacific. For every 15° of longitude, the time changes by one hour—backwards (or behind Greenwich Mean Time), to the west and forwards (ahead of GMT), to the east.

As we now know, the 'unnatural' speed of travel across time zones upsets your body clock. Apart from this problem of jetlag ALWAYS take into consideration the time at which you will reach your destination. It is not fun to arrive exhausted in a strange city in the small hours after a 12-hour flight!

Jetlag

Flying across time zones disturbs not only your sleep patterns, but also your pulse rate, body temperature, reaction time and decision-making abilities. Bowel movements and urination times are usually affected too, possibly leading to temporary constipation and further disturbance of sleep through having to get up in the night to urinate.

It has been proven that jetlag is twice as bad after east-bound flights than westbound, whatever the time displacement. Flying north-south has no effect apart from the normal journey fatigue. Jetlag is not usually dangerous, it's more a case of inconvenience.

MINIMIZING JETLAG

- Time your journey so that you can go to bed as near to your 'old' home bedtime as possible
- DON'T make decisions or enter an important meeting straight after arrival. Your faculties will be greatly impaired
- Try to take it easy for the first 24 hours after arrival
- Use an aperient and/or mild sedative on arrival for constipation and insomnia ONLY if absolutely necessary
- Diabetics should take their meals and administer insulin injections at the 'old' time during long flights, then change their routine on arrival. Always notify the airline in advance

Most people avoid violence. This is not weak, it's sensible. The pressure of city living has led to a dramatic increase in violent crime. Minimize the risks to yourself. It's time to decide how you will defend yourself, if the worst happens.

Self-defence

THE LAW

Our lives are affected, usually improved, by the laws which govern us. After all, we help to make them! Most people only come into contact with the police in order to give evidence as a witness—in which case we must learn how to observe and remember as much detail as possible.

Occasionally, the 'boot may be on the other foot'. Through mistaken identity, foolhardy behaviour (often linked with alcohol) or 'being in the wrong place at the wrong time', we may find ourselves caught up in the processes of arrest and detainment. Such a situation needs very careful handling if it is not to escalate and get out of control.

Much as the police must be trusted to enforce law and order, there is sufficient evidence that mistakes do occur. In a violent struggle, the police may not be able to tell which person is the aggressor. A joke or a gesture from you in a heated situation might be extremely unadvisable. The police ARE people. They will, like most of us, make the occasional error, panic or misinterpret a situation.

YOU may misinterpret the law! You may be carrying a brick to the local builders' supplier to match the colour. It may be seen as an offensive weapon.

SELF-DEFENCE

Understanding a few points of law might help you to avoid trouble. Most of us never have contact with the police or courts. Some of us may, if only by mistake.

Legal limits

You may only do what is 'reasonably necessary' to defend yourself from attack. Each case is different—so what is 'reasonably necessary' will vary. British law actually states that you can 'use such force as is reasonable in the circumstances in the prevention of crime'. If you genuinely believe your life is at stake, 'reasonable force' may be quite extreme, even involving a weapon in some circumstances.

Basically, the action you take to defend yourself or your property must not be excessive. If the law thinks you overdid it, YOU may be liable for prosecution.

The law expects the victim to run away or withdraw when threatened by violence, but this is not always possible.

Offensive weapons

Q What is an offensive weapon?

A A weapon intended or adapted to cause injury and carried for such use by the possessor. A spanner/wrench may be an offensive weapon, unless you are a mechanic travelling home with a kit of tools.

Q Who decides if a weapon is offensive?

A Usually a judge or jury, or the weapon may be obviously offensive, such as a flick knife.

Q Can I carry work equipment and sports kits without fear of problems from the police?

A The law accepts that tool kits and sports equipment are carried by necessity or for a genuine reason. You are allowed to carry objects which form part of a national costume or for religious reasons.

Q Where can I be liable to encounter problems regarding carrying offensive weapons?

A In any public place—street, park, sports event, entertainment centre, public transport.

Q What if the friends I am with are found to possess offensive weapons? Am I liable?

A The court must prove you had a 'common purpose' with the carriers of the weapons.

Q Can I carry articles with points or blades?

A You may be liable to a fine or imprisonment. You may only carry a folding pocket knife with a blade no longer than 75 mm (3 in).

Q What can I use in my house to defend myself?

A If you keep a brick for hitting burglars with, you may be in trouble. If you grab a walking stick and use it as a defensive weapon, you should be within the law—as long as you don't overdo it. The brick was obviously intended to cause harm.

Firearms

Firearms represent more of a risk than a serious form of protection. Having such weapons in your home may pose a severe threat to children—or to all the members of a household if the weapons are misused by an intruder.

Attitudes to firearms vary enormously around the world. In the US, for instance, the 'right to bear arms' is part of The Constitution. Even though there is legislation, there are an enormous number of cases where weapons are misused. The legal requirements in Britain, however, probably give the safest guidelines for owning and storing weapons.

- No one may own a firearm unless he or she is a member of a shooting or hunting club.
- No application for a weapon licence will be considered unless the applicant has been a member of such a club for a set period of time.
- A licence is valid for only three years. Regular inspections are made to ensure that the owner is adhering to ALL the rules governing the possession, use and storage of a firearm.
- The owner will be expected to keep the actual weapon in a purpose-built, steel, lockable cabinet which is bolted to a floor or wall. A safe may not be good enough.
- Ammunition must also be stored under such conditions—but in another location from the weapon.
- This storage must meet the approval of the police—who will check that regulations are not breached.
- If carrying the weapon through the streets, it must be in a locked security case—it must not be loaded. Ammunition may not be carried in the same case.
- In a car it is not sufficient to lock a weapon in the glove compartment or boot (trunk). A lockable steel security box, fixed to the vehicle, must be provided.

REMEMBER

The above regulations cover firearms only. Powerful crossbows are not covered by such rules, but if they were carried through the streets (to a sporting event, perhaps) you would be liable to prosecution for carrying an offensive weapon. Air weapons—guns or rifles—do not fall within firearms guidelines, but are covered by other rules. You should check with the police, since you may be arrested if, for instance, you fire an air weapon near a public place or highway.

THE POLICE

Aside from traffic incidents, most of us have very few dealings with the police. Occasionally we may need to give evidence, seek advice or report crimes. The police are HUMAN and, above all, trying to ensure that the laws of the country are upheld. Their job is stressful and sometimes leaves little scope for humour.

Problems do occur in heated situations, which may lead to arrest or detainment. You should consider a few points about dealing with the police.

◑ Stay calm and cool, quiet and respectful.
◑ Watch what you say—it really will be used 'in evidence'.
◑ Don't argue, don't make jokes and don't wave your hands about. All of these may go wildly 'wrong' and the situation may escalate.

Access/search

◑ If police officers come to your house, even if they are in plain clothes, they carry full identification—usually including a photograph. You have a right, in most cases, to take the identification and telephone the police station for verification. There ARE criminals who masquerade as police officers.
◑ Always note uniforms, insignia and officer number (usually found on the shoulder or epaulette, or on the ID card of plain-clothes officers).
◑ You can refuse to be searched, but the police have a right to search you (or your vehicle) if: they have a warrant; they suspect that you may be in possession of offensive weapons or items intended to help you commit theft/burglary; they believe that you may be in possession of illegal drugs/substances or stolen goods.
◑ The police may enter your home without a warrant (and use reasonable force to gain entry) if they have a warrant to arrest someone or, occasionally, to arrest without a warrant for a serious crime.
◑ If you are not under suspicion, it is up to you whether you let a police officer into your home. If you do, you may ask them to leave at any time.
◑ They may search your home if there is evidence that a serious offence has been committed or that you may be in possession of stolen goods. If you are arrested, they may search for items which may assist escape or which may provide evidence that you have committed an offence.

○ They may take items away from you—for evidence or to prevent damage or injury.

TYPES OF PHYSICAL SEARCH

The simplest type of search is the 'frisk' or outer garment search. It is usually quick and simple. Don't be clever or difficult. A strip search MUST be conducted at a police station. It must be performed by a member of the same sex (or by a qualified doctor or nurse).

An intimate body search, involving inspection of body orifices, must be authorized by a superintendent, and should be to search for 'hard' drugs or weapons/blades which might be used to cause injury. Intimate searches should be conducted under medical supervision.

Arrest

If you are caught up in a situation that involves the police treating you as a suspect, it is essential to know the basics of the laws which govern the process—and your rights. If you can speak/understand a certain amount of police language, you may be able to ask the right questions:

○ A police officer may arrest you without warrant if there are reasonable grounds for suspecting that an arrestable offence is being or has been committed.

○ OR if there are reasonable grounds to suspect that you were about to commit an arrestable offence.

○ OR if you were clearly about to commit an arrestable offence.

○ A suspect MUST be informed that he/she is under arrest—and the grounds (reasons)—as soon as possible. Giving this information to you will be difficult if you are shouting or fighting, for instance.

○ A police officer must take a suspect to a police station as soon as possible.

○ A suspect has a right to inform someone (directly or indirectly) of the arrest.

○ A suspect may consult a solicitor, but if the offence is serious, the police may delay this for 36 hours (the maximum length of time a suspect can be held without going to a magistrates' court, unless suspected of terrorist activity). A solicitor should attend any court hearing.

○ You don't have to answer questions. If you feel you have been wrongly arrested—you have a right to say so. Whatever you say—even to protest innocence—will be noted.

○ If the police decide to charge you, they must tell you so. You need not answer any further questions and can expect to be allowed to consult a solicitor.

- The suspect should be told if he or she is being charged, detained to provide further evidence or released.
- In most cases you have a right to one phone call. If this is to an anxious parent or a solicitor, there should be no problem. Evidence does suggest that it is not always easy to make the telephone call successfully.

> **REMEMBER**
>
> If you are NOT under arrest but only 'helping with enquiries' you do not have to stay with an arresting officer and you should have this explained to you. In most cases you will come to no harm if you comply. You may have a very real need to be somewhere else, but could arrange to visit the police station at a later time.

WARNING

If you do not comply with the police when arrested, they are entitled to use reasonable force! You can be fined or imprisoned for 'resisting arrest' or 'obstructing a police officer'—even if you are found innocent of the offence for which you were arrested.

Statements/confessions

DON'T make or sign statements if you are doing so to bring an end to questioning. There are cases on record of people confessing to crimes they did not commit—sometimes through fear or pressure. This will not help anyone—least of all YOU.

If you do make a statement, confine it to the absolute truth. DON'T sign it until you have read it through and made any changes/corrections. There may be typing errors which could seriously alter the meaning of the statement.

Fingerprinting

In most countries, a record is kept of convictions and the fingerprints of people who have been convicted. Only offences punishable by imprisonment are usually recorded in this way. The police can only take your fingerprints if:

- You give your consent, in writing, at a police station.
- An officer of at least the rank of superintendent authorizes the procedure in writing. Your consent is not required if there are reasonable grounds for fingerprinting.
- The reasons for fingerprinting must be explained to you.
- If you are charged with an offence, fingerprinting may be done without your consent.
- Reasonable force may be used to take fingerprints.

Identity parades

Codes of practice usually require an identity parade where the officer in charge needs to confirm witnesses' reports—or if the suspect demands an identity parade. You may refuse to take part in an identity parade, although (if suspected) this may prolong questioning and detention. You may stand anywhere in the parade line-up.

Identity parades should be arranged by an officer who is not involved in the investigation. There should be at least eight people of the same apparent age, height and general appearance. If there are two suspects, the line-up should involve at least twelve people.

MISTAKEN IDENTITY

If ever you are asked to help out by attending an identity parade (not as a suspect) there is NO possibility of your being implicated or arrested for the offence. The whole idea of the parade is to find out if witnesses can recognize suspects who have been detained. If YOU are selected by a witness, it proves only that the witness does not recognize the suspect.

Bail

If a formal charge is made against you, you must be given a written copy of it by the arresting police station. At this point you will either be granted bail and released, or taken into custody by the police until the date of the court appearance. Even if you have been refused bail by the police, you still have the right to ask the magistrates' court to overrule the decision.

Finding a solicitor

In theory you do not have to have a solicitor—you can represent yourself. However, when faced with judge and jury and the evidence against you, you may find yourself ill-equipped to make your own case. You must seek professional advice if you don't want a prolonged stay at Her Majesty's pleasure.

Finding a criminal lawyer is no more difficult than locating a solicitor to handle a house purchase. Recommendation from a friend or bank manager is always a good starting point, otherwise a consumer advice centre should be able to help. Legal aid is available for anyone on a low income.

BEING A WITNESS

Whatever information you give MUST be as accurate as possible. Look at the following lists and see the choices for describing a person. Try to form a 'word picture' of someone known to you and a friend. Try describing them and see if the friend knows who you mean.

DESCRIBING A PERSON

SEX: Male or female

AGE (or apparent age): This can often be difficult to assess and witnesses of different ages may 'see' the suspect quite differently

HEIGHT AND BUILD: Tall, short, medium height, heavy, light, stocky, skinny, athletic, chubby, fat

COMPLEXION: Skin colour, spots, scars, birthmarks, moles, warts, moustache, beard, sideburns, fresh-faced, dirty

HAIR: Colour, short or long, curly or straight, clean or greasy, receding hairline, balding, bald

EYES: Colour, glasses, cross-eyed, one-eyed

EYEBROWS: Heavy/light, missing, meet in the middle

SHAPE OF FACE: Round, square, long, thin, fat, wide, triangular

CLOTHES: Colours, garments, lettering, patches, missing buttons, shoes, boots, gloves, neat, tidy, scruffy, shoes, trainers, boots, workwear, uniform

JEWELLERY OR TATTOOS: May have names or initials showing, rings, bracelets, earrings, necklaces, badges

VOICE: Loud, soft, deep, high, accent, impediment (speech defect), breathless, angry, calm, use of words repeated, unusual use of words, intelligent, unintelligent

If you can give some of the above details, you may be able to elaborate as you form a mental picture of the person. Did he/she walk with a limp? Was he/she carrying anything? If the person ran off, in which direction?

Vehicle description

You may see a suspicious vehicle, or a 'getaway' car used by a thief or an attacker. Try to remember:
- Type of car/van/lorry/motorcycle/bicycle
- Colour—one or more, writing, stripes, stickers
- Licence plate number
- Roof rack, towbar, foglamps, aerial
- Condition of car—damaged, scratches, dents
- Details about occupants (number/appearance)
- Direction of travel

UNDER ATTACK

The majority of crimes are committed against property, usually homes and cars, but the number of reported physical attacks is on the increase. Contrary to reports you may read in the newspapers, street crime in cities is not a new phenomenon—but it has become more common in recent years to report crimes and for statistics to be recorded and assessed. Statistics help us identify high-risk groups, times of day and places where most crimes take place—and precautions people should take to reduce the risk of attack.

Most attacks on the streets are (statistically) committed by young men—usually on their own. The most dangerous time is from mid-evening to early morning—especially in the summer. A lot of violent crimes are related to alcohol consumption, and happen when public houses and nightclubs are closing.

Violence

It's impossible to explain or understand fully why acts of violence are 'so common. Robbery as a motive is on the increase—cities combine the very rich with the very poor. Unemployment and poor housing accentuate the problems. Sexual crimes, including rape, have always taken place—and often the aggressors are known to the victim already. Members of minority groups may be victims of otherwise 'motiveless' violence—simply because they ARE members of minority groups. It's always possible that there is no real motive. The attacker may be genuinely mentally ill.

Most street crime injuries involve bruising, broken noses, black eyes, grazes and shock. With aftercare and counselling, many victims 'recover'. But the alarming increase of crimes involving knives and firearms may mean an increase in the number of victims who die.

Most attacks take place in a few moments. The police can do little except arrange for the care of the victims and take statements from witnesses. It is up to YOU to make sure that you are not helpless if attacked. It is up to YOU to be aware of your surroundings and to protect yourself. You must look at self-defence as a real necessity and work out how YOU can minimize the risks of attack.

Awareness

Self-defence begins with keeping your eyes open and your mind alert to dangers. Always be aware of your situation and how it might put you at risk. Look ahead when in the street and be aware of dark corners and alleys, of people loitering, of groups of people heading in your direction. Cross the road if necessary to avoid potential risk. Use your ears too—can you hear footsteps or voices? While focusing on one potential attacker, you might not be aware that another is behind you. Try to judge situations realistically.

You mustn't start assuming that anyone who approaches you wants to hurt you—but you shouldn't assume they don't. A person slumped against a wall might be ill or drunk, or could be pretending. A violent argument between a man and woman, followed by cries of help from the woman, would probably make you want to help—but she might be an accomplice. This puts all of us in a serious dilemma. Most people choose not to get involved at all.

There are obvious high-risk groups for violent crime. Statistics show a particularly high incidence of attacks on women, the elderly and children—but everyone is at risk, particularly when alone.

Select from this section whatever suits your needs. Extra advice for women, the elderly and children is given later.

REMEMBER

The knowledge that attacks DO happen is the first step in self-defence. The awareness of a potentially-hazardous situation—expecting the unexpected—takes us further. But the final step is to equip ourselves mentally and physically to deal with attack or confrontation—minimizing the physical and emotional damage it may cause.

ATTACK AT HOME

It's important to understand that not ALL acts of violence occur on the streets. Many occur in the place you feel safest—the home. You MUST be aware of the kinds of risk involved and take sensible precautions. Attacks may also take place at work or in hotel rooms—places where you are not usually in total control of security procedures.

At home, it's up to YOU to protect your house and its contents (see SECURITY) and—much more important—the physical safety of you and your family.

At the door

NEVER let a stranger into your home. If you arrange for someone, such as a plumber, to call (and it is obviously necessary to admit them) arrange for another member of the family or a friend to be with you.

◖ Fit a door 'viewer' or peephole so you can see the caller.

◖ An outside light is essential to see a caller at night.

◖ Fit a door chain or limiter so that the door may be opened fractionally, allowing you a better look at the caller, whatever the caller is carrying and any accomplices who may be out of sight of the door viewer.

◖ Door chains and limiters, and doors for that matter, MUST be strong enough to resist being shouldered or kicked.

Fitting locks, spyholes and door chains/limiters is fairly easy—but not for everyone. Perhaps you could help an elderly neighbour—or anyone especially at risk. Tell neighbours they can always call you if they are bothered about callers and don't want to be alone with them.

◖ Before opening the door at all, if you are on your own, ask the caller what they want. An 'entryphone' is useful—you might choose not to open the door at all.

◖ Never judge by appearance. Even a child or someone in uniform may be used as a ploy to gain entry to your home.

◖ Don't worry about keeping an unknown caller waiting. A door chain or limiter enables you to check the credentials of the caller, by telephone to his/her headquarters.

◖ ALL official visitors carry identification.

REMEMBER

If you're alone, give the impression you have company. When the doorbell rings shout something like: 'Don't worry Tony. I'll go', 'Take the dog into the kitchen please, Mick'. Even during a conversation with someone at the door, especially if you want to get rid of them, you could call out, 'Hold on, Johnny, I'll be there in a minute'.

⚠ WARNING

If you live alone, DON'T advertise the fact on a label beside your doorbell or 'entryphone'. Add a fictitious name. Women who live alone should NEVER put Miss, Mrs or Ms in front of their names at the door OR in telephone directories. Doing so has been known to attract problems. If you use initials, not a first name, no one knows whether you are male or female.

Your keys—the most obvious means of entry to your home—are YOUR responsibility. You must keep them safe at all times. NEVER put an address tag on your keys. Always have them ready immediately when you reach the door. When you move into a new home—you have no idea how many people may have keys. Be prepared to change locks:

■ If you lose your keys/your keys are stolen
■ If you give copies to tradesmen to allow them access. When their job is finished, you don't want them 'visiting' you again
■ If the locks are old and worn
■ If you feel you can upgrade the locks. Mortise locks should have at least five levers. Rim locks should have a deadlock facility on the inside

REMEMBER

Many rim locks have a replaceable internal barrel containing the lock mechanism. This makes changing the locks fairly easy. Keep the old lock mechanism. After a couple of years you may need to change locks again—you could reuse the old mechanism. The levers in some mechanisms can be rearranged and new keys cut to fit.

❗ WARNING

NEVER keep spare keys hidden outside your home. You may be seen retrieving the key. Most 'professional' burglars or attackers are quite capable of imagining the same hiding places.

At night

Lock all accessible doors and windows. Close all curtains, especially when undressing or if you have possessions you don't wish to advertise to everyone who passes by.

If you hear anything which suggests that someone is trying to gain access to your property—or anyone else's, DON'T investigate. Call the police.

TELEPHONE NUISANCE

An ex-directory telephone number is a good idea. DON'T be intimidated by random telephone threats or obscene calls. Your rising fear or panic may be precisely what the caller wishes to hear.

◗ Keep a loud whistle by the phone. It could give a nuisance caller quite a surprise!

⚠ WARNING

Refuse to be intimidated by telephone calls which only make silly obscene statements. They are unlikely to be followed up. If actual physical violence is threatened, call the police—especially if the caller seems to know who you are, where you work or routes you take walking to and from your house.

- **NEVER** give your complete telephone number when answering the phone. Some 'telephone attackers' dial randomly until they find a victim. If any caller claims they have dialled a wrong number and asks you to tell them what number they have reached, **REFUSE.**
- Say instead: 'What number did you dial?'
- **DON'T** be drawn into a silly conversation which may degenerate further.
- Hang up the telephone.
- Unplugging the telephone (in Britain, at least) means the caller hears a ringing tone and thinks no one is in.
- If you have more than one call, hang up **IMMEDIATELY** once you realize what is happening.
- If you suffer from persistent calls, tell the police. It is possible for ALL your calls to be monitored by the telephone company, allowing only genuine calls to come through.
- Special devices are being developed which may be fitted between the telephone and the wall socket to identify a caller's number.
- You may hear sounds in the background which indicate where the call is coming from. This could help the police.
- Change your telephone number.
- Do NOT leave the phone 'off the hook'. This can cause problems with the telephone exchange.
- You could arrange a signal for friends to let you know they are ringing. They could let the phone ring a certain number of times, hang up and ring again.
- Remove the number label from your telephone to avoid it being seen by anyone who calls at your house and has to be admitted.

REMEMBER

Many telephone answering machines can also be used to monitor calls. After your pre-recorded message has been played it is possible to hear the caller. When you recognize the caller's voice you can pick up the phone—or not. The choice is up to you.

Apart from the possibility of people entering your home at night, always remember that thieves or attackers may get in while you are out. They may not be lying in wait, you might simply catch a thief in the act. When returning home ALWAYS check that lights are on or off as expected, windows, curtains and doors are as you left them.

- If you do disturb a thief, DON'T block his exit path. You may get hurt in the process. Which is more important — your possessions or your life?
- Phone the police immediately once the intruder leaves. If alone, go to a neighbour's house — the intruder may return
- Try to make a quick mental note of the intruder's appearance and any 'getaway' vehicle
- If they attempt to physically harm you, use whatever means you can to protect yourself

REMEMBER

Consider installing panic buttons as part of an alarm system. Hand-held screech alarms (see On the streets) may also be useful indoors. If you live alone, a simple doorbell could be wired to a friendly neighbour's house to act as an alarm. A powerful flashlight might dazzle an intruder and could serve as a club for self-defence.

ON THE STREETS

Most advice given applies especially to people who must be out on their own—although a small group of people may be threatened or set upon. Attacks are most common on individuals. Most unforeseeable attacks take place:

◗ During the summer, usually between the mid-evening and early morning
◗ Away from other people, in lonely 'short cuts'
◗ From behind
◗ When pubs or nightclubs have recently closed
◗ On payday—when wage packets might be carried

Living in a city

In most large cities there has been a great decline in people's respect and consideration for one another. Many areas may feel like battle zones to those who don't live there. But even in ordinary day-to-day dealings, people may be very aggressive and self-centred. Many people live with some level of constant fear of attack or intimidation. Stories of gangs, street crime, vandalism and rape abound and feed these fears.

To add fear to the other stresses people may be forced to live with may make life intolerable. It is essential to replace fear with an increased awareness or preparedness. DON'T take a passive stance. **TAKE POSITIVE ACTION!**

Keep yourself fit. Be sensible about moving about on your own. Use your imagination to learn how you might cope in various attack scenarios. Try to learn some basic self-defence techniques and—most important—enjoy your life! Cities have lots to offer all types of people. DON'T let fear make you paranoid and overshadow all your activities.

When out alone

Plan your routes sensibly. Don't go out for the evening without planning how you will return. Arrange to stay overnight or to travel with a friend or two. You should feel relatively free to move around as you choose—or you might as well stay at home and barricade the doors and windows!

Most common sense precautions can become automatic—a normal, necessary part of everyday life just like eating, sleeping or working:

- Plan ahead. Know the safest route. Carry the phone numbers of a reputable taxi company and enough money to cover the fare. Carry this money separately from your other cash.
- Carry money (or a phonecard) for public telephones.
- Don't use short cuts across wasteland, down ill-lit alleyways with lots of hiding places, through underground walkways, along canal paths, through car parks.
- If you are driven home by a friend or taxi driver, ask them to wait until you open your door and go inside.
- Don't display expensive clothes, bags or jewellery—cover them up.
- Walk facing the traffic to avoid cars pulling up behind you.
- Avoid eye-to-eye contact as you pass people, but still try to assess their intentions.
- DON'T look down as if nervous! Hold your head up and look about at all times.
- At night, on your own in an unpopulated area, don't get close to a car when the driver asks for directions.
- Don't let ANYONE who approaches you for any reason get within arm's length.
- Don't accept lifts from strangers—even those claiming to be 'minicabs'.
- Cross the road if you see a group of people coming down the street towards you.

- Walking with a large dog—even if you 'borrow' one from a neighbour—should deter most attackers.
- Look at your shadow on the ground as you walk past a street lamp. If anyone were approaching silently from behind, their shadow might give you a second or two's warning.

If you see someone under threat in the street, and don't feel able to intervene, at least call the police at the first opportunity. A group of people could use their numbers to discourage an attacker.

BEING FOLLOWED

If you think someone is following you, cross the street. See if the suspect does the same. If they do, cross back. If they show any sign of still following you, walk up to a house and fumble with your keys. If it's occupied, ring the doorbell and ask for help. Try to get to a place where there are other people or stop at a phone and call the police. Go into a shop or, at night, a takeaway restaurant or public house. If possible, leave by another exit. It may be a good idea to stop and look in a shop window. You may be able to study the behaviour of the person, reflected in the glass. Try not to panic—don't run unless it is to a place of definite safety nearby.

REMEMBER

Going up to a strange house and asking the occupants for help may also be dangerous—especially for a woman on her own. The owners of the house may not be friendly or may themselves be suspicious or afraid of YOU—an unexpected caller!

If someone comes to your door and says they are being followed or have been attacked, give them the benefit of any doubt. You don't need to let them into your home—but don't just send them away (which is common). Call the police or an ambulance.

Walking on the road side of a pathway—or even in the road—may be safer when there is no traffic. Don't walk close to doorways and bushes where all someone has to do is grab you. Make it more difficult for them!

Try to act as if you are very confident (even if you're terrified). Walk purposefully. Keep your hands out of your pockets—ready to use to defend yourself. Looking confident implies you can 'take care of yourself' and may put off an attacker. DON'T keep glancing nervously over your shoulder. DON'T hang your head and look at the ground. Keep your head UP—looking and listening at all times.

DRIVING

Being in a moving car—except from the dangers associated with using roads—makes you safer than you might be on foot. But if your car breaks down, you may become more vulnerable to attack.

Plan your journeys properly, especially through unfamiliar territory. Always have a map handy. Always make sure that your car is working properly and that you have enough petrol for the journey.

Try to let someone know where you're going, and at what time you can expect to arrive. It would be sensible to join a major breakdown service. Try to park in busy well-lit areas.

In isolated, unfamiliar or badly-lit areas:

◗ Keep doors locked while driving.
◗ Keep windows closed.
◗ Only stop in an emergency.
◗ Don't get out of the car in areas where you wouldn't choose to walk alone—unless you have to.
◗ Have a good look round before getting out—don't pull up near dark doorways or bushes.
◗ Think (more than) twice before giving a lift to strangers.
◗ If you break down, either look for a telephone (unless you have one in your car) or, at night, lock yourself into your car if you feel safer. Offers of help may not be genuine.
◗ Beware large dimly-lit carparks and multi-storey carparks, especially at night.
◗ If you see an accident or someone whose car has broken down—DON'T get out. It may not be genuine.

REMEMBER

It's a good idea to carry a small amount of spare petrol in an approved petrol can. In hot weather, a can of water could prove invaluable.

- If you think you're being followed—take the simplest route to a busy well-lit area or aim for a police station, a familiar public house, a hospital.
- Don't have valuables in sight in your car. When you stop they may attract someone's attention.
- If you carry money or valuables on a regular basis—AVOID always using the same route.
- Never leave any valuables in your car when you park. Lock things out of sight.

If you see an accident or someone having trouble with their car in an isolated spot, you don't need to stop. DON'T get out of the car. With your doors locked and a window only partially open, drive alongside and offer to phone for help as soon as you reach a telephone (or stop further on and use a car phone, if you have one). If you prefer, signal to the driver as you pass that you will call for help from the nearest public telephone.

If you are stopped

Someone may step out into the road and make it necessary for you to slow down—or stop. If you haven't already done so, immediately lock all doors and close windows as you approach. Change to a lower gear—ready to accelerate away if there is any sign of danger. At night, use your headlights to give you a good view of whatever is happening.

If you must stop, do NOT switch off the engine. Do NOT get out of the car until you are sure there is no danger.

If you really are not sure that the emergency is genuine, drive past—swerving if necessary—and phone the police from the first telephone you can find. If the emergency WAS genuine, you have at least summoned help.

HITCH HIKING

There are dangers from both sides—for the driver and the passenger. As a general rule, hitch hiking alone— especially for a woman, especially at night—is CRAZY. Equally so is for a person alone in a car to give a lift to a total stranger. Most advice that can be given works both from the driver's and the hitch hiker's point of view.

Hitch hiking is rare in cities during the day except on exit roads to major destinations. Depending on how you look at life, it's a legitimate way of getting about if you have no other choice. Unexpected passengers may prevent a driver from being lonely or falling asleep on a long journey. ALWAYS assess the risks.

When driving:

○ A hitch hiker or hitch hikers may be planning to rob a driver or damage/steal the vehicle.

○ It has been known for female passengers to attempt to demand money from male drivers by blackmail.

○ If you give someone a lift and can't get rid of them again—pretend the car has broken down, as realistically as possible. Choose a busy location.

○ If threatened with a knife or other weapon, it may be wise to do whatever they ask.

○ A risky tactic may be—if the road surface is good and there are no vehicles behind you—to slam your brakes on and do an 'emergency stop', in a bid to escape.

When hitch hiking

○ It's very common for women hitch hiking alone to be expected to submit to some form of sexual act 'in return for' being a passenger in a vehicle.

○ Men driving on their own have been known to overpower and abuse more than one female passenger at a time.

○ A lot of modern cars have central locking (all the doors can be locked at once by the driver) or childproof locks which can only be opened from the outside.

○ NEVER sit between two strangers in a car.

○ ALWAYS make sure you can reach and open a door.

○ If a car stops to give you a lift and you don't feel at all sure about accepting the lift, say that you've changed your mind.

○ Take the number of a car whose driver threatens you in any way and report the incident to the police.

When ordering a taxi, always ask WHEN the taxi will arrive. If necessary, the taxi company will phone you when the vehicle is outside your house. Ask for the driver's name. NEVER get in a taxi alone if you are not sure it is the one you ordered. All this may sound fairly extreme, but many taxi companies use simple radio systems. It is possible that messages could be intercepted.

REMEMBER

In some cities there are special taxi services for women and children— mostly at night. Most taxi companies will be sympathetic to your concerns if they want your business.

PUBLIC TRANSPORT

Robbery and assault are common on public transport and are becoming increasingly so. Attack reports confirm that the people who commit these crimes are becoming more and more daring: Most cities have buses, underground trains, overground trains or trams. Despite the difficulty of being unable to get off between stops in some cases, there are some attackers who seem quite prepared to try ANYTHING!

Most attacks occur on routes going away from the city or town centre, particularly in the evening. 'Steaming' is a relatively new phenomenon—a gang proceed through an underground train, train or bus and rob everyone along the way. Such gangs are usually armed—often with knives—and as soon as they can, they jump off and 'disappear'.

Avoiding risks

Travelling late at night after an evening out (when you might be expensively dressed), after working a late shift (when you are probably tired and less alert than usual)—for whatever reason—puts you at risk. Attacks DO occur at peak times, but more commonly in the later evening—especially as transport leaves the main central city area. Alcohol and drunkenness are often contributory factors.

◗ If unexpectedly working late, try to arrange a lift with a friend or to travel with another person.

- Sit as near as possible to the driver of trains/buses (NOT upstairs on double-decker buses).
- On trains and underground trains, sit as close to the conductor/guard as possible, when there is one.
- ALWAYS choose a compartment with lots of people in it.
- Be a 'grey' person—don't draw attention to yourself.
- If your compartment empties—move to a busy one.
- Women alone should sit close to other women.

ALARMS

- Look for the emergency alarm. Stay near it.
- On underground trains, DON'T use the emergency alarm between stations. The train is unlikely to stop in a tunnel, making it impossible to get out, but may stop PARTLY in a tunnel.
- Use the emergency alarm on an overground train in a station, if possible.
- If a bus does not have an alarm button, opening an emergency door usually sounds a buzzer to the driver.

- Use interconnecting carriage doors if necessary to avoid an attack when a train is not at a station.
- Choose the busiest sections of station platforms—try to move around (especially when in underground passages) near a crowd of people.

Be ALERT!

- Avoid eye-to-eye contact with ANY stranger who appears to be 'looking for trouble'.
- DON'T keep your gaze nervously fixed on the floor. Keep your head up and appear ALERT.
- Pretend to read—but DON'T get engrossed. You MUST be aware at all times.
- If you wear a personal stereo, keep the volume low enough to allow you to hear what is happening around you.
- If you are not sitting down, stand where no one can approach you from behind.
- At a station or stop, scan new passengers as they get on.

If you see someone being attacked or threatened on public transport, at least raise the alarm or fetch help if you don't feel you can intervene. If someone cries for help, you MUST do something.

The Guardian Angels are a voluntary safety patrol, formed in New York in 1979. By the time a group was formed in London in May 1989, there were groups in 86 cities in the US. The American groups tend to remain fairly detached. In teams, they patrol (mainly) the underground rail network, riding trains—usually one to a compartment—keeping contact with one another to check that all is well. In the UK, they work in much the same way—but prefer to communicate with the public. In their first 18 months in London they handled about 400 incidents.

Although they were met initially with some opposition from the authorities, this relationship has improved with time—now it is understood that the Angels have the interests of the public at heart.

Guardian Angels always wear a red beret and a white T-shirt with a large red logo on the front. Jackets, when worn, are red with a large white logo on the back and a smaller white logo on the chest. To many people who are not used to them, they may at first appear to be intimidating.

HOW THEY HELP

The Angels have a code of conduct, which should be stressed. They NEVER refuse to help and NEVER walk away from a problem. They will stop fights, disperse people and even make 'arrests'—calling the police if a serious injury occurs or weapons are involved. They are happy to give directions or first aid. They carry phone numbers of crisis centres, hostels and help groups.

Their policy is to recruit women and men of all races, who are not necessarily big or strong-looking—but possess 'the Angel spirit'. They are trained in close-quarter combat. They do get hurt—sometimes seriously.

ADVICE FROM THE ANGELS

Most have jobs and work voluntary 'shifts'. They organize self-defence training and are looking towards community projects—but their main advice is more immediate. It concerns your attitude:

- Criminals look for weakness. Let your body language say that you are strong. Stand tall. Keep your head up. Stay ALERT.
- Believe you have a right to travel without fear of attack.
- Prepare mentally and physically to fight back. Get angry. Take the initiative. DON'T be weak and let an attacker call the shots.

DARE TO CARE

The Guardian Angels can be expected to spread further afield, to other cities in other countries—look for them in Sydney, Paris or Amsterdam. Public opinion regarding the Angels fluctuates—but which is more worrying to you: the existence of a group of people who are trying to help OR the fact that there is a need for such a group? As the Angels say—DARE TO CARE.

SELF-DEFENCE ■ UNDER ATTACK ■ PUBLIC TRANSPORT

Stations/terminals

All sorts of crimes from opportunistic bag snatching to abduction take place at main city stations and large travel terminals, especially at night.

Most genuine travellers are preoccupied with the difficulties of travelling with luggage, more so when in unfamiliar territory. They may be tired from a long journey, or, as first-time visitors, bewildered by the city.

Unattended luggage is likely to be stolen or removed as a suspected terrorist device. In some countries 'suspected' unattended luggage is destroyed!

All sorts of ploys are used to separate people from their bags. Keep your wits about you. Offers of help are **unlikely** to be genuine. Someone may ask you if you are looking for a job, or a bed for the night. USE YOUR COMMON SENSE!

If you have run away from home or have decided to try living in a big city for the first time, be observant and try not to get 'involved' with strangers. Above all, DON'T arrive without money or plans for where you will stay or what you will do with your time. There are help and counselling centres. The police will help you locate these.

MUGGING

Muggers may work in many ways. Most prey on victims who are smaller or appear weaker than themselves. A threat is made and money demanded. They may not speak at all—only 'pounce' on a victim and take money, jewellery and valuables by force. If faced, however, by muggers with a knife or a gun—give them your valuables. Which is more important—your, possessions or your life?

❍ Avoid outward shows of wealth—jewellery (especially gold chains which may be snatched from the neck), expensive clothes/bags, cameras.

❍ Don't carry ALL your cash in one place or in your bag.

- When alone in the street or on public transport follow all the rules already given.
- Some muggers seek out victims on likely paydays. Try to avoid carrying your wage packet home.
- Don't carry cheque books/credit cards/cash all together—separate them.
- Take out only as much money as you need—banknotes could be kept in a moneybelt and change in a pocket or bag.
- Keep your keys separate—if you lose means of identification (such as your driving licence) AND your keys, CHANGE YOUR LOCKS as soon as possible.
- Carry 'spare' money you can give to muggers. They will think it is all you have.
- If anyone asks you for change—or if you give money to someone 'begging' in the street, do NOT take out all your money. Carry spare change in a pocket.
- Some muggers operate on public transport by grabbing a necklace or bag and leaping off. DON'T sit close to the door.

BAG SNATCHING

Apart from the dangers of losing unattended luggage, some thieves operate by snatching bags (often in crowded places) and running off. A shoulder bag is a good idea, but wear the strap diagonally across your body so you can always see it—NOT just hanging on one shoulder.

Women are the main targets. The thief (who could be male or female) can expect to find cash, cheque book, credit cards, keys, identification—EVERYTHING. If you carry a bag, keep money, credit cards or keys in your pockets or in a moneybelt.
- Carry your bag on the side AWAY from a road. There are drive-by bag snatchers—even cyclists and motor cyclists!
- If your bag is snatched, DON'T try to tackle the thief—he/she may be armed.
- Shout 'HELP!' and 'POLICE!' and 'BAG SNATCHER!' Someone may help you (if you're lucky).
- If someone picks up your luggage shout 'HEY, YOU!' and run towards them. If the luggage is/looks heavy, the thief will realize they will be unable to make a quick getaway.
- Bag snatchers usually head for isolated back streets to investigate the bags they have stolen. DON'T FOLLOW!
- Some bag snatchers use knives to cut bag straps. DON'T tangle with them.

- ◑ If it comes to a struggle—let the bag go!
- ◑ Bags left by automatic doors on public transport may be snatched—often just as the doors are closing.

REMEMBER

If you carry EVERYTHING in one bag—your money, identification, keys, cheque book, credit cards—then you will lose EVERYTHING in one go!

TOILETS

When using public lavatories, if there is a gap beneath the door, keep your feet tucked in—don't stretch them out. A very rare form of attack is for people's feet to be grabbed and pulled, causing them to fall off the toilet! The door is then kicked in and they are robbed.

- ■ Only use public toilets when absolutely necessary.
- ■ Use toilets in shops, restaurants, public houses, trains—anywhere you can.
- ■ If anything—a coin, a roll of toilet paper, a glove—appears under the partition between cubicles, kick it back with your foot. Do NOT bend down to pass it back with your hand. While you do so someone may reach over the partition and steal your coat or bag from the hook on the door.

PUBLIC PHONES

When you use public telephones, always keep an eye on—or preferably, keep hold of—your belongings. At night, beware of people approaching from behind—especially if you are using a telephone in a kiosk or booth which would restrict your escape. If nothing else is handy, use the phone as a weapon to defend yourself. Go for the attacker's face.

If you're in a phone box and someone appears to be loitering (not just waiting to use the phone), call the police.

PICKPOCKETS

 Most of us have seen pickpockets operating in the movies, or even magicians performing a very clever 'sleight of hand'. Pickpockets may be clumsy opportunists—similar to bag snatchers—or they may be extremely clever and organized thieves. Some work alone, others may have accomplices who distract you in some way. Some are quite capable of removing watches, rings and jewellery—and taking wallets/cheque books/credit cards from obvious pockets is child's play.

Pockets and bags which fasten with zips or buttons are a good idea, although not 100 per cent secure. 'Velcro' (press tape fastenings) are unlikely to be opened without your knowing what's going on. You may hear them being pulled apart.

◑ NEVER 'wear' your wallet in an exposed back pocket!

◑ Avoid crowds—if you can't, make sure you keep your wits about you. If you carry a bag, keep it in front of you.

◑ Don't join crowds around street gamblers or casual street salesmen. Pickpockets may work with them to remove more of your possessions/cash!

◑ If someone bumps heavily and rudely into you, causing you to ricochet off onto a 'kinder' person who smoothes your clothes, touches you very reassuringly and asks if you are OK before melting quickly into the crowd, you have probably just been robbed.

◑ 'Domestic' arguments may be staged—as you go to help (or stare in amazement) you are NOT concentrating.

◑ USE YOUR COMMON SENSE! Be aware of yourself and your surroundings at ALL times.

RAPE

A lot has been written about rape—a very common crime which has been known to ruin lives. Under half of reported cases of rape of women are perpetrated by men who are total strangers to the victim. Most involve men who are slightly or even well-known to the victim— rape can and does occur within 'relationships' and marriage. Many cases of rape are not reported—sometimes through great embarrassment or genuine fear that the authorities may be unsympathetic.

In some countries—the US for example—a rape may be considered 'self-precipitated' if a woman has (for any reason) exposed herself to risk. Rape and the after-effects of rape are terribly complicated.

There are usually two widely different accounts as to how the event took place. Women must try very hard NOT to develop the feeling that 'all men are rapists!' This, like being too frightened of muggers to set foot outside your home, is no way to live. Women must try to take the initiative, by maintaining a positive attitude, staying alert and avoiding situations which involve risk.

A lot of the preceding advice already given in this chapter is relevant here, but there are more points to consider:

- Many rapes are NOT spontaneous. They may be planned, for instance, by a man who knows or casually 'knows' the victim or regularly sees her walking to and from work.
- Rape may occur as a 'last resort' in an argument—even within marriage. Such rape should also be reported, especially if very violent. 'Domestic' violence could easily escalate. If rape is used as a 'last resort'—what is the next stage? Further bodily harm?
- Rape is more common in summer than in winter, and—like any 'street' attack—at night. The chances of outdoor and indoor attack are equal.
- Many attacks occur in the rapist's or the victim's home.
- Always check lifts before entering. If you live in a block of flats, don't travel in a lift at night. If a stranger gets in with you, keep within reach of the alarm button. Avoid eye-to-eye contact. Be polite, but not friendly if forced to speak.
- Many men cannot tell the difference (it seems) between friendliness and sexual invitations. If in doubt, you must avoid the company of men (especially when alone) or try to develop a fairly firm, 'cold' way of dealing with them.

IF AN ATTACK SEEMS LIKELY

You must try to think clearly and act quickly. Assess your potential attacker. Is he armed? Is he drunk? Is he strong? Could you reason with him?

- Use avoidance ploys or simple self-defence techniques outlined later.
- Buy time. Try to adopt a calm, confident manner. Invite him to your home promising whatever you think will please him.
- DON'T actually take him home (but keep him believing that you are). Lead him to a busy area, run to the nearest point of safety—an off licence, a wine bar, a public house. Scream if you think it's necessary to call attention to the potential attacker.
- Arrange to meet him another time. Make him believe you mean it. Give him a false telephone number.
- If he appears to be losing interest, keep chatting to him to distract him from his purpose.

Many women are (understandably) nervous on their own in cities. A woman may be frightened if you stare (even admiringly) or try to strike up a conversation in a lonely spot. If you spot a woman alone on public transport, DON'T sit close to her. You CAN tell if you are making someone nervous by walking behind them. Cross the street to allay their fears. If you give a woman real cause for fear, she may take steps to defend herself.

During a violent assault

◗ You are being hurt already. Use as much violence as you can without making your situation worse.

◗ Bite.

◗ If you can reach the man's testicles—grab, twist and pull. He may at first think you are 'responding' favourably.

◗ Try and tear out hair, pull off a button, tear off a shred of fabric or fluff from a jumper—anything which might lead to the identification of your attacker.

> **! WARNING**
>
> If the assault is particularly violent and the attacker is armed or very strong, it may be safer not to struggle.

REMEMBER

The law allows you to use reasonable force to defend yourself (using some of the simple techniques and improvised weapons outlined later). You may NOT carry weapons for self-defence. In Britain 'mace' sprays are illegal—but there are lots of other things you can do.

If you are raped/sexually assaulted

Seek medical attention as soon as possible. Even though it may be difficult to restrain yourself, do NOT wash or change your clothes until you have been examined. DON'T have an alcoholic drink or a sedative. Try to get a friend to accompany you.

REMEMBER

In some countries, including Britain, your anonymity is usually protected and your sex life should not be the subject of open discussion in court. In the case of rape by a friend or relative it is NOT necessary to go as far as prosecution. A court order can ban the offender (even your husband) from having further contact with you.

There are rape crisis centres in most cities which will help victims through what is generally considered to be the ordeal of giving statements, being examined and questioned.

There is no doubt the police should be involved, but many cases of rape involving friends, husbands of friends or members of the family may have devastating repercussions on many lives when the truth is discovered. Needless to say, many cases of rape go unreported for this reason. Women may live for years in emotional torment and may have physical injuries which should be treated immediately. Never forget that sooner or later you may need to be examined for pregnancy or sexually-transmitted diseases.

MEN AS VICTIMS

Men may also be victims of assault, domestic violence or violent sexual assault. The aggressors may be other men or women. So rarely are these crimes reported fully that investigations are rare and statistics very hazy. Many cases involve more than one aggressor and weapons. Tragically for the victims, most cultures do not even accept the idea of male rape. Evidence does suggest that the same sort of attack scenarios occur as for female rape and that the physical and emotional damage may be just as devastating.

MULTIPLE ATTACKERS

It is virtually impossible for anyone to cope with more than one attacker at a time. You will be overpowered easily. If you can fight, try to hurt the 'ringleader'—if he goes down the others may hesitate, giving you the chance to escape.

Keep moving—don't let anyone hold you. Use your natural weapons and look for anything you can use. Shout for help. Shout for someone to ring the police. Try to run to a busy area.

It may make the attackers give up if you fall to the ground and go limp, especially if you are middle-aged or elderly and they think you have had a 'heart attack'.

If all they want is your money—give it to them. You may escape without injury.

CAUGHT IN A CROWD

When a crowd panics, or is angry, it can be extremely dangerous. You may be a part of the crowd, and its mood may change—especially true of sports spectators. You may visit a part of the city, unaware that a political demonstration is taking place. The main dangers are falling to the ground and being trampled, or of being crushed as the crowd presses in.

Danger areas

Avoid crowds where there may be a risk of unrest or take extra precautions to minimize risks at sporting events (indoors or outdoors), political demonstrations, large queues, industrial disputes. Crowds are usually well managed, but if the road narrows or there are steps or obstructions, try to get away. Any large body of people, in a rock concert, a cinema, a theatre or a sports stadium can be dangerous if panic breaks out for any reason—a fire or a fight, perhaps.

Be prepared

When you arrive at a large event, look for the safest exit route—indoors or outdoors. Stay away from the centre of large crowds. Near the stage or a barrier, you may be crushed if the crowd surges forwards. If there is no obvious exit, look for a safe haven—under stairs in a stadium, under the stage.

Indoors, the way you came in may not be the best way out. Don't leave with the main crowd, especially if 'rival' spectators will mingle as they leave—hang behind or leave early. If leaving a sporting event, beware of displaying the colours of either team.

If you live near a sports venue, memorize the least popular routes taken by visiting crowds to and from public transport.

If you cannot avoid contact

Don't panic. Keep calm. Gradually edge your way out of the crowds. At all costs—stay on your feet. Don't let the crowd push you towards windows, walls, pillars or steps.

Either fold your arms, with your elbows drawn in tightly, or held out in front of you to create a space. In a tight crush, the danger is that you may overbalance if you can't reposition your feet under your centre of gravity as the crowd surges.

It may be worth trying to hold onto an immovable object, to resist the flow of the crowd.

If you fall

Curl yourself tightly into a ball. Pull your head to your knees and keep your hands clasped round it. Crawl or wriggle, if you can, to the base of an obstruction such as a car or a tree where the crowd must separate—or at the base of a wall.

Driving in a crowd

If you are approaching a large crowd on a bicycle or motor-cycle or in a car—stop and change direction immediately. If you are engulfed, when in a car, turn off the engine. Lock doors and windows. There is a danger that members of the crowd may climb onto your car, break the windows or even try to turn the car over. If you have a chance, leave the car and head for safety.

HIGH-RISK GROUPS

Attack and assault, whatever the motive, will always rely on the assumed vulnerability of the victim—and some groups have always been seen as easy targets.
According to statistics, children, young people, women and old people all fall into the high-risk bracket. So whilst the preceding advice applies to more or less everyone, there are other important factors which should be taken into consideration by the more vulnerable.

CHILDREN

Children are at risk both from adults and other children. Most people have difficulty explaining their problems: young children may find it even more difficult to tell someone that they have been a victim of some forms of abuse or violence. They may not understand that the abuse or violence is unusual.

◗ Encourage children to report whatever happens to them—bullying, encounters with strangers (especially if these encounters happen more than once).
◗ Encourage children to say NO to strangers.
◗ Children must NOT approach a car when it stops, even if the driver only asks directions.

REMEMBER

A child or young person may find it particularly difficult to talk about or describe violent cruelty or sexual abuse—through fear (from threats by the perpetrator), through embarrassment or through fear of not being believed. Statistically, children and young people are more likely to be abused by a close relative or a family friend—which can make dealing with the problem extremely difficult.

Abuse can affect the quality of a person's life in later years. It needs to be dealt with very carefully—usually with the help of specialist counsellors. Actual intercourse is not as common as various forms of sexual assault. There may be no physical proof.

If you are a victim of abuse, you MUST talk to someone—a teacher, a Samaritan, a telephone helpline. An anonymous phone call—even to the police or the Samaritans—may help you find a special helpline or counselling organization.

If, as a parent, you are told of abuse it may be difficult to deal with—particularly if your partner is implicated. Most counsellors believe that the safety of the victim is of the highest priority, but will not underestimate the impact that the disclosure may have on the family.

- Children should be encouraged to believe that THEY own their bodies and it is their right not to be kissed or touched if it upsets them. They MUST tell you if someone upsets them in this way, even if it's a relative.
- Children must NOT accept lifts from strangers.
- If a child is bullied, he/she should tell you.
- Tell the child that if a bully demands money or possessions, they should hand things over and you won't be angry—but they must tell you about it.
- Explain that they can hit, kick or scream at a stranger who wants them to do something with them or take them somewhere, even if the stranger offers sweets.

YOUNG PEOPLE

You may not like having to be home at a certain time—or always having to tell your parents where you are—but if you find yourself stuck without the means to get home, or get into other difficulties, how can they help you?

If you go out on your own, use your common sense while out on the streets (see **On the streets**). If you can afford a taxi to get home, ONLY use a registered taxi. Don't accept lifts from 'mini-cabs' or people you don't really know.

Tell your parents where you are going. If, for any reason, you can't get home, call them. It isn't just so that they don't worry—if you are attacked or left unconscious, where are they supposed to look for you?

If you answer an advertisement in a shop window/a local newspaper for Saturday jobs or babysitting, always take a friend with you. A parent would have to be a little careless if they will accept you, a total stranger, as a babysitter. Most babysitting is arranged through friends of your family.

If you are babysitting, don't tell callers (on the phone or at the door) that you are alone. Don't let anyone into the house. Phone home during the evening to let your parents know that everything is OK.

REMEMBER

You are as much at risk of abuse or attack as any child or adult. Some adults may think you are a child and an easy target. Show them you are not. Don't be fooled by offers of lifts or work that seem too good to be true. You have to start thinking for yourself. When in doubt, stick with friends and don't move around the city on your own.

WOMEN

Women, more than men, may find themselves victims of verbal annoyance, sexual harassment or physical molestation at work or in public places.

HANDS OFF!

If you are subjected to physical 'groping' — it may be worth considering a few ways of dealing with it. Most methods rely on knowing EXACTLY who the perpetrator is — or else you may be accused of assault! You need to be feeling fairly confident to use most of these responses:

- Move in such a way as to dislodge the hand which is touching you and give the molester an angry look. Be sure it IS an angry look.
- Say clearly, but only loud enough to be heard by people in the immediate vicinity, 'Please move your hand away from me'. The usual response is embarrassment — the molester may try to plead innocence.
- If you are not sure who the molester is in a crowd, and you are getting upset, use a louder voice and say, 'I would like whoever is touching me to move their hand away NOW'.
- If you are sure who the molester is — stand on their foot. A heel is good for this. If you hurt them (slightly) plead innocence and apologize.
- Pinch the skin on the back of the molesting hand — a small amount squeezed between two fingernails should do the trick.
- Adjust your bag or your clothing — make a fuss, move around quite a lot and dislodge the hand roughly.

REMEMBER

You can come off worst in this sort of encounter. Using violence (an umbrella brought up between the offender's legs, for instance) may lead to violence. Most public-place molesters rely on packed conditions and people being too timid to complain. They are usually easily embarrassed into keeping their hands to themselves.

No job should include sexual favours or sexual pressure — although attitudes towards what constitutes sexual annoyance vary enormously. Detailed notes should be kept about such pressure at work and reports given to employers.

Sometimes the result may be little more than a reprimand given to a boss or co-worker — sometimes the loss of a job, even by the victim, may result. Unfortunately civilization is not as civilized as it should be.

It's important to find a way to deal with harassment — most people are under enough stress without the addition of petty abuse. In public—often on crowded public transport—quite a few people take advantage of cramped conditions. Some people are able to shrug off such physical contact.

THE ELDERLY

If you are reasonably fit and agile, you will obviously stand a better chance. However fit you are, a stout walking stick or umbrella should be carried at all times. Don't assume that an attacker will necessarily be a young man. Attacks have been perpetrated by women, girls and children. They see you as easy prey—particularly if you have been to collect money. Most of the previous advice applies to you too.

You should talk to your local Crime Prevention Officer about the safety of your home. Follow all the advice (see **Attack at home**). A door chain or limiter is essential. NEVER let anyone into your home, whatever yarn they may spin to you about needing access. Any official visitor will carry identification. They WILL hand this over if requested (even through the letterbox). You CAN phone to verify credentials.

Try to keep a good relationship with your neighbours. They may be your first line of defence if they know your habits and can spot when something is wrong.

REMEMBER

- NEVER keep savings at home. ALWAYS put them into a savings account
- YOU may be able to get help in making your home secure. The Crime Prevention Officer will be able to advise you
- NEVER leave your home without locking the door
- DON'T believe everything you're told—it doesn't take two people to read an electricity meter or to check for faulty plumbing
- Instal a 'panic alarm button'. It will help if you are attacked or fall ill. Some connect to the phone and call emergency services for you

---

⚠ WARNING

If you DO admit people to your home, try to arrange a time when someone else can be with you. If you admit more than one person, one may try to distract you while the other is stealing from another room. If you are fooled and you have admitted a thief—'having a go' may be very dangerous.

SELF-DEFENCE

You need to develop some way of defending yourself physically. Some people are naturally strong and aggressive—others, who may be softer, gentler people, need to learn a few simple 'moves' which can be used in a confrontation. Strength is always a major factor unfortunately—the stronger person nearly always wins.

Surprise is very important—and can often be the deciding factor in a fight, but a person who has learned and practised fighting techniques will quite often win. These techniques need to become INSTINCTIVE, if they are to be any use in a defensive situation.

Fighting fit

The fitter you are, the faster you can move, the longer you can defend yourself, the greater the power you have to defend yourself and to recover from a confrontation. Older people may not be very mobile—or have much stamina—so a confrontation MUST be kept as brief as possible. A woman may not be as strong as a male attacker—and must try to keep physical violence to a minimum.

Fear

Fear must be controlled—learning a few defence techniques can help enormously—as power, mobility and coordination may be seriously affected. If attacked, some people who are very afraid may simply 'freeze'.

Everyone is frightened in a confrontation—this triggers the production of adrenalin which is useful for 'fight' or 'flight'. Breathing correctly helps us focus our energy and take charge of our bodies. In times of great stress, adrenalin may give us extra strength—sometimes surprising strength.

MARTIAL ARTS

There is no one style of defence or fighting that fully covers the needs of the average person. All the martial arts are very effective—most 'self-defence' courses use techniques culled from several types of training. Most require a level of dedication and persistence. Some are more adaptable for women. Other styles may have added advantages depending on your size and weight.

The Western styles of fighting have led to formalized boxing and wrestling. The Eastern martial arts were developed

REMEMBER

Violence breeds violence. What begins as an annoyance or a threat to you, may become physical danger if **YOU** strike the first blow. You may cause the situation to escalate to violence — and may be sorry that you did.
Be very wary of hitting, slapping or kicking someone who clearly has a physical advantage over you.

in climates where people either wore minimal clothing or else loose clothing which offers good firm holds. A lot of the techniques may be ineffective because of the padding effect of layers of clothing in colder climates.

If you have knowledge of any of the formal fighting styles, you will automatically use them—boxers punch, karate experts kick, judo experts use throws. Use whatever comes naturally and develop it. Use your natural advantages.

Short stocky people are good at grappling as they have a low centre of gravity. Tall people have a longer reach. Shorter people may be able to get inside the reach of taller people and but with the head. Most techniques can be tailored to suit your size, age, build and weight.

Ju-jitsu involves some karate-like punches and blocking techniques and many highly-effective throws and holds. Ju-jitsu forms the basis of many self-defence courses. It may take years to learn all the techniques or 'tricks'. It does not require great strength, but can be very effective—particularly for restraining an attacker (although escape may be your first priority). Ju-jitsu is not as lethal as many other techniques.

Judo is practised mainly as a sport, although there is no doubt that many judo techniques are very good when used for defence. It basically involves using the force employed by and the weight of your opponent against him or her. It takes a lot of practice and judgement. Ju-jitsu would probably be more useful for self-defence.

Karate is a discipline for the mind and body—if practised correctly. It involves a lot of ritualized movements and many years of training to achieve real results. It is extremely effective—and deadly—but it is not for the person who only wants to pick up a few ideas. The high-powered kicks and strikes could be very useful—but take a lot of practice.

Kung fu includes a variety of types of technique—some harder and more aggressive than others. It attracts many people because it looks so exciting in kung fu movies. Wing chun techniques, involving low kicks and the defence of the centre of the body—plus the speed which kung fu teaches you—could be useful, especially in close confrontations.

You should investigate locally to see if self-defence classes are available. Most teachers will let you watch a session and discuss the nature of the course. You might also investigate tae kwan do, hapkido and aikido. Kendo and Thai boxing are becoming more popular.

> **! WARNING**
>
> Just a few lessons at a self-defence course could be very dangerous—persist! It would be easy to develop a false sense of security—discovering, when it's too late, that you really haven't learnt anything. Courses should be taken seriously, and followed through. Real confidence and real ability to defend yourself are worth achieving.

WARMING UP

As soon as you begin exercising, you will discover your own physical limitations. Once you become fitter your confidence will increase. It really does help to be able to run, to recover from an attack, to be flexible and have some strength—not just for self-defence (where it is essential), but also in most areas of life.

Warming up helps us loosen up in order to practise self-defence skills. It also helps to develop speed, coordination and strength. Planned exercise—whether weightlifting, running or aerobics—is good for you. Vary the routine so that you don't become too bored. Start by jogging.

> **REMEMBER**
>
> Finding the ideal jogging route isn't always easy. A busy traffic route means you get a hefty dose of smog, but a quiet park could pose a risk to personal safety. Never jog alone in empty streets or at night.

Jogging
- Keep it nice and easy and on the toes.
- Punch forwards with each arm at shoulder level, keeping elbows raised in time with your strides.
- Punch upwards from behind the neck.
- Turn full circle as you run—right side first.
- Keep turning—run backwards, always travelling in the same direction.

◖ Turn again, left side first.
◖ Run forwards again.

Once you settle into the rhythm, you soon find jogging easy. Push yourself by varying the rhythm. Choose a target about 50 metres (about 50 yards) in front of you and SPRING to it.

VARIATIONS

- Hop on left leg for ten strides
- Hop on right leg for ten strides
- Jump with both feet together for ten strides
- Take 'giant' strides—as far as you can—for ten strides
- Sprint for ten strides

Interval training

Interval training is excellent for improving circulation and breathing. It helps develop 'explosive' techniques—sudden bursts of energy which are necessary in self-defence.

Mark three lines on the ground about five metres (about five yards) apart. Sprint out to each line in turn—returning to the base line (nearest to you) each time. As you turn, touch the ground. Don't turn the same way each time—turn sometimes to the left and sometimes to the right.

! WARNING

Before you embark on any fitness or self-defence programme, it is wise to have a medical check-up. Suddenly undertaking strenuous exercise may be very dangerous if, for instance, you are middle-aged and lead a fairly sedentary lifestyle. Particular care is necessary if you suffer from any heart or respiratory problems. If you undertake official classes, always let the trainer know about them. Other conditions which may lead to problems include diabetes, haemophilia, epilepsy, recent head injuries (including concussion) and broken bones.

It's never too late!

You need to find a programme of exercises to build your strength and suppleness. It's surprising how quickly we 'stiffen up', if we don't use our bodies. At school we are encouraged to run about—exercise is organized for us. Most adults find it difficult to run for a bus!

Fitness has become 'fashionable'—look for a local course aimed at your type of physique. It helps to learn some basic techniques and exercises. Music makes it more fun—and running or exercising with a friend is less boring.

Yoga provides some very good stretching and bending exercises, with excellent relaxation techniques—but you need speed and coordination too.

BREATHING/RELAXATION

Breathing correctly is very important. It is a way of recovering quickly from fatigue, a way of relaxing and a way to overcome fear. It can also prove highly effective in helping you to control your temper—so that you can make the correct decision in a tricky situation.

Stand upright with the feet apart—the distance between them should be about the width of your shoulders. Put your hands on your hips and keep your elbows slightly forward.

1 Breathe in through your nose for a count of three. As you breathe in allow your stomach/abdomen to rise and fall, NOT your ribcage.

2 Hold the breath for a count of three, relaxing the body. Exhale through the mouth for a count of three.

3 Continue until relaxed and in full control of your breathing. This process—especially holding the breath and then exhaling it—helps us control our bodies.

REMEMBER

Try the breathing exercise lying down in a darkened room. Relax your body, bit by bit, starting with the extremities. Imagine the room is filled with pure clean white light—your body is filled with dark smoke. Each time you breathe in, you take in the white light. Each time you exhale, you release the dark smoke.

AVOIDING VIOLENCE

Most of us would prefer to avoid any kind of physical confrontation. This is natural and healthy. There may be ways you can ward off actual violence or diffuse a dangerous situation.

Assess the situation

Unless someone actually jumps out and attacks you, you may have a chance to prevent a situation from slipping into violence. Look at your potential attacker. Compare their size and weight and apparent strength with your own. Is it likely that the attacker is armed? Look for:

- Long hair and clothing you could grab.
- Heavy boots/shoes which might cause serious injury.
- Friends—yours or the attackers—who may come to your defence or become otherwise involved.
- A red face, flushed with blood, implies that the attacker is not ready for fighting—otherwise the blood would be diverted to the muscles.
- A white, thin-lipped face and 'tight' voice imply that violence is imminent.
- Follow your instincts. If you have a 'feeling' that there is a problem, then there IS one.
- A fist shaken at you or emphatic hand gestures, including a lot of pointing, may pre-empt violence.
- Are you restricted in movement by your clothing—especially by your footwear?
- Is there anywhere nearby which would give you an advantage? Higher ground?

PERSONAL SPACE

Each of us has a personal space all around us. We become uneasy when people enter this space—especially strangers, more so when they stand behind us. Bullies use this invasion of space to intimidate a victim.

If you are stopped in the street for any reason, keep your distance. This should be at least arm's length. The comfortable distance between people who are having a conversation varies from culture to culture. The British tend to keep their distance more than other nationalities.

Keep talking

Can you diffuse the situation by using words? Can you pretend not to speak the language or not to understand a potential attacker—even acting a little 'simple-minded'? Perhaps you could say calmly:

- There's no need to hit me. If you want my money—here it is.
- I know you're angry but I don't want to fight you.

When a mugger demands money, he/she may be very nervous. Usually a short demand or a threat is made. Rapists, similarly, may just make short, sharp statements.

One way of dealing with the situation is to start talking—about yourself, where you're going, about your ill health—which may begin to make you more of a person. The attacker may start to relate to you more as a person and less as a victim and lose concentration—giving you the opportunity to escape.

Staying calm

Ignore verbal abuse—there is no point in fighting over name-calling. Stand in the on-guard position (see **On guard**) and control your breathing. Staying composed will help you assess the situation and anticipate moves. You must focus on the attacker and what his/her apparent intentions may be. Be aware of possible accomplices. Don't freeze up!

It may be that there is nothing you can say or do to avoid a physical confrontation.

Preparing to fight back

The time to decide how and if you will fight back is NOW. No one can make the decision for you. Evidence suggests that people who have at least attempted to fight back bear less psychological scars than those who have not tried to defend themselves.

Obviously every situation is different—only you can judge what you must do. If you DO fight back, get angry and give it everything you have. You don't have to fight fairly—your attacker won't!

During the confrontation, take any opportunity you can to escape, alert the police or reach safety. This isn't weak—it's SENSIBLE!

PERSONAL ALARMS

You should carry a small personal alarm. Choose one that makes the most piercing unbearable noise possible. An attacker might be very surprised and scared away—especially if help is likely to be summoned by the noise.

Gas-operated alarms tend to be louder than battery-operated ones. They work off small cylinders of gas, but most only sound for about a minute. You might not feel that a minute is long enough for your plea for help to be heard. In which case, investigate battery-operated versions.

You must make sure that the alarm you buy can be locked on—it might be knocked from your hand. The alarm must continue to sound if the device is dropped onto hard ground. If you're not likely to lose hold of the alarm—use it in three to five second bursts for maximum effect.

Make sure you operate a gas alarm UPRIGHT. Liquid gas may be released if you operate it upside down and cause cold burns to the skin or render the alarm ineffective.

Keep the alarm ACCESSIBLE. Have it in your hand or in an easy-to-reach pocket. Test it periodically to be sure it works.

⚠ WARNING

A screech alarm will not necessarily keep you safe. In practice it may do more to surprise an attacker, than to summon help. A variety of alarms were tested in a small alley off one of the busiest streets in central London. Disturbingly—NO ONE took any notice! People in the vicinity were questioned to ascertain then reasons for ignoring the alarms. In most cases, they hadn't heard them, or they didn't know what the sound meant, or they thought the sound was a car alarm! It's time to learn to defend yourself!

ACTUAL WEAPONS

You may not, by law in most countries, carry actual weapons to defend yourself. Knives, clubs, guns—even disabling aerosol sprays are illegal. Always remember that a weapon carried BY you could be used AGAINST you by an unarmed attacker.

In most countries, including Britain, you can only use 'reasonable' force to defend yourself. 'Weapons' may include a specially-sharpened comb or penknife—but hairspray, umbrellas, flashlight and keys are not considered to be weapons and may be carried.

Pepper, scissors or do-it-yourself tools are in a 'grey' area. You may be perfectly justified in carrying them, but you should be aware that in certain circumstances they may be viewed as offensive weapons.

PLEASE NOTE

The following self-defence techniques are intended ONLY for self-defence, not for use with the intention to cause real harm. If you employ any of these techniques there is no guarantee that you will emerge unhurt from a confrontation. See **The law** at the beginning of this section, which explains that (under the law) you may only use reasonable force to defend yourself. Always remember—if YOU introduce actual physical violence into a confrontation, YOU may escalate the seriousness of the incident. Violence leads to violence.

BODY TARGETS

EYES

The eyes are very sensitive — even a speck of grit in your eye, as most people know, can be extremely painful and immobilize you completely for a few seconds. Although it is a small target, the socket helps guide a blow into the eye. For a few seconds your attacker may be defenceless — allowing you to escape or strike at another part of the body.

NOSE

The nose is usually prominent enough for you to strike it from several angles. Lifting it from beneath may force the attacker to raise their head, exposing the throat. A hard blow to the nose is very painful and causes the eyes to water. If a nosebleed starts, the attacker may give up.

EARS

The ears are a limited target, but may be grabbed and twisted. They carry a lot of blood, so if torn, will bleed profusely. Cupping your hands and slapping them smartly over the ears can be very painful. Use as a last resort — it may do permanent damage.

THROAT

The throat is a particularly vulnerable area. A blow to the throat can be very painful and also quite dangerous. If the head is lowered, the throat is protected. Expose it by lifting the nose or pulling back on the hair.

NECK

Strangle holds may prevent air from entering the lungs and blood from reaching the brain. Be very careful — too much pressure for too long can kill! A sharp blow on the back of the neck can stun the recipient of the blow.

STOMACH

A large soft area, which is low enough to be attacked by the knee. Trained people keep their stomachs firm, but the majority of people have no muscle tone at all. If punched or kicked, the attacker may be 'winded' and temporarily incapacitated.

SOLAR PLEXUS

At the top of the stomach in the arch made by the ribs is a particularly sensitive area. Punching here will not hurt your hands — but the attacker may be doubled up.

GROIN

The lower abdomen is sensitive, but perhaps most sensitive are a man's testicles. A well-aimed kick or knee will hurt either sex, but a man may be quite literally brought to his knees by the pain. Grab, twist or pull the testicles. Strike them with a fist, knee or foot.

KNEE

The knee joint is very delicate. A well-aimed downward kick might totally immobilize your attacker. A blow to the inside of the knee (between the legs) is very painful. If you miss, scraping a hard shoe down the front of the shin should also do the trick.

KIDNEYS

From behind, any blow in the region of the kidneys is usually effective. It can cause great pain and breathlessness.

There are many other vulnerable parts of the body which can make suitable targets in your defence. All the joints are vulnerable — any hold which puts strain on one or more is disabling. Fingers can be bent, wrists twisted and elbows locked. If you find any of the nerve points on the body, they can cause a great deal of discomfort. Use them to effect release from an attacker.

NATURAL WEAPONS

Now that you know the targets, you need to learn how to use your natural weapons. The legs are the longest, so they can be used to make attackers keep their distance. The hands are the fastest weapons, but have the disadvantage of being easily damaged.

! WARNING

Most people instinctively punch when threatened, but this is not a good technique. If your hand strikes the attacker's head, you could break a bone in a finger. If you hit their teeth, you may cut your hand and get quite a nasty infection. Your hands can swell up very quickly and become unusable. Even boxers, who know how to use their hands, are VERY careful about them. They make sure their hands are protected before a contest. **PUNCHING IS FOR SOFT TARGETS ONLY.**

PUNCH

To punch correctly, you must make a proper fist. Look at your hand, palm up. Starting with the little finger, curl in the fingers one by one, locking the thumb OVER the fingers. **NEVER** curl the fingers over the thumb. **NEVER** let your little finger stick out. If you make a fist incorrectly, you will hurt yourself more than the person you are hitting.

The wrist **MUST** be locked, or you may damage it. The line of the forearm should follow straight to the knuckles.

A punch should use the whole body — it should come from the legs, through the waist, through the shoulder and finally through the arm. All the power comes from the legs and the torque effect of the waist. DON'T draw back to punch — it's too slow, it announces the strike and it's easy to block an expected punch.

HAMMER FIST

The best way to use your fist! Make a fist as before (see **Punch**). Use it as a hammer wherever you can. Aim for the attacker's nose, ear — almost any target. You'll find you can deliver quite a blow.

PALM STRIKE

If you were told to punch a wall, you wouldn't use a lot of force—it would hurt. If you hit the wall with a palm strike or hammer fist you can use quite a lot of force without damaging yourself. The palm strike can be very effective.

Cock your wrist—angle the hand back slightly on the wrist. Spread the fingers and curl them tightly, holding the thumb in to avoid damaging it.

Use the heel of the hand or the bottom of the palm to strike the attacker's jaw from below. Keeping the movement close to the attacker's body means they may not see it coming. If you miss the jaw you will probably hit the nose, which would have been your next target.

When you hove hit the jaw in this way, your fingers can easily find the eyes.

PALM EDGE

The palm edge can be useful for attacking the throat or the back of the neck, although it may not work as well as it does in the movies!

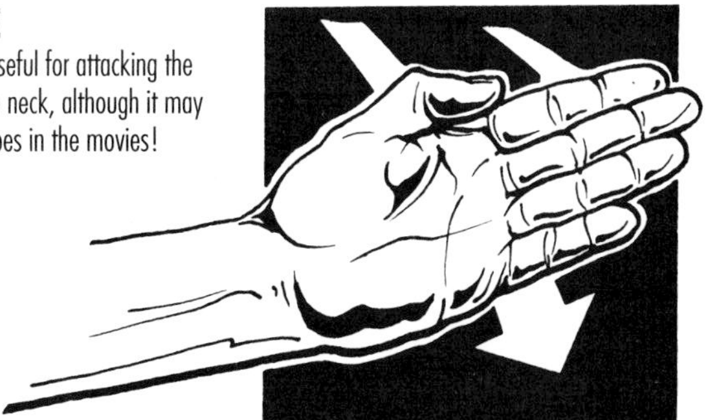

FINGERS

Use these to attack the face and eyes. They can be flicked out with a minimum of effort and used to gouge. Use them as claws. Open your fingers slightly from the palm, strike and scratch. Use a 'swinging' movement like a cat or dig in and drag.

Make the fingers into a chisel point, keeping the thumb tucked safely in, and jab at the philtrum (the upper lip just below the nose), the throat or the eyes.

FOREARM

The forearm is a hard bony structure and is useful for a 'back hander' to your opponent's face or neck. The attacker may think you are turning away. When you lash out with a backswing, they are caught off guard. Almost anything that is banned in boxing is good for defence!

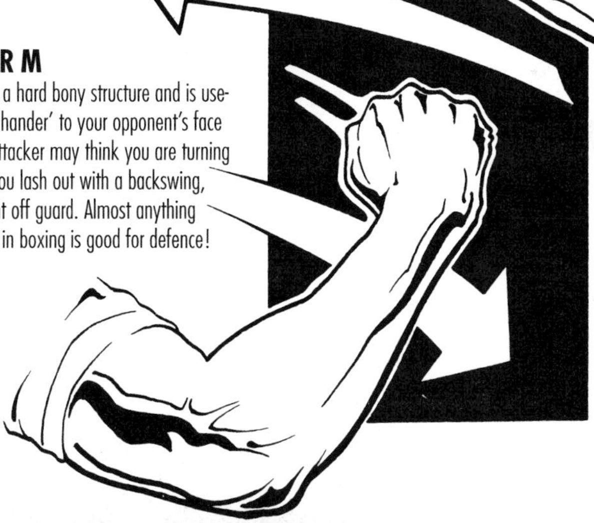

SELF-DEFENCE ■ NATURAL WEAPONS

SELF-DEFENCE

439

ELBOW

The elbow is one of your best weapons. It delivers a considerable blow — often with maximum power and maximum surprise. Snap it up under the jaw or nose. Use it if your attacker is in front, to one side or behind you. Disguise the twist of your body as an attempt to get free and bring the elbow back for all you are worth across the face or under the jaw.

Extra force is added if you make a fist of your hand, cup the other hand round it and literally push the elbow into the attacker's stomach.

TEETH

If you can get any part of your attacker within range — BITE! Any fleshy part will do. Don't let go. Keep on biting. It may be important, especially in the case of rape, to mark an attacker.

FEET

An extended leg could keep an attacker away — especially if you are on the ground. If kicking, don't draw the leg back — this announces the move you are about to make. Take all the weight on the other leg and bend it slightly. Aim the kick to the groin, knee or stomach. Make it a snap action so that the leg is withdrawn immediately. You musn't let your attacker grab your leg and overbalance you (right).

SCREAM!

In tight situations, it helps to scream. It offers several advantages. It may summon help. It drives the air out of your body and helps you focus your energy. It may also shock the attacker. If a real 'bully' finds that you are screaming and lashing out—they may run away.

KNEE

The legs are very powerful and the knee can be used to devastating effect. With practice it can become a very fast weapon. Use for a blow to the attacker's testicles, stomach or even thigh. One advantage of using knees is that they are well below the eye level. If an attacker runs at you—raise a hand as if to defend your face. At the last moment, raise a knee so that the attacker runs right into it at testicle height. Don't look down or you will have 'given the game away'.

The attacker's head may even come into range—especially if 'doubled up' by a blow to the testicles.

HEAD

If grasped face-to-face by an attacker, quickly snap your head forward to hit the nose with your upper forehead. Shorter people can often butt a taller person in the solar plexus or stomach.

If you are grabbed from behind— snap your head back to hit the attacker in the face, preferably on the nose.

REMEMBER

Surprise is all-important. Try to recognize danger and be one step ahead. Run through likely defence techniques if you see a situation developing where you might need them—BEFORE you are required to defend yourself. Move quickly. Don't draw back the hand or the leg to strike—the opponent will be able to see it coming.

Use whatever you have to defend yourself:
■ Coins from your pocket can be thrown in an attacker's face
■ Wrap them up in a handkerchief and use as a club
■ Use your bag, purse, briefcase—aim for the head
■ Umbrellas and walking sticks can be used as clubs or jabbed into feet or stomach—or brought up between the legs to an attacker's groin
■ Hard-soled shoes are essential to be able to kick effectively. Aim for the groin. Scrape your shoe down a shin
■ High heels should be aimed at an attacker's foot or hand. Putting all her weight on a thin heel means an average woman can exert a pressure of nearly three-quarters of a ton!
■ You cannot run in high heels. Take them off and throw them—or use them to strike the attacker
■ A powerful flashlight may dazzle an attacker—and also could make a handy club
■ Grab a handful of dirt, gravel or sand and throw it into the attacker's face
■ Roll up a newspaper and jab it end first into the face or stomach
■ Jab a credit card, comb, hairbrush, anything into the philtrum (the upper lip just below the nose)
■ Scrape a comb across the attacker's face or back of the hand
■ Dig a pen or pencil into the attacker's hand or face—the attacker's impulse may be to defend the eyes
■ Jab or scratch with keys
■ Powder from a compact may temporarily blind an attacker
■ Perfume, hairspray or deodorant can be sprayed into an attacker's eyes

REMEMBER

Disabling sprays are ILLEGAL in most countries, as are knives, clubs and guns. Even a spanner could be considered an offensive weapon if it's not part of the tools of your trade, and could land you in court. If you're not a plumber, builder or whatever, the inference could be that you intend to cause someone harm.

WARNING

Violence breeds violence. An incident that begins with an attacker grabbing your coat and demanding money may escalate into serious assault. If possible, attempt to deter the attacker without being the first one to use real violence. If your attacker is a lot bigger or stronger than you, they may be relying on just frightening you. Once real violence is introduced, you may be way out of your depth.

ON GUARD

To defend yourself against an attacker and use all your natural weapons to the best effect, you need a balanced position from which to fight.

Stand normally—facing the target. The distance between your feet should be about the width of your shoulders. Your favoured leg should be slightly forward and your knees should be bent slightly. Tuck your elbows in, protect your head and body by holding your hands in front of you and lower the chin to protect the throat. You MUST feel comfortable in this position. DON'T strain or stiffen up.

MOVING ON GUARD

Try to make your movements flow. Don't move jerkily. NEVER cross the legs or your stable base is lost and you will be knocked off balance. Practise moving forwards, backwards, to the left and right. Keep the arms up at all times.

Practise with a partner. Don't actually exchange blows. The target is the forehead. If the partner reaches out with a right hand, parry (deflect or block) with your left—and vice versa. Don't stand still—keep moving at all times.

PLEASE NOTE

It is impossible to cover every type of attack and recommend defence techniques to deal with them. The following will give you a clearer picture of the kinds of moves you must make—but no book is a real substitute for taking a self-defence course! Proper training will help you to learn how to channel and use anger—and control fear. You will also have to match the ruthlessness of an attacker.

INTO ACTION!

Don't practise any of the following techniques until you have mastered the on-guard position. Start open-hand techniques with the hands up and open—pleading with the attacker to leave you alone. While this gives them a sense of power or false sense of security, your hands are where you need them—up and ready for the eye jab.

EYE JAB

Flick the hands forward—like a snake striking—extending the fingers. Aim for the eyes. DON'T draw your hand back before you attack. You don't need much force, the eyes are such a delicate target.

Defence against the eye jab is to move the head to one side and to use an arm to parry (deflect) the thrust. ALWAYS parry a straight thrust.

CHIN JAB

Use the chin jab when your assailant has thrown a punch. An untrained person will take their arm back and swing the arm. Step into the swing and parry the blow with your forearm. If the attacker swings with the right hand, parry with your left. Stop the swing and jab upwards under the chin with the palm of your right hand (see **Palm strike**). Push right through the target, pushing the head back. If you are lucky your attacker will fall over backwards and hit the ground.

From this position you can raise a knee at the same time to strike the attacker in the groin and claw the eyes with the hand you have beneath the attacker's chin.

If you miss the chin or the attacker is still standing, grab the hair at the back of the head and pull down hard.

If all these are done quickly, your attacker should go down. You should be able to deliver a chin jab with either hand.

ELBOW STRIKE

The elbow can deliver a considerable blow. With the hand held high in a 'pleading' position an elbow can be lifted suddenly to catch the attacker beneath the chin. Practise this movement so that you could use either elbow. The 'free' hand must be kept up to counter any return blow.

BREAKING HOLDS/STRANGLES

NEVER get in a grappling match with your opponent—it becomes too difficult to use surprise and to strike effectively. Anyone who grabs you from the front usually brings their groin within range. Use your knee! If pulled towards them, you will not be able to get your balance to kick.

ELBOW LOCK

Turn your attacker's hand (with both hands) so that the thumb is downward, locking the wrist at maximum rotation, and the elbow uppermost.

Bring your arm over the elbow and bear down—forcing the arm straight and locking the elbow joint. Exerting pressure now will hurt a lot and the attacker may be frightened that you will break their arm.

FRONT STRANGLE

Use your knee to the groin, head butt and kick. If these are ineffective, swing your right arm over your opponent's hands and bring the elbow back across the arms or chin or strike a 'backhander'. If you can turn to one side in a strangle, you may reduce its effectiveness.

REMEMBER

When in a stranglehold, try to relax — if at all possible! This makes the stranglehold less effective. When straining, the 'Adam's apple' is more pronounced and makes the strangle easier. Sinking down (when in a standing position) may also destroy the attacker's balance.

TWO-HANDED GRAB

If someone grabs you with both hands on your chest, throw your arms over theirs and exert pressure downward so that their arms bend. This automatically makes them drop their head forward — open to a head butt from you.

REAR STRANGLE

If held or strangled from behind, snap your head back into the attacker's face or swing your hips to one side and strike backwards with your hand at the attacker's groin.

Always do the unexpected. Try to destroy an attacker's balance. If they try to push you forwards, into a car or a dark alleyway perhaps, resist initially so that they use extra force. Suddenly take a giant stride forwards. If they fall, kick the groin.

REMEMBER

Try to use an attacker's momentum against them. If someone runs at you, sidestep and use a forearm to the throat. The faster they are moving the greater your blow will be, although it requires careful timing.

KICKING TECHNIQUES

The legs are very powerful weapons, but to deliver this power you must keep your balance. A kick must be fast and effective—it must be a 'surprise'. Don't draw the leg back before you kick.

Using the knee

The knee is an excellent close-quarter weapon and can be used to great effect. The idea is to drive it THROUGH the target—when aimed at the groin you should be trying to lift your opponent off the ground. Aim also for the stomach. Keep your hands up, aiming for the attacker's face to distract them. The opponent's head can be pulled down—perhaps by the hair—and the knee brought up to meet it.

If you are side-on to the attacker, drive the knee hard into the thigh. If you're lucky you will hit a nerve and deliver a 'dead leg'—which is very painful.

If your attacker tries to wrestle you to the ground, try to twist so that YOU land on the top. Aim to land so that one or both your knees are driven into their stomach or abdomen.

> **PRACTISE!**
>
> Stand in the on-guard position. Hold your hands out in front of you, side-by-side at waist height. Transfer your weight to one foot, immediately snap the other knee up into your hands. DON'T lower your hands. Recover at once into the on-guard position. Now adjust your balance to use the other knee. The action should be fast and recovery immediate. Your knee should smack into the palms of your hands.

Using the leg/foot

Kicking also relies on good balance. Surprise and recovery are vital. You must NOT allow the attacker to grab your foot or leg. If you distract the attacker by spitting, screaming and flailing at the face, the kick will be unexpected.

> **R E M E M B E R**
>
> Balance is vital when kicking. All the weight is transferred temporarily to the anchor leg. Use your hands/arms partly for balance and partly to distract the attacker by going for the eyes. The hips and waist give a lot of the turning power and thrust to kicks and strikes. PRACTISE!

MULE KICK

From the on-guard position, transfer the weight onto one leg, pivoting the hips in that direction. Simultaneously raise the other foot, bending the leg. Snap the foot out and down, striking the opponent's knee or scraping down the shin, with the side of your foot. Recover immediately to the on-guard position.

KNIFE DEFENCE

FORCED TO DEAL WITH A KNIFE

Avoid tangling with someone with a knife unless there is absolutely no way to escape. If you are forced to defend yourself:

■ Keep circling away from a thrust to the stomach and other vulnerable areas. Suck your stomach in.

■ Most knife attacks involve diagonal slashes across face or torso. Keep circling to get out of slashing range.

■ If you get into a tangle with your assailant, you're too close. If you MUST grab something, grab the attacker's wrists with both hands.

■ Better still, use mule kick or knee-to-groin to drive the attacker back out of range.

■ Wrap a heavy article of clothing around your hand and forearm to deflect the knife blade.

■ Let the knife impact on a bag or briefcase.

■ Parry the thrust with an umbrella, walking stick or any stout stick (which gives you a longer reach). Go for joints, bones, collarbone, knees, wrists.

■ Don't swing a stick. NEVER pull back before you swing—you have just told the attacker what you are about to do.

■ Jab with the end of a stick.

■ Throw dirt in the attacker's face.

■ Keep shouting for help.

FIREARM DEFENCE

There is very little defence against an attacker armed with a gun or rifle. Some petty thieves rely on the mere sight of a weapon to be sufficient to persuade you to do as they wish. They may not even know how to use a gun. Others may be very proficient and very keen to shoot. You cannot judge whether the gun will be used by accident, in a panic or with cool precision.

Techniques for disarming or grappling with an assailant who is armed with a firearm require a lot of luck, skill and careful timing. Even if you had a lifetime's training in self-defence, you could not be sure of wrestling a gun or rifle from an attacker's grasp.

KEEPING YOUR DISTANCE

The distance between you and your attacker plays a vital part in any confrontation. The distance the attacker needs to punch, kick or use a knife may influence your choice of defence technique. If in doubt—let fly with everything you have.

■ Punching/striking. The attacker is well within range of kicks.
■ Knife attack. You should still be able to kick.
■ Grappling/strangles. Use palm strikes, claw the face, try mule kicks and head butts, use the knees.
■ Stick attack. Back off or step inside swing, block at the attacker's wrist with your forearms and use your hands/knees as well as you can.
■ Firearm attack. Put 50 metres (55 yards) between yourself and the gun— the attacker may be a poor shot. Keep low and don't run in a straight line. Aim to put an obstruction in the firing line—a fence, trees, cars.

ANIMAL ATTACK

 In the urban environment, it's not only other people that give cause for fear—animal attack is a very real possibility, both from domestic pets and urbanized wildlife. Even the humble magpie poses a threat in countries such as Australia, making vicious attacks on unsuspecting passers-by during the egg-laying season.

Dogs, however, remain the biggest threat to the urban dweller. And whilst domestic pets should be reliable, old dogs can become cantankerous, bitches can get defensive when they've recently had a litter, and some breeds are encouraged to be excessively territorial.

Guard dogs and pets which were originally intended as guard or hunting dogs may have an aggressive streak—even the owners may be at risk. Some dogs just don't like strangers. Even a friendly animal might take exception to rough treatment. The 'bark' is quite often worse than the 'bite', but you may be forced to be quite ruthless to defend yourself.

ATTACKED BY A DOG

- Avoid eye-to-eye contact with any 'vicious' animal—especially a dog.
- Stand still—speak firmly (as the owner might) using clear commands such as 'No!' 'Stay there' or 'Sit!'
- Try using a 'good boy' gentle voice to talk to the dog.
- If a small dog goes for your ankles—a soft kick may deter it.
- If the animal jumps up to bite you, offer a well-padded arm—push your arm to the back of the throat. Don't pull back. Give the dog a sharp blow on the nose. If you must, go for the eyes. Twist an ear and pull hard.
- If running or jogging, stop. Walk away before continuing to run.
- If chased by a dog (or other animal) drop an article of clothing or a bag! It may stop to investigate it.
- Don't grab any dog by its tail—it can easily bite you.
- You may be able to pick up a small dog by the scruff of its neck—and deposit it safely on the other side of a fence or wall.

 Not everyone likes dogs and some people—especially children—may be terrified if a large dog approaches. If you own a dog, keep it under control—DON'T let it jump or snap at strangers. NEVER let a dog run free around the neighbourhood. In most countries YOU are liable for the damage or nuisance it may cause.

☠ WARNING

Rabies (hydrophobia) is still common in many parts of the world, although Britain and Australia have more or less eradicated it. The dog or other animal affected may be wild-eyed and foaming at the mouth—but it may not. If you get a large amount of saliva on the skin—or the skin is broken—seek medical attention immediately. It would help if you could identify the animal (see colour pages) as rabies is a horrific and often fatal illness which affects the brain. If bitten, anti-tetanus injections may also be required.

DOG DETERRENTS

There are whistles and electronic devices which are sold, claiming to deter dogs. Dogs have much more sensitive hearing than man—and can hear high frequencies which are beyond our detection. There have always been 'dog whistles' but more recent 'ultrasonic' battery-operated devices claim to stop dogs in their tracks. In some cases they do work, although real guard dogs are trained to ignore the sound. These deterrents could be worth investigating.

URBANIZED WILDLIFE

Depending on where you are in the world, most cities have urbanized wild animals. Most come in to scavenge—and pose little threat unless cornered. Their natural habitat may have been eroded by civilization, and they may have little choice but to scavenge as a means of survival. Racoons, bears, wolves, foxes—even leopards—like to raid waste bins.

On a day-to-day basis, rabies is still a major problem throughout much of the world, and capable of infecting anything from monkeys to dogs to humans. Be very wary of touching a wild animal that appears to be sick or dying and yet is unusually tame. Rabies is a 'clever' virus, altering normal animal behaviour as a means of spreading the disease. Consequently, a wild animal may become 'tame' and a domestic pet savage.

Rabies is just one of the many illnesses transmitted by vectors—in other words, disease carriers (see HEALTH: **Zoonoses**). External and internal parasites also represent attack on a microscopic scale.

SELF-DEFENCE ANIMAL ATTACK ■ URBANIZED WILDLIFE

457

Most dog breeds were intended for specific purposes: guarding, herding, droving, baiting animals or fighting. These origins can be expected to affect their temperament. Powerful dogs need training, discipline and regular exercise, or may become dangerous and uncontrollable. Here's a guide to aid recognition of common types.

1►German shepherd (Alsatian) Versatile guard, used by the police and as a guide dog. Alert and reliable with stamina. Height at shoulder: 66 cm (26 in). **2►Doberman** Developed as a guard in the 1870s. Tough and quick-thinking, but needs careful handling/training. Can be an anti-social nuisance. Height: 68 cm (27 in). **3►Rottweiler** The heavyweight guard, originally a cattle dog. Widely used by police and military. Height: 68 cm (27 in). **4►American pitbull terrier** American-bred fighter, extremely strong and very quick. Bred for a low pain threshold.

Height: 48 cm (19 in). **5▶Bull mastiff** Well-established guard, although the breed is only a century old. Originally used to pin down poachers. Height: 68 cm (27 in). **6▶Giant schnauzer** Bavarian dog bred for droving cattle and pigs, then used as a police and army dog. Height: about 65 cm (25 1/2 in). **7▶Bouvier des Flandres** Bred as a cattle drover, makes a fine guard. Normally calm, but formidable when roused. Height: 68 cm (27 in). **8▶Australian cattle dog** An extremely tough work-ing dog, with great endurance. Needs plenty of activity. Could become difficult if restricted by city life. Height: 51 cm (20 in). **9▶Corgi** Droving dog, bred to nip the heels of cattle to keep them moving. They take an interest in ankles! Height: 30 cm (12 in). **10▶Jack Russell terrier** Bred (as most terriers) for digging out rabbits and other animals. Small plucky dog, unintimidated by the size of an opponent (including humans!). Height: 30 cm (12 in).

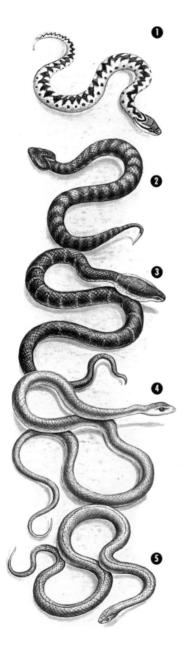

Poisonous snakes are rarely a danger in most cities. Many countries have no poisonous snakes at all. If you are in snake country, always seek advice and take precautions.

EUROPE

1▶Adder *Vipera berus* The only poisonous snake of northern Europe, it varies from olive-grey to reddish-brown, sometimes with a darker zig-zag pattern. Timid, but strikes quickly/repeatedly if cornered or alarmed. Length: 30–75 cm (12–30 in). **Bite rarely fatal.** Larger and more dangerous relatives in southern Europe, the eastern Mediterranean and across northern Asia to Korea.

AUSTRALASIA

2▶Death adder *Acanthophis antarcticus* Thick-bodied and brownish, reddish or grey with darker banding. Well camouflaged, it seldom moves away when approached but waits for prey to investigate then strikes. Length: 45–60 cm (18–24 in). **Highly venomous.** One of Australia's deadliest snakes.

3▶Tiger snake *Notechis scutatus* Thick-bodied, large-headed, tawny-ochre with greenish-yellow, grey or orange-brown bands. Found from southern Queensland down to the populous areas of New South Wales (where it's black rather than brown). Nocturnal on warm summer nights. Length: 130–160 cm (51–63 in). **Aggressive and very poisonous.**

4▶Taipan *Oxyuranus scutellatus* Uniformly light to dark brown, with yellowish-brown on the sides and belly. Found from the Kimberleys through Arubauland to Queensland in northern Australia. Length: up to 3.5 m (11 ft). **Shy, but fierce, when aroused and deadly poisonous.**

5▶Eastern brown snake *Pseudonaja textilis* Found in drier areas of eastern Australia. Slender, yellowish-grey-to-brown with pale belly. Length: 1.5–2 m (5–6 1/2 ft). Active by day. One of several poisonous brown snakes. **Very poisonous, easily roused and when antagonized strikes repeatedly.**

AFRICA & ASIA

6▶Puff adder *Bitis arietans* Thick-bodied, short-tailed and large-headed. Straw-brown with darker markings. Found in savannah and semi-arid areas of Africa and the Arabian peninsula. Well camouflaged, it waits for prey on paths after dark. Length: 90–130 cm (35–51 in). **Highly venomous — causes extensive internal bleeding.** One of many different vipers, which are found all over Africa and Eurasia.

7▶Saw-scaled viper *Echis carinatus* Rough-scaled, pale-reddish to sandy-brown with darker markings and light blotches. Found in arid areas from western North Africa to India. Uses serrated scales on its side to make a threatening noise. Length: 40–55 cm (16–22 in). **Vicious, common, many fatalities. Causes death from internal bleeding.**

8▶Russell's viper *Vipera russelli* Brownish, with three rows of spots formed of white-bordered black rings, each with a reddish-brown centre. Found in most areas from east Pakistan to China and Taiwan. Length: 100–120 cm (39–47 in). **Responsible for the highest number of viper bites in the area. Highly venomous.**

9▶Malay pit viper *Calloselasma rhodostoma* Fawn, reddish or grey marked with geometric patterns. The belly is yellowish or spotted greenish-brown. Found in Southeast Asia and Indonesia. Length: 60–80 cm (24–32 in). **Bites are common. Dangerously venomous. Has many relatives in the area. Avoid ANY that resemble it.**

SAFETY RULES

NEVER tease, pick up or corner a snake.

BEWARE an apparently dead snake, some only move to strike when prey is close.

STAY CALM If you encounter a snake, do NOT move suddenly or strike at it. Back off slowly. In most cases the snake will try to escape.

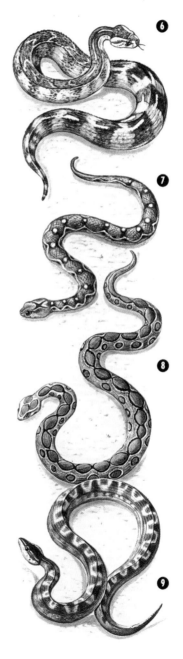

POISONOUS SNAKES

1▶Asiatic Cobra *Naja naja* When alarmed, has easily recognized raised head and spreading hood — often with markings. Occurs in Asia from Middle East to Far East. Common in urban areas. Length: 1.5–2 m (5–61/2 ft). **Highly-toxic venom.**

SPITTING SNAKES

A few cobras spit poison as well as bite. The venom is diluted and therefore not as dangerous, unless the poison reaches an open cut or the eyes. If it does, immediately wash out thoroughly with water. Seek urgent medical attention.

2▶Mamba *Dendroaspis* species Small-headed, very slender, typically with large green or grey-ish scales. Found in Africa south of the Sahara, usually in trees. Length: 1.5–2.1 m (5–7 ft). The large black mambo (*D. polylepis*) is mostly terrestrial. Length: up to 2.6 m (81/2 ft). **Often quick to strike, black mambas deliver large doses of venom which affect the brain and heart. FATAL unless antivenin given quickly.**

3▶Indian krait *Bungarus caeruleus* Small-headed, with black-and-white or black-and-yellow bands down the body. Found in India and Sri Lanka. Length: 1.3–2.1 m (4–7 ft). **Slow to strike, but its venom is especially powerful.** Bite is almost painless. Symptoms may take several hours to develop, by which time it may be too late.

THE AMERICAS

4▶Rattlesnake *Crotalus/Sisturus* species Chunky body, wide head and a rattle at the end of the tail that is usually, but not always, sounded as a warning. Many kinds, widespread in the US and Mexico. The largest are the various diamondbacks, with distinctive diamond-shaped blotches. Length: 45 cm (18 in) to over 2.1 m (7 ft). **Larger species are especially dangerous.**

5▶Copperhead *Agkistrodon contortrix* Stout body, coloured buff or orange-brown with rich brown bands and a copper-red head. Found mainly in the eastern US. Fairly timid, it vibrates its tail if angry. Length: 60–90 cm (24–36 in). **Bites are rarely fatal.**

6►Cottonmouth (Water moccasin) *Agkistrodon piscivorous* Thick brown or brownish-olive body, sometimes blotched, and a yellowish belly, also blotched. The inside of the mouth is white. Aquatic, it is found in or by fresh water in the southern US. Length: 60–130 cm (24–45 in). **Belligerent, do not annoy! Victims have suffered tissue damage and amputations.**

7►Cascabel (Tropical rattlesnake) *Crotalus durissus* Large with diamond-shaped marks, two dark stripes on the neck and a rattle on the tail. Nocturnal and found in drier areas from eastern South America to Mexico. Length: 1.5–2 m (5–61/2 ft). **Aggressive. One of the world's most dangerous!**

8►Common lancehead *Bothrops* species Brownish with paler geometric markings. Length: 1.3–2 m (4–61/2 ft). Its many relatives vary from grey to brown or reddish with similar markings. Various species occur in South America north to Mexico. Some are arboreal. **Have caused many deaths. Tend to loop their body before striking.**

9►Coral snake Slender and strikingly coloured often with bands of black and red separated by bands of yellow or white. Found in forests and grasslands from the southern US into South America. Similar species occur in Southeast Asia. Length: 45–90 cm (18–35 in). **Small-mouthed, somewhat reluctant to bite, but deadly.**

SAFETY RULES
ALWAYS walk on clear paths.
ALWAYS look closely before parting bushes or picking fruit—some snakes are arboreal.
DON'T put hands or feet in places you can't see, use tools or sticks to turn over logs or stones.

BITTEN BY A SNAKE
In the case of snakebite, identify the snake if possible—but don't risk further injury. If the snake has been killed, take it with you so that antivenin can be accurately matched. Get the victim to hospital immediately but avoid them having to walk or expend energy—which speeds the circulation of the poison. Do not apply a tourniquet.

POISONOUS SNAKES

463

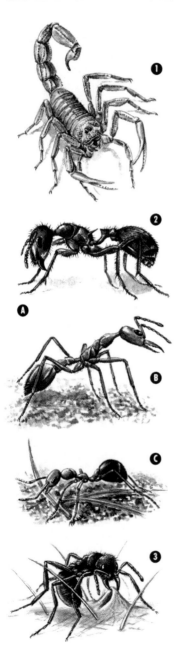

The amount of venom injected is usually tiny, but it contains concentrated doses of fast-acting poisons, similar to those in snake venoms — nerve poisons, stupefying ingredients, a convulsant, enzymes and sometimes formic acid.

Stings can cause faintness, fever, degrees of paralysis, difficulty in breathing, vomiting and diarrhoea. These symptoms usually pass in 24 hours, but an attack of urticaria may follow after a week. Occasionally, severe allergy can cause death.

1 ► Scorpion *Arachnidae* There are many deadly species of scorpion in tropical and subtropical regions, often living in and around houses. Colouring varies from yellow and yellow-green to brown or black, shading from dark body to lighter legs and tail. Almost all are nocturnal. Cornered or crushed scorpions tend to sting repeatedly with the tail — but of the 1500 plus species, only about 50 are dangerous to man.

The most dangerously venomous species in the US is *Centruroides exilicauda*, which is pale brown with dark stripes and about 8 cm (3 in long). In southern Europe/North Africa the most common dangerous species is *Buthus occitanis*. Usually about 10 cm (4 in) long, it inhabits dry areas. The Brazilian *Tityus serrulatus* and the Trinidadian *Tityus trinitatis* cause numerous deaths each year.

Scorpion venom is neurotoxic — death results from respiratory/cardiac failure. Seldom fatal in adults.

2 ► Ant *Formicidae* Found worldwide and closely related to the wasp and bee. In most species the stinger is very small and the venom mild, but the dangerous species possess a combination of potent venom, large stingers and numerous aggressive workers. **The harvester ant** (A) *Pogonomyrmex* is an aggressive species found in the western US and South America. Will attack without provocation, inflicting dozens of painful stings in seconds. Though very painful, the stings have little lasting effect except in cases of allergic reaction. **The giant bulldog ant** (B) *Myrmecia gulosa* of Australia is 2.5 cm (1 in) in length with large toothed jaws and a stinger. Can leap several inches and will attack in vast numbers with little provocation.

Not all species have stingers. The blood-red **field ant** (C) *Formica sanguinea*, of Europe, North Africa and northern Asia, squirts a formic acid solution and then bites.

3►Fire ant *Solenopsis saevissima* Introduced into the US from South America, fire ants are now a major pest in southern states. Dark red in colour and about 2.5 cm (1 in) in length, they build nests up to 90 cm (3 ft) high. The stings are very painful and can cause severe allergic reaction.

BEE/WASP *Hymenoptera*
Many species are social insects that occur in colonies which can number hundreds of workers. They build nests in the ground, in trees, under the eaves or in the walls of houses.

4►Honey bee (Hive bee) *Apis mellifera* A worldwide species, the most highly developed of any commonly-seen social insects.

A honey bee can sting only once and the barbed sting is left in the skin of the victim. The glands at the base of the sting continue to release venom for several minutes, so it should be removed as quickly as possible without pinching the sac. Symptoms: mild pain and swelling. **Severe allergic reactions are rare but may result in rashes, respiratory difficulties or death.**

5►Wasp *Vespula* Common throughout the Northern Hemisphere. Often build nests in or near human habitation. They build football-shaped nests covered with a rough paper made of chewed plant fragments. The chief danger is from swarms, especially when the queen migrates from the old nest to establish a new colony. Painful sting, but usually little effect other than some swelling.

6►Hornet *Vespa* A large member of the wasp family. It is best to stay away from any hornet seen, as the nests are often hidden in the ground or gaps between buildings and the killing of a single worker could disturb the entire colony. They will chase over some distance, inflicting numerous stings.

SPIDERS *Arachnidae*

All spiders possess poison glands at the base of their mouthparts, though few can harm humans. Try to prevent spiders making their home in your house by not letting junk build up in basements, attics, closets, outdoor sheds, and garages.

1 ► Tarantula *Theraphosidae* Despite their menacing appearance, most species are harmless. However they possess special barbed hairs on the abdomen which are released when the tarantula is cornered. These are not venomous but are capable of causing local skin rashes.

2 ► Funnel-web *Atrax robustus* These small 5 cm (2 in) long sleek-black Australian spiders are dangerously venomous. Commonly found in the Sydney area, they spin a web with a deep horizontal funnel. **Their fangs can penetrate clothing and the venom injected is powerful enough to kill an adult in 90 minutes.** There is an antivenin. **Seek urgent medical attention.**

3 ► Black widow (Hourglass spider) *Latrodectus* species Found in North and Central America, and the Caribbean. Most species are 1–2 cm (1/2 – 3/4 in) long with a round abdomen and a red/orange hourglass marking on the abdomen. They build webs in dark secluded places (garages, outbuildings, cellars, attics). Also found in piles of clothing or paper and under lavatory seats.

The venom glands are under voluntary control, so that it is possible for a black widow to bite without injecting any venom. Medical attention should still be sought.

Symptoms: within an hour a dull pain begins to spread from the bite up the limbs and into the back, chest and abdomen. Nausea, cramps, vomiting and increased perspiration may be experienced. **FATAL in young children. Seek immediate medical attention.** An antivenin is available.

4 ► Red-back *Latrodectus hasseltii* An Australian relative of the black widow. Its web is a tangle of dry threads. Often found around buildings, especially outdoor lavatories! **Can be FATAL.** Seek urgent medical attention. There is an effective antivenin.

5►Katipo *Latrodectus katipo* The New Zealand member of the genus, found on beaches of North Island and the western coast of South Island. Lurks under driftwood, stones or well-sheltered plants. **Can be FATAL Seek urgent medical attention.**

6►Brown recluse (Fiddleback) *Loxosceles* species Found in Asia, the US, South America and Europe. Species are all similar: various shades of brown and a faint 'violin mark' on the centre of the mid-section. Common house pests, they inhabit dry dark corners or closets, and often hide in piles of clothes or paper. Most bites occur when people put on clothing or shoes the spiders have crawled into. ALWAYS shake out items of clothing before putting them on.

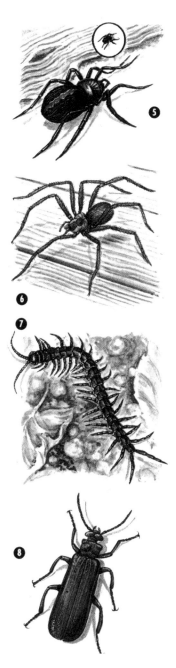

   The venom is necrotic, destroying muscles and fat. At the time of the bite there is a slight burning sensation. Over the next 8 hours the pain increases, the wound turns red and blisters. Bites are slow to heal (usually about 2–3 weeks) and leave a scar — gangrene can set in if untreated, leading to amputation. **Serious complications occur in a small but significant percentage of cases, in which the kidneys are affected and death may result. Seek URGENT medical attention.** Usual treatment involves excision of the tissues at the site of the bite followed by skin grafts.

7►Centipede *Scolopendra* species Found worldwide under stones, in piles of wood and damp places. Most species outside of the tropics are 2.5–7 cm (1–2¾ in) long. The hook-like fangs are located under the head and are used to grasp prey and inject venom. Bites may puncture the skin and cause discomfort, swelling and infections, but they are rarely fatal. DON'T swat a centipede with bare hands. If one gets on you, avoid being stung by brushing off in the direction the animal is moving.

8►Spanish fly *Lytta vesicatoria* Found in southern Europe. Not technically venomous, but possess a chemical called cantharidin which makes them dangerous to handle. Produce intense skin irritation when crushed, causing painful blisters.

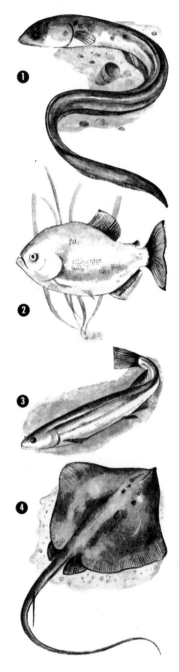

RIVERS

1►Electric eel *Electrophorus electricus* Rounded, olive to black in colour with paler underbelly. Native to Orinoco and Amazon river systems of South America. Often prefer shallow water, where there is more oxygen. Length: up to 2 m (6 1/2 ft). The shock from a large one can be 500 volts, enough to knock you off your feet — lethal to a small child. **Very dangerous if you are shocked while swimming.**

2►Piranha *Serrasalmus* species Found in the Orinoco, Amazon and Paraguay river systems of South America. They vary in size, some are very small, but all are deep-bodied and thickset, having large jaws with razor-sharp interlocking teeth. Length: up to 50 cm (20 in). **They attack in groups and can be very dangerous, particularly in the dry season when the water level is low.**

3►Candiru *Vandellia cirrhosa* Not much thicker than a pencil lead, a scaleless, parasitic catfish of South America. Said to have an affinity for urine and to seek to enter the urogenital openings of the body, their rear-pointing spines causing great pain. Length: 6.5 cm (2 1/2 in). **Wear tight protective clothing over genitalia if tempted to bathe in suspect rivers and streams.**

SEAS/RIVERS

4►Stingray *Dasyatidae* A danger in shallow waters worldwide. There are marine and freshwater species, the latter occurring in tropical South America and West Africa. Very variable, but all share flat shape and long narrow tail with one or more barbed venom-producing spines along its length. Usually lie buried in mud/sand. When stepped on by an unwary wader they swing their muscled tails with enough force to drive the spine into the foot or leg of the victim. The wound is very painful and slow to heal. If untreated, secondary infections can rapidly develop. **Amputation or death may result, seek urgent medical attention.**

In waters where stingrays are known to be found, waders should shuffle their feet along the bottom to avoid stepping on them.

☠ WATER CREATURES

SALT WATER

5▶Tang (Surgeonfish) *Acanthuridae* Found in tropical oceans where they often form large schools. Thin-bodied oval fish, often colourful. Small venom glands in the dorsal spines, but the danger comes from the razor-sharp spines near the tail, which in some species open like a flick knife.

6▶Weeverfish *Trachinidae* Tapering, dull-coloured, they lie partly buried in shallow sandy bays along the coasts of the North Sea, Mediterranean and the Atlantic coast of Africa. Length: 30 cm (12 in). The most venomous fish in temperate and northern seas. Spines on the back and gills inject venom. **Secondary infections and gangrene may develop. Can be FATAL.**

7▶Toadfish *Batrachoididae* Bottom-dwelling fish, found almost worldwide in tropical, sub-tropical and temperate waters — marine and freshwater. All species look much alike — dull-coloured with large mouths. Length: 7–10 cm (2 3/4–4 in). Spines on either side of the head inject venom (though not all varieties are poisonous). Stings are painful but not dangerous.

8▶Zebrafish (Lionfish) *Pterois* species Lives amongst reefs, rocks, or in sand, in tropical and sub-tropical seas, particularly the Indian and Pacific Oceans. Spectacular, usually with bands of reddish brown and white, large filamented fins and long dorsal spines. Length: 10–20 cm (4–8 in). Contact with the spines causes intense pain and swelling. **Heart failure may result. Seek urgent medical attention.**

9▶Stonefish *Synanceiidae* Found in tropical Indo-Pacific waters and the Red Sea — especially common along the northeast coast of Australia. Their drab colours, lumpy shape and sedentary lifestyle make them difficult to see amongst the rocks and reefs they inhabit. Length: 40 cm (16 in). **When stepped on, the dorsal spines inject enough venom to cause agonizing pain, convulsions, unconsciousness and even paralysis of the limb. Recovery may take months. Sometimes FATAL.**

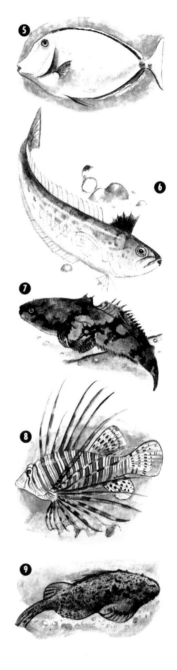

WATER CREATURES

469

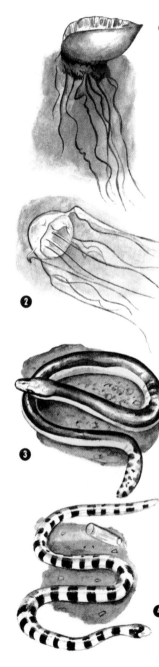

1▶Portuguese man-of-war *Physalia* species Not a jellyfish, but a colony of semi-independent cells. Mainly sub-tropical—commonest in waters of high salinity like those around Australia and Florida—but the Gulf Stream may carry it to more temperate waters. The transparent float or umbrella is harmless—the numerous tentacles with their stinging cells may stream out for 15 m (49 ft) beyond the 20 cm (8 in) float. **The stings are intensely painful and may cause breathing difficulties. Even 'dead' ones can still sting.**

2▶Sea wasp (Box jellyfish) *Chironex fleckeri* A cube-shaped bell, 25 cm (10 in) across, with clusters of tentacles trailing from the corners. These may be up to 9 m (30 ft) long. Large swarms sometimes invade public beaches in Australia between October and March. **Venom can be FATAL in large doses—can kill within five minutes. Stings are intensely painful—there is a risk of victim losing consciousness and drowning. An antivenin is available. Seek urgent medical attention.**

3▶Yellow-bellied sea snake *Pelamis platurus* One of the most widely distributed of all sea snakes and one of the smallest: 72–88 cm (28–35 in). **Highly-toxic venom causes respiratory paralysis and muscular pain and weakness. Can be fatal.**

4▶Banded sea snake *Laticauda colubrina* Range extends from Bay of Bengal, Japan, and around coasts of Australia and New Zealand. Grows to maximum length of 1.4 m (4 1/2 ft). Grey above with creamy underparts, a series of black bands around the body and a paddle-like tail. Partly terrestrial. **The bite is said to be painless, but venom is twice as toxic as cobra venom. Seek urgent medical attention.**

5▶Blue-ringed octopus *Hapalochlaena* species Small, sometimes only fist-sized, with a maximum spread of 20 cm (8 in), found in shallow waters and pools off Australia. Will bite if disturbed. **Venom causes swelling, dizziness and respiratory paralysis. Apply**

artificial respiration. **Seek urgent medical attention.**

6▶Cone shell *Conidae* Sub-tropical and tropical gastropod, found on beaches and in shallow waters, reefs and ponds. Varying patterns. Length: 2.5–10 cm (1–4 in). Venom is injected through a flexible, harpoon-like barb which pokes through the narrow end of the shell. **Can cause temporary paralysis and breathing difficulties, and DEATH within six hours.**

SHARKS
Only a handful of shark attacks on people are recorded every year and few are fatal, but caution is advisable when in waters where sharks occur. Of the many species, only a few of those that swim inshore have been known to attack humans.

Try not to attract the attention of a shark. Head for shore as quickly as possible. Blood, urine and dead fish are known to lure sharks, so avoid carrying fish or urinating in the water. If you see a fin, don't take chances.

7▶Bull shark *Carcharhinus leucas* of the tropical west Atlantic is stout, grey above and white below. It likes shallow water and may swim some way up rivers. Length: up to 4 m (13 ft). **Aggressive and dangerous.**

8▶Nurse shark *Ginglymostoma* species Includes the grey nurse of Australian waters. Heavily built, large-finned, greyish above, white below. Length: 3 m (10 ft). **Often found close to shore.**

9▶Tiger shark *Galeocerdo cuvieri* Found in tropical and sub-tropical waters, white belly, back blotched or barred when young, more evenly grey when mature. The head is wide with squared-off jaws and snout. Length: 3.5 m (11 1/2 ft). **Often found close to shore.**

10▶Sea urchin *Toxopneustes pileolus* Most sea urchins have needle-sharp spines capable of piercing even gloves and flippers. *Toxopneustes* is a venomous variety found in tropical Indo-Pacific waters with three-valved spines, each with its own venom gland. **Seek medical attention.**

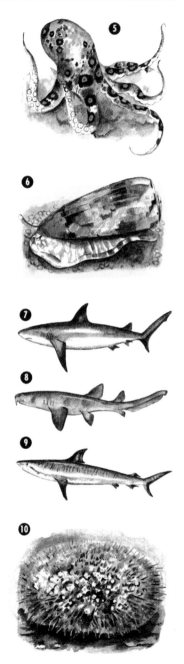

Vectors (hosts or carriers of disease) range from small insects to large mammals. Some play host to insect parasites such as fleas and mites, which are the actual disease carriers. Rabies is carried by a number of mammals, though few invade urban areas. They include the domestic dog and cat, wild dogs that scavenge around settlements, and stray and feral animals that roam urban areas. In countries where rabies has been excluded, there is still a risk of other infections from a bite (see also HEALTH: Pets and your health).

1▶Racoon *Procyon lotor* Long brown fur with tints from grey to yellow, distinctive 'bandit' face mask. Found in the US, parts of Central Europe and Russia, it is well adapted to town and city life. A shy nocturnal creature, regarded as a pest because of its scavenging. An agile climber which can hold small objects and even open windows. In some cities it has learned to nest in chimneys. **It will bite only if cornered or molested, but it can carry rabies. If bitten, seek urgent medical attention.**

2▶Striped skunk *Mephitis* species Sometimes found in city suburbs/towns in the US. It is infamous for its habit of spraying enemies with a strong-smelling secretion from its anal glands — aimed with great accuracy, up to 4 m (13 ft). Look out for the warning signs: before spraying, a skunk will stamp its forefeet, erect its tail, then do a 'handstand' with tail pointed over its head away from you. Don't wait for it to drop back on all fours and spray — get as far away as you can! **Skunks may bite if molested — seek urgent medical attention because of the risk of rabies.**

Control: Mothballs may deter them from making a home under buildings. To avoid long-term problems, make under-house spaces rodent-proof. Seek expert pest control.

3▶Red fox *Vulpes vulpes* Common inhabitant of cities and towns throughout Europe, Soviet Union, Asia, Australia and the US. Length: 60–90 cm (24–35 in). Wary of humans, they will only bite if cornered or molested. **Can carry**

rabies. **If bitten, seek urgent medical attention.**

4▶Coyote (Prairie wolf) *Canis latrans* Smaller relative of the wolf, adapting to town and city life in the western US. Invaluable as a predator of rodents, though attacks on children reported in Los Angeles. Length: 120 cm (47 in). **Can carry rabies. Seek urgent medical attention if bitten.**

5▶Vampire bat *Desmodus rotundus* Found in arid and humid tropical and subtropical regions. Nocturnal, they suck the blood of sleeping victims. Attacks on humans are rare, but do occur. Saliva contains an anaesthetic, so bite might not be felt while you are asleep, and an anti-coagulant so wound may bleed profusely. Body length: 7.5–9 cm (3–3 1/2 in). **Can carry rabies. Seek urgent medical attention if bitten.**

If ANY bats invade a building, seek specialist advice. Many countries have laws concerning these endangered species.

6▶Chipmunk *Sciuridae* A burrowing rodent of the US and Asia, with black-striped yellowish fur. Most commonly found in crawlspaces under houses and in gardens. Length: 25 cm (10 in). **It can carry plague and fevers. Its fleas will bite humans.**

7▶Brown rat (Common rat) *Rattus norwegica* (A), the black rat (ship's rat) *R. rattus* (B) and the smaller house mouse *Mus domesticus* (C), can all be a serious health hazard. Their urine and faeces contaminate food. Their fleas can carry diseases including plague, typhus, hepatitis, salmonella, fevers and worm infections. Their sharp teeth damage woodwork, plaster, metal pipework and electric cables—sometimes causing fires. Lengths: A 19–25 cm (7 1/2–10 in), B 14–20 cm (5 1/2–8 in), C 6–8 cm (2 1/2–3 in).

Control: Block all possible house entry points. DON'T leave food exposed. Indoor bait and traps may be successful for mice—set them where you find droppings. For serious infestations, ALWAYS contact an expert pest controller. NEVER handle a dead rat or mouse—there is serious risk of infection from fleas/lice.

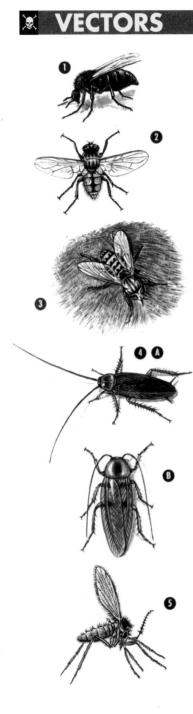

1▶African blackfly *Simulium damnosun* Found near running water in parts of Africa, Central and South America. Will crawl up trouser legs and shirtsleeves and burrow through any possible opening in clothing in their search for blood. Bites produce bloody red spot that will itch for days. Vectors of microscopic tissue worm which causes river blindness (onchocerciasis) which affects more than 20 million people. Length: up to 5 mm (1/4 in). Control: use fine mesh nets. Wear chemically-treated clothing. If bitten, seek urgent medical attention.

2▶Housefly *Musca domestica* Danger comes from their feeding on faeces and human food—they digest by regurgitating a 'vomit drop' of previous meals on food, a residue of which remains to contaminate the food. Known carrier of 40 serious diseases including anthrax, typhoid, cholera, amoebic dysentery, intestinal worms. Length: up to 5 mm (1/4 in). Bluebottles (blowflies) pose similar risks. Control: Cover food and waste containers. Use door and window screens.

3▶Horsefly *Tabanus* species Large brownish or blackish with clear or smoky wings. Some 2500 species are to be found worldwide. The bites can penetrate leather and are very painful. Can transmit anthrax, rabbit fever (tularaemia), worm diseases and forms of sleeping sickness. Length: 1–2.5 cm (1/2–1 in). Control: Destroy breeding sites. Use door and window screens.

4▶Cockroach *Blattidae* Nocturnal, red-brown, well adapted for running, with a flattened shape that allows them to penetrate cracks and crevices. These scavengers are found worldwide in warm places—drains, sewers, kitchens. Their bodies and faeces can transmit a range of diseases including salmonella, typhoid, dysentery, polio and worm infestations. Many asthma sufferers are allergic to dust from their discarded skins. The American cockroach (A) is 5 cm (2 in) long and prefers humid conditions; the smallest and palest household cockroach is the brown-banded cockroach (B) which lays eggs on vertical surfaces such as curtains and gets into TVs and computers, causing damage. Control: Poisons soon become ineffective as roaches build up a resistance. Traps can help. Seek specialist help for serious infestations.

5►Sandfly *Phlebotomus* species Chiefly tropical—common in parts of Asia, Africa, Central and South America and Mediterranean areas. Bloodsucking adults with long slender legs give a painful bite that can transmit sandfly fever, leishmaniasis—an often malignant disease with various forms affecting about 12 million people. Control: Insecticide spray and repellent-treated screens. Seek local specialist advice.

6►Mosquito *Anopheles* species Of all insects, these two-winged flies of the tropics, sub-tropics and warm temperate regions, cause man the most illness, economic loss and discomfort. Carriers of malaria, yellow fever and worm diseases such as elephantiasis. Control: Screens and fumigants. If visiting a risk area, seek medical advice in advance.

7►Human flea *Pulex irritans* One of 2000 different kinds of flea. When vibrations/warmth signal approach of potential host, a flea can leap several times its own length to feed. Most fleas are host specific, but will sample other hosts if the opportunity arises, making them dangerous vectors of some diseases and parasites.

8►Human lice *Pediculus humanus* Sucking lice are living hypodermic syringes, transmitting diseases from infected hosts to healthy ones. A louse feeds almost hourly, making a new puncture and causing a new itch each time. Body lice (A) usually live on underclothing, head lice (B) cement their eggs ('nits') onto the host's hair. A single female may lay 300 eggs in a month. Control: Wash clothes regularly. Use special shampoo. Check and recheck, treat again as necessary.

9►Pubic lice (Crabs) *Phthirus pubis* Usually live in public hair, in the armpits, sometimes in other body hair including the eyebrows. They are an irritant, but not a vector.

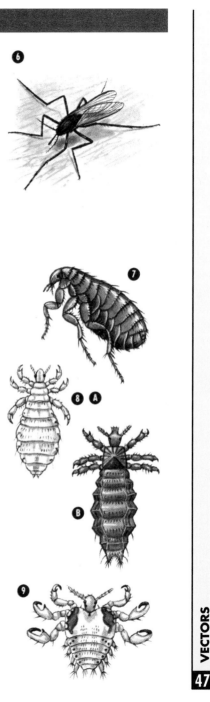

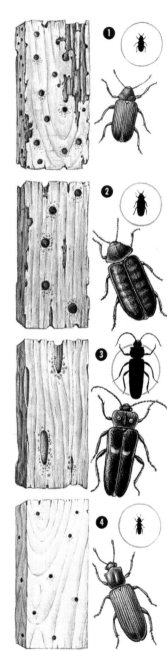

1▶Woodworm (Common furniture beetle) *Anobium punctatum* Common name for a small flying beetle or its larvae, which bore through wood. The exit holes are 1.5–2 mm (1/16 in) across, surrounded by small lemon-shaped pellets. Fresh clean holes and dust (in warmer weather) indicate current activity. See also SAFETY FIRST: **Rot/infestation.**

2▶Deathwatch beetle *Xestobium rufovillosum* Found in Europe and parts of the US. Larvae tunnel into dead or partly decayed wood, structural timbers and furniture. By the time they are detected, by presence of 3 mm (1/8 in) exit holes with coarse rounded pellets or clicking sound adults make by banging their heads against the wood, serious damage may already have been done. Not common.

3▶House longhorn (Old house borer) *Hylotrupes bajulus* A very destructive beetle found worldwide in warm conditions, named after its two long antennae. The larvae bore through seasoned softwoods. It is claimed that their activity is audible. Oval exit holes, 10×6 mm (3/8×1/4 in) are surrounded by a cluster of fairly large, compact wood-dust pellets. Larval stage lasts 11 years, allowing time for considerable damage.

4▶Powder post beetle (Auger beetle/Shot hole borer) *Lyctidae.* Found worldwide, so called because of the dry, powdery dust extruded from burrows when adults emerge. Exit holes circular, 1.5 mm (1/16in). Tends to attack seasoned or dry timber. New timber may be affected.

5▶Termite *Reticulitermes* species Typical of the numerous species found worldwide in warmer regions. Some species cause widespread damge to homes, outbuildings, fences and posts and are viewed as a serious problem in (at least) the US, Africa and Australia—although at least two species have become established in parts of Europe, particularly around the Mediterranean. Most species have the same or similar members — *Reticulermes* are typical with soft pale-bodied workers (A), which process the wood. Other members of the colony usually only eat 'secondhand'.

WOOD-BORING INSECTS

The soldiers (B) may have very large jaws (some species have smaller jaws and a horn-like extension on top of the head). Either type is capable of exuding a repellant, possibly caustic, secretion to deter attackers. Winged adults (C) have darker bodies and swarm in great numbers in the mating season. They fly off to start new colonies (often only over a short distance). When they land, their wings become detached but may be seen glistening on the ground.

Depending on species and the size of the colony, they may nest in dry, damp or rotten wood or build large galleries underground (sometimes with large 'ventilation' tubes above ground). Some, reliant on moisture and darkness, may construct easily-visible extension tubes as tunnels to reach wood above ground without having to expose themselves to sunlight.

Termite infestation should be taken very seriously. Treatment should be done by specialists, and may involve spraying of timbers and soil, digging in and destroying nests and replacing timbers.

6▶Carpenter ant *Camponotus* species Large brown/block ants found in the US, continental Europe and Asia. Live in wood and chew out burrows. Easily mistaken for termites. Reproductive winged members of the colony leave in a swarm to form new colonies. Piles of broken-off wings a clue to their presence — other indications: slit-holes in wood with faeces resembling sawdust outside. They leave nests to scavenge for food. Partial to sugar.

7▶Carpenter bee *Xylocopa virginica* A tunnelling bee, it resembles a very large bumblebee. It makes a single large hole, 2 cm (3/4 in) in diameter, turns through 90° and excavates extensive galleries (which run with the grain of the wood), in which to lay eggs. Length: 2.5 cm (1 in).

READING THE SIGNS

We are surrounded by symbols which try to instruct us or alert us to danger. They are usually accompanied by an appropriate message or explanation of a specific hazard.

In general, safety signs fall into four categories. A blue sign indicates that the message is mandatory and must be obeyed. Green symbols offer instructional guidance on safety and safeness. A yellow and black triangle is an alert to danger—a warning of a slip hazard perhaps. A red symbol with a diagonal line through it means 'Oh no you don't'!

Most of the following examples can vary in meaning—always read the accompanying message. Before you read the explanatory notes, see how many you understand.

WORK HAZARDS

Workers, visitors and even the general public may be at severe risk.

1▶ An exclamation mark is a general hazard warning. Read the accompanying message. **2–9▶** All warnings that protective gear is required must be obeyed, or preferably exceeded. There may be a risk of permanent physical damage. **10▶** Wear a safety harness. This may apply on scaffolding, on a rig, on the framework of a tall unfinished building. Don't take chances! **11▶** Now wash your hands. Sometimes seen on the exit door from a lavatory—but may indicate that you should wash toxic/harmful substances from your hands before going to the lavatory or eating. **12▶** Keep locked. May restrict access for security or safety reasons. **13 & 14▶** If your machine has a fixed or removable guard, use it! **15▶** Sound your horn. Forklift and other vehicle drivers should alert pedestrians and other drivers of their presence. **16▶** Stock correctly. Heavy boxes or crates may come crashing down on you. There may also be a danger of goods lower down in the 'pile' being crushed.

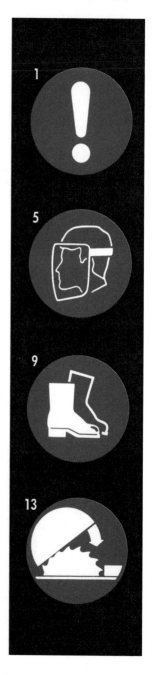

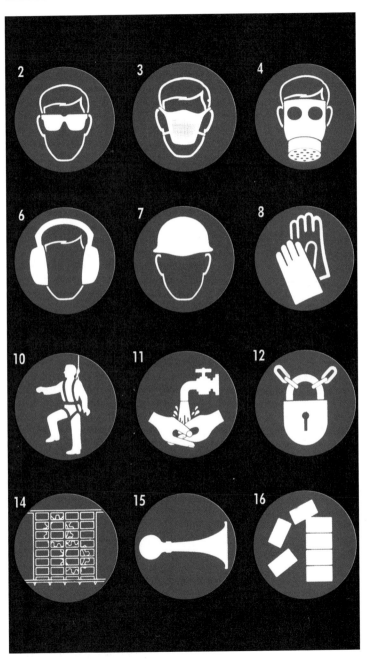

READING THE SIGNS

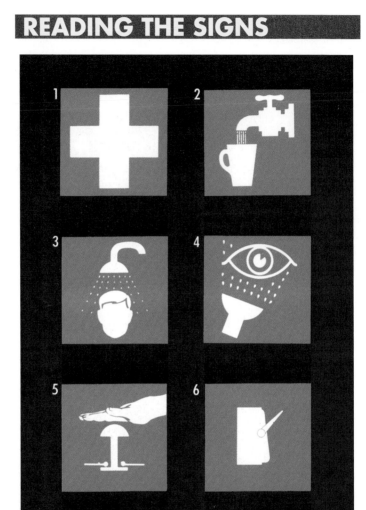

WORK GUIDANCE

You MUST know the dangers of your job — and how to deal quickly with accidents.

1► First-aid symbol. Should be clearly displayed. **2►** This is drinking water. If cancelled by a diagonal stroke or a cross, the meaning is reversed — do NOT drink. **3►** Emergency showers — or simply showers, if a full wash is required after working with hazardous materials. **4►** Emergency eyebath/eye-washing station. **5►** Emergency stop. Hit the button! **6►** This is the main switch. It will deactivate the machine/a group of machines/a whole department/the whole workplace.

All green symbols offer information and/or guidance and are there for your safety. They are often combined with directional arrows, telling you how to reach a first-aid station or emergency showers for example. Fire exits and fire escape routes are usually shown in green.

READING THE SIGNS

480

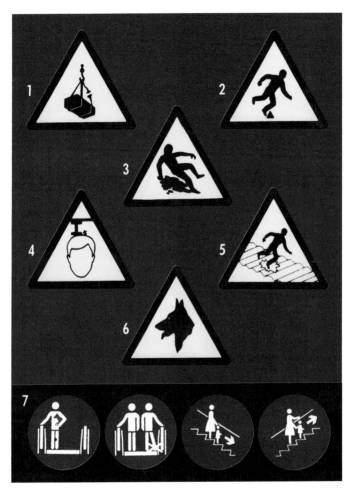

PHYSICAL RISKS

These symbols are aimed primarily at the workforce, but visitors and members of the public are also required to take note.

1▶ Beware! Objects or heavy loads may come crashing down on you. **2▶** Trip hazard. This indicates that there is a change in ground level, an uneven surface or there is debris lying about. **3▶** Slip hazard. Ground or flooring may be greasy or wet. All people are at risk unless wearing non-slip footwear. **4▶** Overhead danger. There are sharp projections or obstacles which may cause severe injury. **5▶** Roof covering is unsafe — do NOT attempt to walk on it without crawlboards. Risks may be very severe. **6▶** Guard dogs on patrol. They don't ask questions! **7▶** Escalator signs to remind you to carry dogs (and small children!), not to stand with your foot near the edge of the step and to face the direction of travel. Even if YOU feel safe on escalators, there is always the danger that someone behind or in front of you may fall and take you with them.

READING THE SIGNS

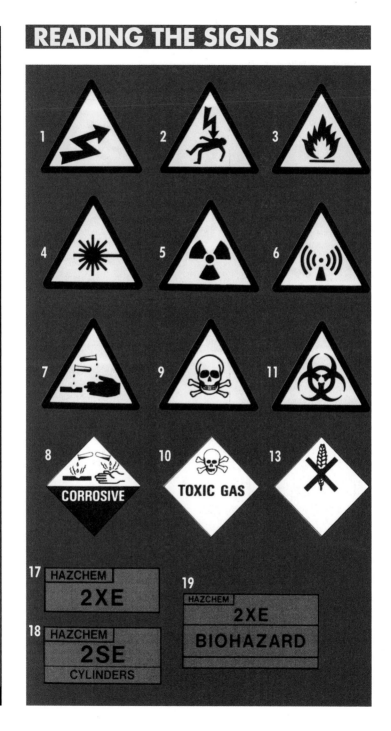

READING THE SIGNS

SUBSTANCE HAZARDS

Risks may be very serious indeed. Usually, accompanying warnings spell out the dangers and the emergency action required. These signs may indicate that you should not even be in the vicinity of the substance, or may instruct you to wear proper protective gear.

1▶ There is a risk of electrocution—possibly involving very high voltages above normal domestic or three-phase power. **2▶** Voltages may be high enough to arc across a gap—even getting close could mean risking death. **3▶** There is a severe risk of fire—perhaps from a flammable substance or a heat-producing process. **4▶** Lasers in use. Degrees of danger usually indicated, from class 1 (low) to class 4 (high). The higher the class, the greater the risk to eyes, skin and tissues. **5▶** Radioactive hazard. Risks may be slight (class 1) or severe (class 3). Read the warning which will indicate the type of radiation—gamma rays or x-rays, perhaps, which are ionizing radiation. **6▶** Non-ionizing radiation. Risks are low but exposure should be minimized. **7 & 8▶** Corrosive substance likely to burn skin and tissues. There may be other qualifying notices. **9▶** A skull and crossbones is another multipurpose sign, often indicating toxic dangers, as in **10. 11▶** One of the more recent signs—biohazard. These may be viral, fungal, bacteriological, disease or genetic hazards. **12▶** Biohazards are usually explained more fully. **13▶** Harmful substance. Keep away from foodstuffs. **14▶** Usually accompanied by the words harmful or irritant. Common on domestic products. **15 & 16▶** Intended for the workplace, but serve as a reminder to all that when CO_2 or halon fire extinguishers have been discharged, the area MUST be evacuated and ventilated. **17–20▶** These are the most common variations of hazchem (hazardous chemical) warning signs. The codes indicate the measures the fire brigade must take to deal with a fire involving potentially dangerous substances (see WORK & PLAY). These symbols are commonly seen on haulage vehicles.

READING THE SIGNS

FIRE

1▶ A typical example of a fire drill notice—intended for the workplace, but could be adapted for the home. **2▶** Escape door or emergency exit this way. There may be colour-coded arrows to follow if the route is complicated. **3▶** The tick indicates an assembly or muster point, where a personnel check can be made to ensure that a building has been fully evacuated. **4, 5 & 6▶** Self-explanatory because of the added wording. Now you know where the extinguisher and hose are, do you know how to use them? **7▶** Fire hazard. **8▶** Risk of explosion. **9 & 10▶** Obvious! No smoking. No naked flames. If you see these, DON'T ignore them. **11▶** If a fire breaks out here, DON'T use water or a water extinguisher. **12▶** Extinguishers are colour-coded—use the right type for each fire (see FIRE!).

GENERAL WARNING

13▶ Any message preceded by an exclamation mark means: Read this ... you are at risk. **14▶** Standard symbols to indicate: Keep dry. Fragile (treat as glass). This way up. There are hundreds of variations on these worldwide. **15▶** Litter may be a nuisance, but it may also be a severe hazard or a fire risk. **16▶** Smile! You're on closed-circuit television (CCTV)! **17▶** NOT drinking water. **18▶** No pedestrians. Can also mean 'keep out', which might be for your safety. **19▶** No crash helmets! Take it off before entering a building/area, especially where money is handled—or you could be mistaken for an armed robber. **20▶** No dogs. **21▶** No comment!

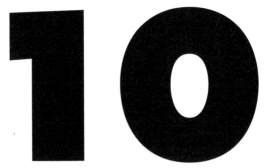

10

Who knows where terrorists may strike next? Everyone is a potential target, now that innocent 'bystanders' are used as pawns in attempts to apply political pressure. High-risk targets must toughen up their security.

Terrorism

TERRORIST THREAT Are you at risk? • Prime targets

BOMBS Detection • Save lives • If you receive a bomb threat • Hoax or real? • Letter bombs • Vehicle bombs • Expert searches

HIJACK/KIDNAP Hostage! • Rescue • Kidnap! • Escape • Ransom • Tied up/gagged

UNDER FIRE Hold-ups and robberies • Holding an attacker • Post-event stress • Body search

RISK LIMITATION Preventing intrusion • Staff training • Alarms • Evacuation • Bomb search • High-security lifestyle

BULLET/BLASTPROOFING Glass • Clothing

Terrorism is not new. The destruction of property, murder, assassination, kidnap and hostage-taking have long been the tools of those who believe that violence is a means of achieving political and criminal goals. What makes modern terrorists distinct from their historical predecessors is the technology at their disposal and the greater range of opportunities for terrorism which modern life provides—aeroplane hijacking being just one example. More value is now placed on the lives of ordinary people and it is they—not just the rich and powerful—who have become targets.

The global nature of media coverage and the publicity it can give has also fuelled the use of terrorism. There is no doubt also that some nations finance and support the activities of terrorists in other countries with which they have political differences.

The more innocent the victim, the better for terrorism. The reality is that we are ALL potential targets and that we are ALL vulnerable.

How terrorists work

Injury, torture and murder have been tools of terrorists both to destabilize a community through fear and to deter opposition—whether the perpetrators are acting against an established government or in the name of a political cause. Bomb attacks and assassination, however, tend to bring the greatest publicity. They draw attention to a cause and seek to weaken authority. By taking hostages, terrorists may bargain for political concessions or the release from prison of fellow terrorists and sympathizers. Bank raids, kidnap and blackmail may be employed to raise funds to finance their activities.

Not all terrorists are motivated by political aims, the same tactics may be used for direct financial gain or personal revenge. Terrorism has also been adopted by certain 'extremist' groups, who feel a need to use violence (as opposed to peaceful demonstration and political pressure) to further their cause. But high on the list of motivation remains racial or religious hatred and intolerance.

The threat of terrorism is no longer limited by distance. Terrorists, if unknown, can use any form of transport and move as easily as any innocent traveller. In recent years, terrorist attacks have included shootings, hijacks and bombings. Bombs mailed as packages or hidden in cars are probably the most common. They may be detonated by movement, a timer mechanism or by remote control.

Some countries, and particular cities, have a high incidence of terrorist activity. Although terrorists can strike anywhere and at any time, it is wise to be aware of world affairs and in which countries there is likely to be a terrorist threat to YOU. If you are going on holiday to a country which has a history of political instability, be aware of how this could affect you and if necessary alter your plans and defer the trip to a safer time.

Military personnel and government employees are always at risk, but if you are connected to a group/company/organization whose activities may attract terrorist attention—YOU could be a target. If you only live or work near a likely target you could also be at risk.

IF YOU BELIEVE YOU ARE A POTENTIAL TERRORIST TARGET, SEEK ADVICE FROM THE POLICE

Even if you, your workplace or your home are unlikely terrorist targets we are all vulnerable to attacks in public places. In many cases terrorist attacks have been directed at military and government targets, but things have changed. Department stores, airports, railway stations and underground stations—almost any busy, public place—could now be 'high risk'. Body and bag searches are common during known periods of terrorist activity.

PRIME TARGETS

Although attacks on the general public are fairly indiscriminate, statistics help us identify and protect certain groups of people who are always at risk from some form of terrorism. It's not possible to account for the effects of sudden political change or war—either of which could suddenly tip the balance against you.

HIGHEST RISK:

- Government officials
- Industrialists
- Service chiefs
- Celebrities

LOWER RISK:

- Civil servants
- Embassy officials
- Members of armed forces
- Families of all the above

If you are a prime target

○ Follow security advice (see SECURITY) regarding home, workplace and telephone.
○ DON'T advertise your presence or your movements—even a wedding announcement could tell an assassin where you are going to be at a certain time.
○ Book hotel accommodation under an assumed name.

TERRORISM TERRORIST THREAT

489

- DON'T stick to a pattern. Vary transport, times and routes to work. Avoid roadworks and congested streets. Change vehicles during longer journeys.
- Make a habit of looking out of the window before leaving a building. Watch out for people loitering on foot or sitting in parked vehicles.
- Glance over your shoulder frequently—but unobtrusively, you DON'T want to attract attention—and use reflections in shop windows to see if you are being followed.
- Be 'friendly' with neighbours. You may need them to let you know if anyone is asking questions or taking an unusual interest in you—but DON'T divulge your business or movements to them.
- Telephone ahead when making journeys and telephone base when you arrive.
- Arrange a coded phrase to be used if you are in danger or under duress and need help. Choose an arrangement of words that can be fitted into normal conversation. It can be anything which sounds innocent—'I didn't bring a raincoat', 'The roads were clear', 'Sorry. This is a really bad line'.

BOMBS

The most feared and most commonly used weapon in the terrorist's arsenal is the bomb. Modern technology serves the terrorist well with ever-more reliable and complex detonation systems and increasingly-powerful explosive substances. A bomb does not care who it kills—this knowledge and the fear that it creates in innocent people, is why the bomb is the ultimate weapon of terror.

Bomb detection

Dogs, with their very powerful sense of smell, have been trained to detect the presence of most explosives. Police forces and anti-terrorist organizations around the world use sniffer dogs. There are some types of explosive, particularly plastic ones, which are odourless and very difficult to detect.

Metal detectors can be used to scan packages and luggage—metal is an essential part of bomb-making—but it is also found in many everyday items. The most effective application is screening for parcel and mail bombs. Hand-held detectors are suitable for small offices and mailrooms. For rapid scanning of large numbers of items a larger fixed apparatus will be needed. Some are capable of handling 9000 packages an hour and taking parcels up to 5 cm (2 in) thick.

X-rays can be used to 'see' into packages and luggage. X-ray machines are the most efficient way of screening for bombs—they are used at all international airports. The operator must be expertly trained to identify suspect devices quickly, since thousands of pieces of hand luggage must be scanned per day.

Save lives!

Few terrorists set out on suicide missions. Bombs, the most common form of attack, are usually planted and detonated once the bomber has escaped. By being constantly aware and keeping an eye out for abandoned/suspect cases or packages (especially on transport systems and in public places), you may save yourself and others from injury or death. If you see a suspicious package or container:

◐ DO NOT touch it—moving it may trigger an explosion.
◐ DO NOT shout 'Bomb'. This could cause dangerous panic.
◐ Move away and tell others to do the same—but tell them quietly and calmly.
◐ If you are on a moving underground/subway train, do NOT activate the alarm between stations. In some countries the train will stop immediately, which could jolt the package. Usually the emergency signal only alerts the driver, who will stop the train at the next station.
◐ On trains with doors between carriages, tell other people to move away and then alert the guard/conductor.
◐ If you see someone leave a bag or package behind, call after them. If they deny any association or just run off, treat the object as suspect.
◐ Make a mental note of the appearance of the person. Your description of them may prove vital later (see SELF-DEFENCE: **Being a witness**).
◐ If you have a camera with you, take a photograph—even a back view may be helpful in identification.
◐ Inform a police officer or other responsible official or telephone a police emergency number.

If you receive a bomb threat

Most threats or warnings are given by telephone. If you take such a call it is VITAL to remember all that is said accurately. Write points down as they occur. Try to get as much information as possible. This may be possible by direct questioning or by pretending that you think the call is a practical joke—more details may be given by the caller to persuade you.

Telling the caller that children are at risk or otherwise appealing to his/her humanity—'surely you don't want to kill innocent children'—may gain time or information on how to prevent an explosion. Give details of the call to the police or a security officer IMMEDIATELY. If facilities are available, threat calls should be recorded.

Information needed

ALL information about the caller and about the call is VITAL for anti-terrorist and police action. A check list should be available by EVERY telephone in the workplace—whoever takes the call may be frightened or flustered. It will help to ensure that you get the following information:

CALL CHECK LIST

1 The message (word-for-word if possible) including any codewords
2 Gender of caller: male/female
3 Age group of caller: child/adolescent/adult/elderly
4 Speech style of caller: rapid, slow, calm, excited, accented, unusual or frequent use of words
5 Background noises
6 Time
7 Date
8 Your name

Hoax or real?

A hoax threat (without a bomb being placed) causes considerable disruption. It is illegal, dangerous and STUPID. Fear and panic are the obvious results. Knock-on effects may disrupt the lives of thousands of innocent people, through chaos on public transport or loss of working time for example. There is always the danger that a hoax may be part of a pattern to distract attention from a real bomb placed elsewhere—or a series of hoaxes may be designed to encourage a real warning to be ignored. There are many more hoaxes than genuine threats, but no threat can be dismissed until it has been very carefully assessed. People at risk MUST, of course, be protected. Always ask yourself:

❍ Does the call fit into a pattern of previous terrorist attack?
❍ Is it likely that you or your workplace could be at risk from terrorist attack?
❍ Is there anything in the message which could identify it as a genuine threat, including details of the type of bomb?

This last question highlights why you MUST make an accurate record of the conversation. Certain phrases, even specific code-words, are known to be used by terrorist groups in their warnings. The caller may have privileged knowledge of the interior layout of a building—and say so.

Letter bombs

A letter bomb is a clumsy terrorist weapon. Although the likelihood of the bomber being identified is reduced to a minimum, most intended victims don't open their own mail. Letter bombs cannot easily be directed at exact targets. All mailroom staff, secretaries, family members and others who handle correspondence for a potential target (individual or organization) should understand the risks and relevant procedures. Instructions on screening should be clearly displayed wherever packages are opened.

There is no easy way to recognize a letter bomb—a terrorist will probably make it look as ordinary as possible—though hoax devices are sometimes made to look suspicious in the hope of causing maximum panic.

WATCH OUT!

LOOK! READ! SMELL! THINK!

Beware of any unexpected mail, especially if no sender's name and address is given. If not addressed to you, find out if the addressee is expecting a package. Telephone the sender and check whether the package is genuine. Look out for:

■ Padded mailing bags—one of the most widely used and ideal ways of disguising an explosive device
■ Packages that are unexpectedly heavy for their size
■ Grease marks/sweating on packages—it may be caused by old or unstable explosive substances
■ A smell similar to almonds or marzipan
■ Wire or tin foil protruding from package
■ Packages that feel spongy or seem unbalanced in weight
■ Excessive sealing tape—it could be there to hold down a spring detonator
■ Inaccurate addressing or sub-standard typing
■ Excessive postage paid
■ Packages containing other sealed containers

REMEMBER

Packages may include a series of trigger devices. They do not detonate the explosive immediately, but trigger a time delay in the hope of getting the bomb through the mailroom to its actual target.

If it might be a bomb

◑ LEAVE IT undisturbed.

◑ DON'T cover it or place it in water.

◑ Inform police/security staff.

◑ WRITE DOWN all details of packages which have come through the mail—postmark, stamps, how addressed, sender's name and address, condition. It may help later. If the device detonates, all information may be lost.

◑ Clear the room/area.

◑ Open windows. If you cannot, tell colleagues and passers-by to stay well clear of them in case of flying glass.

◑ Lock the door so that no one can walk in.

Do NOT open or attempt to defuse. Leave it to the experts!

Vehicle bombs

Terrorists may turn a vehicle itself into a traveling bomb to be driven into or at a target. Alternatively, a car may be parked near a target or left ready to detonate by timer/remote control. Defence against this form of attack can only rely on refusing admittance of unauthorized vehicles to high-risk areas, preventing parking and constant vigilance by potential targets and people who protect them.

Bombs may be fixed to the target's own vehicle. To do this, terrorists must obtain access. Constant surveillance and secure garaging of vehicles when not being driven deny terrorists access (unless they have infiltrated the target's security). This may be possible for VIPs or the wealthy, but in cities many vehicles have to be parked in the open and left unattended.

The target or the driver of the vehicle MUST have a thorough knowledge of it to be able to spot any changes. You need an expert knowledge of what the vehicle looks like inside and out, underneath, in the boot and under the bonnet. Always lock your vehicle, with windows and sun roof closed. Leave things arranged inside so that you can see if they have been disturbed in any way. Better still, keep the boot, glove compartment—all obvious hiding places—CLEAR.

Hubcaps are a favourite bomb concealment location—the same make of hubcap is bought, fitted with a bomb and substituted. Discreetly mark your hubcaps and wheels—a tiny dot or scratch will serve the purpose. Line the marks up—if they are out of alignment they must have been removed and replaced by someone.

Fill up keyholes with soft wax. This will do no harm to the lock, but will indicate if anyone has inserted a key. Thin strips of clear tape across door, bonnet and boot openings will leave tell-tale breaks if they are opened.

Before you get into your vehicle ALWAYS carry out a search. A detailed search takes hours, but if you are a high-risk target, a regular search must be done every time the vehicle has been left unattended. The techniques given above save search time—a search is good practice for any vehicle user (especially if you park your car near any potential target) during a period of terrorist activity.

VEHICLE SEARCH

- Walk round the vehicle to inspect it, WITHOUT touching it
- Look for objects under the vehicle and under the wheel arches—crouch down if necessary
- Look for any signs of adhesive/insulating tape, pins, wire, or other suspicious items
- Look for any signs of entry. Are the tape seals still intact on the door, bonnet and boot?
- Look for fingerprints or smudges on the bodywork. Keep your vehicle clean and polished so that signs of tampering show up.
- Check out the interior through the windows. Is everything inside the vehicle EXACTLY as you left it?
- Look behind bumpers, inside spoilers and around the underside of bodywork—these are common bomb concealment areas
- Check interior again after you have opened the door

REMEMBER

A visual check of the underside of your vehicle—places like wheel arches, spoilers and the bodywork—can be made easier and quicker with a simple tool made from a mirror attached to a stick. In some countries such tools are sold by security equipment suppliers. A flashlight can be fitted to give better visibility.

WARNING

If ANYTHING makes you suspicious, do NOT touch any part of the vehicle. EVACUATE the area and call the police.

Expert searches

Detailed searches should ONLY be undertaken by trained specialists. They involve thorough inspections of all parts of a vehicle and must be carried out whenever the vehicle has been serviced, repaired or left unattended.

In addition to the checks already listed, the underside of the vehicle must be inspected thoroughly. ALL parts of the engine and exhaust system must be inspected by someone who is completely familiar with the particular make and model of vehicle. A device may be heat sensitive and will detonate when the vehicle warms up. The boot, spare wheel and tool box should also be emptied/stripped down and searched. The electrical system must be thoroughly checked—bombs are often linked up to the wiring of a vehicle so that they explode when the ignition key is turned or when a particular electrical circuit is operated. ALL wires must be identified.

Once the mechanical/external checks have been made and those areas have been cleared, a thorough internal search must be made. All removable parts of the vehicle interior—upholstery, carpets and seats—must be taken out and searched. Door panels, dashboard and all linings must be checked for tampering—where necessary removed. All storage compartments must be emptied and searched.

REMEMBER

When a vehicle has been given a detailed search and cleared, make future quick checks easier by using anti-tampering devices (see VEHICLE SEARCH panel). ALWAYS keep the vehicle in a secure area. High-risk target vehicles should NEVER be left unattended.

HIJACK/KIDNAP

 Terrorists have shown themselves to have little or no regard for human life—this is graphically illustrated by hijacks and kidnappings of recent years. By kidnapping someone or by taking hostages, in a hijack, the terrorist gains or hopes to gain the advantage or publicity for a 'cause' by putting people's lives at risk.

Hostage!

The terrorists AND hostages are likely to be in a highly-emotional state in a hijack situation. If you are innocently caught up in a hijack, you MUST remain as calm as possible. Everyone is at great risk in such a stressful situation where emotions are at their most volatile.

❍ DON'T be aggressive.
❍ DON'T make yourself stand out. By drawing attention to yourself, you increase the risk of being 'singled out'.
❍ Hide or dispose of documents or other items which might increase hostility towards you.

○ Cooperate with terrorists in preparing meals, tending the injured and generally looking after others—including the terrorists themselves.

○ Avoid eye-to-eye contact with the terrorists.

○ Try to reassure any hostages showing signs of strain—make allowances for behaviour caused by stress.

○ If a rapport can be built up with the terrorists, ask if conditions can be improved for everyone—blankets to counter cold, for example.

○ Drink plenty of water—you need over a litre (two pints) a day. Food intake can be reduced dramatically but water is essential.

○ Avoid alcohol—it dehydrates your body AND you need to keep your wits about you.

○ Be prepared for difficulties with sanitation, particularly in an aircraft/train/bus, where toilets are not designed for extended use.

○ If you understand the 'politics' of the situation, do NOT imagine that you could dissuade the terrorists from their cause. It would be safer to agree with them.

○ Keep your mind occupied.

○ Talk to your captors about personal matters, show them photographs of your children. The more they begin to see you as an individual with a life and relatives or a family of your own, the less you may represent a 'victim'. The more they identify with you personally, the more difficult it becomes for them to harm you.

Rescue

If a rescue attempt is made, get to the floor and use your arms to protect your head (see **Under fire**). Do NOT move until you are told by your rescuers that it is safe to do so. DON'T try to be a hero—rescuers will not have time to make positive identification before shooting to kill. Their priority is minimum casualties—DON'T risk accidental injury. Follow instructions exactly and make an orderly exit as quickly as possible—the aircraft or building may have been wired with explosives. Vehicle/aircraft fuel may be a fire risk.

Kidnap!

Kidnap can only be guarded against by careful and comprehensive security measures. Avoid struggling—you may be injured—but if you can attract attention, witnesses can raise the alarm and give valuable information to the police. Once abducted, stay as calm as possible. Do NOT get aggressive.

Try to work out where you are being taken, look out for identifiable landmarks. If blindfolded or in a windowless vehicle, listen for sounds that might tell you where you are. You can also try to work out your route from the movement of the vehicle. It might help you keep your mind occupied. At best, if your sense of direction is accurate, it might help you if you get a chance to escape.

REMEMBER

Kidnap situations vary widely from incarceration in a windowless darkened cell, to almost normal living conditions. It is impossible to predict/prepare for the experience. One thing to remember at all times is that as a hostage you are more valuable ALIVE.

Getting to know your kidnappers may help you to assess how to behave, but BE COOPERATIVE. The more they can relate to you, the better. You may receive better treatment.

Escape

Escape may be worth considering, but should be attempted ONLY if you are extremely confident of success—though inventing plans may be a way of keeping up morale. If you are denied reading material and other ways of passing the time, you MUST keep yourself mentally alert and your mind off your situation as much as possible. Recite poems or songs, recall movies scene-by-scene, set yourself sums and puzzles, invent stories and commit them to memory—whatever works.

WARNING

Plotting an escape MAY make you feel better, but think again. Are the kidnappers armed? How many of them are there? Where are you? Where will you go? In some cases of kidnap, victims may actually be taken to other countries! If you don't seem to be at great risk physically, it may be safer to stay put.

Ransom

ALWAYS inform the police if you are asked to pay a ransom (or subjected to any form of blackmail). Payment will not necessarily obtain safe release of a kidnap victim—kidnappers could go on asking for more money or inflict injury anyway. Get as much information from the kidnappers as possible. If contact is made by telephone, a check list similar to the type you need to take details of a bomb warning would be useful (see **If you receive a bomb threat**). Keep any written communications clean for forensic analysis by the police.

In many kidnap/hostage situations you might find yourself tied up and gagged. Being tied up, even for a short time, can be very painful and may even cause serious injury by cutting off blood circulation. To avoid injury and make the restraints less effective, do the following:

- 'Relax' as you are being tied to a chair. Slump down, keeping the small of your back away from the chair. When you straighten up, there may be enough slack in your bonds for you to escape.
- Expand your chest and keep it expanded until the tying is finished.
- Try to keep your knees or ankles slightly apart.
- Keep your hands slightly apart—but don't overdo it—you need just enough to wriggle out of your bonds.
- If you are on the ground with your hands tied behind you, try to pass your hands under your rear. With your hands in front you can then use your teeth to undo knots.
- If gagged, try to catch the gag in your teeth so it is not forced all the way back into your mouth.
- Rub a blindfold against a shoulder or any convenient edge to push it UP— don't try to lower it, your nose will get in the way.
- Use any sharp edge to wear through ropes.
- If you are tied up with other people, teamwork can achieve more—try to undo a fellow captive's knots with your teeth or fingers.

Instructions for the handover of cash may be set up by a series of telephone calls or letters—both are ways for the police to trace the kidnappers. Kidnappers may invent highly complex procedures for the transfer of ransom payments to reduce the risk of capture. Recent schemes have involved a complicated network of bank accounts and bank self-service cash machines. However, banks will usually be ready to cooperate with the police.

Although the police must be involved, it is often stated very early on by kidnappers that you must NOT involve them. At all costs, the press must be prevented from hearing of the kidnap—media coverage will give the game away.

UNDER FIRE

Firearms, particularly automatic weapons and hand guns, are commonly used by terrorists. They can only be used at relatively-close range, and therefore increase the risk of an attacker being captured or injured by returned fire. Nevertheless, any precautions against terrorism MUST include an awareness about how to behave when threatened by an armed person and what to do if you are shot at.

> # REMEMBER
> The romantic impression given about guns and their effects in the movies has very little bearing on reality. DON'T think about being a hero if you are confronted with a gun, and NEVER use one unless you have had training in gun handling—you are likely to do yourself more harm!

Shots being fired without warning can quite often go unnoticed—you may not evens hear a bang! Unless you are close to it, the sound of a shot can be muffled in a crowd and difficult to identify. You may not SEE the gun or its effects. If you have the slightest suspicion that shots are being fired, drop to the ground IMMEDIATELY and cover your head with your hands—tell anyone near you to do the same. 'Hitting the deck' like this makes you less of a target and also protects you from deadly ricochet and flying debris.

If there is any cover USE IT, but on no account move if an attacker has told you not to. If an attacker has singled you out or you are in range—in an armed robbery, for example—your best defence is to do exactly what you are told to do. An attacker, particularly one who has already wounded or killed someone, is likely to be very highly strung. The slightest movement may be interpreted as a threat. All it takes is the squeeze of a trigger.

Anti-terrorist and security personnel are usually trained to disarm an attacker. Without training, you are likely to put yourself and those around you at more risk by taking on an armed person. Remain as calm, still and close to the ground as possible. This is VITAL if gun fire is returned by police or military personnel. You could easily get caught in crossfire, if you get up and try to run for it. Even if your attacker has been shot and seems to be out of action, do NOT move until you are specifically told to do so.

Hold-ups and robberies

The risk of being present during an armed robbery is surprisingly high in cities—whether the robbers are 'only in it for the money' or terrorists trying to raise funds for their activities. If you work in a bank or with large quantities of cash, that risk is even greater.

Wherever money changes hands or is loaded into vehicles (especially where large sums are involved) the process must

ALWAYS be screened from the public by effective barriers and partitions. Doors through these should ONLY be opened to known staff, when appropriate codes/credentials are given and confirmed. Locks should be operated from the inside, NOT by an external key.

If a staff member is being threatened or is under duress, they should use a code to indicate the situation to their colleagues. It should be easily recognizable and phrased as a request, like: 'Can you ask Johnny to let me through please'. 'Johnny' could be the code for 'I am being threatened with a gun'. 'Let me through please' could be a signal to activate an alarm system.

REMEMBER

DON'T risk your or anyone else's life for the sake of money. Cash can be recovered—life cannot! Tills may have automatic locking/timing devices, which would make handing over cash impossible. Without that provision there is NO alternative but to hand over the cash if faced with a gun.

IN A HOLD-UP

- Sound alarms which alert security staff/police, without signalling to robbers. Alarm triggers should be undetectable, except to trained staff
- Stall for time if you can, without raising suspicion
- DON'T take any direct action against robbers
- DO as you are told
- DON'T try to run away—unless the robbers are unaware of your presence and you can escape safely
- DON'T draw attention to yourself
- DON'T shout or scream
- DON'T argue—either with the robbers or others caught up in the robbery
- DON'T volunteer information
- MAKE a mental note of everything you see and everything that happens—what the robbers look like/voices/behaviour/sequence of events
- As soon as possible telephone the police and report everything to them Write your evidence down as soon as you can, the stress of the occasion often makes exact recall difficult later. Pass it on to the police as soon as possible (see SELF-DEFENCE: Being a witness).

Holding an attacker

In the unlikely event that YOU may have to search a suspect or restrain one, follow this procedure. There is always a possibility that rescuers may not be able to tell (initially) terrorists from hostages, of course, so this could happen to you!

If you see any violent incident do **NOT** enter the building or area where it is happening. Telephone the police at once. If you can still see what is happening, **DON'T HANG UP** when you have given the alarm. Stay on the line so that you can report events to the police as they happen. Make a written record of what you saw as accurate evidence before details are forgotten. Date and sign it before delivering it to the police.

If it is impossible to get to a telephone, observe carefully, make notes and report to the police as soon as possible (see **SELF-DEFENCE: Being a witness**). If you are carrying a camera, take photographs.

POST-EVENT STRESS

Kidnap, hijack, armed robbery and other terrorist action can be deeply traumatic with long-lasting effects. Hostages and hijack victims may be physically and mentally exhausted, sometimes with an irrational sense of guilt. They may sleep badly and have nightmares. Help from professional counsellors should be sought to help cope with these problems. In many cases, victims should be treated for shock (see **HEALTH: Save a life!**).

Body search: Pull off any outer garments or coat and make suspect stand facing a wall. Lean them against it with their arms up, palms flat against it and legs apart. The feet should be far enough from the wall to force the suspect to lean hard against it. Hook your foot round the suspect's ankle—ready to pull them off balance if they try to escape. As you make the body search for concealed weapons/devices, feel clothes for ANY unusual object but be careful of blades and syringes.

❍ Turn suspect's head to one side and open mouth to check whether anything is concealed inside.

❍ Starting at right hand, feel up right arm to armpit, across chest—checking pockets—back along shoulders, down right side of abdomen, around waist and up left-hand side of abdomen, then along left arm.

❍ Start at right foot, up right leg—checking pockets—groin and down left leg to foot.

To secure a suspect: Blindfold, then tie hands behind back, securing them to a fixed object through the legs. Use shoelaces (if nothing else is available) to tie thumbs tightly together behind back.

Once suspect is secured, search through his/her outer garments, and any bags. Reassure them they are at no risk and that they should not struggle.

REMEMBER

This is very unlikely to take place and is obviously not something you could do by yourself. Once an attacker has been restrained or tied up, you must NOT sink to their level by being violent or causing pain. Let the police deal with them.

RISK LIMITATION

Normal home security devices are inadequate if you are combatting terrorism, rather than break-ins or burglary. In the workplace, high security measures are essential (see SECURITY) for companies/organizations at every level of risk.

General security and fire precautions will provide basic anti-terrorist measures but, if premises are a potential target for terrorist attack, more extreme measures are necessary—specific strategies must be devised. Managers who have to balance risk to property against the expense of precautions must first consider the 'cost' of possible damage to people.

Confidence in security procedures will encourage better morale and greater efficiency among staff. They should be introduced to anti-terrorist measures in a positive way—generating paranoia will be counterproductive.

Preventing intrusion

In the office or enclosed industrial site, a multi-layer system to prevent entry and detect would-be intruders, coupled with physical checks, will be effective, provided security guards are vigilant (see SECURITY).

Visitors and clients may find such measures off-putting, so it will help to warn them in advance and remind them that inspections are for their own safety.

Bombs do not need to be carried into a building, they can be fired—as a rocket or mortar bomb—or placed near the outside of the building. Seal letterboxes and fit grilles across vents, flues and other places where objects could be inserted. Seal accessible windows and instal shatterproof glass (or cover with metal grilles). Secure manhole covers and in a very high security situation use gates to bar access via sewers and other service channels.

Set up a perimeter barrier 10 metres (11 yards) or more (if possible) from the building. Do not allow unauthorized vehicles to park within this boundary. Never allow parking close to a high-risk building and erect physical barriers to prevent it. These may take the form of banked flower beds or terraces and need not be obvious barriers. All authorized cars should be locked and a guard posted to ensure they are not tampered with in any way.

Inside the building

The ingenuity of terrorists means that a bomb can be hidden ANYWHERE, but you can make it harder for them. Reduce the risks greatly by not providing:

◑ False ceilings
◑ Air-conditioning vents and pipes
◑ Oversized light fittings
◑ Accessible lavatory cisterns
◑ Any other locations with removable covers

Use furniture that provides minimum concealment—glass topped tables, no shelving, except in locked cupboards. Blinds or shutters are preferable to curtains. Plant pots and troughs provide further places to hide explosive devices. They may be 'attractive', but they are particularly hazardous in entrance lobbies and public access areas.

Enforce rules against untidy desks and work areas—don't allow work to be piled up on them. It may be easy to slip a bomb under a pile of papers or in a document tray. An orderly and tidy working environment will make anything unusual or suspicious more noticeable to staff.

Similarly, scrap materials in industrial premises should not be allowed to accumulate and provide scope for concealment. Boxes, canisters and raw materials should not be stacked or stored in areas to which visitors have access. A device could easily be placed among them.

REMEMBER

When building from scratch, do not exclude terrorism when it comes to considering general security. Avoid architectural features, fittings and equipment that could be used as hiding places. Use sloping ledges inside and out, rather than flat surfaces (for example) and consult a security expert from the outset. Changes on the drawing board will cost much less than later alterations.

In some public buildings, such as department stores, it is virtually impossible to impose more thorough security checks on the public than a bag search on entry. It is VITAL, therefore, that staff and customers are vigilant and look out for suspicious packages or odd/questionable behaviour. Staff should also be well versed in procedures for reporting suspect devices, evacuation and further action.

Damage limitation

Good staff training and well-planned evacuation routes will counter the risk of panic and quickly clear the building. Apply protective film to windows to reduce the risks from flying glass. Fit anti-blast curtains made of tightly-woven polyester with a weighted hem. These will absorb the blast and glass fragments. Fire-fighting equipment should be readily accessible at all locations (see FIRE!).

Staff training

Every company and organization should already have fire drills, staff training—and regular practices! These procedures can be extended to cover the threat of terrorism. The police or terrorism experts can give advice on how to draw up antiterrorist plans to suit individual circumstances.

All staff must be aware of potential risks and the recommended measures that should be taken to combat them—though information that could be useful to terrorists should be disclosed on a 'need to know' basis only. No member of staff should EVER discuss security, the layout of a building or the use of individual areas, with ANY outsider.

Staff should also be made aware of the fact that 'careless talk costs lives'. A conversation between two authorized members of staff in a public place may be overheard by terrorists—even seemingly inconsequential topics of conversation may be useful. If you do not have canteen or refreshment facilities, staff may regularly meet in cafés, restaurants and public houses. When they are together, discussing 'work' may be unavoidable. The news of the impending visit of an important person, for instance, may spread like wildfire.

If the company is not large enough to have its own security officers, a member of staff must be given this responsibility and become the contact with police and other organizations. At least one, preferably two, members of staff should be appointed as

TERRORISM TERRORIST THREAT ■ RISK LIMITATION

bomb-threat officers. They MUST be trained in the recognition of explosive devices. Neither they, nor any other member of staff, must ever take charge of bomb disposal—that is a job for the experts.

All staff members must know any coded warnings (see **Hold-ups and robberies**) and be familiar with the company's evacuation procedures. Each must know their exact responsibilities. In an evacuation, some may be responsible for seeing visitors or customers off the premises whilst others may supervise staff muster points.

> **! WARNING**
>
> The safest way to deal with a bomb or a suspect device is to get as far away from it as possible—as quickly as possible. Unless you have VERY good reason to believe that a bomb is a hoax, evacuate the area.

> **REMEMBER**
>
> Lines of communication must be established for any emergency situation and some means of contacting all staff quickly devised. Public-address systems or internal telephones may be available, but some reserve system should be considered, perhaps loud-hailers.

> **! WARNING**
>
> Walkie-talkies are not suitable for contact in bomb emergencies—electronic detonators are sometimes designed to be activated by radio frequencies. Unless an expert has seen the bomb, you cannot be sure if it is to be detonated by a trigger, by a timer or by remote control.

Staff emergency codes

In shops, hotels, airports, conference and entertainment centres, or anywhere where the public gather in large numbers, it is VITAL to have a means of alerting responsible staff to a terrorist situation quickly WITHOUT causing panic. The best way of communicating is by a public-address system and using a series of codes, devised to cover all possible emergencies. These should be known and understood by members of staff as part of their regular training.

If the emergency can be contained and the public are not at risk, then the rest of the staff should know what to do, without causing general panic. If action is needed, staff can implement emergency drills and be ready to give assistance unhindered.

Emergency codes will vary according to the circumstances, but they could follow the style 'Mr Black report to reception', for example. To the untrained ear this is a normal enough message, but to staff the name 'Mr Black' would mean a bomb threat and 'report to reception', carry out emergency evacuation procedure.

Alarms

All alarm systems should be regularly tested. They need to be sounded during working hours, unannounced, to test evacuation procedures. People in high-risk buildings may be nervous enough without being subjected to alarms, which could mean anything. Establish an alarm language. Three short blasts, pause, three short blasts could mean that the alarm is for testing evacuation procedure only. A continuous alarm could indicate a real fire emergency. A continuous stream of short blasts could indicate that there is a bomb in the building.

Evacuation

There may be no way to tell initially whether a bomb warning is a hoax or not, but you CAN'T take chances—the only course of action is to evacuate. In workplaces, or at school, everyone on the premises should know the evacuation drill. This will be much the same as a fire drill (see FIRE!: **Equipment & drills**).

Where members of the public are present (places such as shops, theatres or sports stadia), it is up to the staff to organize evacuation—calmly and quickly. Providing a building has adequate emergency exists, the public should not be at risk.

Planning an evacuation procedure depends on a building's layout and use. In most countries there are laws governing the number and location of exits, often depending on how many people use the building. If you are responsible for safety in a public building you may be liable to prosecution if regulations are not followed. Equally, if you are just an employee, and you suspect that safety rules are not being maintained, you MUST report this to the relevant authorities—people's lives may depend on it!

If you are evacuating the public from a building in the event of a bomb threat, there are special instructions which you must give to people. Some contradict instructions you may be used to in fire drills.

◗ DON'T leave belongings behind, especially bags. This reduces the number of possible suspect packages that bomb disposal experts have to deal with.

- In a theatre or similar venue with tilting seats, make sure that your seat is lifted so that a search is made easier.
- Once clear of the danger area, report any suspicious object and its location to police or security officer.
- Once out of danger, DON'T leave the scene—if a bomb has been planted the police will need to eliminate from suspicion as many people as possible. If you were in the vicinity, be prepared to give your name and address so that you can be interviewed later.

Bomb search

If a bomb warning has been given and evacuation has taken place, the only people who should re-enter the evacuated area are the police, security officers and the bomb squad. However, the best people to search each area are the people who work in it. Without their knowledge, a search is likely to be time wasting and possibly ineffective. For this reason it is VITAL that once a bomb alert has been given, staff have a quick-search procedure which they can put into action as they evacuate.

There is every possibility that more than one bomb may have been planted—in more than one location. A bomb may be intended to target people who have just evacuated from the area where a warning has been given, or to kill police or anti-terrorist forces who are at the scene to investigate. The aim of a search, therefore, is to gather information over as wide an area as possible—not just in the location given in a warning.

Search procedure

Search strategy must be worked out in advance. Staff or individuals must understand their functions in a search and how to use any relevant equipment.

> ## REMEMBER
> The key to an efficient and fast search is familiarity with the area—if YOU know what belongs where, then you are more likely to spot something out of place! Keep the work area tidy!

The size of an individual's search area depends on the number of available searchers and how familiar they are with the territory. In a complex environment—an oil refinery is a good example—each person can only search the limited area they are familiar with. If you have a work station—where the majority of your daily working time is spent—then this should be your search area.

If you are responsible for a wider working area or a group of staff, work out the minimum number of people necessary to carry out work-station searches. In an office this might mean two or three people searching desktops and floors for suspect devices before or while evacuating. Meanwhile, all other members of staff leave quickly and calmly.

If you are a bomb-threat officer or the person with overall responsibility, it is up to you to search other possible sites such as stairwells, lavatories, reception/meeting rooms. That done you can evacuate the building, taking with you a detailed floor plan for the police to refer to.

Once reunited at the designated muster point, staff must report the findings of their search:

◑ Nothing sighted—state area searched
◑ Possible sighting—give description and location
◑ Definite sighting—give description and location

By the time the police or bomb squad arrive a senior member of staff should be able to account for ALL personnel and visitors—the building should be FULLY evacuated. Give a good assessment of the presence of suspect devices. It should be possible to provisionally rule out some areas as being low-risk and direct the bomb squad to more likely locations.

WARNING

DO NOT re-enter the building/area until the all-clear has been given by the bomb squad. NO muster point for staff should ever be near enough to the building for people to be at risk from flying glass or debris, if the bomb explodes.

SEARCH RULES

■ Any unauthorized person found during a search should be treated with suspicion. The police will want to talk to them
■ Search with your eyes ONLY—a search is visual NOT physical
■ DON'T open cupboards/drawers/doors
■ DON'T touch anything that looks suspicious
■ Remove ONLY your own personal belongings
■ Remain quiet—you should also be LISTENING for unusual noises (ticking, perhaps—although that's a rather old-fashioned idea)

High-security lifestyle

The risk of attack or kidnap for a potential terrorist target can be reduced if the location and style of their accommodation are considered carefully. NEVER give the terrorist the advantage.

Densely-populated cities allow good communications with emergency services, usually have street lighting and may offer a choice of routes for journeys. It is easier to keep a low profile and to mingle in crowds—but this environment can favour the terrorist as well. Suburban areas may offer larger properties with private grounds. With adequate resources a perimeter wall/fence can offer good protection to a suburban home, but road routes into the centre of the city may be more of a problem, especially at rush hour.

Flats in apartment blocks usually offer limited access. The main building door can be equipped with an 'entryphone' and closed-circuit television (CCTV), but there may be more than one entrance. 'Visitors' might be admitted by other occupants of the building. Terraced houses have limited access but a detached house is preferable. If it is multi-storey, it offers better observation points. Ground-floor doors and windows, at least, MUST be made secure (see SECURITY).

A house in grounds should have a high perimeter wall, with the minimum number of entrances/exits. A drive or approach road should have speed bumps to slow cars. Garages/car-parking areas should be well away from the house, well lit at night, with limited access. There should be no bushes, trees, or other obstructions within 50 metres (about 50 yards) of the house. Grounds should be floodlit at night. ALL security and alarm devices should be used (see SECURITY).

Personal bodyguards should supervise identification and admission of all visitors. Immediate access to potential targets should not be given to a visitor. A small delay—an offer of refreshment may be an excuse—allows security staff time to double-check visitor identity and assess possible terrorist risk. In a contact situation like a meeting or interview, the layout of the room is also important. The visitor should be offered a low, padded chair, with no arms. This makes it difficult to get up quickly. The potential target should sit behind a heavy desk which offers a protective barrier.

If a meeting is being held in private, a hidden panic button should be accessible in case of an emergency.

REMEMBER

To maintain security all household employees must be totally reliable, so great care must be taken when choosing staff. Strict vetting procedures MUST be followed, references checked and private life investigated. It is VITAL that staff are treated well and feel appreciated. Dissatisfaction with wages, conditions or any resentment towards the person they are protecting could lead to a breach of security.

BULLET/BLASTPROOFING

Bulletproof glass

This is made of several layers of glass and plastic. The laminate can be made as thick as required, and will stop all small arms fire. It is ideal in high-risk areas such as banks, post offices—anywhere where large amounts of cash are handled.

The glass is very heavy, which is partly why bulletproof cars/vans need special suspension. Bulletproof glass will also perform well if an explosion occurs. Like conventional laminated glass, there is little fragmentation.

Bulletproof clothing

Nowadays this is usually made from kevlar—an almost unbreakable man-made fibre. Its very long strands can be woven into almost any garment to protect the heart, lungs—all vital organs. Several layers may be used, depending on the seriousness of the threat—six layers will stop a bullet from a handgun. More layers are needed if the garment is to be any use against a magnum or special ammunition. If high-velocity rifles are likely to be involved, a ceramic plate is placed over the kevlar, but the disadvantage is that this makes the armour very bulky and inflexible—and detectable.

Kevlar's enormous strength (and long fibre length) stops a bullet penetrating, but can't stop the trauma of impact—a 'trauma pack' beneath a bulletproof vest is worn to dissipate the shock. It will still feel like you've been kicked by a mule.

All heads of state and high-risk personnel wear kevlar. In the US, in areas where there are spates of drive-by shootings or local battles going on, even children are sent to school in bulletproof clothing. It sounds extreme, but in one period in 1990 in Boston, 90 per cent of gunshot fatalities were innocent women and children caught in the crossfire of street fights.

Although kevlar will also stop grenade and bomb fragments from penetrating, it has been known to fail under impact from a knife or ice-pick. The head will always be vulnerable—there are ballistic helmets, but an important dignitary is unlikely to want to be seen wearing one to a function!

Bulletproof clothing has saved hundreds of lives, but doesn't actually guarantee safety. Although the bullets won't penetrate, there is likely to be bruising and even broken ribs. A company in the US (who claim to have saved many 'important' lives) actually demonstrate a bulletproof vest by having a man shoot himself. For the purposes of the demonstration he places two telephone directories under the vest to absorb the shock!

A lightweight vest (**A**), which can be worn inconspicuously under a shirt and jacket, may be more comfortable—but cannot offer full protection. Adding layers of 'protection' means adding bulk (**B**), but such a vest may still be worn under a coat. If subtlety's not the aim, (**C**) as an outer garment offers good protection and mobility. Uniforms, jackets and overcoats are all available with varying levels of 'protection' built in.

Blastproofing

Most blastproofing is tailormade to the location—where blast-proofing is possible at all. Cars tends to have layers of kevlar built into the bodywork or steel and plastic shielding. The underside of a car—vulnerable to bombs—can be made strong enough that a blast will lift the car but cause no actual damage within the vehicle.

Any building designed to withstand or contain a blast should have a bund—a blast wall—constructed around it. The roof should be designed to collapse inwards, not to fragment and fly off. The blast wave takes the route of least resistance, so partition walls and glass will be the first and most likely barriers to be blown away. ALL glass should be laminated to reduce the risk of flying fragments, which is the cause of a majority of casualties.

Blast blankets made of kevlar may be draped over an explosive device and weighted with sandbags to cut down the damage done by a blast.

REMEMBER

The majority of injuries caused by a bomb blast are from flying glass and debris. A muster point for the staff of a large building should ALWAYS be well away from this risk. If a bomb explodes near the general area where you are, do NOT go to inspect the damage. Terrorists sometimes use more than one device—to catch spectators and emergency services.

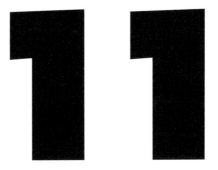

11

Environmental changes may have increased the threat of natural disaster. The restlessness of the planet may manifest itself in many ways, from floods to earthquakes—sometimes resulting in huge death tolls. Understand the dangers.

Disasters

NATURAL DISASTERS

Routine is the key to everyday living. It's about getting up, remembering to take the dog for a walk, paying the electricity bill, going to work, and coming home to switch on the news and watch what has been happening to other people. Yet every day, somewhere in the world, disaster strikes, throwing normal life into chaos and reminding us that no matter how scientifically advanced our society, nature will always have the upper hand.

All disasters, from earthquakes and volcanoes to hurricanes and tornadoes, from flooding to drought and severe winter storms, have the power to devastate our safe urbanity, and force us to face the fragility of ordinary life. Suddenly, with no warning at all, you may have to cope with disasters totally outside your experience. Emergency services are designed to step in on such occasions, but you may well have to rely on your own instincts to survive—at least for the first few days.

By being prepared, you can reduce the fear, panic, inconvenience and loss that usually surround a disaster. Know the dangers and act accordingly. Know how to protect yourself, feed yourself, and prevent needless injury until help arrives. But most of all, be a survivor!

DON'T PANIC

No matter how severe or life-threatening the disaster, panicking will get you nowhere. It leads to muddled thinking and puts you at even greater risk. Take a few deep breaths and calm yourself down, that way you'll be able to make a more logical assessment of the dangers.

Morale

Never underestimate the psychological effects of losing a loved one, a home or treasured possessions. Everything you know and rely on may be swept away from you. Disaster victims will be subject to an enormous amount of stress and may be terrified, irritable and exhausted. Try to boost morale by concentrating on the positive aspects of being alive and the future. Survival is about THE WILL TO SURVIVE.

Predicting disaster

Meteorological stations around the world use satellites, radar, observation boats and planes to study weather conditions 24 hours a day, and play a major role in warning of predictable disasters such as flooding. Global communications mean that

Britain's Met Office will be informed of a hurricane in Miami, as it's happening. You can find out about foreign and local weather conditions by calling one of the telephone information services such as Weathercall or Snowline.

The Met Office is also responsible for faxing councils with local weather information, so that emergency and social services have time to go on the alert. Every county council has an emergency planning officer responsible for coordinating services and voluntary organizations in the event of a disaster. Contingency plans usually include a 24-hour phone line for those who need help. Know the emergency procedure for your area. Details can be found in your library or consumer advice centre or can be obtained from your local council.

Personal safety

Just as disaster brings out the best in some people, it can also bring out the worst. Society can break down in the most ordered of cities, and looting and violence become as much of a threat as the disaster itself. In 1965 and 1977 New York suffered blackouts, where power and communications throughout the city were suddenly inexplicably cut off. All hell broke loose (according to reports) with lootings, muggings and arson occuring on a horrendous scale. Be aware of the dangers from other people who may try to take advantage of the situation. Always put life before possessions.

Emergency supplies

An emergency supplies kit will be your lifeline. Store it in a cool, dark, dry place and make sure everyone in the home knows where it's kept. Keep a regular check on the state of supplies—DON'T wait until a disaster occurs before you find out your batteries are flat. Keep the kit with a list of last-minute things you're likely to need, such as contact lenses or specialist medical supplies—you might not be able to think quickly enough at the time of emergency.

Purifying water

The body can survive up to three weeks without food. Without water, death may occur between three and eleven days. If supplies are cut, domestic cold water cisterns/tanks are a good source of water (you may need a flexible connecting hose), since they can hold as much as 180 litres (about 40 gallons). Do NOT use water in toilet cisterns for drinking.

If water is contaminated, strain through paper towels or clean cloth to remove any sediment. If a heat source is available, boil at a rolling boil for at least five minutes. Alternatively, use water-purifying tablets. If neither is available and you are absolutely desperate, you can use liquid household chlorine bleach (NEVER the granular form, which is poisonous) or two per cent tincture of iodine. Only purify enough water to last a maximum of 48 hours—that way you minimize the chance of recontamination.

☠ WARNING

If water is severely contaminated — possibly by sewage — or if there are chemical contaminants which may 'cancel out' the effects of bleach or iodine, take great care. Listen to warnings on the radio concerning drinking water. DON'T take chances!

Add two drops of chlorine bleach to one litre (about two pints) of clear water. If the water is cloudy, double the 'dose' to four drops. A plastic bucket for domestic use holds about nine litres (about two gallons) of water—increase the 'dose' to 16–18 drops for clear water, 30 drops for cloudy.

If using tincture of iodine, use three drops for one litre (about two pints) of clear water and 24 drops for a bucketful. Double the 'dose' to six drops if the water is cloudy.

EARTHQUAKE

Major earthquakes can cause death and destruction on a massive scale. No other natural phenomenon is capable of such devastation over so large an area in such a short space of time. In most cases the actual movement of the earth only lasts about 15 seconds (it may last several minutes), but it is usually the after-effects which wreak the greatest havoc.

Buildings and bridges collapse, crushing people under the debris. Power and telephone lines are destroyed, causing fires which rage out of control because of building failures and broken gas pipes. In the California earthquake of 1906, the resultant fire lasted four days before it burnt itself out, leaving a path of destruction that left 300,000 homeless. In 1985, a massive earthquake in Mexico City reduced to rubble the homes of 100,000 people. As recently as 1988, a 6.5 magnitude earthquake in Armenia killed 25,000 people as well as making thousands more homeless.

Just as devastating, earthquakes can set off a chain of other natural disasters such as landslides, mudslides, avalanches, volcanic eruptions and tsunamis—gigantic tidal waves reaching 30 metres (100 feet) high and travelling great distances, before crashing into coastal areas at speeds of up to 645 kph (400 mph).

It's a common misconception that earthquakes only occur along geological fault lines—recognizable 'cracks' in the rock structure which indicate past movement. San Francisco is inextricably linked with violent earthquakes, built as it is on the San Andreas fault—a zone of faulting and cracking that extends for hundreds of miles. Yet in the majority of earthquakes, fault rupture never even reaches the surface and so is not directly visible.

The Richter Scale

The definition of an earthquake is any abrupt disturbance within the earth's structure resulting in the generation of elastic or seismic waves. It is the passage of these seismic waves through the earth that usually causes the violent shaking at its surface, the earthquake. The result can be anything from a slight tremor to a full-blown rupture, and is usually expressed on the Richter Scale (although other magnitude scales are also used by earthquake observatories).

Based on seismograph recordings, the Richter Scale has no maximum or minimum levels—the biggest earthquakes so far measured have reached magnitude 9. Each unit recorded represents a tenfold increase in terms of size, which makes an earthquake measuring 8 on the Richter Scale 10,000 times bigger than one of magnitude 4. An 8 would undoubtedly cause major devastation, whereas a 4 would probably result in only slight damage.

1976—THE EARTHQUAKE YEAR

As many as 300,000 people are believed to have died in 1976 as a result of earthquakes, aside from the hundreds of thousands of casualties and the enormous losses suffered.

February 4: Guatemala Approx 23,000 deaths
May 6: Italy 900 deaths
June 25: Irian Jaya 6000 deaths
July 28: China 250,000 deaths
August 16: Philippines 2000 deaths
November 24: Turkey 4000 deaths

Earthquake zones

The major earthquake zones or areas of seismic activity are most common around the edges of the Pacific Ocean—up the western coast of South and North America and back down the western rim through Japan, the Philippines and the South Sea Islands—and, less actively, along the Mediterranean Sea. But earthquakes can occur anywhere.

Even Britain suffers tremors—in 1984, Newton in central Wales experienced shaking houses, cracking walls and overturning furniture. In 1990 another tremor in the south of England shook pictures on walls. In the US, scientists estimate that 70 million people in 39 states could be classed as high risk.

Despite scientific advances and seismological research in China, Japan, the US and the Soviet Union, no method has yet been devised to predict the time, place or magnitude of earthquakes with a high degree of accuracy and consistency. Small tremors, also known as foreshocks, may be the only advance warning, which is why earthquakes are so deadly. This means that year-round preparations, particularly for those in known unstable areas, are an absolute necessity to plan for and survive a major earthquake.

Getting prepared

Frightening though it is, the movement of the earth itself is seldom the direct cause of injury or death. The earth does not open up and swallow cities whole! Earthquake injuries are usually caused through building collapse or structural damage and flying debris. If you live in a high-risk zone, check for potential hazards and take immediate steps to make your home as safe as possible (see SAFETY FIRST).

◑ Check for cracks in your home's walls and ceilings. Cracks wider than 3 mm (1/8 in) could indicate potential weakness.

◑ Heavy light fittings should be anchored solidly to joists above ceilings.

◑ Defective electrical wiring, leaking gas and inflexible utility connections are potentially hazardous—have them checked out by specialists. Know where and how to shut them off at the mains in case of damage during the quake.

◑ Water cisterns/tanks are a potential danger if they fall over or burst in an earthquake. What's more, the water they contain could save your life if services are cut off. Make sure they are secure and in good repair.

◑ Ensure shelves are safely fastened to walls. Place large or heavy items on lower shelves to reduce risk of injury. Put breakables in cupboards that can be fastened shut.

◑ Children's play areas should be away from brick or concrete walls which are not steel-reinforced and could collapse during earth tremors.

◑ Ensure everyone in the home knows what to do in an emergency. Locate safe spots in each room—under sturdy tables or in strong interior doorways. Identify danger areas such as windows which may shatter or top-heavy furniture such as bookcases.

◑ Keep a regular check on your emergency kit. Store water in airtight containers and replace every six months. Reckon on at least 14 litres (about three gallons) of water per person for a 72-hour period or longer. Ensure medication and food supplies are not past their use-by dates. Make sure portable radio, flashlight and extra batteries are working. It doesn't take long for old flat batteries to leak and possibly damage other emergency supplies.

○ Devise a plan for reuniting family and friends after the disaster. You may be at work. Children may be at school. Transportation and communication may be disrupted.

During the earthquake

Stay where you are. Take cover against an inside wall or strong internal doorway. Avoid exterior walls, doors, heavy furniture or appliances. Do NOT rush outside—you may be injured by falling glass or masonry.

○ If you are outside, get into the open away from power lines, buildings and lampposts.

○ If you are in a crowded public place, do NOT make a dash for the exit—everyone else will have the same idea. Take cover, away from the risk of falling objects/debris.

○ If you're in a tall building, such as an office block, get under a heavy desk away from exterior walls and windows. Do NOT use stairs or lifts in an attempt to get to lower floors—most people are injured when moving around.

○ Extinguish candles and naked flames in case of fire or gas explosion. If you smell gas, DON'T smoke.

○ Do NOT take refuge in cellars, subways or underground tunnels. Exits could cave in or become blocked.

○ If you're driving, STOP. Do NOT get out—your vehicle will offer some protection. Crouch below seat level if possible. Do NOT stop on bridges, near buildings, or under trees, electrical power lines or large signs in case they collapse.

○ Aftershocks have been known to occur anything from one minute to a year later! Most happen 24–48 hours after the main tremor and cause further damage to already weakened structures. Wait until you get the all-clear from police or rescue teams before leaving safety.

After the earthquake

Check for injuries. Apply first aid if necessary or wait for help. Leave severely damaged buildings in case of further collapse. Only return when authorities deem it safe to do so.

○ Wear sturdy shoes to prevent injury from fallen masonry and glass.

○ Clean up spilled household chemicals and potentially-harmful substances.

○ Gas leaks and damaged electrical wiring are major fire hazards. Shut off at the mains if damage is suspected and do not turn back on until checked by specialists.

○ Do NOT smoke or operate electrical appliances if a gas leak is suspected—even a telephone could create a spark sufficient to cause an explosion.

TSUNAMI

Often more devastating than an earthquake itself is the tsunami (pronounced soo-name-ee) or 'tidal' wave which can follow as a result of underwater disturbance. More than 200 tsunamis have been reported in the Pacific in this century alone, some of which have resulted in waves more than 30 m (100 ft) high smashing into the coast with enormous destructive power. Earthquakes don't always cause tsunamis, but they may also be generated by underwater landslides and major volcanic eruptions.

Tsunamis are capable of travelling enormous distances. Although they move quickly, there is usually time for adequate warning. A tsunami generated on the coast of Chile, for example, would take about ten hours to reach Hawaii.

A major problem associated with a tsunami warning is that people allow their curiosity to overcome their common sense. When Berkeley, California received a warning during the Alaskan earthquake of 1964, the Chief of Police later complained that not only did local people go to the shoreline to watch for the great wave, so did some of his staff!

It's VITAL to understand that a tsunami is not usually ONE wave but a series. People can be fooled into returning to the danger area after the first wave, only to meet their deaths under the onslaught of a second or third. At Crescent City, California, during the same Alaskan earthquake, seven people went back to a pub in the flooded area to collect their valuables after the first and second waves had passed. Assuming the danger was over, they stayed for a drink and were hit by a third wave. Five of them were drowned.

◑ If you can see that water pipes are damaged, shut off the supply. Check with local emergency officials whether sewage pipes are intact before flushing toilets. Plug bath and sink drainholes to prevent sewage back-up.

◑ Weakened chimneys can topple during an aftershock—approach with caution. Do not use a damaged chimney. A major fire could start.

◑ Beware of items falling out when you open cupboard doors.

◑ Do not use the telephone except in an absolute emergency. You could tie up much-needed lines for the rescue services.

◑ Check food and water supplies. Take great care over hygiene and sanitation.

◑ Do NOT use your car except in an emergency—roads may be impassable. Keep streets clear for rescue vehicles.

◑ Do NOT go out sightseeing. Apart from the danger to yourself, you could hinder rescue efforts.

◑ Listen for up-to-date emergency information on your portable radio.

Earthquakes can cause mudslides and landslides which may necessitate evacuation in your area. People living in coastal areas should be particularly aware of the possibility of tsunamis (sometimes called, incorrectly, 'tidal' waves). When a warning is issued, stay away from the shoreline and prepare for evacuation. Do not attempt to save possessions when your family, friends or other people are in danger.

REMEMBER

It is essential to stay as quiet as possible if rescuers are searching rubble for survivors. They need to be able to hear even the faintest cries for help. Unless you are told to evacuate, you should help search the immediate area—but only when you KNOW it's safe to go outside.

VOLCANO

When the volcanic island of Krakatoa in Indonesia suddenly exploded in 1883 the blast was so violent that the sound could be heard 3000 miles away across the Indian Ocean. Clouds of ash and dust were ejected into the atmosphere with such force that it took almost two years for it all to fall back to earth.

Altogether there are 1343 known active volcanoes around the world, Indonesia boasting the highest number—200. There is also a high concentration around the Pacific, though Europe has a few such as Mount Etna in Sicily, Vesuvius in Italy and Hekla in Iceland.

Volcanic eruptions of the earth's interior core matter, usually where the crust is deeply fissured, can cause fierce explosions of lava, gases, dust and rock—particularly in a 'central' volcano, which has the recognizable cone. Why a volcano should decide to erupt, perhaps after hundreds of years of apparent dormancy, is usually linked to viscous lava clogging up the cone of the volcano and causing a massive pressure build-up.

There are also instances where local earthquakes have triggered off a volcanic eruption—a bit like shaking up a bottle of fizzy drink. Such was the case in the eruption of Mount Saint Helens, Washington, in 1980, where 65 people were killed—including a professional volcanologist!

Given the cataclysmic nature of an erupting volcano, and the speed at which molten lava can travel, you would be forgiven for thinking that survival would be in the lap of the gods. However there are usually visible warning signs weeks before an euption and, even if prior evacuation has not been possible, there are emergency precautions you can take.

Advance warnings

A volcano may 'grumble' on and off for years, letting everyone know that they can expect an eruption sooner or later. On the other hand, the eruption may take place in a few hours or days. DON'T take chances! DON'T climb a volcano for a closer look. Pay attention to:

- Audible rumblings from the volcano or the ground
- Ash and gases appearing from the cone, the sides or round the volcano
- Earth movement, whether faint tremors or earthquakes
- Presence of pumice dust in the air
- Acid rain fall
- Steam in clouds over the mouth of the volcano
- Rotten egg smell near rivers, betraying presence of sulphur

Emergency procedure

- Leave the area immediately. Do NOT waste time trying to save possessions.
- Be prepared for difficult travelling conditions. If vehicles get bogged in deep ash you may have no choice but to abandon them, in which case run for the nearest road out of the area. You may be able to hitch a lift.
- Avoid areas downwind from the volcano if ash is being expelled. Use a scarf or wet handkerchief as a mask—the combination of ash and acidic gas can cause lung damage.
- Protect eyes with any snug-fitting goggles.
- Beware flying debris—any hat stuffed with newspaper might help prevent head injury. Wear thick padded clothing for body protection.
- Always check for mudflows when approaching a stream channel. A mudflow can move faster than anyone can run— even buildings may not stop one.
- Shelter in buildings (other than emergency refuges) ONLY as a last resort. Walls can be crushed by rocks and lava. Roofs are subject to collapse, even from just the weight of ash and debris.

NUEE ARDENTE

A nuée ardente is the scientific name for a rapidly-moving, incandescent cloud of gas, ash and rock fragments accompanying a volcanic eruption. Faced with a red hot cloud travelling at more than 160 kph (100 mph), you only have two choices. Either shelter in an underground emergency refuge or hold your breath and submerge yourself underwater—in a river, a lake, the sea. The danger may pass in 30 seconds or so.

HURRICANE

A hurricane is a severe tropical storm capable of releasing as much energy in one day as a one megaton hydrogen bomb. Also known as a cyclone (literally 'coil of the snake') or typhoon, a hurricane can travel up to 300 kmph (about 200 mph)—bringing with it storm surges, flash flooding and tornadoes and leaving a phenomenal trail of destruction in its wake.

Hurricanes have caused more damage in the United States than any other type of natural disaster. As many as 50 countries with a total of 500 million population are exposed to the risk every year.

True hurricanes never form in cooler European waters. From June through to November, the US, Hawaiian islands

MEASURING HURRICANES

International classification of hurricanes is usually based on wind speeds measured over one, three or five minutes or in single gusts. Wind speeds are often expressed on the Beaufort Scale, although in the US the common classification is simply:
1 Tropical storm
2 Severe tropical storm
3 Hurricane

STRONG GALE
Wind speed: 75–88 kph (46–55 mph)
Beaufort scale: 9
Likely effects: Leaves and small branches stripped off trees. Structural damage to buildings unlikely.

VIOLENT STORM
Wind speed: 103–114 kph (64–71 mph)
Beaufort scale: 11
Likely effects: Large branches snapped off trees. Shallow-rooted trees may fall. Roof and other structural damage. Fences blown down. Lorries blown over. People blown over.

HURRICANE
Wind speed: 117 kph plus (73 mph plus)
Beaufort scale: 12 and over
Likely effects: Widespread damage. Weak structures flattened, roofs blown away. Trees blown down. A large amount of airborne debris. Windows blown in. Shelter for people essential.

and Caribbean are at their most vulnerable. In the steamy tropics the general weather conditions are favourable to hurricanes virtually every day.

No one can explain exactly why a hurricane develops. It starts at sea as a centre of low pressure. Given the right combination of atmospheric conditions, it can gather energy from evaporated water in the ocean, whipping into a whirling windforce. The result is a spiralling catherine-wheel effect, with a calm 'eye' in the centre, which may exceed 40 km (25 miles) in diameter.

A hurricane could cover 1.3 million square kilometres (500,000 sq miles) and last up to three weeks before blowing itself out, usually over land. Small wonder that anything caught in the eye is virtually helpless. The eye of Hurricane Carla in 1961 was packed so tightly with trapped birds that observatory weather planes were unable to fly through it!

Enormous destructive power aside, the hurricane has another deadly secret weapon—unpredictability. Despite world observatories' facility to spot and monitor hurricane activity, thanks to satellites and radar, there is still no way to predict when and where a hurricane may strike.

Hurricanes are able to 'loop the loop' without warning—Typhoon Wayne doubled back on itself three times before hitting China in 1986. International collaborations such as Spectrum (formed in 1991) have been set up to conduct research into hurricane behaviour, but it will be years before reliable warnings are possible.

These days, the catastrophic effect of hurricanes is being reduced as a result of better designed housing in affected areas. On the Florida coast, for example, the hurricane death toll has fallen over the last 20 years, even though the area is more densely populated.

When Cyclone Tracy hit Darwin, Australia, in 1974, only 400 of the city's 11,200 houses survived relatively intact. Darwin has been almost totally rebuilt with hurricane-proof structures but (fortunately) they have yet to be tested.

Naming hurricanes

Since 1953 hurricanes in the Atlantic, Caribbean and Gulf of Mexico have been given personal names devised by the World Meteorological Organization as a means of identification. Originally they were named only after women, supposedly because of their unpredictable and capricious behaviour! These days they are named alternatively with male or female names. Revised every six years, the names run alphabetically. 1991 hurricane names, for example, are Bob, Claudette through to Victor and Wanda. In the event of an extremely damaging hurricane a name is retired permanently.

Early warnings

Tracking stations around the world constantly monitor hurricane activity and provide up-to-date information to weather services. There are two types of hurricane alert:

Hurricane watch This means that a hurricane MAY threaten coastal and inland areas, but it does NOT mean that a hurricane is imminent. **Action:** Listen for further information on your local radio and television stations and be prepared to act promptly.

Hurricane warning This means that a hurricane is expected to strike WITHIN 24 HOURS. **Action:** Keep listening for advice from local authorities and act according to recommended emergency procedures.

Getting prepared

Before hurricane season:
- If you do not have a storm cellar, know where your nearest evacuation shelter is—it could save your life.
- Check your house is sound, particularly the roof. The strongest winds in a hurricane occur immediately after the transit of the eye and blow in the OPPOSITE direction. It is then your house will be most vulnerable.
- Keep your property clear of junk—a hurricane can turn it into missiles.
- Check emergency supplies.

After a hurricane warning

- Keep the radio/TV on for more information and instructions. Watch out for warnings of tornadoes—often spawned by hurricanes.
- Fill car with petrol and be prepared to evacuate if told by the emergency services to do so.
- Leave low-lying/coastal areas that might be swept by high tides or floods.
- Leave mobile homes for substantial shelter. Caravans are prone to overturning in strong winds.
- Secure outside objects. Even fridges have been known to take flight in hurricanes.
- Board up windows to prevent breakage through debris or wind pressure.
- Store drinking water in clean bathtubs, sinks and containers. Water systems may be damaged or contaminated in the storm.
- Lock up pets.

During a hurricane

❍ If you are unable to evacuate and you do not have a cellar, shelter in the strongest part of the house—under the stairs, perhaps. Take emergency supplies with you.

❍ Do not shelter near an internal chimney breast. Even in Britain's gale of October 1987 people were killed when chimneys collapsed through ceilings.

❍ If the house starts to break up, protect yourself with mattresses and blankets or get under a strong bed.

❍ Stay inside—high winds and flying debris can be lethal.

❍ Do NOT drive—cars are no protection in a hurricane.

❍ Do NOT be fooled by the calm eye! If the storm centre passes directly overhead there may be a lull lasting anything from a few minutes to an hour, before the wind picks up to hurricane speed again in the OPPOSITE direction.

After the storm

❍ Apply first aid to any casualties.

❍ Stay where you are until given the all-clear.

❍ Listen for information on radio/TV stations.

❍ Do NOT go sightseeing—you may interfere with rescue work and put yourself in danger.

❍ Drive ONLY when necessary—roads may be littered with debris and should be kept open for emergency services.

❍ Do not use telephone unless an emergency. You could tie up lines needed for rescue services.

❍ Report broken power cables, water, gas and sewage mains.

❍ Prevent fires—fire-fighting could be impaired because of low water pressure.

❍ Eat perishable foods first if electricity is cut off—you don't know how long you may have to rely on your existing food supplies.

TORNADO

The low, train-like rumbling of a tornado approaching is indeed an awesome sound, for it brings with it one of nature's most violent storms. With whirling winds of up to 300 kph (about 200 mph), a tornado has a funnel-type effect causing everything in its path, apart from the most solid structures, to be sucked up as if into a giant vacuum cleaner. The pressure is sufficient to lift light buildings—at the very least, doors and windows can be sucked out.

Fortunately, the base of a tornado is usually only 25–50 m (82–164 ft) across, which reduces the level of destruction to a smaller area. Also, unlike hurricanes, tornadoes normally only travel up to 16 kilometres (ten miles) before petering out. That is not to say that their power should be underestimated! Tornadoes can strike at any time of the year and often occur on the fringe of hurricanes.

Even Britain has had its share of tornadoes, albeit on a small scale. The Newmarket tornado of 1978 caused £1 million of property damage. In 1981, 58 tornadoes were reported in one day in an area from Anglesey to the eastern coast.

Tornado precautions

Emergency procedures before, during and after a tornado warning are much the same as for hurricanes, though there are a few extra precautions which should be taken:

◑ The best protection is an underground shelter or basement of a steel-framed or reinforced-concrete building. If none of these is available, take cover on the lowest floor in the centre of the house, under a strong table, making sure there are no heavy appliances above. This should reduce the risk of injury if walls collapse. Stay away from windows!

◑ As the tornado approaches, CLOSE all doors and windows facing the storm and OPEN all those on the opposite side. The aim is to equalize pressure inside and outside to prevent the roof being sucked away or the building collapsing.

◑ If outside, do not attempt to escape—not even in a car.

◑ If there is no shelter nearby, lie flat in a ditch or ground depression and cover your head. You will stand less chance of being sucked up by the tornado or hit by flying debris.

FLOOD

Flooding may not have the same dramatic impact as an erupting volcano, say, or the effects of a monumental earthquake, but it is probably one of the most common and widespread natural disasters worldwide. The statistics are horrendous.

Flooding can occur for many reasons—heavy rainfall (particularly in valleys after prolonged spells of hot dry weather), snow melting on mountains, rivers changing course, dams collapsing and high tides in coastal areas. The Thames Barrier in London was constructed to prevent flooding during a North Sea surge. This is where high tides combine with low pressure and north or north westerly winds, as far away as Scotland, to push waves down the eastern coast of England, gathering force and raising sea levels.

Living close to the sea is immensely popular, despite vulnerability from eroding sea walls, but the squeeze for land has meant housing estates may be built in (frankly) unsuitable areas, such as the flood plains of rivers. If 'global warming' has the effects predicted, it is estimated that by the year 2050 sea levels could rise by as much as 50cm (20in). It may not sound much, but it could mean that many areas of the world could be drastically affected by the constant threat of floods.

Getting prepared

Fortunately, weather centres can usually predict flooding and so give early warning to local authorities and the media. Advance warning can be anything from a few hours to a few days, so there's normally time to organize yourself before water starts heading your way.

Before the flood

○ If you live in a known risk area, find out the official flood height which will affect your house (from your local council) to help you act accordingly.
○ Listen constantly for radio and TV reports.
○ Turn off electricity as threat becomes more immediate. Do NOT wait until water has entered the house and risk electrocution.
○ Turn off gas in case of potential damage and leaks.
○ Block gaps outside doors with sandbags, usually provided by the emergency services. If not available, improvise with plastic bags or pillowcases filled with earth or gravel, or stuff cracks with old blankets and carpets.
○ NEVER stack sandbags round the outside walls to keep water out of a basement. Water seepage could cause pressure that could damage the walls or even raise the basement out of the ground. Better to fill the basement yourself with clean water if you think flooding will occur. It will equalize pressure to prevent structural damage and will make cleaning up afterwards easier.
○ Sandbag exterior windowsills if it looks like water will rise that far.
○ Move to the upper floors, taking emergency supplies, pets and any valuables/papers you simply cannot live without. REMEMBER: Lives are more important than possessions.
○ Take spare clothes and shoes with you—the flood may be around for days.

Flash floods can be deadly and highly unexpected, unless you know the danger signs. They are common when torrential rain follows a long dry spell. The ground cannot absorb the rainfall quickly enough and the flood begins.

Alice Springs in Australia's 'red centre' suffers from an incredibly arid and desert climate with summer temperatures around the 45°C (113°F) mark. So permanently dry is the town's Todd River that they hold an annual Henley-on-Todd Regatta. Competitors run up and down the river bed in bottomless boats with their legs sticking out (like 'Flintstones' cars)! Yet every couple of years surprise storms transform the river bed from a barren desert to a raging torrent. Many people have drowned under the force and speed of the first onslaught of water.

Flash floods occur all over the world. If a flash flood is imminent or leaves you stranded outdoors:

- Seek shelter on higher ground immediately. REMEMBER: You don't have to be at the bottom of a hill to be on low ground!
- If your car stalls, abandon it. You and your vehicle could be swept away
- Do not try to swim—the currents are deadly
- Do not try to cross a fast-moving stream unless you are POSITIVE it is no more than knee height. It may look safe, but the current could still be strong enough to sweep you off your feet

➊ Fill baths and containers with water—supplies may become contaminated.
➊ Brightly coloured cloth is useful for signalling for help.

During the flood
➊ Keep listening to reports on your portable radio.
➊ Keep a constant watch on what's happening outside.
➊ Stay where you are, unless told to evacuate—travelling could put you at even greater risk.
➊ NEVER drink floodwater—it may carry disease.
➊ Do NOT use telephone except in dire emergency.
➊ Obey officials if told to evacuate.
➊ If life is endangered, get onto the roof and signal for help.
➊ Only in extreme emergency should you attempt to make a raft—flood water moves very fast and you may put yourself at greater risk.

After the flood
➊ Do NOT go outside until you know it's safe—more storms could be on the way.
➊ Boil or purify ALL water, until local authorities declare municipal supply safe. Flood water could have contaminated the cold water mains supply.

○ Do NOT use fresh food or water that has come into contact with flood water.
○ Do NOT turn on electricity or gas until told to do so.
○ In likely areas, beware poisonous snakes or spiders seeking refuge in your home.
○ Beware contracting diseases from paddling in flood water.
○ Wear strong shoes when walking through subsided flood water—you could step on sharp debris or broken glass.
○ Check drains carefully for blockage—they could pose a hygiene threat if not cleared properly.

LIGHTNING

 Every year there's an average of six lightning strikes per square mile in Britain alone—that's about 4200 just over London. Amazingly enough few people are actually killed (around 12 per annum is the national average), though many are seriously injured or burned. If you are outdoors in a thunderstorm and you feel your hair start to stand on end, that's a sure indication that lightning is about to strike.

○ Drop to your knees IMMEDIATELY. Bend forward, putting your hands on your knees. That way if lightning strikes the electrical charge may pass through your limbs to the ground, bypassing your heart.
○ Do NOT hold metal objects (such as golf clubs) and keep away from metal structures—they will attract lightning.
○ Do NOT stand under a tall isolated tree in an open area—it could act as a lightning conductor.
○ Do NOT stand on a hilltop where you would project above the landscape—YOU could act as a lightning conductor!
○ Seek shelter in a low area in dense shrubbery.
○ Even rubber-soled shoes are no guarantee of safety.

DROUGHT

 Recent years have seen an unparalleled awareness of drought in Britain and many other countries. No longer is it something that happens to other people! From hosepipe fines and car-washing bans, even tap water cannot be relied on for drinking

as reservoirs sink to new lows. Yet for many countries, drought is a simple fact of life. In parts of the Middle East, for example, the only drinking water comes from the sea, having undergone processing in huge desalination plants. In many other areas of the world, drought survival is about water conservation, eking out every precious drop in the hope that supplies will not run out before the next rainfall.

The city of Perth in Western Australia is a case in point. Every household is allotted a certain amount of water, and anyone exceeding it has to pay for their excesses. Television news bulletins report overall weekly consumption compared to existing water supplies, and set new low 'targets' in an effort to reduce expenditure.

Fire, sanitation and disease are the main problems associated with drought, and if the current climatic changes are attributable to the much-vaunted 'greenhouse effect', we may have no choice but to learn to live with them.

THE 'GREENHOUSE EFFECT'

Thanks to the burning of fossil fuels, use of CFCs and intense cultivation and deforestation, man is increasing the concentration of 'greenhouse' gases in the atmosphere. This, say scientists, could change the global climate, causing floods in some regions and droughts in others. It is estimated that the average global surface temperature increased by 0.5°C (0.9°F) from 1900 to 1989. Unless we change our ways we may accelerate the trend.

Living with drought

❍ Do NOT waste water. A hosepipe BAN means exactly that, and could signal the start of severe water shortage.

❍ Do NOT drink tap water if local authorities issue warnings to that effect. Shrinking water levels can result in contamination and, in extreme drought, dead animals may pollute water sources. Boil, purify, or buy bottled water.

❍ If your water supply is cut off and mobile water tanks or standpipes are installed in streets, this is designed for drinking water ONLY. The situation is too severe to waste water on other uses.

❍ Do NOT use the toilet, but leave enough water in the bowl to act as a barrier to prevent smells and possible disease spreading from sewers up the pipes.

❍ Buy a camping chemical toilet. Alternatively, you may have to face making an outdoor latrine.

❍ Re-use water as much as possible.

❍ Ensure food is always covered—flies could prove a problem.
❍ Try to practise good hygiene, despite lack of water, especially when preparing food. Hot unsanitary conditions are a breeding ground for germs.
❍ Try to eat foods with a high moisture content (such as fruit) and which need little preparation or clearing up.
❍ NEVER throw cigarette ends casually out of cars or anywhere else. Grass could be tinder-dry and fire-fighting services severely hampered.
❍ If driving any distance, carry your own water in case the engine overheats. In Western Australia, it's quite common to see the following sign in petrol stations: 'Do not ask for water as a punch in the mouth often offends'!
❍ Watch out for structural damage to houses, particularly those built on clay. You may have to fell trees too close to the house to prevent the roots causing damage to foundations.

THE BIG FREEZE

 The picture postcard romance of a snow-laden landscape soon begins to wear thin when the harsh reality of winter sets in—blizzards, snow drifts, arctic winds at more than 55 kph (35 mph) and freezing temperatures which, with the wind chill factor, can be the equivalent of −40°C (−104°F).

Even areas which normally experience mild winters can suddenly be hit with blizzards and extreme cold, rendering entire cities powerless for days. The result is always human suffering and, all too often, death.

Unlike Britain, where two centimetres (less than an inch) of light snow can bring the nation to a standstill, many countries such as Norway and Finland have had to invent ways of coping with the extreme winter conditions as part of everyday survival. In Alaska and regions of the Soviet Union, for example, houses are built on stilts because conventional foundations would buckle in the permafrost soil.

Canada sees 140 days of snow every year, and there are many lessons to be learnt from their approach to winter living. House insulation is paramount, pipes are submerged below the frost line, a second set of 'storm windows' is the norm, and people can even buy plastic draught excluders which fit into electrical sockets on exterior walls to prevent every tiny bit of heat escaping! Cars are 'winterised' with snow chains, gasahol (anti-freeze for petrol), plug-in block heaters to keep the engine warm overnight. People never drive with less than half a tank of petrol in case moisture in the air in the tank freezes and causes the vehicle to stall.

Winter complacency, underestimation of the severity of conditions, lack of preparation and, often, lack of common sense are all contributing factors to unnecessary deaths—from what is after all a seasonal event. But with a little preparation before the cold starts to bite, you can be ready for the worst that winter can bring.

Getting prepared

◑ Ensure your house is well-insulated, especially the loft. Lag pipes to prevent freezing and insulate the hot water tank to prevent heat loss. Don't continue loft insulation under a cold-water cistern—the small amount of heat beneath it may keep it from freezing.

◑ Check for draughts around exterior doors and windows. A heavy curtain over the front door can make halls warmer. Plastic sheeting taped over windows is a simple, cheap alternative to double glazing.

◑ Service central heating—it has a habit of going wrong when it's most needed.

◑ Kitchen foil, fixed shiny side out on walls behind radiators, will reflect heat.

◑ Check electrical fires are in good working order. Many winter deaths are caused through occasionally-used electrical fires which prove to be faulty.

◑ Electric blankets should be serviced ANNUALLY.

◑ If you have an unused fireplace, get it cleaned and unblocked and stock up on fuel. It could prove a last-resort.

◑ Check emergency kit and supplies (see beginning of this section). A camp stove could be vital, especially if you have an electric hob or oven.

◑ Food is fuel for the body. Ensure you have enough supplies for at least three days, but resist the temptation to stockpile. You don't really need 15 loaves of bread to see you through a winter emergency in a town or city and you could cause hardship for other people.

◑ Make sure you have enough winter clothing.

During bad weather

◑ Listen to radio/TV for weather reports and emergency information. Call social services if you need help for yourself or a relative living alone.

◑ Have emergency supplies at hand in case of power failure.

◑ Live in one room, if you can't keep the whole house heated.

◑ Do not block all ventilation—avoid build-up of potentially toxic fumes from fires and heaters.

- Drink plenty of hot drinks to make you feel warmer.
- If your pipes freeze, shut off water at the mains and turn on all taps to drain the system in case of burst pipes. Drain water into containers to ensure an adequate supply.
- If there is power failure, do NOT open freezer. A closed freezer should stay frozen for up to 48 hours.
- If central heating does fail, turn it off as a safety measure.

If you must go outside

- Dress accordingly. Several thin layers are warmer than one thick one. Mittens are warmer than gloves, and hats will prevent heat loss. Frostbite and hypothermia are SERIOUS hazards.
- Avoid over-exertion. The combination of excessive physical activity and cold can KILL.
- Do NOT drink alcohol. It lowers the body temperature.
- Do NOT dry wet clothes on or too close to heaters—it's a major fire risk.

HYPOTHERMIA

Hypothermia is a killer, particularly for elderly people. Initial signs are clumsiness, disorientation and drowsiness. As cooling of body temperature progresses, the individual becomes less active, paler and may become deeply comatose. Rewarm the casualty's body around the ribs and heart first to increase the 'core' temperature—NOT the extremities. If casualty is conscious, give a warm sweet drink. Seek urgent medical attention (see also HEALTH: Save a life!).

12

Maintaining your health is largely down to common sense, but may be difficult against constant city stresses. Judging the seriousness of an injury or illness and knowing how to apply first aid may help you to save a life.

Health

TAKE CONTROL

The prime responsibility for your health rests with you. True, not everyone has an ideal start in life—poverty, poor housing, family break-up, and the possibility of inherited disease or handicap may mean that from birth onwards your health has been affected by factors that are totally beyond your control.

In addition, you are at the mercy of an environment that includes atmospheric pollution, passive smoking, and over-processed foods contaminated by pesticides and additives. You may have little or no control over where you live and the kind of work you do.

Health advice always makes it sound as though all you have to do is follow some simple rules and you'll be fit, healthy, and live to a ripe old age. All the same, everyone can do a great deal to improve or maintain the best possible level of health. Inform yourself about the way your body works. Recognize and act upon any changes you observe, which might indicate that something is going wrong.

Self-neglect—whether caused by an unwillingness to recognize the possibility of disease or the mistaken assumption that your symptoms are too trivial to 'bother the doctor' with—may result in a worsening of the condition, serious permanent disability, or even death.

A positive approach—taking advantage of the many sources of information and advice provided by the media, coupled with a determination to seek professional help without delay when you're worried about your health—can reduce the risks of serious illness and premature death. Such an approach may also slow down the aging process, enabling you to look forward to an active and independent life well beyond the age of 70.

It's up to you

Taking responsibility for your health may involve some quite drastic changes in your lifestyle! Change is never easy. The important first step is to decide that you WILL adopt a healthier diet, take more exercise, stop smoking, cut out or cut down on alcohol—but don't try to do it all at once. Gradual change is far easier to accept than a 'blitzkrieg' approach. Each change will bring its own reward, encouraging you to embark upon the next.

It's far better to admit that it might take a year or so to reach your target of optimum health, than to try to change habits of a lifetime—depriving yourself of some of the props that have seemed to keep you going—and become a 'new' person overnight. A rushed approach can lead to failure, discouragement and unwillingness to try again. Ill-advised dietary changes, sudden exercise or stopping regular medication may have serious side effects. Always seek medical advice before making a change in lifestyle.

Looking after yourself is no longer an option. It's a necessity. Around the world delays in hospital consultations, the expense of treatment, shortages of beds and medical supplies have made us realize that most health systems just can't cope.

HEALTH CHECK LIST

There are many very basic ways in which you can influence your own health. You've seen them hundreds of times, to the point where we cease to pay attention. Read them again!

- ◑ Eat sensibly. Make sure you have a balanced diet, limiting fats and increasing fibre. Medical opinion does not favour vitamin and mineral supplements unless they have been prescribed.

- ◑ Don't drink alcohol or keep your alcohol consumption to a minimum. Alcohol affects your general health and commonly leads to irreversible damage to the heart, liver, brain and nervous system.

- ◑ Don't smoke! It is a major cause of general ill health and commonly leads to lung cancer, heart disease and chronic bronchitis, among other disorders. Your cigarette smoke can also harm other people.

- ◑ Take exercise. It will improve and maintain physical and mental fitness. Match it to your age and condition and choose a form that you enjoy. Seek medical advice.

- ◑ Watch your weight. Join a club if you find self-restraint is difficult! Seek medical advice.

- ◑ Reduce stress. Stress is linked to many serious illnesses/disorders. Talk through your problems with family, friends, employers or professional counsellors. Learn relaxation techniques. Make whatever adjustments you can to your lifestyle.

Gone are the days of the traditional family doctor with 'all the time in the world' to solve your health and personal problems. The doctor's receptionist alone may put you off making an appointment. Now is the time to decide that you will take some responsibility for your own health. Your first response must NOT be to totally abandon hope that your doctor can and will help you. Nor should it be to rush to pharmacies and 'health' shops to dose yourself up with over-the-counter medicines!

PLEASE NOTE

This section is intended to provide guidelines, advice and information. It can't offer actual diagnoses or prescribe treatments, but it may help you recognize and deal with some common conditions and serious emergencies. DON'T use it as a hypochondriac's handbook!

PREVENTATIVE MEDICINE

In recent years medical and technical advances have been made in the early diagnosis of illness—in many cases the earlier a problem is treated, the more successful the treatment. Immunization against certain illnesses has become routine, so that many life-threatening conditions have been almost eliminated.

Screening

Some types of screening, such as checking your weight or blood pressure, may be undertaken by your doctor. Certain groups of people are routinely specifically screened—for instance, pregnant women, newborn babies, people over 75. Those in particular racial groups, people who have relatives with certain serious disorders, and people who work in hazardous conditions should receive special screening.

You can carry out some types of screening on yourself. Some cancers in their early stages appear as small 'lumps'—self-examination of breasts or testes (for instance) for lumps or growths at regular intervals is ESSENTIAL. A doctor or nurse will demonstrate the techniques.

Women have been encouraged, for many years now, to check their breasts for any abnormalities and to report vaginal or abdominal discomfort. Many men do NOT realize that testicular cancer is increasingly common. The best time to examine the testes is in a warm bath—the scrotum is relaxed, allowing free movement of the testes.

Basic screening options

Antenatal screening

- Blood group
- Anaemia
- Haemophilia (high-risk groups)
- Rubella
- Spina bifida (and other congenital defects)
- Hepatitis B, HIV infection (high-risk groups)
- Syphilis
- Chromosomal abnormalities (high-risk groups)
- Sickle-cell disorders (high-risk groups)

ANAEMIA

Anaemia is surprisingly common, yet people may suffer for years without knowing what is wrong. The name describes a condition where the number of haemoglobin molecules in red blood cells is below normal. Haemoglobin carries oxygen from the lungs to the tissues of the body. Anaemia is usually caused by iron deficiency — iron is an essential component of haemoglobin.

Minor bleeding, such as menstrual blood loss, causes levels of haemoglobin and iron to decrease, but is not usually harmful. However, although pregnancy stops menstrual blood loss, greater drains are made on iron levels by the developing foetus. Pregnant women are particularly at risk from anaemia.

SYMPTOMS: Lethargy, tiredness, headaches, brittle nails, pallor, breathing difficulties during exercise, dizziness and pain in the chest.

ACTION: Seek medical assistance. Foods with good sources of iron are fruit, green vegetables, wholemeal bread, lean meat and beans. Iron supplements are available, but should be taken under medical supervision. An excess of iron in the body is also undesirable.

WARNING

Adult-strength iron tablets, which look like 'sweets', can be extremely dangerous if swallowed by a child. Seek medical attention at once, if this occurs.

Infant and child screening

- Heart disease (congenital)
- Hypothyroidism
- Cataracts/visual defects
- Deafness
- Sickle-cell (high-risk groups)
- Testes (displaced or undescended)
- Hip dislocation
- Dental caries (tooth decay)

Adult screening
- ◑ Cervical cancer
- ◑ Breast cancer
- ◑ Testicular cancer
- ◑ Lung cancer
- ◑ Blood pressure
- ◑ Cholesterol
- ◑ Glaucoma/visual defects

Immunization

It is possible to induce or boost immunity to certain serious diseases. All that is usually required is one or more injections of appropriate vaccines. Immunization against polio is even simpler—the vaccine may be swallowed on a lump of sugar. Immunization has brought many diseases under control. Smallpox has actually been eradicated!

Routine immunization is given to babies and young children at various ages. Your doctor or clinic will advise on this (see **Childhood diseases**).

Seek advice when you intend to travel abroad. A doctor/travel adviser will be able to tell you which serious diseases you need to be immunized against in any particular region (see **Tropical diseases** and IN TRANSIT).

The following vaccines are routinely available:
- ◑ **MMR:** Mumps, measles and German measles (rubella)
- ◑ **DPT:** Diphtheria, whooping cough (pertussis) and tetanus
- ◑ **Polio**
- ◑ **Influenza** Only for two out of three strains
- ◑ **Hepatitis B**
- ◑ **Meningitis** Only for two out of three strains
- ◑ **Cholera**
- ◑ **Typhoid**
- ◑ **Rabies**

REMEMBER

Immunization for influenza is normally only given to the elderly or those at occupational risk. Cholera and typhoid vaccines offer only partial protection. Other precautions, such as food and water hygiene, should be observed in areas where there is a risk of contracting the diseases.

▌ WARNING

Vaccines may be dangerous if you are pregnant, if you are HIV positive, if you have cancer or any prolonged fever or infection. Always state whether you are taking any medication.

YOU AND YOUR DOCTOR

✚ **Many doctors would much prefer to be educators, helping you to adopt a healthier lifestyle, suggesting positive ways of self-help, encouraging and monitoring rather than treating. Whether or not such an ideal is attainable, use your doctor's knowledge, help and advice to prevent ill health in the first place. The traditional idea that a visit to a doctor is only necessary if you are actually ill is gradually changing.**

Preventative medicine must involve a two-way discussion between you and your doctor on ways you can spot trouble before it becomes serious. More important, you need advice on how to make sure that problems are less likely to occur.

However, most people still only see a doctor when they experience illness, pain or discomfort—to seek reassurance, advice or treatment—or worse still, when their condition is so serious that emergency action (such as instant hospital admission) is required.

REMEMBER

When you visit your doctor, it is VITAL that he/she understands what is troubling you. Although doctors are trained to communicate with their patients, you might find your doctor difficult to talk to or understand. Don't feel inhibited about discussing personal matters—he/she is professionally bound to keep your conversation confidential. You should be free to talk about anything that is bothering you. It is possible that you and your doctor might not 'get on'. Find another doctor.

Get organized

The responsibility for describing and pointing out symptoms lies with YOU—your doctor cannot read your mind. There are some important ways in which you can prepare for a visit to the doctor, to make the best use of the time:

◗ Check in your own mind why you want an appointment.
◗ Think over how you have been feeling. Have there been any significant changes in your eating habits, your bowel habits or sleeping patterns?
◗ Think about any particular worries that you may have.
◗ Make notes—your mind might go blank during a consultation or you might forget something important.
◗ Always be totally honest.

How doctors work

A consultation with your doctor usually involves several processes. The first is history-taking—the doctor asks a series of carefully selected questions, which are designed to build up a picture of your problems. While you are talking, a good doctor will not only be listening but also seeing whether your appearance gives clues to your state of health. He/she is checking to see if you look relaxed and generally fit and healthy. Your skin, hair, eyes and teeth all can give indications of your general condition.

If your doctor thinks it necessary he/she may move on to the next process—a physical examination.

Diagnosis

It is not always possible for a doctor to make an immediate diagnosis and it would not be proper for him/her to guess at your condition. If the doctor has taken a blood or urine sample, for example, there may be a delay until the results of tests have been returned.

If the doctor is able to diagnose your condition at once, it is VITAL that you understand the explanation the doctor gives you, and the prescription and the treatment he/she recommends. Write down any instructions and ask for an explanation of any part of the diagnosis that is too 'technical'.

Finally, FOLLOW THE DOCTOR'S ORDERS—but if the suggested treatment does not appear to be effective, make another appointment.

❗ WARNING

If you suspect serious illness or you are in considerable pain seek medical assistance—go to your doctor, call your doctor or (if necessary) call an ambulance. Of course it may be a false alarm, but a good doctor prefers a patient who is worried about a symptom to raise the alarm, rather than suffer in silence.

You may hold back because you don't want to 'bother' a doctor, but you are not qualified to decide whether a problem is serious or not. If there is any risk at all that the condition could be serious, speedy diagnosis and treatment are VITAL.

If you need a doctor to visit you at home, telephone early in the day, so that the doctor can plan the day's visits'.

If you, or a child, may be suffering from an infectious disease, don't carry the infection to the doctor's waiting room—ask for a home visit.

If a child's temperature is raised beyond 39°C (102°F), it is definitely advisable to ask a doctor to visit.

WARNING

X-rays are invaluable and — compared with recent inventions — employ quite basic technology, but they must be used with care. The x-ray 'image' is achieved by passing a form of electromagnetic radiation through the body. An image is created on a photographic plate behind the body.

The use of this radiation has recently caused great concern. At high levels, it can cause neurological disorders and cancers. The dosage of radiation necessary to take an x-ray photograph is usually small, but doctors are worried that too many unnecessary x-rays are given. The effect is cumulative. One report revealed that in the UK at least a fifth of all x-ray examinations did not actually provide new or useful information. ALWAYS check with the doctor that an x-ray is essential. Keep a personal record of your own x-rays — when they were taken and why.

With heavy workloads in many hospitals, radiologists and occasionally their patients have been exposed to unnecessarily high levels of radiation from x-rays, as a result of ineffective protective and screening procedures. In some countries x-ray procedures may have to be closely examined.

Seeing a specialist

Depending on your particular problem, your doctor may not be able to make a diagnosis from history-taking and a physical examination. You may be referred to a specialist or consultant, perhaps at a hospital. Consultants and specialists usually focus their training on particular types of problem and have access to a wide range of high-tech procedures. 'Imaging' techniques allow doctors literally to 'see' into their patients.

Ultrasound, computed tomography (CT) scanning, magnetic resonance imaging (MRI) and positron emission tomography (PET) scanning sound like science fiction. These techniques use various forms of electro-magnetic rays, sound waves and radioactive substances to create 'images'. There's usually little or no discomfort.

Endoscopy, the use of special viewing instruments that can be inserted into the body, has greatly improved since the advent of fibreoptic technology. If absolutely necessary, minute samples of possibly diseased tissues can be painlessly removed for examination (biopsy) at the same time. Quite often, if there is an incision at all, it's tiny.

Ultrasound

Ultrasound is a diagnostic method in which high-frequency sound waves are transmitted into the body. Their reflections show images of body organs, observable on a screen. It is thus possible to detect abnormalities and pinpoint their position. It is claimed that the technique is completely safe. There is growing concern that the use of ultrasound during pregnancy should be limited to one 'screening', unless an abnormality is found. No one yet knows whether long-term damage to the baby might be caused by too many routine screenings.

TREATMENT

✚ Even the relatively short space of ten years has transformed some areas of modern medicine. While the basic practices of the doctor's surgery remain unchanged, the treatment of illness has seen major advances. Like other areas of modern life, computer technology has speeded up developments, allowing medical science to discover and understand more about the human body.

Drugs

Discoveries have also been made in the development and production of new drugs. More is now known about the ways in which many drugs work—this means they can be made safer and more effective. Modern biochemistry has made it possible to create synthetic drugs, but we still have an enormous amount to learn about the usefulness of substances found in 'nature'. More and more naturally-occurring substances are being 'discovered' and put to use.

A better understanding of how the body works has led to the ability—again only recently achieved—to 'tailor-make' drugs for specific purposes. This fine-tuning can result in fewer harmful side effects.

Surgery

Many surgical procedures have been superseded by less-invasive techniques.

Lithotripsy This is a non-surgical treatment that uses ultrasound to pulverize deposits such as kidney stones in the body. Repeated pulses of high-energy ultrasound waves are focused on a stone to break it down into particles small enough to be passed out of the body in urine or faeces.

Laser treatment The benefit of the laser—a narrow beam of intense energy—is that it can be used to make highly exact and fine cuts in tissue. At the same time, it cauterizes severed blood vessels, thus sealing them and stemming blood flow. Extremely delicate operations, such as eye surgery, have been made safer and faster by such exact treatment.

Endoscopy Not only does endoscopy have a diagnostic use, it has also made it unnecessary to 'open up' patients for certain surgical procedures. Special tools can be attached to thin endoscopes, allowing surgeons to carry out surgery inside the body—sometimes through only tiny incisions.

When more traditional surgery is required, it needn't be at all dangerous or frightening. There have been significant discoveries and improvements—a greater understanding of body tissues and the development of new materials, such as special plastics and carbon fibres. Surgical implants can be carried out with a much greater degree of success. Where arthritis has led to joint damage, it is often possible to replace an affected joint with an artificial one. Hip-joint replacement now lasts for ten or more years in 70 per cent of cases.

Transplants

There have been remarkable developments in the field of transplant surgery. In the early days, only organ transplants between identical twins were possible, because the body would only accept an organ that essentially matched the original. These days the problems of rejection have been reduced by the development of various drugs. Kidney, heart and lung transplants are carried out every year, with varying degrees of success. However, finding a suitable donor remains the first hurdle. Carrying a donor card means your organs could be used to give another person life in the event of your death.

ALTERNATIVES

Orthodox medicine is not the only way of treating ill health. There have been other methods in use for centuries. Many people have become suspicious of orthodox medicine. There have been instances where great damage has been caused by 'unsafe' drugs prescribed before sufficient testing has been done. Side effects of many major drugs may be as bad as or worse than the symptoms they were supposed to cure. It is recognized that no medication is without side effects, however slight.

Most alternative medicines and therapies are seen as 'kinder' or 'gentler' treatments. It's no longer fashionable to sneer at alternative medicine—more properly called complementary or holistic medicine. It is usually the intention of such practitioners to treat the whole person or the root cause of the disease—not only the symptoms.

Many so-called 'primitive' societies evolved complex systems of medical treatment, which have helped them for centuries. Most of these were based on plants—valuable medical resources that are now at risk with the destruction of the habitat in which they grow. Herbalists created many of the orthodox medicines used today—common drugs such as aspirin, some heart medicines and birth control pills are based on 'natural' substances. However, bear in mind that many off-the-shelf herbal medicines in health-food shops have not been offically tested or are not regulated because they are allowed to pass as 'food supplements'. Some have proved dangerous.

Practitioners of alternative medicine often spend more time on diagnosis than is practical in the orthodox doctor's surgery. This can be a particular bonus when patients are suffering from stress-related disorders.

It is difficult to make sure that any alternative practitioners you consult have been properly trained—there are some charlatans about. Ask about their qualifications and membership of the appropriate organizations. There is usually a national or regional 'headquarters' which keeps a register of approved members. Look in your phone book or ask at a public library. Some forms of alternative medicine have little or no scientific basis. This, however, does not mean that they won't work for you. Some of the most popular are:

- **Osteopathy** and **chiropractic** treat joint and muscle problems by manipulation.
- **Homeopathy** treats diseases and disorders by administering small amounts of substances which, in healthy persons, would produce symptoms similar to those being treated.
- **Acupuncture** uses the insertion of needles through the skin to stimulate nerve impulses to treat the patient. Some conventionally-trained doctors now also use acupuncture.
- **Acupressure** uses pressure and massage in the same way as acupuncture uses needles (also known as shiatsu).
- **Hypnotherapy** uses hypnosis in the treatment of emotional and psychogenic problems. It is sometimes used to 'cure' the cigarette addict!

Many people choose these 'alternatives' over 'conventional' medicine. Others may investigate the possibility of these forms of treatment only when they have a persistent, 'incurable' or terminal condition. This can sometimes lead not necessarily to a miracle 'cure', but at least to the alleviation of some symptoms. There is often a therapeutic value in just receiving caring attention.

When a serious illness has been treated, or an operation performed, you may be weak and 'delicate'. If you have been told that your condition is progressive or incurable, you may be deeply depressed or shocked. You may have been told—or have deduced—that your lifestyle will have to change.

Even a minor operation might require lengthy recuperation that could interrupt your life and make demands on relatives and friends—the people who care for you. In these situations you must seek support, you cannot face this crisis on your own. Other conditions—multiple sclerosis, for instance—may pose long-term problems, with varying levels of assistance required as the illness progresses.

Professional counsellors are trained to help people deal with the changes in lifestyle that a health crisis can cause. Your condition may cause you to feel acutely depressed. This can also be an 'illness'—slowing recovery. There are support groups for almost every condition—seek them out.

The spread of AIDS has helped to publicize the psychological effects of being a victim of illness. AIDS sufferers, but also victims of other diseases, often find that they are rejected by the community—yet in most cases they pose no health risk from normal daily contact with people around them.

People with illnesses or longer-term diseases need and must have physical assistance and emotional support, and may look to you to give it to them. If your relative or friend is in this position, find out all you can about the illness, join a support group and DON'T be afraid to ask for the help of others.

Self-treatment

Do NOT attempt to 'self-prescribe' with over-the-counter medicines or vitamin and mineral supplements on a regular basis. Most preparations have side effects of one sort or another, or can cause allergic reactions—either at once or by 'acquired sensitivity' over a period of time. If your diet is adequate and varied, supplements should be unnecessary. If you regularly take unprescribed medication for pain (of any sort), for recurrent coughs and colds, or to regulate bowel movements, you may be harming yourself or masking the symptoms of a more serious problem. You MUST seek medical attention (see POISONS: **The medicine cupboard**).

RECOGNIZE ILLNESS

Although the best protection against illness is to look after yourself and stay fit, some risk of disease can never be eliminated. You may still have to deal with a health crisis. While you may be able to cope at home with minor upsets (colds and flu, for instance), obviously there are more serious conditions that will require medical assistance.

It would require several medical textbooks to cover every possible serious or worrying medical condition. This is only a general guide to recognizing and dealing with medical emergencies. Without proper training, you cannot be expected to assess the seriousness of a condition, or to make an accurate diagnosis—although some medical emergencies may be obvious.

REMEMBER

Reading and committing to memory some of the information in this section could help you to save a life—at least help you to prevent unnecessary suffering, if you recognize an illness quickly. Often, someone may be ill for days before a doctor is called!

WARNING

EMERGENCIES: Many illnesses, diseases and disorders may become life-threatening—some slowly, others very suddenly. If a person you know is suddenly in great pain, collapses, convulses, vomits uncontrollably or runs a high fever, you shouldn't hesitate to seek urgent medical assistance or call an ambulance. They may have a previous history of illness or even a permanent health problem or allergy.

If the 'victim' is a stranger, it may be difficult to deduce what is wrong with them. If they are conscious, ask them if they know what is wrong. They, too, may have a 'permanent' illness or condition—they may even carry a card or wear a bracelet to let everyone know they have a serious health problem or are on regular medication. If the 'victim' is unconscious, search them quickly for any evidence to suggest that they have a particular problem of this sort.

Unsteadiness, irrational or abusive behaviour, or collapse may suggest a person is drunk. It can also result from a serious emergency, such as may occur in someone who is diabetic. If someone collapses, obviously from alcohol, place them in the recovery position (to keep the airway open). Don't forget that anyone who is unconscious is at risk and that alcohol can lead to severe poisoning and death.

The intention in this section is to help you know when to call for help. If in any doubt—especially if someone is obviously very ill—don't waste time thumbing through a book. Pick up the telephone instead!

Seek medical attention indicates that the sick person should be seen by a doctor. You'll have to decide whether to take them to the doctor, to call the doctor to visit or to drive them to the casualty department of a hospital.

Seek urgent medical attention indicates that the condition may be more serious or even very serious. Don't waste a single moment!

CALL AN AMBULANCE means just what it says. Do so, and then monitor the 'patient's' condition or apply first aid (see **Save a life!**) until the ambulance arrives.

CHILDHOOD DISEASES

As soon as a child starts mixing with other children, at playgroup, nursery or school, he/ she inevitably comes into contact with all the common childhood infections. For two or three years there will be not only colds, tonsillitis, ear infections and so on, but the risk of catching the well-known 'infectious diseases'.

Many of these can be prevented if the child has been immunized against them. Parents should make sure to take their child to the clinic or doctor at the correct ages, and that they keep a record of all the routine immunizations. Some of the diseases—for example, whooping cough in babies and young children, polio at any age—can be very dangerous.

Immunization, whether by injection or by swallowing a drop of the vaccine on a lump of sugar (polio immunization) usually has few after-effects. However, some may cause slight pain and swelling at the site of the injection, followed by mild fever (something like flu) and irritability. Painkillers such as paracetamol (NEVER give aspirin to a child under twelve) can help. Sometimes, as in the case of measles, there may be mild symptoms of the disease, which will pass quickly.

Some parents have been worried about the whooping cough vaccine because in a tiny minority of cases there has been brain damage following this immunization. If you are concerned about this, talk to your doctor. Medical opinion favours immunization because whooping cough in very young children can be a VERY serious disease, with much greater risks than those posed by the innoculation.

AT 2, 3 AND 4 MONTHS
- DPT (one injection)

For diphtheria, whooping cough, tetanus
- Polio (by mouth)

AT 12–18 MONTHS

Usually before 15 months
- MMR (one injection)

For mumps, measles and German measles

AT 5 YEARS

About the time of starting school
- Diphtheria/tetanus (one injection)
- Polio (by mouth)

Schedule in the UK, as of May 1991

CHICKEN POX

Common infectious childhood disease. Not usually serious, but rare cases involve complications. Once you've had chicken pox, you are immune. The virus is spread in airborne droplets. Infectious from about two days before a rash appears, to approximately one week after. The same virus *Herpes zoster* can cause shingles in later life.

SYMPTOMS Two to three weeks after contact, a rash appears behind the ears, in the mouth, armpits, on the trunk, upper arms and legs. There may also be a dry cough. Within a day or two, rash resembles clusters of itchy red spots, which turn into fluid-filled blisters. These soon dry out to become scabs. There may be a raised temperature.

WARNING In rare cases there may be complications, with coughing, vomiting and a high fever. If this occurs, or if the spots become inflamed, if the eyes are affected or a severe headache starts, seek urgent medical attention.

ACTION Seek medical attention—a home visit is advisable. Treatment includes low doses of paracetamol to reduce fever and calamine lotion to relieve itching. The child's nails should be kept short. Scratching of the blisters and scabs should be discouraged, to prevent infection and scarring.

WARNING Adults who have never had chicken pox should stay away from anyone who has got it (or away from any adults with shingles). Women in the final stages of pregnancy should be especially careful—newborn babies can develop severe/dangerous attacks of the disease. Anyone who is being treated for leukaemia or is HIV positive could be at great risk.

SHINGLES An adult person, who has had chicken pox, may contract shingles.

SYMPTOMS Pain in the area of skin affected, then, about five days later, a rash. This starts as small, raised, red spots that soon become blisters. After three days the blisters turn yellow, then dry and crust over. Within a couple of weeks the crusts fall off, often leaving small scars.

ACTION The sooner medical treatment with anti-viral drugs is given, the better the chances are of reducing the severity of the disease and nerve damage.

WARNING Once the disease has passed its earliest stages, little can be done to relieve post-herpetic pain. On the face, shingles is usually confined to the eyelids and forehead, but can be extremely serious if the eyes are affected.

MEASLES

Viral illness which may be serious. Usually affects children, but can also occur in adults. In most cases, once you've had measles, you're immune.

SYMPTOMS Fever, rash, runny nose, sore eyes and a cough. In four days, rash develops on the head and neck and spreads to the whole body. Lymph glands may become enlarged. Symptoms start to disappear after three days.

ACTION Seek medical attention—a home visit is advisable. In rare cases complications develop. Give low doses of paracetamol (NOT aspirin) and plenty of drinks. Antibiotic drugs will take care of any minor, bacterial infections if these occur.

PREVENTION Important, because of possible complications. The MMR vaccine is now routine. See **Immunization schedule.**

WARNING Those with impaired immune systems, such as anyone with leukaemia or who is HIV positive, should keep away from children with measles. This also applies to pregnant women—20 per cent of pregnant women with measles miscarry—though there is no evidence that the illness causes any birth defects.

MUMPS

Serious viral illness. Usually affects children, but may also occur in adults. Once you've had mumps, you are immune. 25 per cent of boys/men who contract mumps (after puberty) experience inflammation of the testes (orchitis). If this occurs, one testis may be swollen and painful for a few days and may shrink slightly when the illness passes. If both testes are affected, there is a slight chance of subsequent infertility. Women may develop inflammation of the ovaries. In either sex, inflammation of the pancreas can cause abdominal pain.

SYMPTOMS Usually develop gradually. A raised temperature, difficulty swallowing and pain behind the ears, leading to swelling of glands beneath the jaw bone on one or both sides of throat. Swelling usually disappears after seven days.

ACTION Most cases last a few days only. A doctor will make sure there are no complications. Give low doses of paracetamol (NOT aspirin, for children under twelve) and plenty of fluids. Patient should rest in bed until symptoms have gone. If chewing is painful, soup is a good idea. The main priority in severe cases of orchitis is to reduce the pain and inflammation.

PREVENTION MMR vaccine is part of the routine immunization schedule. Boys (after puberty) and men should avoid infection if they have not had mumps or been immunized against the illness. If symptoms of mumps do start to develop, an immunoglobin injection may offer a degree of protection against orchitis.

POLIO

Infectious viral disease, which mainly (but not exclusively) affects children. Three closely-related polio viruses have been identified. Infected people pass the virus in their faeces. Scrupulous hygiene is VITAL.

SYMPTOMS Sore throat, mild fever, headache, vomiting and muscle ache are early signs of the minor form of polio and last for two or three days. Only about 15 per cent of those infected go on to develop the major form of polio, in which the early symptoms are followed by chronic head-ache, high fever, intolerance of light, a stiff back and neck, muscle pains and perhaps twitching, all caused by inflammation of the membranes covering the brain and spinal cord (meninges). In rare cases, paralysis of some muscles occurs. This may lead to breathing difficulty and may even cause death.

ACTION Seek urgent medical attention or call an ambulance. Although the majority of those who suffer the most serious kind of polio make a full recovery, cases where paralysis sets in can be fatal and MUST be treated straight away. Hospitalization is VITAL, in case artificial respiration may be needed. There is no effective drug treatment, but physiotherapy can aid recovery.

REMEMBER Polio can be prevented by immunization at specified ages during childhood. As a result of routine immunization, polio is very rare in industrialized countries, but still a risk in the Third World.

GERMAN MEASLES

Also called rubella. Common viral illness in children. Only a few similarities to measles. It is of importance as a severe risk to women in the early stages of pregnancy. Once you have had rubella, you are immune.

SYMPTOMS A rash of pink, slightly raised spots on the face, behind the ears and on the trunk and limbs for a few days. Glands, especially near the back of the neck, may become swollen. There is often a fever and swollen—sometimes painful—joints.

ACTION Paracetamol should reduce the fever, and fluids may be given to drink. If there is pain from the joints or glands, or the patient's temperature rises, call a doctor.

WARNING A pregnant woman should avoid ALL contact with anyone who has rubella. If she is exposed to the virus there could be severe consequences for the foetus. Women who have not had rubella, and wish to have children, should be immunized against the virus. Pregnancy should not be considered until at least two months after the immunization.

SCARLET FEVER

Infectious disease, caused by an airborne bacteria. It is transmitted when the patient coughs. These days it is no longer a dangerous disease in the UK.

SYMPTOMS Sore throat, fever, headache and a rash, which begins as tiny red spots on the neck and upper body and soon spreads over the face. White coating and red spots cover the tongue. A few days later the tongue is bright red. As the fever and rash start to subside, the skin on the hands and feet may be flaky or peeling.

ACTION Seek urgent medical attention. Antibiotic drugs, such as penicillin, may be prescribed. Ensure plenty of rest and fluids. Paracetamol will ease the fever.

WHOOPING COUGH

Caused by an airborne bacterium, whooping cough (pertussis) can be fatal. Spread when the sufferer coughs. Usually affects small children and babies. Most infectious before the 'whoop' of the cough has begun.

SYMPTOMS Sneezing, mild cough, runny nose and sore eyes. Soon develops into bouts of coughing, usually with the characteristic 'whoop' sound. Symptoms may last for up to ten weeks, if not treated immediately.

ACTION Seek urgent medical attention, especially if a child vomits after coughing or appears to turn blue. Keep the patient warm. Give plenty of fluids and small regular meals. Avoid exposure to cigarette smoke and draughts.

PREVENTION Prevention is important, because of possible complications. These include dehydration and pneumonia, which are life-threatening to very young children and babies, especially if premature. Infants should be immunized.

TONSILLITIS

Infection of the tonsils, causing throat and tonsils to become sore and visibly inflamed.

SYMPTOMS Difficulty swallowing, fever, earache, headache, swollen and sore lymph nodes in the neck, and a bad taste in the mouth. Temporary deafness may occur, or an abcess beside the tonsils (quinsy).

ACTION If the symptoms do not clear in a day or so, or there is a severe infection, seek medical attention. Antibiotics may be necessary. Ensure plenty of rest in bed, lots of drinks. Low doses of paracetamol (NOT aspirin) may relieve the fever.

HEART DISEASE

In many countries, heart disease is the leading cause of death among adult males—the incidence among women is increasing. About eight babies in 1000 are born with a heart defect, usually as a result of the mother contracting German measles (rubella) during pregnancy (see **Childhood diseases**) or as a complication of Down's syndrome.

The most common form is ischaemic heart disease, leading to a narrowing or obstruction of the coronary arteries which supply blood to the heart muscle. Chest pain (angina pectoris)—caused by lack of blood supply to the heart muscle—is brought on by severe stress or exertion, but may be relieved by rest. Unremitting chest pain may not be due to angina, but to a myocardial infarction ('heart attack').

See a doctor immediately if you suspect any sort of heart trouble. Early treatment is VITAL. Prevention and treatment of heart disease include improving the diet—eating less animal fat and salt—exercising and stopping smoking/drinking.

HEART ATTACK

When a section of heart muscle 'dies', it is known as a myocardial infarction.
SYMPTOMS Overpowering, relentless pain in the chest, shortness of breath, giddiness, profuse sweating, irregular pulse, blue skin and lips, vomiting and possibly collapse and unconsciousness.

Cardiac arrest means that the heart has stopped pumping. It may be caused by myocardial infarction, hypothermia (dramatic heat loss), drug overdose, electrocution, severe blood/fluid loss, or anaphylactic shock (acute allergic reaction).

SYMPTOMS Sudden collapse, unconsciousness, no pulse, but initially would still be breathing. Monitor breathing closely and be prepared to apply artifical respiration.

WARNING The response to both a myocardial infarction and a cardiac arrest should be to CALL AN AMBULANCE. It is VITAL to act quickly to avoid the risk of possible brain damage. While you are waiting for medical help to arrive, give cardiac compression and artificial respiration (see Save a life!).

HYPERTENSION

Unusually high blood pressure (measured in the main arteries), even when resting. Most people's blood pressure is increased as a response to exercise and in times of stress, but decreases when they are resting.

SYMPTOMS There may be no symptoms. If the hypertension is severe, there could be shortness of breath, headaches, dizziness and problems with the eyesight. Antihypertensive drugs can be used for severe cases or as a last resort, but self-help measures are the preferable option.

PREVENTION Many people have hypertension without knowing it. For this reason, it is important to have regular medical check-ups. The condition affects up to 20 per cent of adults, particularly middle-aged men. To some extent it can be avoided by stopping smoking, stopping drinking, losing weight (if overweight), decreasing animal fat and salt intake, and exercising regularly. Unchecked, hypertension may lead to heart failure, heart disease, damage to the kidneys or the eye retinae (retinopathy).

CANCER

Over one in four people are affected by cancer during their lifetime. It is known to cause over 140,000 deaths every year, in the UK alone, and (after heart disease) is the second most common cause of death in the Western world.

An abnormal growth or lump in the body is not necessarily cancerous (malignant)—most growths are benign. When cancerous cells in organs or tissues do occur, they may destroy or disturb circulation, bones, nerves and tissues. Cancerous cells sometimes travel rapidly through the system to form new tumours elsewhere in the body.

Whether or not a person develops cancer is largely determined by their genetic make-up. Cancer-causing agents may trigger the disease. Cancer has been linked to many factors, including diet, smoking, drinking alcohol, occupational hazards and the use of food additives. Sexual/reproductive history can also prove an influencing factor in the incidence of cancer in women. Cervical cancer has been linked to number of sexual partners, for example, whereas motherhood whilst comparatively young may lessen the likelihood of breast cancer.

Australia has a high incidence of people with skin cancer. Those most at risk in the northern hemisphere are pale-skinned holiday makers, who expose themselves to the sun in short bursts (see IN TRANSIT).

What to watch for

The earlier the diagnosis, the higher the chance of a cure—80 per cent of breast lumps turn out to be benign. The cure rate for testicular cancer in its early stages is between 95 and 100 per cent. Cancer produces a range of symptoms. Any of the following signs should be reported to a doctor immediately, if they persist for several days or more.

◑ Changes in breast shape or 'lumps'.
◑ Unexplained nipple discharge.
◑ Menstrual bleeding, unrelated to periods or menopause.
◑ Development of lumps or changes in the scrotum.
◑ Constant unremitting pain in the abdomen.
◑ Bleeding (painless) during urination.
◑ Persistent problems with bowels/defecation.
◑ Throat problems (especially when swallowing).
◑ Unexplained grating or husky voice.
◑ Heavy cough with blood in the phlegm.
◑ Regular headaches.
◑ Mysterious and swift loss of weight.
◑ Sores or ulcers which do not heal.
◑ Moles which grow or constantly itch.
◑ Lumps or patchy changes in skin condition.

In recent years, the means of diagnosing cancer at an early stage and the range of treatments available have dramatically improved. Screening tests, microscopic examination of cells by biopsy, scanning and imaging techniques all give VITAL diagnostic information and cause little discomfort to the patient. Almost half of all cancers may now be cured. The survival rate, for the years after diagnosis, continues to increase—although it does depend on the part of the body affected.

LEUKAEMIA

Name for forms of cancer in which there is an overproduction of white blood cells from the bone marrow. This inhibits or disturbs the manufacture of normal red and white blood cells. As these invade various organs in the body, other problems may develop—especially in the spleen, lymphatic system, liver or brain.

SYMPTOMS 'Sore' bones, bleeding gums, recurrent headaches, unexplained bruising, pale complexion, fatigue, swollen lymph nodes (in groin, neck and armpits), recurrent infections (including coughs and sore throats).

ACTION Seek urgent medical attention. Depending on the type of leukaemia, antibiotic or anticancer drugs, radiotherapy, or bone marrow transplant may be used.

A progressive disease which affects the central nervous system, causing the destruction of areas of the covering (myelin) of the nerve fibres in the brain and spinal cord.

MS has an enormous variety of symptoms, ranging from numbness and tingling in the limbs, to paralysis and incontinence. General symptoms include fatigue, apparent clumsiness, 'loss of balance', visual disturbances and muscle weakness. There are often dramatic and unpredictable improvements and relapses—a patient may be severely disabled one week and seem normal the next.

Relapses are more common after injury, infection and physical or emotional stress. The patient may be left with mild symptoms for long periods of time, or even no symptoms at all. Tragically, some people may be severely disabled or even paralysed in their first year of illness.

Research to find a cure continues. Some people with MS believe that changes in their diets, including reducing the intake of starch and increasing sunflower or evening primrose oils, can help. As yet, no drug has been found to help all MS victims, although some drugs alleviate specific symptoms. Perhaps the most beneficial treatment is physiotherapy, which strengthens muscles and restores mobility and independence.

EPILEPSY

A disorder which can lead to occasional losses of consciousness. Convulsions may occur. The condition may be 'inheritable'. Seizures may be triggered (for instance) by injury or severe illness. There are two main types of epileptic seizure—general (grand mal) and absence (petit mal).

GENERAL SEIZURE

SYMPTOMS Having fallen to the ground unconscious, the victim will firstly go rigid, then start to twitch involuntarily. Their breathing may be erratic or stop totally. After the victim's muscles relax—bladder and bowel control may be lost. He/she may be confused, may feel very drowsy, and usually cannot remember the event.

ACTION Carefully loosen tight clothing around the neck. Move anything which might cause harm—especially to the victim's head. NEVER restrain the victim or place anything in the mouth. NEVER attempt to move the victim unless in danger of injury. When the attack is over, place the victim in the recovery position (see SAFETY FIRST) and allow him or her to regain consciousness.

WARNING Seek urgent medical attention. CALL AN AMBULANCE if seizure continues for more than five minutes, if another seizure follows or if consciousness is not regained a few minutes after the seizure has ended. Prolonged seizures can be fatal without immediate emergency treatment.

REMEMBER Epileptics should advise their colleagues on what to do if a seizure occurs and carry a special card or bracelet stating that they have epilepsy. Statistics suggest that the condition may be worse in childhood and adolescence. Many people 'grow out' of the condition, or find that medication helps to keep it under control.

ABSENCE SEIZURES

A temporary loss of consciousness for up to thirty seconds during which the epileptic is oblivious to his/her surroundings, appearing to be daydreaming. The attack may even pass without anyone noticing. There may be many 'attacks' in a day, especially in children, and schoolwork could suffer.

DIABETES MELLITUS

A disorder in which the body does not manufacture the hormone insulin in sufficient quantities, if at all. Insulin controls the way the body regulates sugar levels in the blood. This condition may be present from birth, although it may manifest itself at any time during a diabetic's life.

SYMPTOMS Excessive thirst. Constant need to pass urine. Weight loss, fatigue and hunger, caused by the body's inability to use or store energy-giving glucose. Blurred vision, muscle weakness, tingling and numbness in the hands and feet. Recurrent skin infections, candidiasis (thrush), and urinary tract infections such as cystisis. Most diagnosed diabetics can usually treat themselves—but may need help.

WARNING If a 'diabetic' or someone suspected of having diabetes seems in distress and unable to help themselves, seek urgent medical attention or CALL AN AMBULANCE. Without regular insulin injections, the sufferer may lapse into a coma and die.

DIABETIC EMERGENCY

Two main conditions can occur. Hypoglycaemia — too little sugar in the blood — can rapidly cause a life-threatening condition. Hyperglycaemia — too much sugar in the blood — normally develops slowly and the diabetic is prepared for the problem.

SYMPTOMS Low blood sugar may cause the diabetic to appear dizzy or drunk and aggressive. He/she may look pale and sweating. Breathing is shallow. There may be trembling and a rapid pulse, possibly with rapid loss of consciousness. The diabetic may or may not be aware of the problem.

ACTION If you are quick, it may only be necessary to give the casualty a sweet drink, sugar or chocolate to raise the blood sugar level. If the diabetic lapses into unconsciousness, seek urgent medical assistance.

WARNING Diabetics who have been taking insulin for quite a few years may have few warning symptoms that they are slipping into hypoglycaemia. An overdose of insulin may be extremely dangerous. Diabetics should not miss meals or eat only tiny snacks. Low blood sugar levels, if allowed to persist, can cause brain damage. It is possible for the condition to lead to coma and death. If you know a diabetic, discuss the situation so that you will know how to help if a problem arises.

GENETIC DISORDERS

There are many diseases and disorders which are passed on from generation to generation through families. Some genetic defects may not be apparent until the person is mature—others may be evident at birth. Some traits, such as colour blindness, do not usually cause any real disability. Others, including haemophilia or sickle-cell anaemia, may be life-threatening.

Traits for genetic disorders may be carried by either parent. Sometimes, through mutation of the genetic material, a child can be born with a disorder that has never been seen in the family before. Whatever the origin of the condition a person may inherit, it may directly affect their health, their fitness and susceptibility to disease throughout their life.

When considering having children, couples with known genetic disorders, or who have genetic problems in their family history, should seek medical advice about the likelihood of disorders occurring in their children.

The couple may find professional counselling of great help. This is obviously an emotionally-traumatic situation and professionals are able to help with the short- and long-term problems that may arise. They are also in a position to provide support over time and instigate communication with many counselling groups.

This is by no means a comprehensive list of 'inheritable' diseases—but indicates how serious and complex they may be.

CYSTIC FIBROSIS

Cystic fibrosis may be evident from birth, although it may not be detected for months or even years. It is caused by a defective gene. The main manifestations are repeated infections of the lungs and the failure of the body to make use of food. A child may 'fail to thrive'. A sticky mucus secretion, may be present in the nose and throat, the air passages and sometimes in the intestines. Faeces may be pale, shapeless and strong-smelling. Males and females may later suffer from infertility.

Early diagnosis and treatment are VITAL to avoid complications like pneumonia, bronchitis and lung damage. Antibiotic drugs and physiotherapy can be of great help.

A special protein-rich, high-calorie diet should be devised with added vitamins to encourage the digestive system to function more normally. Current medical care means that, although there is still a risk of permanent lung damage, the quality of life for those affected is now far higher than it would have been a few years ago.

HAEMOPHILIA

A disorder caused by a deficiency of a blood protein, known as factor VIII, which is essential to blood clotting. Haemophilia is the result of a defective gene, carried by females but almost always passed on to males.

Haemophiliacs tend to experience recurrent bleeding, not always caused by injury, usually into the muscles or joints. Bleeding episodes may be painful, although the severity varies greatly.

Early diagnosis and control of severe bleeding episodes is vital to avoid crippling deformities of the joints. Bleeding is usually controlled with concentrates of Factor VIII, either as a preventative measure or as treatment when bleeding has just started.

Being a genetic disorder, it's quite likely that a haemophiliac will have a close male relative also suffering from the condition (although in a third of cases there is no history of it). Haemophiliacs and their female relatives should seek genetic counselling before starting a family. Haemophiliacs should avoid strenuous contact sports and opt instead for gentler forms of exercise, such as walking and swimming.

HUNTINGTON'S CHOREA

A rare inherited disorder affecting the central nervous system, manifesting itself in middle age. There is a 50–50 chance that a parent may transmit the relevant gene to a child.

When the condition becomes apparent, walking, speaking and hand movements gradually become difficult. Spasmodic and jerky movements may escalate, possibly followed by mental deterioration and severe behavioural problems—often requiring hospitalization.

There is no known treatment, though tranquillizers may calm the nervous system. Genetic counselling is essential for those at risk, if they intend to start a family.

MUSCULAR DYSTROPHY

A muscle disorder of unknown cause. Muscular dystrophy may appear at birth, in infancy or as late as the child's fifth or sixth year. The most common form is duchenne muscular dystrophy, the result of a recessive gene that is carried by females and passed to males.

Once the disorder becomes apparent, there may be a slow but progressive degeneration of the muscle fibres. It is usually diagnosed by the patient's appearance and movements.

As yet there is no 'cure' for muscular dystrophy, although many cases benefit from surgery to the heel tendons to improve walking ability. The affected person should be encouraged to remain physically active, to keep healthy muscles in 'working order'. They should avoid becoming overweight. The parents and brothers or sisters should seek counselling before having more children or starting a family.

SICKLE-CELL ANAEMIA

A blood disease, occurring mainly in black people and, sometimes, in people of Mediterranean origin. It causes the red blood cells to become abnormal, resulting in a life-threatening form of anaemia. In the UK alone, about one in 200 people of West Indian origin and one in 100 of West African origin have the disease. Approximately one in ten has the trait (in other words, carries the disease), which may be passed on when two carriers have a child.

The first signs—shortness of breath after exertion, jaundice, pallor, headaches and fatigue—may appear after the age of six months. The symptoms may be triggered by a childhood illness, dehydration as a result of prolonged diarrhoea and vomiting, or very cold weather.

Sickle-cell anaemia attacks various parts of the body. Children may experience pains, 'tender' bones, and blood in the urine from kidney damage. Brain damage may cause seizures, unconsciousness or stroke. Some children may develop a life-threatening form of anaemia. From puberty there is a risk of blood poisoning, if the spleen stops working. There is also an increased risk of pneumonia. Treatment includes immunization against pneumonia, and penicillin, antibiotic drugs, painkillers and folic acid supplements.

People who do not know whether or not they carry the sickle-cell gene MUST find out. A couple who both carry the gene or have the disease should obtain counselling before having children. Although the survival rate in under five-year-olds is still poor, improved methods of treatment have increased the chances of longer-term survival. Thirty years ago the disease was usually fatal to children. Today, some sufferers are able to have children of their own.

BACK/JOINT PAINS

The most common causes of 'time off work' are musculoskeletal strains, which include back pain and sports injuries. These warrant a visit to the doctor and may be soothed by painkillers and alleviated by physiotherapy. Some resolve themselves, spontaneously. Most back strains can be avoided, by not overstraining when lifting heavy weights—keep your back straight and bend your legs when you must move weighty objects.

RHEUMATOID ARTHRITIS

More common in women than in men, it may begin insidiously and can lead to joint deformity. Most of the connective tissues may be affected, but the worst damage is usually done to the joints.

SYMPTOMS Stiffness, pain and swelling in the small joints of the hands and feet, followed by problems with the larger joints, such as hips and knees. There may be diminished movement of the joints and muscle wasting. Tissues round the

joint may become inflamed. Many of the internal organs may be affected, including the lungs, heart, kidney and eyes. The central nervous system is also at risk.

ACTION Seek medical attention to reduce the pain. Anti-inflammatory drugs, physiotherapy and (very rarely) steroids may help. The best advice is to start treatment early to head off the risks of disability.

OSTEOARTHRITIS

Affects almost everyone past the age of 65. There is a degeneration of the joint surfaces, especially where there has been a previous

injury or deformity. The incidence increases with age and obesity. Women tend to be more prone to the disease than men.

SYMPTOMS Pain is induced by movement, particularly in the evenings. Small joints of the hands are the first to become affected, in most cases, but may be followed by pains in the spine, knees and hips. As the condition progresses, there may be stiffness, immobility and even joint deformity.

ACTION Reducing the pain is the first priority. It may be necessary to use a walking stick or walking frame to reduce pressure on the joints. If the feet are affected, there are special shoes which may make walking easier. If overweight, weight loss would be advisable. In serious cases, joints may be fused or replaced surgically.

RESPIRATORY PROBLEMS

Respiration is the process by which oxygen reaches body cells and carbon dioxide is emitted from the body. Although it may seem simple, it involves a highly complex set of organs which are prone to disease, damage and infection. Respiratory disorders can affect the air passages, causing obstruction, or affect the lung tissues.

If there is no apparent external cause for severe breathing difficulties—airborne dust, fumes or a crushing injury—there are several other immediate possibilities. If the victim is not actually choking (see **Save a life!**) and does not have a history of respiratory illness, seek urgent medical attention, especially if blood is coughed up. The cause may be:

Oedema A dangerous build-up of fluid in the lung. This is most likely if the person is already ill with heart disease or kidney failure. Certain drugs, such as steroids, or hormone therapy may also cause fluid misplacement in the body. Seek urgent medical attention.

Pulmonary embolism Severe difficulty in breathing, dizziness, sharp pain in the chest when breathing, blood coughed up, rapid pulse rate may indicate that a small blood clot is restricting the flow of blood between the heart and lungs. Tends to affect women—after pregnancy or after surgery. Seek urgent medical attention.

Ventricular failure Most likely to occur when there is a severe problem with the heart, causing fluid to build up in the lungs. The heart would be treated first in this sort of emergency. Seek urgent medical attention.

ASTHMA

Bronchial asthma is a very distressing condition, normally associated with childhood—although it may persist into adulthood. Characterized by recurrent attacks of breathing difficulty, usually with wheezing. Most asthmatics have their first attack when they are small children—but it is possible for asthma to develop later in life.

SYMPTOMS Include mild or acute breathlessness, panicky wheezing, apparent inability to breathe in, coughing, sweating, inability to speak, rapid pulse and great anxiety. In rare cases, the attack may be severe enough to cause a blue facial pallor. Such an attack may be life-threatening. Seek urgent medical attention.

ACTION Most asthmatics carry medicines, usually in the form of an inhaler. They may need help to use the inhaler. It consists of a small aerosol with a mouthpiece. It is usually necessary to take two full puffs. The drug should dilate the constricted bronchioles in the lungs in 15 minutes — allowing air to enter the lungs. If the attack has been unusually severe, seek medical attention.

REMEMBER Some forms of asthma may be due to external causes, such as breathing in a harmful substance or a substance to which the person is particularly sensitive. There may not be any history of attacks. Many asthmatics have common allergies, such as hay fever or sensitivity to house dust. Attacks may be brought on by anxiety, stress or sudden emotional upset. Most 'grow out of' the condition by their early twenties. Even if the condition persists, there are very effective therapies and treatments.

PNEUMONIA

Inflammation of the lungs due to an infection, usually carried by viruses or bacteria.

SYMPTOMS Chills and fever, breathlessness and a cough with greeny-yellow phlegm or sometimes blood. It may be difficult or painful to take air into the lungs. This may be due to inflammation of the lining of the chest and lungs (pleurisy).

ACTION Seek urgent medical attention. Mild pneumonia usually responds well to antibiotic or antifungal drugs, depending on the cause of the infection, with paracetamol to reduce the fever. In some cases, particularly if the victim is elderly, hospitalization may be necessary.

TUBERCULOSIS (TB)

Tuberculosis is still a major problem in the Third World. It is an infectious bacterial disease — the bacterium is transmitted when the person sneezes or coughs. The immune system usually stops the infection, but scar tissue may be left in the lungs. In a small minority of cases the infection is more serious and may spread to other parts of the body. In these cases, further tissue damage may occur — especially to the lungs.

SYMPTOMS Pain in the chest, a heavy cough, breathlessness, loss of appetite, loss of weight and a sweaty fever which may be worse at night.

ACTION Seek urgent medical attention. If not treated early, the bacteria occasionally lie dormant in the lungs and other organs of the body, to be reactivated years later. Provided the full course of treatment is followed, most victims fully recover.

HYPERVENTILATION

Usually caused by extreme anxiety, hyperventilation takes the form of excessively deep or rapid breathing. Can also be caused by kidney failure, oxygen deficiency and diabetes.

SYMPTOMS Over-breathing reduces the carbon dioxide in the blood, making it more alkaline, which tends to increase feelings of anxiety — making the problem worse.

ACTION Breathing into a paper bag for a while should help to restore carbon dioxide to the blood. Even when the attack has passed, seek medical attention.

PANIC ATTACK Sudden, often unexplained, attack of acute anxiety. The main outward signs are often hyperventilation, heart 'palpitations', sweating, shivering, fainting and tearfulness. Given the effect hyperventilation can have on anxiety levels, the victim should be encouraged to breathe into a paper bag for a few moments while being calmed and reassured as much as possible. Attacks may be related to work, the home, crowds, public transport, being outside. Recurrent panic attacks may be symptoms of certain phobias and it is these phobias which should be treated. Attacks usually pass after a few minutes and, although very unpleasant, usually do no actual physical harm. Seek medical attention. There may be a help-group or counselling group which could provide support.

BRONCHITIS

Inflammation of the bronchi, the airways that connect the windpipe (trachea) to the lungs. Acute bronchitis may start suddenly but not last very long. Chronic bronchitis lasts longer and may recur over several years.

SYMPTOMS Persistent coughing, large quantities of yellow or green phlegm, wheezing, breathlessness, discomfort in the chest and, in acute bronchitis, fever.

ACTION Seek medical attention. Antibiotic drugs will help to treat or prevent a lung infection. In the case of chronic persistent bronchitis, a doctor will check that any blood coughed up is not a result of a more serious lung problem

PREVENTION The most common cause of bronchitis is a history of smoking. Those who are subject to bronchitis, especially chronic bronchitis, would be extremely foolish to continue to smoke. Passive smoking, that is, being in an environment of cigarette smoke, may also lead to bronchitis.

EATING DISORDERS

Every time we switch on a television or open a magazine, we are confronted with images of slim healthy people with perfect teeth and perfect lives. Young people, especially teenage girls, are particularly susceptible to this onslaught. At a time when their bodies are changing in a way they cannot control, young people may find that they can control their figures. They may begin to starve themselves for long periods of time (anorexia nervosa), or overeat in binges and then force themselves to vomit (bulimia nervosa). Both are dangerous. Seek medical attention.

Obesity is in many ways the opposite of anorexia nervosa. Very overweight people consume far more in calories than their bodies are able to convert. If they do not have a network of friends or family to offer emotional support, people may resort to 'comfort eating', or deliberately gain even more weight as an expression of unhappiness.

An overweight person, as well as someone who appears to be healthy, may be suffering from malnutrition. A diet of 'fast' or convenience foods, chocolate bars and potato crisps, coupled with a social life based around restaurants and public houses, will not provide the balanced diet necessary for good health. The health of anyone, whatever their weight, will suffer. If you do live on pre-packed sandwiches and convenience foods or skip meals altogether, a multivitamin and mineral supplement may be advisable. Consult a doctor.

ANOREXIA NERVOSA

SYMPTOMS Severe weight loss, with great lengths taken to avoid food. Anorexics may lie to their families, saying they have eaten out so that they can miss meals. Extreme choosiness over food, excessive exercise, use of laxatives, 'baby' hair on body, thinning of head hair and absence of menstrual periods. Anorexics often refuse to or are unable to recognize their thinness.

ACTION Anorexia nervosa is difficult to treat and in rare cases may prove fatal. Hospital treatment is usual to help the person return to normal weight. He/she will often require psychotherapy for some time. Full recovery is possible. Under stress, relapses may occur. Doctors should be able to suggest self-help organizations which can be very supportive of former anorexics.

BULIMIA NERVOSA

Bouts of overeating followed by self-induced vomiting. Bulimia nervosa often starts as anorexia nervosa—in both there is a fear of becoming fat. Mainly affects girls and women between the ages of 15 and 30.

SYMPTOMS Bingeing and vomiting may occur more than once per day. In extreme cases this may lead to tooth decay from gastric acids, dehydration and loss of potassium causing weakness

and cramps. Bulimics may take laxatives to help 'drive' food from the body. They are not always very thin and may be of normal weight. They may be distressed about their compulsive behaviour, depressed or, in extreme cases, suicidal.

ACTION Treatment is required to supervise and regulate eating habits. Sometimes psychotherapy and antidepressant drugs are used. Full recovery is possible, but there is a risk of relapse.

FOOD HYGIENE

In recent years there have been well-publicized scandals connected with the food we eat— salmonella in eggs, for instance, and listeria in soft cheeses. Mass production of food may sometimes put profit before quality. Food-hygiene legislation is not always strict enough or properly enforced. The result is that food poisoning has increased—particularly in urban areas.

Take care to store and prepare food carefully in your home. Bacteria need food, warmth, moisture and time to multiply. They cannot grow on dried or dehydrated foods, which are safe until rehydrated. Bacteria thrive at 37°C (98.6°F)—body temperature—multiplying every two to 20 minutes. Danger zones are between 5°C and 63°C (40°F to 145°F). Normal room temperatures are within this range, so avoid leaving food out of the fridge for too long. A fridge thermometer will give a very accurate temperature reading.

BE SAFE, NOT SORRY

Simple food hygiene measures can virtually eliminate the risk of poisoning or contamination:

- Always wash your hands before and DURING food preparation— especially after handling meat and poultry, rubbish bins and after using the lavatory.
- Always rinse cutting boards and utensils after each type of food, to avoid cross-contamination, especially of raw and cooked meats.
- Always wash fruit and vegetables thoroughly in clean water.
- Throw away anything that smells or looks spoiled, bulging or rusting tin cans, any food that has passed its sell-by/use-by date, mussels that do not open when cooked.
- Make sure that frozen poultry is completely thawed before use, and then well-cooked.
- Keep refrigerators between 2°C and 5°C (35°F and 40°F).
- Keep freezers at – 18°C (–0.4°F). Check the temperature of refrigerators and freezers regularly with a thermometer.
- Heat food such as soups and pies to a temperature of at least 63°C (145°F).
- Do not reheat food more than once.
- If using a microwave, check that food is cooked all the way through.
- Never leave foods such as butter, cheese and meats at room temperature.
- Never keep cooked and uncooked foods together.
- Change dishcloths and teatowels every day.
- Never allow pets, their hairs or cat litter on to work surfaces, especially if there is a pregnant woman in the house.
- Do not wrap cheese and other fatty goods in cling film (it contains a possibly dangerous chemical which may leach into fats).
- Babies, toddlers, pregnant women and the elderly should not eat eggs, unless these are cooked until both the whites and yolks are solid.
- Never brush or spray hair in the kitchen or near food; if your hair is long, keep it covered or tied back.
- Never smoke whilst preparing food.

Animals and food hygiene

Animals and food hygiene do not go together. Domestic pets and pests can cause diseases (see **Pets and your health**). Mice or rats can be detected by droppings, damaged packaging, spilled foods and broken eggs.

Mice are not usually a serious threat to health, but they are unhygienic. They are attracted by fallen scraps of food, so floors should be kept clean. Humane traps, which capture mice alive, can be used to reduce numbers, but mice tend to move around houses under floors, through wall cavities and roof spaces, and are therefore difficult to eradicate completely. Call your local environmental health officer or a specialist if the problem persists.

Rats pose a serious health risk and are increasingly prevalent in cities. They can be killed using anticoagulant poisons, but should be dealt with only by a specialist.

Bats are generally harmless to humans in temperate zones. Indeed, they are a becoming increasingly rare and are a protected species. It is illegal to interfere with them, or their nests. Call your local authority for advice.

Cockroaches may pose serious health risks—they may carry salmonella and hepatitis in their faeces. A severe infestation may be difficult to remove. Seek expert assistance.

FOOD POISONING

When a naturally poisonous substance or contaminated food has been swallowed, it is VITAL to pinpoint the source. There are many poisonous mushrooms, toadstools and poisonous plants. Vegetables and fruit that are normally safe may be contaminated by high levels of pesticide or fertilizer. Food may have been incorrectly stored or prepared. Sometimes, symptoms may develop rapidly or they may appear over a couple of days.

SYMPTOMS Nausea, vomiting, diarrhoea and stomach pain. Possibly shivering and fever. If infection is severe, there may be a risk of shock or unconsciousness.

ACTION Mild cases of food poisoning can be treated at home. Give plenty of liquids, with a small amount of salt and sugar — but nothing to eat. Most cases clear within three days. If in any doubt, seek medical attention.

WARNING If diarrhoea/vomiting is especially severe, or if victim collapses or has difficulty in swallowing/speaking, seek urgent medical attention. Do not induce vomiting as a way of removing infected foodstuff. If the poisoning is severe, it is helpful to keep samples of the suspected meal or vomit for tests help pinpoint the cause. Other people may be at risk.

BACTERIAL POISONING

SALMONELLA This name covers a group of bacteria. Commonly associated with raw chicken, eggs, raw meat and seafood. Symptoms usually develop within 24 hours. The young and elderly are particularly at risk. If a baby is affected, it may be vital to replace

lost body fluids. Anyone who has had salmonella food poisoning should bear in mind that — for several months after the symptoms disappear — they may excrete the bacteria in their faeces. Consequently scrupulous hygiene should always be practised.

TYPHOID Caused by one of the salmonella bacteria—*Salmonella typhi*. Associated with food or water contaminated by sewage. Alternatively flies or cockroaches may carry the bacteria to food. Shellfish exposed to sewage may be another risk. Any case of typhoid MUST always be reported to the local health authorities in industrialized societies—where the disease has become very rare (see **Tropical diseases**).

LISTERIA *Listeria monocytogenes* is a very common bacterium. Humans may be exposed to it from salads, chilled food which is heated for eating, badly-cooked meat, soft cheeses. It can multiply in chilled food in a refrigerator if the temperature is not low enough—one important reason for observing sell-by and use-by dates. The infection—known as listeriosis—may be very serious and can kill very young children, the elderly and anyone with a depressed immune system. Pregnant women may be at risk.

STAPHYLOCOCCUS Even proper cooking may not destroy toxins caused by various strains of 'staph'. The most likely cause is from septic sores on the hands of the person preparing food. Symptoms usually occur within two to six hours.

BOTULISM Rare, but associated with incorrect canning or preserving of food. The spores of the bacterium are resistant to cooking. Although rare in most developed countries, this is a very serious form of poisoning. Symptoms (apart from the symptoms mentioned earlier) usually develop within two days and include difficulty swallowing and disturbed vision. Early treatment is essential.

TROPICAL DISEASES

Many diseases of tropical areas are linked to inadequate hygiene and geographical factors such as temperature and humidity. Malnutrition may contribute to the spread of disease, because it lowers the body's ability to fight infection. Many of these diseases were once also common in temperate zones, but have been eliminated by improvements in public health and hygiene—in particular keeping water and soil free from contamination.

It's worth bearing in mind that many tropical diseases are very localized. Sleeping sickness, for example, an extremely serious parasitic disease spread by the tsetse fly, only occurs in areas of Africa. The malaria map of the world is constantly changing and, what's more, mosquitoes can become resistant to the preventative drugs. That's why you should always get your doctor to check the most up-to-date medical information for the region you intend to visit.

Vaccinations should be obtained before travel to tropical areas, but note that these only offer partial protection—other precautions, such as food and water hygiene, must also be observed scrupulously. Avoid uncooked food. Take insect-repellent sprays, a mosquito net to use at night and long-sleeved clothing and socks to protect exposed skin. For information on risk zones and special immunizations see IN TRANSIT.

CHOLERA

Infection of the small intestine, caused by food or water contaminated with the bacterium *Vibrio choleroe*. Cholera is prevalent in Africa, the Middle East and Far East. Cholera can only be controlled worldwide by improving sanitation and ensuring that contaminated water does not come into contact with drinking water. Major epidemics tend to follow natural disasters, especially in crowded conditions or if few people are left to nurse the sick.

SYMPTOMS Profuse watery diarrhoea and vomiting. Can lead to dehydration.

ACTION Seek urgent medical attention. If the diarrhoea is severe, the victim may dehydrate rapidly and die. Lost fluids should be replaced by plenty of liquid, to which has been added salt and sugar at one level teaspoon of salt/eight level teaspoons of sugar in one litre (about two pints) of water. Most patients make a full recovery, but some may need hospital treatment if dehydration continues despite fluid replacement.

PREVENTION Travellers to risk areas should obtain a cholera vaccine (which usually lasts for six months). Boil all drinking water or buy it from reliable sources.

DENGUE

Viral disease spread by the mosquito *Aedes aegypti*. Southeast Asia, Africa, the Pacific region and South/Central America are all risk areas. Outbreaks known in Mexico, Puerto Rico and the US Virgin Islands.

SYMPTOMS Rash, headaches, fever, extreme muscle and joint pains. These initial symptoms tend to subside after a couple of days, only to reoccur. It may take some weeks for a full recovery to be made, although serious complications are rare.

ACTION There is no 'cure' for dengue and no vaccine. Painkillers may be needed to relieve the symptoms. Steps should be taken to avoid being bitten by mosquitoes in the first place — wear protective clothing (particularly at dawn/dusk) and use insect-repellent sprays and mosquito nets.

MALARIA

Serious disease spread by the bites of *Anopheles* mosquitoes. Malaria is a major health problem in the tropics. It affects 200 to 300 million people worldwide and kills over a million people every year in Africa alone.

SYMPTOMS Symptoms usually appear within two weeks of being bitten, but may not appear for up to a year if the victim has taken antimalarial drugs inconsistently. Symptoms are fever with chills and shaking, with three stages. At first, the patient may be cold, with uncontrollable shivering. The temperature may then rise to 40.5° C (105° F). There may be profuse sweating, which helps bring the temperature down. All

these symptoms may be accompanied by severe headaches, malaise and vomiting. After the fever attack the patient may be weak and exhausted.

ACTION Seek urgent medical attention. This is an emergency and may require admission to hospital. Antimalarial drugs and, in severe cases, blood transfusions may be given.

PREVENTION Antimalarial drugs should be taken by all visitors to the tropics, including pregnant women. Wear clothing that covers legs and arms in the evening. Insect repellent sprays and mosquito nets all reduce the likelihood of being bitten.

Acute, highly-infectious disease affecting the digestive tract, caused by the bacterium *Salmonella typhi*. Paratyphoid is milder and caused by the bacterium *Salmonella paratyphi*. Both are contracted from contaminated food or drink, and are more common in hot climates than in temperate areas.

SYMPTOMS Headache, aching limbs, fatigue, fever lasting several days, then subsiding and returning a few days later. Constipation and mental confusion accompany the fever. Later diarrhoea, bloating of the abdomen and red spots on chest and abdomen may occur.

ACTION This is an emergency. Seek urgent medical attention as soon as salmonella poisoning is suspected. Treatment is likely to include antibiotics, rehydration and hospitalization — usually in isolation.

PREVENTION Scrupulous hygiene in handling food. Avoid shellfish from possibly contaminated waters. Eggs, which have recently been identified as carrying the bacteria in some cases, improperly thawed frozen poultry and some 'cook-chill' foods must all be thoroughly cooked to destroy the bacteria. Immunization against typhoid and paratyphoid lasts several years.

TYPHUS

A group of infectious diseases caused by microorganisms that are similar to bacteria (rickettsiae). Usually spread by insects such as fleas, mites and ticks.

SYMPTOMS Headache, constipation, collapse, pain in the back and limbs and coughing Followed by fever, mild delirium and a rash of small red spots. There may also be a weak heartbeat.

ACTION Seek urgent medical attention. Prompt treatment may help to avoid death from septi-

caemia (blood poisoning), heart failure, kidney failure or pneumonia. Part of the treatment may include antibiotic drugs. Convalescence is often slow, especially for elderly people.

PREVENTION Insecticides to control louse infestation, vaccination and protective clothing may all discourage bites from tyhpus-carrying insects. Preventative measures are important in crowded tropical conditions and after a natural disaster.

YELLOW FEVER

Infectious disease transmitted in urban areas by *Aedes aegypti* mosquitoes. Some types may be spread from monkeys to humans. Yellow fever is prevalent in Africa, Central and South America.

SYMPTOMS Nosebleeds, headaches, nausea and fever. Heartbeat may be slow, despite the fever. Recovery usually takes place within three days. In severe cases there may be pain in the back, neck and legs. Rapid liver damage may lead to yellowing of the skin (jaundice) and kidney failure. In untreated cases, these may

be followed by mental disturbances or delusions, then coma and eventually death.

ACTION There is no effective drug. Treatment involves maintaining the volume of blood in the body, often by transfusion. Most people recover before the disease progresses. Once you've had it you're immune.

PREVENTION Vaccination should be sought by travellers to risk areas, except for babies under a year old. A vaccination certificate is required for entry into or exit from some countries where the disease is prevalent.

Plague killed 25 million people, in Europe alone, in the 14th century. Today it is responsible for up to 50 human deaths each year, mainly confined to parts of Africa, South America and Southeast Asia. Caused by the bacteria *Versina pestis* and spread by the bites of fleas that have lived on dead rats. There are two types—bubonic, characterized by swollen lymph glands or 'buboes', and pneumonic, which affects the lungs. Both must be treated IMMEDI-ATELY or will almost always prove fatal.

Travellers to high-risk areas should NEVER touch a rat or the carcass of any animal or bird. A vaccine is available for people in high-risk occupations.

STDs

✚ **Sexually-transmitted diseases are infections spread primarily, in a majority of cases, by sexual intercourse, heterosexual or homosexual. Confidential diagnosis and treatment is given at STD clinics, where a doctor will assess the infection or infections—usually prescribing a course of antibiotic drugs. Tests are usually made after the symptoms have gone, to ensure that the patient no longer has the infection.**

From the late 1970s it became clear that some STDs, such as herpes (which can become chronic) and hepatitis B (which can be fatal), could not be cured with drugs. With the recognition of the HIV virus and AIDS in 1982, STDs became even more life-threatening.

All recent partners of an infected person should be traced by an STD clinic, especially since these contacts could be infected with a type of sexually-transmitted disease which shows no obvious symptoms. The confidential tracing and treatment of anyone who has had recent sexual contact with the person being treated is essential to the control of STDs.

SAFER SEX

Sexual intercourse is only completely safe if both you and your partner have never had sex with anyone else and if neither has ever had an STD (which could have been contracted non-sexually). Safe sex was traditionally practised by anyone wishing to avoid contracting any STD, but is now IMPERATIVE with the recognition of HIV and AIDS. To reduce the likelihood of contracting any STD you should:

○ Never have unprotected sexual intercourse, except with a monogamous partner whose sexual history and HIV status is known to you.
○ Always use a condom and spermicidal jelly. Novelty condoms (coloured, flavoured, or textured) tend to he less reliable than the thicker varieties. Remember that oil-based lubricants like petroleum jelly and baby oil can damage the rubber. Water-based lubricants should be used.
○ Reduce the number of your sexual partners.
○ Remember that prostitutes have a high rate of infection. They should ensure that their clients use a condom, to protect both the client and themselves.

High-risk sex

Transference of body fluids may lead to transmission of STDs or HIV, if one or more partners carry them:

○ Unprotected vaginal or anal intercourse. Remember that secretions will still be exchanged even if the unprotected penis is withdrawn before ejaculation.
○ Any sexual activity that draws blood from the genital area or breaks or chafes the skin.
○ Putting the fingers (aggressively), hand or object into the vagina or rectum before, during or after sexual intercourse. This may damage the lining of the vagina or rectum and increases the risk of STD/HIV transmission.

Low-risk sex

Little or no risk of STD or HIV transmission:

○ Oral/genital contact. The risk is increased if the woman is menstruating or the man ejaculates into his partner's mouth.
○ Oral/anal contact. Infections such as gonorrhoea can be passed on in this way.
○ Sharing 'sex toys'.
○ Putting one or more fingers (gently) into the vagina or anus.

Very-low-risk sex

Minimal transfer of body fluids:

◐ Penis/body contact or vagina/body contact.
◐ Mutual or group masturbation.
◐ Sex toys used with, but not shared with, a partner.
◐ Urination in sex.

No-risk sex

No transfer of body fluids:

◐ Masturbation on your own.
◐ Sex toys used on your own and not before vaginal or anal intercourse.
◐ Touching or massaging, not in the genital/anal area.
◐ Bondage or beating. Breaking the skin increases the risk.

AIDS

Acquired immune deficiency syndrome is not a single disease, but a group of infections and cancers caused by HIV (human immunodeficiency virus). AIDS has an extremely high fatality rote, but the number of people infected with the HIV virus who go on to develop the condition is statistically very low. HIV is believed to be a life-long infection, which can manifest in a variety of ways—from no symptoms at all to severe damage to the immune system. This means that the body has only limited protection from the diseases and infections it may come into contact with.

SYMPTOMS Minor symptoms of HIV infection include inflamed skin, particularly on the face, rapid weight loss, diarrhoea, fever and oral condidiasis (thrush). Diseases that tend to be more acute in HIV-infected patients include shingles, tuberculosis and herpes simplex infections. The brain may also be affected, causing disorders like dementia. Fullblown AIDS includes various cancers, diseases of the immune system, pneumonia, diarrhoea, candidiasis and chronic or persistent herpes simplex.

ACTION Diagnosis for AIDS is based on a positive result to an HIV test (note that there is no such thing as an 'AIDS test'), along with the presence of characteristic infections and tumours. So far, there is no cure or vaccination against the disease, although research contin-

ues. According to the form their illness takes, patients respond in varying degrees to antibiotic drugs, antiviral drugs, anticancer drugs and radiotherapy.

REMEMBER The most useful treatment for anybody with HIV infection or AIDS is practical help with their everyday lives, if and when they need it, and reassurance and companionship. Bear in mind that it is the person with AIDS who is in danger—infections such as measles and chicken pox have serious consequences because of the deficient immune system. Normal non-sexual human contact poses NO risk to the non-infected person.

AVOIDING AIDS There are only three ways' to contract HIV infection:

■ Sexual contact, especially unprotected vaginal and anal intercourse. Not all sexual activities carry a risk.

■ Mother to baby, before and perhaps during childbirth. Very rarely, HIV antibodies can be transmitted by breastfeeding.

■ Shared needles and syringes. This includes sharing needles to inject drugs, use of medical equipment for more than sue patient and, in rare cases, the re-use of acupuncture, tattooing and body-piercing needles without proper sterilization.

In the past, HIV is known to have been transmitted through blood transfusions, and blood products to treat haemophiliacs. Today, all donated blood, semen and organs are meticulously screened for HIV and any othe infections and blood products are heat-treated. Nevertheless, people in high-risk groups should NOT donate blood, semen or organs such as kidneys.

REMEMBER The HIV virus is not transmitted through insect bites, coughing or sneezing, cuddling, dry kissing, mouth-to-mouth resuscitation, nor by sharing lavatory seats and household utensils, such as drinking glasses, crockery and cutlery.

HIGH-RISK GROUPS People most at risk of contracting HIV infection are:

■ Promiscuous men and women, particularly if they have numerous sexual partners without using condoms.

■ Men and women who believe they are not at risk and have casual sexual intercourse without using condoms.

■ Prostitutes, and their clients, if they do not use condoms.

■ Subsequent partners of both of the above.

■ Intravenous drug users and their partners (male or female).

■ Bisexual or homosexual men and their partners, if they have causal sexual intercourse without using condoms.

■ Hospital patients who receive unscreened blood transfusions or blood products — including haemophiliacs — although treatment of blood and blood products should now rule this out.

■ People who undergo surgery in some Third World countries, where equipment may be scarce and must be re-used.

■ People who have unprotected casual sex in areas where there is a very high incidence of HIV infection, such as central Africa and Haiti (and most major cities).

REMEMBER Anyone who thinks they may have been exposed to HIV can request a blood test. However, it would be wise to obtain counselling to determine whether you really need to have the test done and how to prepare yourself for the news of a positive result. Also bear in mind that you may have to wait **at least** three days (perhaps a lot longer) for the result.

Because the HIV virus takes time to manifest itself in a detectable form, someone who has only recently come into contact with the disease may have a negative result in an HIV test and consider themselves free of the disease — but they may be positive. Those in high-risk groups should follow a negative result with another test about six months later. Some people who are HIV positive do not develop full-blown AIDS for several years — others remain HIV positive, but do not go on to develop AIDS.

If you are HIV positive or have AIDS, you will have to be monitored every three months or so to check that the T-cells (the helper white blood cells) are o a safe level. You can help to keep the number of T-cells high by eating well, exercising and making sure that you get plenty of rest. For practical help, advice and friendship contact your local AIDS help group.

CANDIDIASIS

Also known as thrush. A fungal infection caused by *Candida albicans*. It affects moist body tissues such as mucous membranes, especially in the vagina but also penis, anus or mouth. Candidiasis, as an STD, is transmitted by sexual contact with an infected person. When antibiotic drugs are taken or when the body's resistance to infection is lowered (after or during a serious illness, for example), the fungus may multiply rapidly. *Candida albicans* is usually found in moist body tissue — it's only when it gets out of control that a problem develops. It can also spread from one moist body area to another and can sometimes affect the gastrointestinal tract.

SYMPTOMS **Vaginal candidiasis** commonly appears as a thick white discharge that takes the appearance of cottage cheese. May be accompanied by itching and discomfort on passing urine, although some women have no such symptoms. **Penile candidiasis** is quite rare, and is more likely to infect an uncircumcised penis, causing inflammation and soreness under the foreskin. **Oral candidiasis** may appear as creamy-yellow sensitive patches anywhere in the mouth. **In babies** candidiasis takes the form of an irritating rash with white flaky patches in the general nappy area.

ACTION Seek medical attention. Vaginal pessaries or creams containing anti-fungal drugs are usually required. These should be applied to the affected a reas as directed. It is recommended that sexual partners are treated at the same time, to prevent reinfection. Women who suffer repetitive bouts and who take oral contraception should consider changing to another, non-hormonal type. Oral candidiasis should be treated with a special mouthwash. In most cases, condidiasis is relieved or 'cured' very quickly and simply, but tends to return. Areas prone to infection should be kept clean and dry.

CYSTITIS

By no means classed as an STD, but may be linked to infections caused by sexual intercourse — or may be mistaken for an STD. An inflammation of the lining of the bladder, usually caused by bacterial infection. In women, bruising of the urethra caused by over-enthusiastic intercourse can also lead to cystitis. Recent research has cast doubt on the safety of vaginal deodorants and bubble-baths. Both may lead to irritation and cystitis. Links are being explored with bladder cancer.

SYMPTOMS Burning or stinging pain when passing urine, frequent urge to pass urine but with only small amounts passed, cloudy or bloody urine, chills, fever and nagging pain in the lower abdomen.

ACTION Seek medical attention. Drink large quantities of water to flush out the bladder, but not excessively to the point where you feel ill. Drink-

ing water with a teaspoon of sodium bicarbonate about four times a day will increase the alkalinity of the urine — anyone with high blood pressure or heart trouble should consult their doctor first. Painkillers may help, but do not take over-the-counter remedies for more than 48 hours. If symptoms persist seek medical assistance again, as cystitis may spread to the kidneys. Pregnant women are particularly at risk from kidney infections. Prompt treatment with antibiotic drugs will usually settle the infection within 24 hours.

REMEMBER Women who have repeated attacks of cystitis should urinate as soon as possible after sexual intercourse. Sometimes similar symptoms may be caused by a sexually-transmitted infection — if you have had a recent change of sexual partner or if your partner develops similar symptoms, see your doctor or visit an STD clinic.

HERPES

Not necessarily an STD but the general name for a group of viruses, of which herpes zoster (see **Chicken pox**) and herpes simplex are the most significant. Herpes zoster is also the virus which causes shingles. Most people carry HSV1 (herpes simplex virus type 1) and may develop cold sores

and, in some cases, genital herpes. HSV2 more commonly leads to genital herpes.

SYMPTOMS Pain, burning or itching in the affected area, followed by the development of sore red skin eruptions or blisters, which burst to form small painful ulcers. In a severe attack, glands in the groin may become swollen and painful, and there may be a fever. There may be pain on passing urine if the tip of the penis or the vagina are affected. The sores may appear anywhere on the body, provoked (it is believed) by stress, poor health, illness with fever, even by overheating in the sun.

ACTION Seek medical attention. There is little or no treatment, except to avoid disturbing the sores — the virus is highly contagious. If the sores become infected, antibiotic drugs may be prescribed. Anti-viral drugs may be prescribed in severe cases.

Kissing or sexual contact should be avoided during an attack, and for a week afterwards.

WARNING A severe attack may be life-threatening, especially if the eyes or the brain become affected. Seek urgent medical attention. A baby might become infected, if the mother has an attack at the time of childbirth. People who are HIV positive should avoid all forms of herpes, if possible. The consequences could be severe.

GENITAL WARTS

Soft warts which grow in and around the anus or the vagina, or on the penis. Genital warts are caused by a virus and spread through sexual contact with an infected person. However, it may take as long as twelve months after the initial contact for the warts to appear.

ACTION Seek medical attention. Genital warts are usually painless and can be removed by simple surgery or the application of ointment,

but tend to recur. Some may spontaneously disappear. Sometimes, however, clusters of warts may develop, causing great discomfort.

REMEMBER Links are suspected between genital warts and cervical cancer (see **Cancer**). This makes doubly important the need for cervical smear tests for any woman who has had genital warts.

GONORRHOEA

Also known as 'the clap', it is spread primarily by sexual intercourse (including vaginal, anal and oral sex) and from mother to baby during childbirth. It is one of the most common STDs and, once detected (incubation period is less than two weeks), can usually be 'cured' very quickly and simply. Current safe sex practices (using condoms) should reduce the incidence statistics — if they are followed!

SYMPTOMS Discharge from vagina/penis and pain on passing urine. More than half of women infected have no symptoms. If the anus is infected, the rectum and anus may become inflamed and a discharge may be apparent on the faeces. After oral sex a 'sore' throat may develop. In babies it causes an inflammation of one or both eyes.

ACTION Seek medical attention. Antibiotics, usually penicillin, are administered. After treatment, further tests confirm eradication of the infection. Sexual partners are normally traced confidentially through STD clinics and told they may have gonorrhoea to prevent reinfection or spreading of the disease.

WARNING Gonorrhoea must be treated as early as possible to prevent it from spreading to other parts of the body. In men, it may cause inflammation of the testes and prostate, affecting fertility. In women it can spread to the fallopian tubes and cause pelvic inflammatory disease, with a risk of a misplaced (ectopic) pregnancy. It can also affect female fertility. If untreated gonorrhoea is allowed

to spread through the system, a form of arthritis may develop—leading to painful and swollen joints. There may also be chronic blood poisoning which (very rarely) can affect the major organs (brain and heart) and lead to death.

NONSPECIFIC URETHRITIS (NSU)

An inflammation of the urethra, which may not have an identifiable cause. Most cases are caused by bacteria or viruses. This is the most common type of STD worldwide, affecting men and women.

SYMPTOMS In men, there may be a clear or pus-filled discharge and pain on passing urine. Symptoms are sometimes absent. In women, there are usually no symptoms unless there are complications.

ACTION Tests are made to identify the cause of the infection, but this can be difficult as there are so many possible causes. For women, diagnosis often relies on the fact that a partner has the condition and may have passed it on. Both partners should be treated with antibiotics. Most cases can be 'cured', but infection/symptoms commonly recur. Follow the full course of medication and seek further attention if the problem persists.

PUBIC LICE

Also known as crab lice or 'crabs'. An infestation of *Phthirus pubis*—a small insect that lives at the base of pubic hairs and feeds on the host's blood. They are very easily transferred during sexual contact.

SYMPTOMS Pubic lice are usually felt (as pinching or itching) and seen (as small dark 'spots'). The females lay tiny eggs which are also visible to the naked eye. On hairy men they may also be found in body hair, on the legs, arms and even on the face or head. Parents may transmit lice to their children. It is said that, if exposed to cold air or water, one can feel the lice 'tighten their grip'!

ACTION Seek medical attention. Insecticide lotion containing benzyl benzoate or lindone (to kill eggs as well) should be applied. More than one treatment may be required to kill all the eggs. Sexual partners should also be treated. Clothes, sheets and blankets should be washed in very hot water.

REMEMBER Other body parasites such as scabies, and fungi such as ringworm, may be spread by sexual contact.

SYPHILIS

Usually transmitted sexually during intercourse, less commonly by kissing, mutual masturbation or by contact with items soiled by an infected person. Traditionally viewed as life-threatening, syphilis responds well to antibiotics—although the sooner it is recognized, the better. Syphilis is caused by the bacterium *Treponema pallidum,* which has access through breaks in the skin or through membranous tissues of the genitals, anus or mouth. Since mothers-to-be are now screened, congenital syphilis is rare.

There is a 30 per cent chance of contracting the disease after contact with an infected person.

SYMPTOMS Three to four weeks after contact, a painless ulcer with a wet base, up to 1 cm (1/2 in) across, appears on the site of the infection—genitals, anus, rectum, lips, throat, and sometimes on the fingers. This heals within four to eight weeks. At six to twelve weeks, there may be a skin rash. In white people this takes the form of noticeable crops of pinkish round spots. In black people, the rash is pigmented and appears darker than the surrounding skin. Other symptoms include fever, fatigue, headaches, 'aching bones' and loss of appetite. Scalp hair may fall out. Infectious pinky-grey patches can develop on moist skin areas.

Latent syphilis: This stage of the disease may persist for years or even indefinitely. The infected person appears to be quite 'healthy', but a few cases (if untreated) may lead in time to tertiary syphilis.

Tertiary syphilis: This stage can take as many as 25 years to develop—perhaps as few as three. The effects will vary from one individual to another. Tissue death or break-down of any parts of the body may occur. Other effects of tertiary syphilis include cardiovascular syphilis (which can lead to heart valve disease) and neurosyphilis (with progressive brain/spinal cord damage and general paralysis).

ACTION Seek medical attention. All stages of the disease respond to treatment with penicillin, but any organ damage (which only occurs in the later stages) is irreversible. As treatment begins, there may be flu-like symptoms and tender lymph nodes. Regular check-ups should follow to confirm that the infection has been dealt with. It will show up (harmlessly) in the blood for many years.

REMEMBER The first and second stages of the disease are infectious, the latent and tertiary stages are not. Measures to avoid contracting syphilis include practising safe sex. Confidential contact tracing means that others who may have become infected can be informed and treated. Some may have no symptoms and may spread the disease, unwittingly, to others.

HEPATITIS B

Viral disease, resulting in inflammation of the liver. It may be extremely serious, with damage to or death of liver cells. People in obvious high-risk groups—those with many sexual partners, intravenous drug users and their partners, prostitutes and their clients, doctors and nurses—should seek advice. There is a vaccine available.

SYMPTOMS Begins as a flu-like illness (which may be severe and debilitating) and leads eventually to jaundice—yellowing of the skin. In rare cases, if not treated, liver failure may occur, leading to death.

ACTION Treat as a very serious illness. Seek urgent medical attention.

WARNING A baby born to a carrier mother may be at severe risk. Antenatal screening is advisable. There is no risk from contaminated blood products. Blood is screened and/or heat treated to remove the risk.

PARASITES

Numerous organisms may live in or on humans. Not all spend their entire life with the host—some are just 'passing through'! Some bring serious diseases, whereas others may cause few noticeable symptoms.

About 30 **protozoa** may 'inhabit' humans—many causing serious conditions such as amoebic dysentery, malaria and toxoplasmosis. **Fungi** are fairly simple parasites and may only cause simple skin infections, such as ringworm. Others may invade the lungs or other body tissues and leads to serious illness or even death.

Many **bacteria** are essential within the human body, others are 'invaders' causing diseases such as pneumonia, meningitis and skin disorders. Many are well-known and deadly—salmonella, for example.

Even **viruses** are parasites. There are many types of virus—the smallest known disease agents (about 100th the size of the smallest bacterium)—leading to some very well-known illnesses from the common cold to warts, from rabies to viral hepatitis, from AIDS to polio—and all the common childhood diseases from mumps to measles. Immunization programmes exist to combat many viruses.

WORMS

Many people are embarrassed by the very idea of having 'worms', but most humans have at least one type of worm in their lifetime. These may live in the gut, in bile ducts, or in blood vessels. Many produce eggs which are excreted by the host.

Liver flukes infest the bile ducts of the liver and have a bizarre life cycle. They normally infest sheep and lay eggs which are excreted in the faeces. The eggs are then eaten by snails, hatch and emerge as immature flukes which are passed on to water flora—including watercress. In the Far East, liver flukes may be ingested by humans in raw or undercooked freshwater fish. In the early stages of infestation there may be rashes, night sweats, liver enlargement and tenderness. Later, jaundice may develop.

The largest types of **tapeworm** usually occur in regions where there is inadequate sewage disposal and unhygienic preparation of meat for food. Animals may play intermediate host, humans ingesting the larvae from undercooked meat. The largest tapeworms may be 9m (over 29 ft) long! They attach themselves to the wall of the intestine and trail throughout it.

The **dwarf tapeworm** is only 2.5 cm (1 in) long and found worldwide, especially in tropical regions, and is usually caught from an infested person or in insanitary conditions. The eggs pass out in faeces and may be ingested by hand-to-mouth contact. In central and Western Europe, Australia and New Zealand there is a rare but serious illness involving cyst formation in the lungs and other tissues. The cause is the larva of a small tapeworm, which relies on sheep and dogs to complete its life cycle.

Roundworms such as threadworms and hookworms are fairly common worldwide. Many cases of infestation don't cause any noticeable symptoms, unless there are large numbers of worms.

Threadworms are very common, infecting as many as one in five children. They are about 1cm (less than 3/8 in) long. Their eggs are laid in the skin around the anus, causing severe itching. Scratching may detach the eggs, which are then transferred back to the mouth. Keep children's fingernails short and encourage good hygiene. Wash underclothes and nightclothes at high temperatures to kill eggs and worms, and seek medical attention.

Hookworms are more common in tropical regions, infesting an estimated 700 million people. The larvae burrow into the skin of the feet, travel to the lungs, from where they are coughed up and swallowed, to mature in the gut to about 12 mm (1/2 in) long. They may cause anaemia.

EXTERNAL PARASITES

External parasites, such as lice and mites, may be passed from one human to another. Ticks may be picked up from animals or vegetation. Seek medical attention in all cases. Eradication of most parasites is usually simple and quick.

There are three main types of **lice**—all are about 3 mm (1/8 in) across—infesting specific areas: the head, the body, the pubic region. **Head lice** (see colour pages) suck blood from the scalp, causing itching and soreness. ANYONE can get lice, particularly children. Nowadays lice are NOT to be associated with poverty or lack of body hygiene. In fact, lice find it easier to cling to clean hair. Eggs (known as 'nits') are commonly attached to the base of hairs.

Body lice are (unfortunately) associated with dirty clothes and can (in warmer countries) transmit serious diseases such as typhus. Clothes should be washed in very hot water or burned.

Pubic lice seek the warm soft pubic areas but may migrate to other hairy body areas, including the head. Known as 'crabs', they are most frequently transmitted during sexual contact (see **STDs** and colour pages).

Bedbugs are flat, wingless insects about 3 × 5 mm (1/8 × 1/4 in). They live in furniture and bedding, or cracks in walls and floors nearby, and emerge at night to feed on humans. Their bites may be visible and itchy, and may lead to infected sores. Seek medical attention.

The numerous types of **mite** are all tiny. Some transmit serious diseases such as typhus, others are fairly harmless to most people. A severe house dust allergy can usually be traced to an allergy to the faeces of **dust mites**—millions live in EVERY home in carpets and bedding, where they feed on tiny flakes of human skin.

Others, such as the **scabies mite**, live IN human skin—they are highly contagious. The burrows look like small scaly grey lumps—often on the hands, in the armpits, around the genitals, leading to reddened patches on the limbs and torso. Usually all members of a household need to be treated at once—but treatment is almost always effective immediately, although the irritation may take a couple of weeks to subside.

Mites known as '**chigoes**' may be picked up in grassland, scrubland or beaches in Africa or tropical America. Also known as 'jiggers' or 'chiggers', the mites usually penetrate the skin of the feet or lower leg. They are actually sand flies and harvest mites. The females burrow into human skin to lay their eggs. Treatment usually involves removing the creatures with a needle and the application of antiseptic ointment.

PETS AND YOUR HEALTH

Zoonoses are infections or diseases of animals that can be transmitted to humans—some zoonotic organisms are able to adapt themselves to a variety of different species. Zoonoses are most commonly associated with pets and livestock (obviously because these animals coexist closely with humans), but are also known to be transmitted to humans by wild animals, such as rats.

In addition to zoonoses, there are many examples of human diseases which can be spread by vectors—animal carriers. Rat fleas were responsible for spreading the Black Death, the 14th-century plague that claimed 25 million lives in Europe alone. Nowdays, rats (or their fleas) still spread diseases such as plague, typhus, Weil's disease and Lassa fever.

There are known to be about 30 animal diseases which may be transmitted to humans—a few of these from household pets. Some have extremely serious consequences.

Pets

Pets and domestic animals are generally beneficial to our health. Cats and dogs provide companionship and it has been proven that stroking an animal can be therapeutic for the owner, reducing tension and, in some cases, lowering blood pressure. Dogs have to be taken for walks, providing exercise for their masters (whether you like it or not!). Specially-trained dogs can help the deaf and blind.

Any pet that is properly cared for and in good health presents no threat, provided these guidelines are followed:

- Wash your hands whenever you have touched pet food, faeces, urine, or cat litter.
- Provide separate dishes, utensils and can-opener for pets.
- If a pet has diarrhoea or seems 'off colour', be scrupulous about hygiene.
- Take a pet to the vet whenever it is ill.
- Arrange regular worming and flea treatments.
- Wash your hands before you eat, drink or smoke, if you have been playing with or stroking an animal.
- NEVER allow an animal to lick your face, however much it may seem to be an expression of 'love'.
- DON'T allow young children to play with soil—especially if dogs or cats use the area as a 'lavatory'. This includes public play areas, where dogs are exercised.

Pregnancy and pets

Infections from domestic animals have access through the placenta to the foetus. This can result in miscarriage or severe malformation. If a woman becomes infected with certain diseases in late pregnancy, the child may be subject to disorders of the nervous system or blindness—although symptoms may not show for a few years. Pregnant women should:

- NEVER change cat litter trays or come into contact with animal faeces
- NEVER handle dog or cat food—especially raw meat
- NEVER stroke stray cats or dogs
- NEVER handle soil

Allergies

Some people develop allergies to dog or cat dander, the tiny scales from animal skin and fur that are present in the air and make up a large percentage of household dust. This can cause asthma or urticaria (a type of skin rash), sneezing, itchy eyes—all may be slight or severe. REDUCE contact with the animal concerned and seek medical attention.

 WARNING

TOADS ARE DANGEROUS: NEVER allow small children to play with toads. The toad's skin secretions are highly poisonous and can enter the body through the mouth, by sucking fingers with poison on them or through a break in the skin. The poison can affect the heart, cause convulsions and even lead to death.

ZOONOSES

CAT SCRATCH DISEASE

Not a common occurrence. A cat bite or scratch may lead to blistering and swelling, occasionally to an absess. The infection progresses possibly causing headaches, fever and swelling of the lymph nodes. The exact cause of this infection has only been identified recently. The disease does not imply that the animal is ill. Seek medical attention.

CRYPTOSPORIDIOSIS

A protozoic infection leading to nausea, fever, abdominal pain and watery diarrhoea—in children, the elderly or anyone with a suppressed immune system (as is the case with people suffering from AIDS) it may be life-threatening. Can be transmitted by animals or humans. Treatment includes rehydration. Seek medical attention.

HOOKWORM

Mainly confined to tropical and sub-tropical regions, this parasite can be transmitted from infected pets by skin contact with larvae. There are no long-term health risks, but an itchy rash may occur on the skin, especially on the feet. Serious infestation may lead to lung problems, abdominal pains and anaemia. High-risk areas: places where cats and dogs defecate—playgrounds, beaches, parks. Seek medical attention.

LYME DISEASE

Only identified in 1975, this condition is caused by a bacterium which is transmitted by a tick which lives (normally) on small birds, rodents and deer, but which will bite dogs and humans. Symptoms include joint inflammation, lethargy, fever, aches. Unless treated, these may persist for several years. In rare cases, serious complications may develop, affecting the heart and the central nervous system. High-risk areas where ticks may be 'picked up' include scrubland, woodland, areas of long grass—especially if deer are present. The condition is known to occur in Europe and the northeastern United States. If Lyme disease is suspected, seek urgent medical attention.

PET THROAT

Over a third of all domestic pets carry a streptococcal bacterium which causes sore throats. Persistent or recurrent sore throats in a household with pets should always be investigated. Seek medical attention. DON'T allow pets to lick your face.

PSITTACOSIS

Usually transmitted by caged birds and causing a variety of symptoms, including fever, headaches, a bad cough, sore throats, lethargy, breathing difficulties, and rashes.

The bird may show no signs of ill health, or may show signs of ornithosis. Avoiding dust from droppings, feathers and actual contact with birds is essential. In rare cases, if untreated, psittacosis can be debilitating or even fatal. Anyone who has contact with birds as pets, as a hobby or for a living, should monitor their health and seek urgent medical attention if a problem develops.

NOTE: In recent years, links have been investigated between bird-keeping and a greatly increased rate of lung cancer.

RABIES

Transmitted by infected saliva from animal bites and caused by a virus that attacks the central nervous system. Over 12,000 cases are reported annually worldwide. Rabies is extremely rare in Great Britain, Australia, Scandinavia and Japan. It is endemic in countries where quarantine laws are relaxed, especially in parts of Europe, Asia, Africa, the United States and South America, where it may be carried by wild dogs, foxes, wolves, badgers and vampire bats. Cases from domestic dogs are rare.

SYMPTOMS May not appear until months after infection. Fever, agitation, delirium, weakness or paralysis, extreme salivation, difficulty swallowing and brain inflammation. Fear of water (hydrophobia) may set in, followed by coma and death.

ACTION If a bite does occur, especially in these regions, the animal should be isolated for observation. The bite must be inspected by a doctor IMMEDIATELY. Seek URGENT medical attention. A vaccine is available, given as a course of injections over several weeks. Other infections may also be transmitted by an animal bite.

RINGWORM

Although the name suggests a worm, this is a popular name for a group of common fungal skin infections, which can be transmitted by pets, from one human to another, or from infected soil or household objects. Symptoms include ring-shaped, red, scaly or blistery patches on the skin. Seek medical attention.

Eradication involves the use of anti-fungal powders, creams and drugs. Reinfection is common. Hygiene must be scrupulous.

TOXOCARIASIS

Potentially VERY serious infection caused by infestation by the larvae of the common dog roundworm (also lives in cats). Eggs of these parasites are passed by dogs (and cats) into soil. The eggs can remain viable for several years. Humans — most commonly children — are infected when playing with an infected dog or in infected soil.

Once the eggs hatch in the body, the larvae migrate to the liver. Early symptoms may include abdominal discomfort, mild fever, throat infections, nausea, coughing or wheezing — often not severe enough for infestation to be suspected. Complications can develop, leading to convulsions, behavioural changes, and — if larvae migrate to the eye — impaired vision or eventually **blindness**. If this condition is suspected, **seek URGENT medical attention.**

WARNING Dogs MUST be wormed on a regular basis (consult a vet). Young children who play with dogs or in gardens/ parks where dogs may defecate are AT RISK. Recent surveys have found that soil in most public parks is severely contaminated. The eggs are NOT destroyed by cold weather. Care should be exercised by EVERYONE who handles soil. When gardening, for example, avoid hand-to-mouth contact.

TOXOPLASMOSIS

Potentially VERY serious infection. Humans may be affected (and often are) by eating infected pork or lamb, although the protozoa responsible for the infection may live in several species. More seriously, the organism can multiply in a cat's intestines and eggs are then excreted in cysts within the faeces. Humans may become infected by touching the cat, or handling a litter tray or soil contaminated by cat faeces.

In healthy humans the infection manifests itself only with flu-like symptoms, although eye inflammation may also occur. **However, in** pregnant women, particularly in the early stages of pregnancy, the risk of infection must be avoided AT ALL COSTS. There are severe risks to the foetus, including abortion and stillbirth. Babies may be born with abnormalities ranging from anaemia (and other blood disorders) and enlargement of the liver or spleen, to an enlarged head, brain damage, blindness, deafness. In some cases the baby may appear normal but may develop severe problems in the first two years—including eye, ear and brain diseases.

Anyone with a suppressed immune system (especially anyone suffering from AIDS) must also take great care to avoid infection. The consequences could be devastating—possibly leading to complications involving the lungs, heart and brain.

WARNING Always wear gloves or wash your hands thoroughly after handling a cat, a litter tray, cat's faeces, raw meat or soil ALWAYS dispose of cat litter trays carefully and promptly. Toxoplasmosis is an EXTREMELY common infection worldwide.

WEIL'S DISEASE

Properly called leptospirosis. Human infection may occur from contact with the urine of infected rats—a possibility when swimming or enjoying other watersports in city rivers, dockyards or canals. Dogs may also be infected. Symptoms of human infection include fever, chills, headache, muscular pains, rashes, eye inflammation. The kidneys and liver may be affected. In many cases, recovery is slow—the nervous system could be affected, producing symptoms similar to meningitis. Seek urgent medical attention.

BITES/STINGS

Many species of animals and insects bite or sting. Cat and dog bites are fairly common—less so, are bites and stings from smaller creatures such as hornets, scorpions and jellyfish. It may not always be clear what has caused a sudden pain, especially if a sting occurs while swimming. If in any doubt, seek medical attention.

Some people may have extreme allergic reactions to bites and stings (see Anaphylactic shock). In rare cases, this can be life-threatening. Unfortunately a person has no way of knowing whether they are particularly sensitive until a problem develops.

ANIMAL BITES

Always have an animal bite inspected and treated. There is a high risk of infection from bacteria or viruses in the animal's saliva. Depending on where you are in the world, this is particularly important if there is any possibility of rabies. Cat bites more commonly become infected than dog bites—this is thought to be because the puncture wounds are smaller, preventing natural seepage of fluids. It may be advisable to report any dog bite to the police. The animal might prove a danger to other people. ALWAYS seek medical attention.

HUMAN BITES

Whatever the reason for the bite, it should be remembered that infection of the wound is very likely. Tetanus booster shots may be advisable. Certain viruses may be transmitted in the saliva of infected people, including hepatitis, rabies, herpes—HIV transmission may even occur. A human bite must ALWAYS receive medical attention.

BEE/WASP/HORNET STING

Get medical help immediately if anyone is stung more than once or stung in the mouth or throat. The latter is dangerous because the swelling caused by the sting could obstruct breathing. If possible, give the victim ice cubes to suck.

TREATMENT Bee stings should be carefully removed by stroking the sting with the side of a needle and then extracting it with tweezers. Do not squeeze or prod the poison sac, as this will only release more venom. Apply bicarbonate of soda in a cold compress.

WARNING Some people are allergic to insect stings. Seek URGENT medical attention if anyone shows signs such as:
- Facial and throat swelling
- Wheezing/breathing difficulties
- Vomiting
- Dizziness
- Severe itchy rash
- Collapse

SCORPION STING

May be no more serious than a bee sting. Some species may be particularly venomous and a sting could cause diarrhoea, vomiting, sweating and severe pain. The heart may be affected. It is likely to be more serious for the elderly or young children. ALWAYS seek urgent medical attention. It may be necessary to use an antivenin to control the effects of the sting.

SPIDER BITE

A tarantula bite is generally about as serious as a bee sting. Other species including the black widow and the brown recluse (US), the funnel web and red-back (Australia) may occasionally cause death. If you have been bitten by any spider, seek urgent medical attention. It may be necessary to use an antivenin — these are available for all dangerous species.

JELLYFISH STINGS

May range from a mild itchy rash to incapacitating pain. DON'T touch any fragments of 'tentacle' adhering to the skin — vinegar (acetic) acid should be used to neutralize their stinging capability. Calamine lotion will soothe. Many dangerous species are found in tropical waters. The effects of the Portuguese man-of-war (not strictly a jellyfish) may be severe. Stings are most dangerous while swimming far from shore. The pain may make it difficult to move — let alone swim. After any sting, ALWAYS seek medical attention. If there is sweating, vomiting, breathing difficulties, convulsions or loss of consciousness, SEEK URGENT MEDICAL ATTENTION.

FISH 'STINGS'

Mainly in tropical waters, some species of fish such as scorpion fish and stone fish may be extremely venomous. The blue-ringed octopus, sea urchins and stingrays can cause severe injury, too. Even if you have not identified what type of creature has caused the pain (by a bite or poisonous sting) SEEK URGENT MEDICAL ATTENTION.

SNAKE BITE

Victim may feel dizzy, nauseous, hot and faint. Effects vary according to species of snake, its size, the amount of venom injected and the age, health and size of the victim. SEEK URGENT MEDICAL ATTENTION. With prompt treatment, most victims recover fully. Antivenin, antibiotic drugs and tetanus injections should be given in the event of any bite, poisonous or otherwise. If you kill the snake, keep the body to aid identification so that the antivenin can be matched accurately. NEVER apply a tourniquet.

REMEMBER The chances of serious injury or death as a result of being bitten by a snake are slight. Nevertheless, you should avoid all possible confrontations with snakes. Never corner a snake. Never try to pick one up with a stick. Never approach one to take a photograph — the snake could interpret your behaviour as a threat. Given the enormous variety of poisonous snakes and the enormous variety of potential victims (in terms of age, sensitivity, health), the results of a bite can be highly unpredictable. Never hurry a snake-bite victim—it will speed the passage of poison around the body.

SUBSTANCE ABUSE

PLEASE NOTE

Substance 'abuse' is a very misleading term. It should refer to the abuse of one's body with a substance—after all, it's not the substance that's being abused. More correctly, we should say substance 'use' or 'misuse'.

Heavy drinking, tobacco smoking, taking certain 'drugs' for non-medicinal purposes or inhaling solvents, are all types of substance use or misuse. They are all very dangerous and can lead to death in many ways—through poisoning, accident, disease, choking, anaphylactic shock (severe allergic reaction) or heart failure. Intravenous drug users, who share needles, are at increased risk of contracting HIV and hepatitis B.

The only way to reduce the health risks of tobacco smoking is to STOP SMOKING. Habitual alcohol and drug users must also try to stop but, like smokers, may need medical and emotional support. Confidential self-help groups like Alcoholics/Narcotics Anonymous, worldwide organizations, offer practical help and companionship for those who wish to break free.

The use of 'drugs', the misuse of prescribed drugs and the misuse of other substances to relax or to achieve out-of-the-ordinary or 'mystical' experiences are widely publicized. In most countries, governments provide leaflets in health centres, schools and colleges drawing attention to the risks. Many young people 'experiment'. Some are more vulnerable to peer pressure. Most will lose interest after a short time, with no lasting ill effects. This is NOT a new phenomenon (remember 'reefers' in the 1950s?). Parents should look back to their own teenage years—the first cigarette (and how disgusting it tasted!) and the first under-age alcoholic drink.

Many anti-drug campaigns are aimed directly at heroin, cocaine and crack. The popular belief that cannabis use can inevitably lead to use of these drugs has been largely discounted. However, the use or misuse of certain drugs/substances may cause distress to the user's family and friends—it may alter behaviour patterns, increase the risk of accident and may cause general ill health or actual physical harm. The 'market' is sometimes flooded with impure or contaminated products, which could kill—especially if injected. There is a danger with mixing drugs/substances, or mixing drugs/substances with alcohol. The biggest risk is that of overdose, especially with sedative drugs.

Sedative drugs induce sleep and reduce muscle activity. If they are taken to excess, the whole body system is slowed down and this could result in coma. Breathing may stop, the tongue may fall back and block the windpipe, or the unconscious user may choke on his/her own vomit. The user can die.

The drugs/substances that most often result in coma are heroin, morphine, strong tranquillizers, barbiturates, glues, inhaled solvents, gases and lighter fuels, and ALCOHOL. Coma is less likely to result from hallucinogens such as LSD, 'magic' mushrooms, cannabis or mild tranquillizers.

Injecting a drug directly into a vein is the most dangerous way to take it, as the whole dose acts immediately and the 'hit' is more intense. Blood may be 'backwashed' and stored in the syringe, which is why intravenous drug users who share needles are particularly at risk from HIV and hepatitis B.

The body usually develops a tolerance for increasingly large doses of any drug/substance. Just as a drinker quickly finds that he/she has to drink more to feel the effects of alcohol, so a drug/substance user has to take more to get a 'hit'. If the user of a powerful drug, such as heroin, decides to cut down, the tolerance gradually reduces. A single injection of the former high dose can then cause death.

OVERDOSE SYMPTOMS

Depending on the amount and type of drug/substance involved, one or more of the following symptoms may occur:

■ Vomiting
■ Difficulty breathing
■ Sweating
■ Hallucination
■ Dilation or contraction of the pupils
■ Unconsciousness
■ Coma

ACTION This is a general guide for all types of drug/substance overdose:

■ Seek urgent medical attention/CALL AN AMBULANCE if the victim is unconscious or having breathing difficulty. This is a life-threatening situation.

■ Collect a sample of vomit and any evidence —syringe, pill, bottle or container—near the victim, which may help with medical diagnosis.

■ If the victim is unconscious but breathing, check for any obstructions in the mouth and throat and place in the recovery position (see **Save a life!**).

■ If the victim stops breathing, you must begin artificial respiration immediately (see **Save a life!**).

■ Monitor the pulse. If the victim's heart stops beating (and ONLY if it does) apply cardiac compression (see **Save a life!**).

■ Stay with the victim and keep him/her calm. Make sure no more drugs/substances or alcohol are taken. Keep the victim away from bright lights and loud noise which may intensify the problem. Prevent any accidents. A 'violent' person may have to be physically restrained to prevent self-injury. Be as gentle as possible.

■ If solvents/gases have been used, remove the source and open doors and windows. Take care to avoid the risk of fire, especially if flammable substances have been used.

■ With milder drugs such as amyl nitrate and cannabis, it may only be necessary to wait for the effects to wear off provided that no other drugs have been used as well.

■ NEVER induce vomiting, unless the victim has DEFINITELY taken tranquillizers or barbiturates very recently. The only safe way to do so, is to insert your fingers GENTLY into the mouth towards the throat until the victim's 'vomit reflex' takes over. NEVER do this with the victim on his/her back. If the victim is semi-conscious or unconscious, there is a severe risk of choking, or of fluid entering the lungs. Always call an ambulance FIRST—don't waste time!

■ NEVER try to keep the victim awake by walking him/her around or by giving black coffee to drink. This will only speed the effects of the drug.

AMPHETAMINES

Swallowed in tablet or capsule form. Amphetamine sulphate powder is sniffed, swallowed, dissolved or injected. Traditionally known as 'pep' pills or 'speed'.

OVERDOSE SYMPTOMS The user will be restless and irrational, with outbursts of delusions and hysterical behaviour, even fits. There will be rapid breathing/pulse rate, twitching and increased body temperature.

AMYL NITRATE

Originally available only as glass ampoules containing a clear liquid. The vapours are inhaled. It is more common as 'poppers' in small, usually brown, glass bottles. Smells like dry-cleaning fluid.

OVERDOSE SYMPTOMS Excessive doses will cause a severe headache and flushed skin,

staggering and incoherence. The victim may collapse and have difficulty breathing.

BARBITURATES

Capsules and tablets of barbiturates are swallowed, often with alcohol, which can prove fatal. Pentothal, though not common, is injected. Barbiturates are 'downers'.

OVERDOSE SYMPTOMS The victim may be drowsy, semiconscious or comatose. If Pentothal is injected breathing may stop, especially after a large dose.

COCAINE

A fine white powder that may be sniffed ('snorting'), dissolved in water and injected ('mainlining'), mixed with tobacco and smoked ('freebasing') or chewed in the form of coca leaves. Most available cocaine can be presumed to be impure. It is often combined with amphetamine sulphate ('speed'), which is cheaper. Particularly unscrupulous 'dealers' may extend the drug with almost ANY white powder, sometimes with disastrous consequences.

OVERDOSE SYMPTOMS Cocaine overdose causes hysteria, delusions, physical tremors and fits. There may be an increase in the pulse, body temperature and breathing rate. The user could suddenly collapse from anaphylactic shock.

CANNABIS

Smoked in 'joints' or pipes, on its own or mixed with tobacco. It may also be added to cake mixtures and eaten. The effects of the drug are rarely dangerous. When taken to great excess, may impair judgement for several days. The main risk is that the drug may be contaminated with poisonous plant matter or insecticides.

OVERDOSE SYMPTOMS Drowsiness, disorientation and mild hallucinations. In many cases, the user will fall asleep before toxic levels are reached.

SOLVENTS/GASES

Freely available in the form of lighter fuel, glues, dry-cleaning agents, petrol, nail varnishes and removers, all of which give off an intoxicating vapour. Glue or liquid is poured into a paper bag or onto a cloth and inhaled. The gases halon (used in some fire extinguishers) and butane (used for lighter fuel) are also abused.

OVERDOSE SYMPTOMS Solvents/gases often lead to coma. Breathing may stop, particularly if the victim falls forward into the bag or cloth and continues to inhale the fumes. Extreme exertion after prolonged use may put the heart at risk. Halon affects the nervous system and leads to convulsions and even death.

HEROIN/MORPHINE

Both are usually smoked, but can also be swallowed or injected directly into a vein ('mainlining') once dissolved in water. Both are highly addictive.

OVERDOSE SYMPTOMS Vomiting, coma and contracted pupils. Breathing may stop.

METHADONE

A synthetic drug resembling morphine. In liquid or tablet form, it is usually swallowed. Otherwise, methadone may be injected into a vein. Widely used in drug clinics to reduce the heavy dependence and cravings felt by heroin or morphine addicts. Methadone is not, however, without side effects.

OVERDOSE SYMPTOMS The same as for heroin/morphine.

OPIUM

Raw opium is smoked in a pipe or swallowed. Refined opium is swallowed, smoked, injected, or heated and the fumes inhaled.

OVERDOSE SYMPTOMS Similar to heroin/morphine, but with less likelihood of breathing difficulties or coma.

LSD

Most hallucinogens, including LSD (lysergic acid diethylamide) and magic mushrooms, are swallowed. LSD is also known as 'acid'.

OVERDOSE SYMPTOMS Overdoses are rare, but accidents may occur through a 'bad trip' where the user experiences illusory terrors and tries to escape them. Occasionally the user may become violent, especially if very frightened.

TRANQUILLIZERS

Some tranquillizers, like Librium and Valium, are mild, whereas others like Largactil may have a profound sedative effect and can induce drowsiness even in normal doses. They are swallowed or occasionally injected.

OVERDOSE SYMPTOMS Dangerous if taken with alcohol — can result in coma. Coma is less likely if tranquillizers are used on their own. Most tranquillizers used for prolonged periods (six weeks or more) are addictive or may lead to dependence. Even when prescribed by a doctor, they should only be used for short periods.

ALCOHOL DEPENDENCE

Habitual and compulsive heavy drinking commonly leads to hypertension, heart disease, liver disorders and brain disorders. Alcohol over-use is also closely linked to numerous social problems — traffic/industrial accidents, physical assault, marital breakdown, domestic violence and child abuse. Alcohol IS a poison — if the drinker collapses or becomes unconscious, there is a possibility of alcohol poisoning.

SYMPTOMS Heavy drinking. In more serious cases, nausea, vomiting, shaking in the morning,

lapses of memory, abdominal pain, facial redness, unsteadiness, confusion and irregular pulse. A habitual drinker may hide bottles, become aggressive or grandiose, be drunk for long periods of time, neglect his/her appearance or food intake, constantly promise to give up and have uncontrollable personality changes such as anger, violence and selfishness. After a heavy drinking session, the sufferer may be in a stupefied state. There is a severe danger that the drinker may fall into a coma and stop breathing.

WARNING Never assume that a person who looks drunk IS drunk. Other conditions such as diabetes mellitus and some drugs can produce similar apparent effects. Always check for the smell of alcohol.

ACTION In the event of collapse or unconsciousness, seek urgent medical attention. If breathing stops while you are waiting, give artificial respiration. Do NOT induce vomiting. Loosen tight clothing, check that the mouth is clear. Place casualty in the recovery position to keep air-way open (see **Save a life!**).

REMEMBER Treatment to help compulsive drinkers give up drinking altogether involves detoxification to help them get over withdrawal symptoms, which may include — in extreme cases — hallucinations (delirium tremens), severe shaking and convulsions. Longterm counselling and support may be needed to help them rebuild their lives. People who think they have an alcohol problem should contact their doctor or a help group such as Alcoholics Anonymous as soon as possible.

TOBACCO SMOKING

Nicotine, a constituent of tobacco, contains several cancer-causing (carcinogenic) substances. There is a direct link between the amount smoked and the likelihood of cancer of the lung, mouth, lips, throat, kidney, pancreas and bladder. The majority of people with head and neck cancers have a history of heavy alcohol use and smoking.

Another harmful effect of smoking is coronary heart disease, which kills more middle-aged men in industrial societies than any other disease. Smokers are about three times more likely to contract the disease than nonsmokers. Smoking also damages the arteries of the legs and brain and can lead to strokes. A number of other diseases have been linked with tobacco smoking.

WARNING Smoking and pregnancy: Smoking is extremely harmful during pregnancy — the foetus is likely to be smaller with less chance of survival. After birth, children of smokers are more likely to develop asthma and other respiratory diseases.

A woman who smokes during pregnancy increases her child's risk of getting cancer. Other risks include spontaneous abortion, stillbirth and premature birth, as well as a baby of impaired growth and increased susceptibility to infection.

PASSIVE SMOKING People who live, work or spend a lot of time in the company of smokers are at risk of contracting smoking-related illnesses. Passive smokers often suffer the immediate discomforts of coughing, wheezing and watery eyes. Evidence shows that the children of smokers take in the equivalent of 80 cigarettes each year and are more likely to smoke when they are older. It is also thought that those who live with smokers have an increased risk of developing lung cancer.

STOPPING SMOKING If you want to stop smoking, the first and most important thing to do is to decide that you definitely will give up. Cutting down is usually only temporary. Acupuncture, hypnosis and chewing gum containing nicotine may help. Many people do not like the idea of gaining weight if they stop smoking. This is NOT a foregone conclusion, although it may occur temporarily because smoking increases the metabolic rate (the rate at which the body uses up food). Those who give up tend to eat more at first — sweets and 'comfort' foods may satisfy 'cravings'. However, it is far safer to be slightly overweight than to smoke. Smoking is an addiction and a smoker wishing to give up may need help and support from family, friends, or a help group.

Some people may be surprised to find these common beverages classed as dangerous substances. But both contain tannin and caffeine, which have adverse effects on the body. Both can become addictive.

Coffee contains caffeine in larger amounts than tea or cola drinks. Caffeine is a stimulant drug, which affects all the organs and tissues of the body. In the short term, caffeine combats drowsiness and fatigue and can help the drinker to concentrate. In larger quantities (more than five cups per day) it may cause over-stimulation, anxiety, irritability and restlessness. If you are suffering from or being treated for any anxiety-related condition, the effects may be dangerous.

Research shows that people who drink coffee throughout the week (at work, perhaps) often suffer withdrawal symptoms, such as irritability, headaches and fatigue, at the weekends if they go without a cup of coffee for just a few hours. Caffeine overdose causes muscle twitching, heart palpitations and abdominal pain.

As with other drugs, users tend to build up a level of tolerance and have to drink more before they are able to feel the effects. Caffeine can improve short-term athletic performance and is therefore banned in high doses during sports competitions.

Tea also contains caffeine, but in smaller doses than in coffee. More importantly, it contains tannin which was once used for medicinal purposes but has now been found to cause liver damage and constipation in large doses. Tannin is also suspected of having links with mouth cancer.

Oil of bergamot, used in Earl Grey tea, contains agents that are suspected of being cancer-forming (though this is unproven). Even herbal teas, which can contain impurities, may have side effects. Some people may develop an allergy to camomile tea. Comfrey tea has been reported to cause liver toxicity and even liver damage.

Any kind of tea is best taken in moderation. It is wise not to drink the same variety of scented tea over long periods of time.

EMERGENCIES

There are numerous health emergencies that can occur or develop which might be hard to recognize. Some are much more common than others, but ALL of the following may be life-threatening.

ANAPHYLACTIC SHOCK

Uncommon dangerous allergic reaction to a substance — penicillin, antitetanus, local anaesthetic, poison, drug, a particular food, an insect sting/bite, a snake bite. Look for the possible cause.

SYMPTOMS Pain in abdomen, swelling of the face, tongue or throat, diarrhoea, constriction of the airways to the lungs, itchy rash and a sudden lowering of blood pressure.

ACTION Lay the victim down and raise his/her legs to ease the flow of blood to the heart and brain. If unconscious, place in recovery position. If breathing or heartbeat cease, give artificial respiration and/or cardiac compression (see **Save a life!**). In most cases, an injection of Adrenalin is required. Seek URGENT medical attention.

REMEMBER ALWAYS carry a card which describes your particular allergy. People who are known to have a severe allergy which could lead to anaphylactic shock are encouraged to carry a syringe of Adrenalin at all times. If you know such a person, familiarize yourself with the procedure in case help is needed at any time.

APPENDICITIS

Acute and painful inflamation of the appendix. The cause is usually unknown, but may be due to an obstruction of the appendix by faeces. Without surgery, the appendix may become further inflamed and infected, which may cause gangrene (tissue death) in the appendix wall. The appendix may burst. Appendicitis is the most common cause of abdominal surgery in developed countries. However, it is rare in the very young or the elderly.

SYMPTOMS Discomfort in the central abdominal area. This commonly develops into a more intense pain in the lower right-hand side. It is usually accompanied by loss of appetite, vomiting, nausea, coated tongue and unpleasant breath.

ACTION Seek URGENT medical attention.

GALLSTONES

Stones that develop in the gall bladder when a chemical imbalance occurs in the composition of bile.

SYMPTOMS Most cases of gallstones do not have any symptoms, unless a stone becomes lodged and blocks the duct leading from the gall bladder. This can cause severe pain in the upper abdomen or between the shoulder blades, nausea and vomiting, flatulence, even jaundice.

ACTION Seek medical attention, urgently if there is great pain or distress.

RISK GROUPS Gallstones are unlikely to occur until after puberty. More women are affected than men — particularly if they have had several children. Being overweight adds to the risk.

HEPATITIS A, B & C

Inflammation of the liver, which is usually caused by viral infection. Hepatitis C (also known as non-A/non-B hepatitis) is caused by more than one virus.

Hepatitis A is transmitted by food and water that has been contaminated by faeces. Most of the world's population has been infected by the virus at some point in life and a natural immunity has been established. Only those people with no immunity are at risk from the virus.

Hepatitis B is commonly transmitted sexually or by contaminated blood. Intravenous drug users are particularly at risk. Carriers may have no apparent symptoms and infect others unknowingly. A vaccine is available, but is only recommended for high-risk groups — people with many sexual partners, prostitutes and their clients, intravenous drug users, doctors and nurses. Preventative measures include practising safe sex, not sharing needles, and making sure that equipment for tattooing and ear or other piercing is properly sterilized.

Now that blood to be transfused is made safe by heat treatment, there is no danger of Hepatitis B or C being caused by blood transfusion in most countries. Transfusions and surgical procedures/equipment in underdeveloped countries, however, carry a greater risk to the patient.

SYMPTOMS Slight to very severe flu-like symptoms, which do not pass. Weakness, lethargy, nausea (especially at the thought of food). Urine may be dark, while faeces may be pale. Jaundice tends to develop later.

ACTION If a form of hepatitis is suspected, seek urgent medical attention. There is no direct treatment for types of viral hepatitis, but plenty of bed rest, a balanced diet and avoidance of alcohol all help.

KIDNEY FAILURE

Occurs when one or both kidneys become partially or totally unable to filter the blood (normally waste products are passed into the urine). Kidneys also help control blood pressure and salt levels. The condition may occur suddenly (acute) or over a period of time (chronic).

SYMPTOMS Acute kidney failure is characterized by a reduced need to pass urine, lassitude, vomiting and shortness of breath. It may occur after serious illness, severe injury or physical shock, such as heart attack, heavy bleeding or burns.

Chronic kidney failure usually begins with nausea, loss of appetite and weakness. It may be a result of diseases that affect the kidneys such as diabetes mellitus. Diabetics should be aware of this risk. If the condition is not treated immediately it will lead to fatigue, vomiting, weight loss, headaches, bad breath, itchy skin and to collapse, coma and even death.

WARNING Acute kidney failure can be cured if treated without delay. If it is allowed to develop, chronic kidney failure cannot be cured. Seek urgent medical attention.

LIVER FAILURE

May be as a complication of acute/chronic hepatitis, but also associated with cirrhosis of the liver, which may be caused by alcoholism.

SYMPTOMS Weakness and loss of appetite, vomiting, aching muscles, restlessness, fatigue and possibly coma. Jaundice may develop. The brain may be affected.

ACTION Seek urgent medical attention.

MENINGITIS

Inflammation of the membranes that cover the brain and spinal cord usually caused by a bacterium or virus. A vaccination is available, but so far it is of only limited use. Bacterial meningitis is the more dangerous. Most cases are young children.

SYMPTOMS Unremitting headache, fever, vomiting, and a need to keep away from light. Babies may have high-pitched cry, blotchy skin, fretfulness and difficulty with waking and feeding. Viral meningitis (fairly common in winter) is less serious and has influenza-like symptoms.

ACTION Seek URGENT medical attention. Viral meningitis usually requires no actual treatment. The worst symptoms usually clear in a week or two, but some headache and lassitude can persist for as long as three months. Bacterial meningitis needs immediate treatment, usually with intravenous antibiotics. There is a slight risk of some damage being caused to the brain and the condition can be life-threatening.

PEPTIC ULCERS

These affect one in eight people at some time in their lives and may occur in the duodenum, stomach or oesophagus.

SYMPTOMS May include a burning pain in the abdomen which is relieved by eating but returns after a few hours. A common complication is bleeding from the ulcer, which may result in blood being vomited and the production of black faeces. Such bleeding may cause anaemia and is life-threatening.

ACTION If you think you may have a peptic ulcer, seek medical attention. Antacid medication will neutralize the excess acidity and help the ulcers to heal. Drugs to treat peptic ulcers are now commonly available.

REMEMBER Some ulcers respond to self-help. Stop smoking, avoid alcohol, tea, coffee and aspirin, eat several small meals a day rather than two or three large ones. As psychological stress is a factor, investigate the causes of stress in your life, and look into relaxation techniques.

RUPTURES/HERNIAS

Ruptures/hernias are caused by an organ or other tissues pushing through a weak area of muscle. Usually abdominal, they have been known to occur where incisions have been made for an operation.

Groin ruptures are common. In men, these appear as a bulge in the groin or scrotum; in women, as a swelling at the top of the thigh. Sometimes the bulge can be pushed back, but when this is impossible, there may be great pain.

Ruptures and hernias are often caused by lifting heavy objects — especially in older people with weaker muscles, people who are overweight and people with a congenital abdominal weakness. Other causes are persistent coughing, straining to pass faeces and sporting injuries.

WARNING 'Strangled' hernia: A rupture which is accompanied by vomiting may

indicate that the swelling has cut off the blood supply. Lay the victim in a half-sitting position with head and shoulders supported. Do not attempt to reduce swelling. Seek urgent medical attention. Surgery may be necessary.

SEPTICAEMIA

Serious, possibly life-threatening condition — otherwise known as blood poisoning. Caused by the multiplication of bacteria and the subsequent release of toxins. Septicaemia can result from an existing infection or an infected burn or wound. In some cases, septic shock may occur. This involves tissue damage and a dramatic drop in blood pressure, which interferes with the circulatory system and can damage the kidneys, heart and lungs.

SYMPTOMS Shivering, fever, 'panting', headache and (usually) confusion or delirium. Hands may be hot. There may be rashes or jaundice. In septic shock, a feeble rapid pulse signals an emergency.

ACTION Antibiotic drugs and intensive treatment mean that the likelihood of death through septicaemia has been greatly reduced. The survival rate for septic shock, however, is still as low as 50 per cent. Seek URGENT medical attention.

TETANUS

Disease of the brain and spinal cord (central nervous system) caused by the bacterium *Clostridium tetani*. Infection can occur through cuts and wounds. Tetanus vaccination is given as a matter of course to all children, in the UK at least. Booster vaccines are recommended every ten years. NEVER neglect any wound or animal bite, particularly if it is deep. Bathe wound immediately. ALWAYS seek medical attention.

SYMPTOMS Contraction of the facial muscles resulting in a fixed grimace, stiffness at the hinges of the jaw (lockjaw) and aches in the back and abdomen. There may also be fever and heavy sweating. If untreated, there may be painful spasms in the muscles. Rarely, these lead to suffocation.

ACTION Tetanus can be fatal, but the majority of people who are treated sufficiently early make a complete recovery. Seek URGENT medical attention.

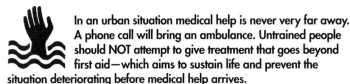

SAVE A LIFE!

In an urban situation medical help is never very far away. A phone call will bring an ambulance. Untrained people should NOT attempt to give treatment that goes beyond first aid—which aims to sustain life and prevent the situation deteriorating before medical help arrives.

There are many cases where prompt action can prevent worsening of an injury, ease discomfort and sometimes even save a life. In cases of cardiac arrest, or when breathing has stopped or when it is impaired, first aid can be of vital importance, if administered properly while waiting for an ambulance to arrive.

EVERYONE should have a knowledge of basic first-aid procedures so that they are equipped to deal with any emergency. Take a course and learn how to do it properly. No book can compete with qualified people in terms of demonstrating the correct techniques and procedures.

The techniques given here should enable you to provide urgent help in emergencies in the city.

❗ BE AN EXPERT!

Do YOU know how to apply first aid—even just the basics? WHY NOT? It is absolutely ESSENTIAL that one person in every home and one person in every department of every workplace knows how to deal with medical emergencies. Don't be the parent or the friend who watches helplessly while a loved one dies.

Further danger?

Once the immediate threat to life has been dealt with and emergency medical assistance is on its way, it is YOUR responsibility to reduce any further threat/danger to the victim or yourself. Under most circumstances you should try NOT to move or carry the victim—you may cause further injury to the victim and to yourself.

It may be essential to move him/her away from a source of danger—a fire or a collapsing building. In the case of a road accident, stop the traffic—if the victim is inside a vehicle, ONLY move them if there is the threat of further injury from fire or explosion.

Post-traumatic stress

After any accident, attack, injury or extremely stressful event in a person's life, there may be a period of anxiety, fear or depression. It may begin immediately, or over a period of

REMEMBER

DON'T PANIC
Panic affects judgement and coordination. Stay calm, you may need to give the casualty constant reassurance. Knowledge of first aid is itself an antidote to panic.

IMPROVISE
You can't prepare for every emergency and each requires a different response. Be ready to use a T-shirt to control heavy bleeding or an ordinary credit card to close a sucking wound over a punctured lung until medical help arrives.

ASSESS RISKS
Consider the dangers of each possible course of action. Choose the one you believe to be the best.

If there are other people who can help, send someone to call an ambulance while you give the casualty urgent aid. Tell them what to say so that the emergency service know what to expect when they arrive and will have things ready to help the casualty's needs. If there are more people than are needed to help, send them away. Crowding round can increase the casualty's distress and obstruct emergency services when they arrive. If you witness an accident and see the injured are already being well-cared for, move on. Don't get in the way.

If you suspect that the casualty may have a disease and that there is a risk of infection, protect yourself by wearing gloves or putting your hands into plastic bags. The AIDS crisis has made people very frightened of coming into contact with other people's blood. If someone is bleeding severely, you MUST do something!

! WARNING

NEVER induce vomiting, unless you are absolutely certain that the casualty has swallowed barbiturates/tranquillizers/poisonous plant material very recently.
NEVER use tourniquets.
NEVER remove large pieces of glass or metal from a wound.
NEVER give up. Always do all you can to help and wait for medical help before assuming that someone is dead—drowning victims can sometimes be resuscitated for 25 minutes after being presumed dead.

months—often as the person 'relives' the event in their mind. YOU can lessen the trauma at the onset—reducing the embarrassment, fear and panic of a casualty during the event by giving as much caring reassurance as possible. Hold a hand. Stroke a forehead.

Do everything you can to keep 'onlookers' away. You may need help to care for a casualty and you WILL need someone to call an ambulance. DON'T permit people to crowd round a casualty, particularly if they are saying things like: 'Is he dying?', 'Isn't she bleeding a lot!' and, appallingly thoughtless as it may seem, 'Look, her foot's come off'.

Often called 'shock', this post-traumatic stress may need to be treated by support and counselling as part of general aftercare.

First-aid kit

Normally a small first-aid kit is intended for minor accidents, but equip yours to be able to apply first aid to more serious injuries. Keep it in a sealed plastic container, out of reach of children and in a cool dry place. Your kit should contain:

- ○ Antiseptic lotion and antiseptic wipes
- ○ Cotton wool for cleaning wounds—NEVER use cotton wool or any 'fluffy' fabric as a dressing. Even a clean T-shirt or a pad of clean lavatory paper would do in an emergency
- ○ Blunt-ended scissors for cutting bandages/dressings
- ○ Painkillers such as aspirin/paracetamol—NEVER exceed the stated dosage
- ○ Antihistamine cream for insect stings/bites
- ○ Thermometers (oral and rectal)
- ○ Tweezers to remove large splinters
- ○ Eyebath/eyewash solution
- ○ Adhesive dressings for small cuts and grazes
- ○ Roll of fabric adhesive tape
- ○ Triangular bandage and safety pins for making slings
- ○ Sterile dressing for covering wounds
- ○ Sterile eye dressing and bandages for eye injuries
- ○ Crepe bandages for sprains and wounds in awkward places such as elbows/knees/ankles
- ○ Tubular gauze bandages and applicator tongs for whole-finger dressings

FIRST AID

You are in an emergency first-aid situation, an ambulance has been called and it's on its way, but do you know what to do next and shouldn't you be doing it already? The concept of saving a person's life may seem daunting but, in fact, life can often be sustained by a few simple procedures.

The instructions given in this chapter may seem complex, but to save a life all a casualty needs is an unobstructed airway with adequate breathing AND blood circulation. It's that simple. Knowing this helps to work out when a first-aid situation is life-threatening.

You must act at once if a casualty has stopped breathing, their heart has stopped beating or if they are bleeding severely. If they are unconscious, the airway may become blocked and they may suffocate. In any more serious first-aid situation, your first priority is to check the casualty's vital functions:

Open airway

A blocked or obstructed airway can be caused by a foreign body—such as food or vomit—or by a constriction of the air passage—often as a result of unconsciousness and by the tongue itself (which can fall back inside the mouth). To check, with the casualty on their back:

Lift up casualty's jaw to dislodge tongue, tilting the head back. If breathing is heavy or noisy, check if there is any blockage in the mouth. Sweep a finger around inside, but DON'T push any matter back into the throat. Place casualty in RECOVERY POSITION ☞

Check breathing

If casualty is unconscious, detecting breathing can be difficult—use your eyes, ears and touch. Hold victim's airway open as in ☞ OPEN AIRWAY and place your ear over mouth and nose. Look at line of victim's chest for any movement, while listening. Feel for breath on your cheek.

Pulse

The heart pumps blood around the body, so it is VITAL to know whether this function is still being carried out. Putting your ear to the chest and listening will not always work—feeling for blood flow in arteries just below the skin, a pulse, will. To check:

The most reliable pulse can be felt at the neck—the carotid pulse. Turn face to one side. Slide fingers from voice box into depression alongside and press gently.

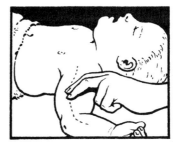

With a baby check brachial pulse, located on inside of upper arm midway between shoulder and elbow. Press index and middle fingertips lightly towards bone.

Check bleeding

Major external bleeding will be obvious, but a dramatic loss in body fluid (including blood) may not be apparent if the injury is internal. Check for:

- ◑ Pale/grey skin, particularly in and around lips
- ◑ Cold skin
- ◑ Shallow/rapid breathing
- ◑ Weak/rapid pulse
- ◑ If victim is unconscious, clear airway and treat for SEVERE BLEEDING/SHOCK ☞

INTO ACTION!

If a check tells you that a vital body function is impaired, act IMMEDIATELY. If you are alone, stabilize the casualty before leaving to find a telephone. If you have help, send someone to phone for an ambulance while you attend to the casualty.

BREATHING

If a casualty has stopped breathing, oxygen cannot enter the body and they will die in minutes—irreversible brain damage can occur after four minutes. If the airway is clear, but there is no breathing, begin artificial respiration. The best method is mouth-to-mouth, although mouth-to-nose and mouth-to-mouth-and-nose are also used.

Mouth-to-mouth

Also known as the kiss of life, the aim is to get air into the lungs. Position the head to ☜ OPEN AIRWAY. Check for blockages in the mouth/throat. Remove loose dentures (leave fixed ones). Loosen restrictive clothing around casualty's throat and chest.

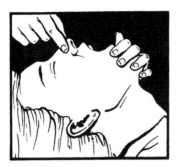

Make sure the head and chin are tilted well back. Hold the nostrils closed with a finger and thumb.

Seal your mouth over casualty's mouth and blow two quick breaths (inflations), watching the chest. It should rise. Remove your mouth and chest will fall. If chest does NOT rise, the airway may still be blocked. Check airway is open. You may need to treat casualty for CHOKING ☛

Check ☞ PULSE. If there is no pulse, use CARDIAC COMPRESSION ☞ Continue mouth-to-mouth at a rate of 12–16 inflations a minute until breathing is restored. DON'T GIVE UP. You may need to assist the casualty's breathing with further inflations. When the casualty is breathing independently, place in RECOVERY POSITION ☞

REMEMBER

If it is not possible to use casualty's mouth for artificial respiration, use nostrils instead. Seal casualty's mouth with thumb and put your mouth over nostrils. Proceed with artificial respiration. This method is useful if the mouth is injured or in cases of suspected poisoning, where you may be at risk of contamination.

BABY/CHILD

Infants need special techniques for artificial respiration:
- Seal your mouth round baby's/child's mouth AND nostrils. Don't tilt a baby's head back too far.
- Breathe gently into lungs at a rate of 20 inflations a minute.
- Check pulse after two inflations—see ☛ PULSE.

CIRCULATION

Your  PULSE check will tell you if the casualty's heart is beating. If it has stopped, there is no point in continuing with only artificial respiration because blood circulation is necessary to take oxygen to the brain. The heart must be squeezed between breastbone and spine, effectively operating it like a hand pump. Cardiac compression must be applied **in conjunction with** mouth-to-mouth.

☠ WARNING

Compression should ONLY be carried out by someone who is trained in first aid, who has had practice in the technique and who can establish conclusively whether or not the casualty's heart has stopped. NEVER give compression if the heart IS beating—even if only a very faint pulse can be felt. You could stop the heart.

Cardiac compression

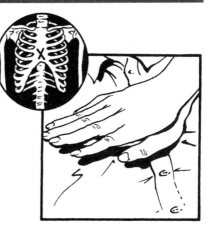

Lay the casualty on their back on a firm surface and kneel at their side. Place heel of one hand on lower half of casualty's breastbone (sternum) — the central chest bone between the ribs. Make sure your hand is about an inch above where the ribs meet, NOT on the end of the breastbone or below it. Place the heel of the other hand on top. Keep fingers off casualty's chest.

With arms straight, rock forward and press down about 4 cm (1 1/2 in) 15 times. The rate should be at about 80 times per minute — that's more than once per second. Press smoothly and firmly. Erratic or rough pressure could cause injury.

Give two lung inflations, mouth-to-mouth, and continue the sequence—15 compressions, two inflations. Make the first check for victim's pulse after one minute, thereafter at three-minute intervals. DON'T GIVE UP.

If two first aiders are present, give five compressions followed by one deep inflation on the upstroke of the fifth compression. Repeat. First aider giving mouth-to-mouth should also check pulse. DON'T GIVE UP.

As soon as a pulse is detected, STOP compressions, but continue mouth-to-mouth until casualty is breathing unaided. Place victim in RECOVERY POSITION ☞

BABY/CHILD

Use less pressure and more compression. For a baby/toddler, light pressure with two fingers is enough, at a rate of 100 compressions per minute. Only depress the chest a out 2.5 cm (1in). For older children up to ten years, use the heel of one hand only and push lightly 90–100 times per minute to a depth of about 3.5 cm (1 3/8 in). Give five compressions to one lung inflation.

CHOKING: UNCONSCIOUS

Breathing has stopped, but artificial respiration fails to raise casualty's chest. The airway is blocked. You MUST remove the obstruction. First clear casualty's mouth by turning head to one side and sweeping the inside of the mouth with two fingers. DON'T push any object further down the throat! Take particular care when doing this to a baby. If this fails, use:

Back slaps
Roll UNCONSCIOUS casualty onto side and support chest with your thigh. Position casualty's head well back to open airway. Slap casualty smartly between shoulder blades with heel of your hand. Repeat up to four times if necessary. Check the mouth to see if the blockage has been dislodged.

Abdominal thrust

If back slaps fail, try abdominal thrusts. The aim is to use the air in the casualty's lungs to dislodge a blockage by a series of thrusts to the upper abdomen. Such action could damage internal organs. Use the abdominal thrust ONLY as a last resort.

Place UNCONSCIOUS casualty on back with head in 🡒 OPEN AIRWAY position. Kneel astride casualty's thighs—if this is not possible, kneel alongside.

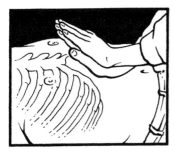

Place your hands, one on top of the other, with the heels of your hands resting above navel. Keep fingers clear.

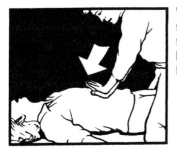

With your arms straight, make quick thrusts upwards and inwards as if up into the centre of the rib cage. Thrusts must be strong enough to dislodge blockage. Repeat up to four times if necessary.

BABY/CHILD (UNCONSCIOUS)

For a child, technique is the same as for adults—but use only one hand and don't press quite so hard. For a baby, put two fingers of one or both hands between the navel and bottom of breastbone. Press downwards/forwards quickly. Repeat up to four times, if necessary.

CHOKING: CONSCIOUS

Symptoms
◑ Casualty may be panicking and clasping throat
◑ Casualty cannot speak or make any sound
◑ Veins of the face and neck may be 'bulging'
◑ There may be blue discolouration of the lips
◑ Casualty obviously cannot breathe or
◑ Casualty may stand up and thrash about
◑ If there is any breathing, it may be very noisy

IS IT A HEART ATTACK?

■ Victim clutches chest, not throat
■ Casualty (if able to speak) complains of chest pain
■ Rapid breathing, usually quiet and shallow
■ Casualty may sit down or slump
If you suspect a heart attack, CALL AN AMBULANCE. See HEART ATTACK ☞

Encourage casualty to cough out the blockage. If they cannot, help them to a chair and bend their head over between their knees so that it is lower than lungs. Slap them sharply between the shoulder blades, using the heel of your hand. Slap four times, if necessary. Check casualty's mouth to see if blockage is dislodged. Repeat slaps. If blockage is not dislodged, use 'conscious' abdominal thrust—HEIMLICH MANOEUVRE ☞

BABY/CHILD (CONSCIOUS)

Lay a child over your lap with head hanging down, supporting child under chest. Use back slaps with the heel of your hand. If blockage is not dislodged, use HEIMLICH MANOEUVRE ☞
   For a baby/toddler, use a lot less pressure when slapping. Be prepared to use ☞ ABDOMINAL THRUST if blockage is not dislodged. If you use your fingers to try to remove a blockage from the mouth or throat of a baby/toddler, be VERY careful not to force the blockage further in.

Heimlich manoeuvre
With a CONSCIOUS choking casualty, stand behind them and put your arms around them. Make a fist of one hand and press it thumb inwards above the navel, but below the breastbone. Clasp your other hand round the first.

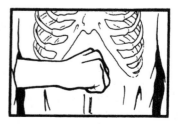

Pull upwards and inwards with a quick movement up to four times. Pressure used should compress the upper abdomen. If there is no response, repeat ➤ BACK SLAPS and return to Heimlich manoeuvre. Check the mouth to see if the blockage has been dislodged.

BABY/CHILD (CONSCIOUS)

For a baby, use conventional ➤ ABDOMINAL THRUST. For a child, depending on height, stand or kneel behind them or sit them on your lap. Support the back with one hand and apply the Heimlich manoeuvre with one hand only. Don't use nearly as much pressure as for an adult, but the pressure must be sufficient to force out the blockage.

REMEMBER

If you are alone, adapt the Heimlich manoeuvre by positioning yourself to use the back of a chair, the post at the bottom of a staircase or any other blunt projection.

SEVERE BLEEDING

An adult has around 6 litres (11 pints) of circulating blood. The loss of 0.5 litre (1 pint) can cause mild faintness. Loss of 1 litre (2 pints) causes faintness, increased pulse rate and shallow breathing. 1.5 litres (3 pints) leads to collapse, and more than 2.25 litres (4 pints) can be fatal. Immediate steps must be taken to stop blood loss.

Bleeding may not be apparent—internal bleeding may occur, particularly after a fall, a blow or a crushing injury. Severe external or internal bleeding often leads to SHOCK ☞ Shock can kill.

Blood releases its own first-aid agent when bleeding occurs. Particles in blood form clots which plug up the wound and stop the flow. If bleeding is heavy, clotting cannot take place but you can help by restricting the flow of blood from a wound. If the wound is small restrict blood flow by:

Direct pressure

Place your finger, hand or a dressing over the wound and press. This squeezes the blood vessels around the wound and cuts the blood supply. Pressure MUST be kept up for at least five to 15 minutes to let clotting take effect.

Ideally you should cover the wound with a sterile dressing, but the priority is to prevent blood loss so use any clean non-fluffy cloth. DON'T lift the dressing up to look underneath.

If no dressing is available, use your hand. Squeeze the edges of a gaping wound together. If the wound is on a limb, raise it above the level of the heart—lay the victim down and prop up head or limbs.

If YOU are wounded and alone, use a free hand to apply direct pressure to a wound—do NOT wait for assistance.

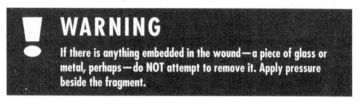

WARNING

If there is anything embedded in the wound—a piece of glass or metal, perhaps—do NOT attempt to remove it. Apply pressure beside the fragment.

If there is more than one serious wound, pads/dressings to control bleeding may be fixed in place with bandages or improvised bandages—but NOT tied so tightly that circulation is restricted.

PRIORITIES

When bleeding is coupled with cessation of breathing, treat both at the same time. This is a double emergency.

Pressure points

If there are multiple lacerations on a limb and a large amount of blood is being lost OR if there is major arterial bleeding (blood is bright red and spurts rhythmically), you MUST act very quickly.

You must find the place where the relevant artery crosses a bone, and apply pressure with your fingers to slow or cut off the flow of blood.

WARNING

Do NOT apply pressure at a pressure point for more than 15 minutes. You are cutting off the blood supply to the tissues.

BLEEDING ARM/HAND

Feel between the muscles on the inner side of the upper arm. Push the artery against the bone, with an inwards/upwards pressure. Watch the bleeding and adjust the position of your fingers until stops.

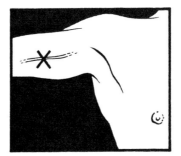

BLEEDING LEG/FOOT

With the casualty lying on the ground, raise the knee of the bleeding limb and apply pressure to the centre of the fold, where the thigh joins the groin. If you can't find the bone with your fingers, use the heel of your hand. Watch the bleeding and adjust the pressure until it stops.

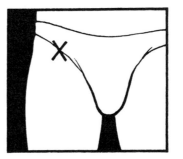

REMEMBER

ALWAYS elevate the bleeding limb, above the level of the heart, to make it easier to staunch serious bleeding.

Stab wounds

Most knife wounds involve slashing injuries and may lead to serious bleeding. Stab wounds caused by a knife or any other sharp object can be extremely serious. The actual puncture may be small, but internal injuries are likely.

If the knife (or other sharp object) is still in the wound, do NOT attempt to remove it. Treat ☞ SEVERE BLEEDING by applying pressure beside the knife or around it. If there are other cuts or wounds, apply direct pressure to them. If bleeding is extensive, check for signs of SHOCK ☞ Keep the casualty as calm as possible while waiting for help to arrive.

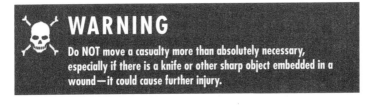

WARNING

Do NOT move a casualty more than absolutely necessary, especially if there is a knife or other sharp object embedded in a wound—it could cause further injury.

Gunshot wounds

If the 'bullet' passes straight through, the exit wound is often larger than the entry wound. Serious damage to tissues, organs, blood vessels and nerves may result from any gunshot wound. Internal bleeding may be more extensive than external—watch for signs of SHOCK ☞

ALWAYS check to see if there IS an exit wound. Calm and reassure casualty. Deal with ☜ SEVERE BLEEDING while waiting for the ambulance to arrive. Monitor ☜ BREATHING and ☜ PULSE until help arrives.

SHOCK

Shock is caused by a serious and dangerous reduction in the blood flow or fluid levels in the body. It should NOT be confused with post-traumatic stress disorder, which follows an emotional or physical trauma.

Severe bleeding, loss of body fluids from severe burns or from persistent/prolonged vomiting or diarrhoea commonly lead to shock. Shock may also be caused by a heart attack/ failure of an artery or electrocution.

The priority is to encourage the supply of blood to the vital organs—the heart, brain and lungs.

Symptoms
- Skin cold and clammy
- Casualty weak/dizzy/faint
- Pulse is shallow and rapid
- Casualty may be thirsty
- Vomiting/unconsciousness
- Skin paler than normal (greyish)
- Loss of colour in lips

You MUST act quickly. Do NOT excite the casualty in any way or move them more than absolutely necessary. Shock is life-threatening. CALL AN AMBULANCE and apply first aid to BURNS ☞ or BLEEDING ☜

Action

Reassure the casualty as much as you can. Lay them on their back (if conscious) with legs elevated about 30 cm (12 in). Loosen tight/restrictive clothing round neck, chest and abdomen.

Do NOT give anything to eat or drink. Cover to keep warm, but do NOT add heat—warming the surface of the body will draw blood away from the vital organs which need it most.

Treat and monitor injuries, especially check breathing and pulse. If there is loss of consciousness, impaired breathing or signs that vomiting may occur, place the casualty in the RECOVERY POSITION ☞ (bearing injuries in mind), until help arrives.

> # REMEMBER
>
> **Your attitude and actions are very important in treating shock. If you appear to be calm and in control of the situation, the patient will feel cared for and respond. Stay with them if you can—NEVER leave a shock victim on their own.**

HEART ATTACK

 This is a general term which is often used to cover any sudden, painful heart condition. An artery supplying blood to the heart may become blocked or severely restricted. Part of the heart may die or the heart may stop.

Symptoms
- Sudden, severe gripping pain in the chest (like extreme indigestion). Pain may spread to the shoulders, throat and down one or both arms.
- Casualty may sit down and seem withdrawn.
- Casualty clutches chest.
- Face, lips, hands and feet may lack colour.
- Sweating is likely.
- Breathing may be rapid, shallow and quiet, or may stop.
- Pulse (heartbeat) may be rapid, weak, erratic or may stop.
- Casualty may lose consciousness.
- Symptoms similar to SHOCK.

It is VITAL to act quickly to reduce the amount of work the heart has to do. **CALL AN AMBULANCE.** TELL THEM A HEART ATTACK IS SUSPECTED.

Reassure and support the casualty, sitting down—if necessary on the ground against a wall. Raise the knees slightly. DON'T excite, worry or move the casualty unnecessarily. Loosen any restrictive clothing. Monitor breathing and pulse until help arrives.

If the casualty stops breathing, loses consciousness or the heart DEFINITELY stops, be prepared to use ☜ MOUTH-TO-MOUTH and ☜ CARDIAC COMPRESSION. When 'stable', place in the RECOVERY POSITION ☞ until help arrives.

WARNING

Compression should only be carried out by someone who is trained in first aid, who has had practice in the technique and who can establish conclusively whether or not the casualty's heart IS beating—even if only a faint pulse can be felt. You could stop the heart.

STROKE

In some cases, symptoms may be confused with the effects of alcohol (but no smell of alcohol will be present). Symptoms may develop suddenly over a period of hours or, more rarely, days.

Symptoms
Depending on the severity of the stroke:
- Headache, dizziness and confusion
- 'Thumping' pulse
- Unconsciousness or slipping into unconsciousness
- Drooping mouth, slurred speech, drooling
- Flushed face
- Pupils of the eyes unequal in size
- Casualty may urinate or defecate

You must act quickly. CALL AN AMBULANCE AND TELL THEM YOU SUSPECT A STROKE.

Reassure and calm the casualty. Lay them down with head and shoulders supported. Make sure saliva drains from the mouth—mop it up to make the casualty feel 'better'. Loosen any restrictive clothing.

Monitor the casualty's breathing and pulse. If the casualty becomes unconscious, place in RECOVERY POSITION ☞ Be prepared to apply ☜ MOUTH-TO-MOUTH if breathing stops, and ☜ CARDIAC COMPRESSION if heart stops.

BURNS/SCALDS

Extremes of heat/cold, electrocution, chemicals or radiation can cause burn injuries. Scalds tend to be caused by 'wet' heat such as steam or hot liquids. The severity of the injury depends on the total skin area affected and the depth of the burn/scald.

Severe burns

⭕ Lay casualty down and protect burn/scald area from contact with ground.

⭕ Cool affected area IMMEDIATELY with clean, cold water.

⭕ Remove any rings/watches/belts or constricting clothing from injured area before swelling starts. NEVER remove anything that is sticking to a burn/scald.

⭕ Remove any clothing soaked in chemicals—take care not to burn yourself.

⭕ NEVER cover burn/scald area in fluffy dressing—use sterile dressing

⭕ DO NOT break blisters/remove loose skin or interfere with injured area. Try not to touch burns directly.

⭕ Do NOT put ointments/lotions or fat on burn/scald area

⭕ If casualty becomes unconscious, place them in the RECOVERY POSITION ☞

REMEMBER

The conscious casualty with serious burns/scalds MUST be reassured and comforted and also treated for fluid loss. Give sips of water—half a cup over ten minutes for adults. Children should sip water continuously.

⚠ WARNING

The risk of infection through damaged skin is VERY high in burn/scald injuries. Keep infection to a minimum by using sterile/clean dressings and clean water to cool injured area. Shock can also develop in cases of large-scale burns/scalds owing to loss of body fluid. Watch for signs of shock and treat as for ☞ SHOCK.

RECOVERY POSITION

When a casualty's heart is beating and the casualty is breathing, place them in the recovery position to keep the airway open. This is particularly important if the casualty is unconscious. If left on their back, a casualty is at risk of suffocation or choking by the tongue falling back into the throat and blocking it, or by choking on their own saliva and vomit. Spectacles/loose dentures should be removed.

Face down, with the head in the correct position, the tongue will fall forward and fluids will drain from the mouth. You may not see any reason WHY a casualty should vomit but it is common, especially on regaining consciousness.

WARNING

NEVER attempt to roll a casualty into the recovery position if a head, neck or spinal injury is suspected. It is better to support the casualty's head in the position in which it was found. If breathing becomes difficult and there are at least six other responsible first aiders, it is possible **WITH GREAT CARE** to attempt to turn the casualty over (see below). Try opening the airway first.

ALWAYS monitor a casualty's breathing—if it becomes laboured or noisy, the airway may not be properly open. Check the position of the casualty's head.

ALWAYS use the recovery position, unless you are examining the patient or applying first aid to other injuries. While doing so, if there is breathing difficulty the recovery position is essential.

Draw the 'upper' arm and leg away from the body, to stop the casualty lying flat. Turn the head in the same direction, angling it back on the neck (with jaw jutting out) to keep airway open. Monitor ➤ BREATHING and ➤ PULSE.

SPINAL/NECK INJURIES

If there is a suspected spinal or neck injury and the casualty's breathing becomes noisy or laboured, it is VITAL to try to open the airway. ONLY try to do so if absolutely necessary! The ideal position for a conscious casualty is lying flat on their back with head supported at each side. Try to lift the lower jaw with the head stationary. If this does not work, angle the head back and lift the jaw VERY GENTLY.

If breathing is very difficult, and there are at least six other responsible first aiders, attempt to turn the casualty. One person (you) should hold the head, while three people position themselves on either side of the casualty.

VERY VERY CAREFULLY roll the casualty onto their side. The spine and neck must be constantly supported in the neutral position and NOT allowed to twist or arch in any direction. Continue supporting the head until help arrives.

> # ❗ WARNING
>
> Spinal/neck injuries: If you suspect that a casualty has suffered a severe back/neck injury, extreme care should be taken in moving them. The casualty should ONLY be moved if they are in more extreme/immediate danger — in a burning building for example. If possible, do NOT move the casualty. Signs of severe back/neck injuries include a loss of feeling/movement below the injured area. A tingling/pins-and-needles sensation in hands/feet often denotes a neck injury. There may be an inability to feel stimuli to skin, and breathing may be weak or laboured.

HEAD INJURIES

A heavy blow to the head, possibly causing bruising or bleeding, may also cause a skull fracture. A skull or brain injury is possible without external signs. After a fall or other blow to the head, a casualty's condition MUST be monitored for several days IN ALL CASES. Seek medical attention.

Concussion
Concussion is likely after a severe blow. Unconsciousness may or may not occur. Often a very brief period of unconsciousness may be taken for granted—in other words, it's over with so quickly that you tend to assume all is OK. But, the brain has been shaken and could be damaged.

Symptoms
❍ Shallow breathing
❍ Loss of facial colour
❍ Cold, possibly sweaty, skin
❍ Rapid weak pulse
❍ Possible nausea and/or vomiting
❍ Possible confusion/loss of memory regarding the 'accident'
❍ Possible unconsciousness
If unconscious, there is a strong possibility of vomiting. Place casualty in the ☞ RECOVERY POSITION (if no spinal/neck injuries are suspected). Monitor breathing and pulse. If consciousness does not return, check level of response by pinching some

> If the casualty has been caught in the blast of an exploding bomb — or even a domestic or workplace explosion involving gas or chemicals — there may be several types of injury.
>
> While waiting for help to arrive, look for signs of fractures, head injuries, burns and lacerations from flying glass and debris. The lungs and other internal organs may also be affected.
>
> Reassure and calm the casualty. DON'T move more than necessary. Treat all injuries according to their seriousness.

skin on the casualty's arm or hand. Look at your watch and note WHEN the casualty fails to respond to pinching or commands to open their eyes. It could be important later on. There is a great likelihood of brain compression. **SEEK URGENT MEDICAL ATTENTION.**

WARNING

Compression of the brain may develop hours or days after the initial injury. SEEK URGENT MEDICAL ATTENTION.

Compression

If blood/fluid is building up inside the skull, causing pressure on the brain, symptoms may include:
- Strong slow pulse
- Numbness, tingling or paralysis of part of the body
- Flushed face
- Noisy laboured breathing
- Uneven pupil size
- Possible rise in body temperature

This is an emergency. **CALL AN AMBULANCE.** TELL THEM YOU SUSPECT COMPRESSION FROM A HEAD INJURY. If the casualty loses consciousness while you are waiting for the ambulance, place in recovery position and monitor breathing and pulse.

Skull fracture

After a severe blow to the head—in a fall or motoring accident for instance—a skull fracture may result. Symptoms include:
- Similar symptoms to **concussion** and **compression.**
- There may be blood or straw-coloured fluid leaking from the nose, ears or eyes.
- There may be obvious surface signs on the head itself—a soft area, a depression or obvious fracture.

If no spinal/neck injury is suspected, place casualty in the ☜ RECOVERY POSITION. If fluid is leaking from one ear only, turn the head to allow the fluid to leak directly downwards. Be very gentle. Monitor breathing and pulse and look out for symptoms of ☜ COMPRESSION.

This is an emergency. **Call an ambulance.** TELL THEM YOU SUSPECT A SKULL FRACTURE.

CHEST INJURIES

A chest wound can give rise to a life-threatening situation, which will require immediate action. Lung function may be impaired, affecting ☜ BREATHING. A wound may cause ☜ SEVERE BLEEDING. Danger signals of a severe chest injury include bright-red frothy blood coughed up by the victim, the sound of air being sucked into the chest and bloodstained liquid bubbling from the chest.

Seal open wound with hand or a credit card. You may need to use quite a lot of pressure and maintain it until help arrives. Place victim in a half-sitting position, support head and shoulders. Turn victim so that unaffected lung is uppermost.

Cover wound with sterile dressing—try to make dressing airtight by covering with plastic sheet or metal foil. Be alert for symptoms of ☜ SHOCK. If victim loses consciousness place in ☜ RECOVERY POSITION.

DROWNING

Not all drowning victims have water-filled lungs. In most cases, quite small quantities of water may enter the lungs. Occasionally the airway goes into spasm as water tries to enter, thereby causing asphyxiation. When water does enter the lungs, it impairs their ability to transport oxygen into the blood.

☠ WARNING

ALL people who have 'nearly drowned', but have been successfully revived, MUST SEEK URGENT MEDICAL ATTENTION. Delayed 'drowning' may occur up to 72 hours later, due to the presence of water in the lungs.

Other effects of immersion may include hypothermia—which is life-threatening—and, if the casualty has been in the water a long time, post-rescue collapse.

Post-rescue collapse is most common when a person has been in water for a long period and is suddenly lifted out (by air/sea rescue, for instance). The water pressure around the body is suddenly removed, causing the heart to work harder. In many cases this has led to tragic death after a person has been 'saved', although rescue techniques are being modified to cater for this. Lifting the casualty in a horizontal position has been found to lessen the effect of sudden removal from the water.

Action

In most cases there is little water in the lungs (if any)—don't waste time trying to get it out, GET AIR IN! You MUST act as quickly as possible—if you can start resuscitation in the water, do so. Remove any debris in the casualty's mouth and apply ☜ 'MOUTH-TO-MOUTH. If ☜ CARDIAC COMPRESSION is also necessary, get the casualty out of the water—quickly.

Keep the casualty warm while waiting for help. When 'stable', place in the ☜ RECOVERY POSITION.

COLD WATER IMMERSION

Starting from a normal body heat of 37°C (98.6°F), a drop in body temperature to 35°C (95°F) is enough to lead to hypothermia. The swimmer may become confused and have difficulty breathing. At around 32°C (89.6°F), the heartbeat may become slower and irregular. Unconsciousness is usual at about 30°C (86°F). If the core temperature of the body drops down to as low as 25°C (77°F), death is likely.

WARNING

The cooling of the body's core temperature will be greatly accelerated by the consumption of alcohol. It is estimated that a quarter of all deaths by drowning are also partly due to the lowering of the body's temperature that alcohol causes.

REMEMBER

People have been resuscitated after 'drowning' up to 25 minutes later. NEVER give up attempts at resuscitation until help arrives. If the water is extremely cold, the casualty may fully recover—the dramatic reduction in body temperature can actually protect the brain.

CHILDBIRTH

If you are pregnant, it is VITAL that you seek regular medical attention. Follow any guidelines or instructions given by your doctor in preparation for childbirth. This may include changes in lifestyle—avoid physical and mental stress which can lead to premature birth/miscarriage. Prepare an emergency procedure for getting help and medical attention.

Even though emergency medical response time in cities is short, pregnant women in difficulty MUST receive first-aid assistance. Slight vaginal bleeding does NOT necessarily mean that a miscarriage is imminent—but medical attention should still be sought.

Miscarriage

Spontaneous abortion of foetus before week 28 of pregnancy. Risks include severe loss of blood (leading to ☜ SHOCK), possibly aggravated by the retention of some birth matter in the womb. Symptoms may include cramp-like pain in the lower abdomen, as well as vaginal bleeding and perhaps passing of birth matter.

◑ Reassure and keep the woman warm.
◑ Position her with head/shoulders supported and knees slightly raised, supported by a cushion/blanket to take strain from abdominal muscles.
◑ Monitor pulse and breathing.
◑ Place clean towel or sanitary pad between her legs.
◑ If bleeding is severe, treat for ☜ SHOCK.
◑ Save birth matter for medical examination later.

Labour

It is quite possible for a woman to go into labour at any time approaching the expected date of birth. In most births, there is no threat to mother or baby—but the warmer and cleaner the surroundings the better. There is usually time for the mother to be taken to a hospital, or for the emergency services to arrive and 'take over'.

Normally the birth process follows clear stages—but there is no 'rule book'. Every birth is different.

! WARNING
If birth seems imminent, do NOT attempt to delay the emergence of the baby in any way.

This stage may take 12–14 hours for a first child, but less in a subsequent pregnancy:
- Cramp-like pain in abdomen/lower back
- Contractions of the womb every 10–20 minutes accompanied by a small amount of blood-stained mucus (a sign that the cervix has begun to dilate)
- 'Waters' break — the membrane containing the baby ruptures, discharging amniotic fluid

Once the 'waters' have broken — there may be a sudden rush of fluid or a prolonged 'trickle' — this means that the baby is being born. **Prepare for emergency childbirth.**

The mother is likely to be anxious and frightened, particularly if it is her first baby. Help her to lie down with her head and shoulders raised—use pillows, coats or have someone cradle her from behind. Restrictive clothing should be removed, especially from the mother's lower half.

Do what you can to arrange as much privacy as possible. Encourage the mother to draw up her knees, place a pad of towels, clothing or blankets underneath her hips for comfort. Keep her warm as much as possible.

Hygiene is VERY important to mother and baby but, in the back of a taxi (for instance), there is only so much you can do. If possible, wash your hands and use clean towels.

Stage two

This stage can take up to an hour. Keep the mother warm, calm and reassured.
- During a contraction, encourage mother to grasp her knees and pull her head towards them, pushing downwards.
- She should relax until the next contraction.
- Contractions normally become stronger and last longer as labour progresses.
- The area around the vagina and anus will begin to bulge as the baby's head approaches.
- The baby's head will begin to appear—get ready, birth is about to take place at any minute.

Stage three: Delivery

Support baby's head as it begins to appear and hold clean towel over anus (if bowel movement occurs, wipe faeces AWAY from birth canal to avoid infection). Support baby's head during each contraction. When crown (widest part) of baby's head has passed through lower end of birth canal, tell mother to stop pushing and start panting.

○ Still supporting the baby's head to prevent it emerging too quickly, clear membrane and birth matter from the baby's face and check that the umbilical cord is not caught around the baby's neck.

○ Still supported, the baby's head will turn to one side as the body moves into position to travel down the birth canal.

○ By lowering the baby's head, you can help the 'upper' shoulder emerge.

○ By gently lifting the baby, you can ease out the 'lower' shoulder and arm.

○ Hold the baby gently around its chest, lift it up and out of the mother, placing it on her abdomen.

WARNING

DON'T pull on the umbilical cord—and BEWARE: The baby may be very slippery.

○ Gently clean out the baby's mouth and clean mucus away from its nostrils. It should cry and breathe naturally. If it does, wrap it up to keep it warm and lay it on its side with its face down to allow fluid to drain from the airway.

○ If baby does not breathe, apply resuscitation techniques. It is NOT advisable to slap the baby!

Stage four

Afterbirth matter and the umbilical cord are normally released from the womb up to 30 minutes later, accompanied by mild contractions. The mother can help by 'pushing' this matter out. DON'T attempt to speed up the process by pulling.

Leave the umbilical cord and placenta, wrapped, beside the baby. When medical examination is possible, the afterbirth matter is needed to check whether anything has been left inside the mother—a potential cause of complications.

Clean the mother up and cover her. There may be some residual bleeding, but it's not normally serious. Place a clean towel or absorbent material between the legs. If bleeding seems more excessive, DON'T worry the mother. Massage her abdomen gently and reassure her.

Editorial/design director
TONY SPALDING

Chief sub-editor
MANDIE RICKABY

Design/symbols
MIK BAINES/TONY SPALDING

Harvill editor
ANNE O'BRIEN

Production controller
LUCY RUTHERFORD

ACKNOWLEDGMENTS

Black-and-white illustrations **CHRIS LYON**
Colour illustrations **SOPHIE ALLINGTON** (Bites & stings, Vectors)
NORMAN ARLOTT (Plants) **CRAIG AUSTIN**/GARDEN STUDIO (Snakes 2)
GRAHAM AUSTIN/GARDEN STUDIO (Dogs) **SEAN MAYERS**/GARDEN STUDIO (Fungi)
ANDREW RILEY/GARDEN STUDIO (Snakes 1, Wood-boring insects)
SIMON THOMAS (Water creatures)
Special thanks to **STOCKSIGNS LIMITED** for supplying signs artwork

Thanks also to:

Janet Ahmed
Tim Shelford Bidwell
Matthew Biggs
Judy Bugg
Ronald Clark
Ian Devey
Paul Diamond
'Gabriel'
Victor Stuart Graham
Adrian Louvain
Howard Loxton
Diana Miller
Janice Nurski
Johnny Pinfold
Kate Sekules
Bill Spalding
David Squire
Kym Turner
Christopher Winter

British Museum of Natural History (Dept of Zoology) UK
British Red Cross Society UK
Building Research Establishment/Dept of Environment UK
Canadian High Commission UK
Emergency Preparedness Centre Ottawa Canada
Federal Emergency Management Agency Washington DC
Friends of the Earth UK
Greenpeace UK
Guardian Angels UK
Health and Safety Executive UK
St John Ambulance Association UK
London Fire and Civil Defence Authority UK
London Underground Limited UK
The Met Office Berkshire UK
National Research Council Ottawa Canada
National Union of Journalists UK
New Scotland Yard/Metropolitan Police UK
Personal Protection Products UK
PSI Co. London UK
Queensland Museum Australia
Royal Horticultural Society UK
The School of Survival Hereford UK
SEE RED London UK
State Emergency Service Queensland Australia
The Terrence Higgins Trust UK
Total Security Systems Limited UK
Trades Union Congress UK
United States of America **Embassy** UK